GLYCOCONJUGATE RESEARCH

VOLUME I

Members of the Organizing Committee
Endre A. Balazs
John F. Codington
John D. Gregory, Cochairman
Roger W. Jeanloz, Treasurer
Renée K. Margolis
Richard U. Margolis
Lennart Rodén, Cochairman
Charles C. Sweeley

Sponsored by the Society for Complex Carbohydrates

Proceedings of the Fourth International Symposium on Glycoconjugates,
Held in Woods Hole, Massachusetts, in September 1977

GLYCOCONJUGATE RESEARCH

Proceedings of the Fourth International Symposium on Glycoconjugates

VOLUME I

Edited by

JOHN D. GREGORY
The Rockefeller Institute
New York, New York

ROGER W. JEANLOZ
Harvard Medical School
and Massachusetts General Hospital
Boston, Massachusetts

ACADEMIC PRESS New York San Francisco London 1979
A Subsidiary of Harcourt Brace Jovanovich, Publishers

Academic Press Rapid Manuscript Reproduction

ACADEMIC PRESS, INC.
111 Fifth Avenue, New York, New York 10003

United Kingdom Edition published by
ACADEMIC PRESS, INC. (LONDON) LTD.
24/28 Oval Road, London NW1 7DX

Library of Congress Cataloging in Publication Data

International Symposium on Glycoconjugates, 4th, Woods Hole, Mass., 1977.
Glycoconjugate research.

1. Glycoproteins—Congresses. 2. Glycolipids—Congresses. I. Gregory, John D. II. Jeanloz, Roger W. III. Title [DNLM: 1. Biochemistry—Congresses. 2. Glycosides—Congresses. W3 IN918NR 4th 1977g / QU75.3 161 1977g]
QP552.G59I55 1977 574.1'924 79-15164
ISBN 0-12-301301-1 (v. 1)

PRINTED IN THE UNITED STATES OF AMERICA

79 80 81 82 9 8 7 6 5 4 3 2 1

Dedicated to the memory of Ward Pigman

CONTENTS

Asterisks denote invited speakers

CARTILAGE PROTEOGLYCANS: SYMPOSIUM IN HONOR OF MARTIN B. MATHEWS

STRUCTURE–FUNCTION RELATIONSHIPS

CONTRIBUTORS

Numbers in parentheses indicate pages on which authors' contributions begin.

ABE, TOSHIAKI (701), *Department of Biochemistry, University of Tokyo, Bunkyo-ku, Tokyo 113, Japan*

ALBERSHEIM, PETER (415), *Department of Chemistry, University of Colorado, Boulder, Colorado 80309*

AMAR-COSTESEC, ALAIN (1007), *International Institute of Cellular and Molecular Pathology, Avenue Hippocrate 75, B-1200 Brussels, Belgium*

AMINOFF, DAVID (1011), *Departments of Internal Medicine and Biological Chemistry, Simpson Memorial Institute, The University of Michigan, 102 Observatory, Ann Arbor, Michigan 48109*

ANDERSON, JOHN C. (493), *Department of Medical Biochemistry, University of Manchester Medical School, Oxford Road, Manchester M13 9PT, England*

ANDO, SUSUMU (79), *Department of Neurology, Yale University Medical School, New Haven, Connecticut 06510*

ARNOLD, WILFRED N. (1015), *Department of Biochemistry, University of Kansas Medical Center, Kansas City, Kansas 66103*

ARNOTT, STRUTHER (321), *Purdue University, West Lafayette, Indiana 47907*

ASPINALL, GERALD O. (83), *Department of Chemistry, York University, Downsview, Toronto, Ontario, Canada M3J 1P3*

BÄCKSTRÖM, GUDRUN (713), *Department of Medical Chemistry, Royal Veterinary College, Biomedical Center, S-751 23 Uppsala, Sweden*

BAERT, F. (317), *Laboratoire de Physique des Solides et Equipe de Dynamique des Cristaux Moléculaires, Université des Sciences et Techniques de Lille I, B.P. 36, 59650 Villeneuve d'Ascq, France*

BAKER, JOHN (115, 313, 329), *Institute for Dental Research and School of Medicine, University of Alabama in Birmingham, Birmingham, Alabama 35294*

BALDUINI, CESARE (157), *Institute of Biological Chemistry, Faculty of Sciences, University of Pavia, 27100 Pavia, Italy*

BANKS, JOHN R. (95), *Department of Biological Chemistry, Milton S. Hershey Medical Center, Hershey, Pennsylvania 17033*

BASU, SUBHASH (873), *Department of Chemistry, University of Notre Dame, Notre Dame, Indiana 46556*

BEARPARK, THERESA (937), *Department of Biochemistry, Queen Elizabeth College, Atkins Buildings, Campden Hill, London W8 7AH, England*

BEAUFAY, HENRI (669, 843), *Laboratoire de Chimie Physiologique, Université Catholique de Louvain 75-39, Avenue Hippocrate 75, B-1200 Brussels, Belgium*

BEECK, HANNELORA (951), *Westfälische Wilhelms–Universität, Physiologisch-Chemisches Institut, Waldeyerstrasse 15, 44 Münster (Westf.), West Germany*

BERANEK, JIRI T. (87), *Department of Biochemistry, New York Medical College, Valhalla, New York 10595*

BERGER, ERIC G. (637), *Medizinisch–Chemisches Institut Universität Bern, CH-3000 Bern, Switzerland*

BERNACKI, RALPH J. (809), *Department of Experimental Therapeutics, Roswell Park Memorial Institute, Buffalo, New York 14263*

BERROU, ÉLIANE (737), *Laboratoire de Biochimie, Université de Paris VI, Faculté de Médecine Saint-Antoine, 27 rue Chaligny, 75571 Paris, France*

BETTELHEIM, FREDERICK A. (545), *Department of Chemistry, Adelphi University, Garden City, New York 11530*

BEYER, THOMAS (247, 641), *Duke University Medical Center, Durham, North Carolina 27710*

BHATTACHARYYA, SAMBHUNATH (91), *Department of Biochemistry, Duke University, Durham, North Carolina 22710*

BHAVANANDAN, VEERASINGHAM P. (95, 499), *Department of Biological Chemistry, Milton S. Hershey Medical Center, Hershey, Pennsylvania 17033*

BINETTE, J. PAUL (149), *Department of Biochemistry, Boston University School of Medicine, Boston, Massachusetts 02118*

BISHOP, STEPHEN H. (105), *Department of Biochemistry, Baylor College of Medicine, Houston, Texas 77030*

BLACIK, LAWRENCE J. (561), *Research-in-Aging Laboratory, Veterans Administration Medical Center, North Chicago, Illinois 60064*

BLOK, CORRIE M. (681), *Laboratory for Medical Chemistry, Vrije Universiteit, van der Boechorststraat 7, Amsterdam, The Netherlands*

BOAT, THOMAS F. (503), *Department of Pediatrics, Rainbow Babies and Children's Hospital, Case Western Reserve University, Cleveland, Ohio 44106*

BOBBITT, THOMAS F. (245), *Department of Biochemistry, University of Massachusetts, Amherst, Massachusetts 01003*

BOER, PIETER (759), *Laboratory for Physiological Chemistry, State University of Utrecht, Vondellaan 24 A, 3521 GG Utrecht, The Netherlands*

BOERSMA, ARNOLD (99), *Unité de Recherches n° 16 de l'I.N.S.E.R.M., Place de Verdun, 59045 Lille, France*

BOUCHARD, MICHELINE (881), *Centre National de la Recherche, Centre de Biophysique Moléculaire and Laboratoire de Chimie Biologique, Université d'Orléans, 45045 Orléans, France*

BOUCHILLOUX, SIMONE (645), *Laboratoire de Biochimie Médicale, Faculté de Médecine et de Pharmacie, Boulevard Jean-Moulin, 13385 Marseille, France*

BOUQUELET, STÉPHANE (933, 937), *Laboratoire de Chimie Biologique,*

Université des Sciences et Techniques de Lille I, B.P. n° 36, 59650 Villeneuve d'Ascq, France

BOURRILLON, ROLAND (567, 1037), *Centre de Recherches sur les Protéines, Faculté de Médecine Lariboisiére Saint-Louis, 45 rue des Saints-Pères, 75006 Paris, France*

BRADY, ROSCOE O. (855), *Developmental and Metabolic Neurology Branch, National Institute of Neurological and Communicative Disorders and Stroke, National Institutes of Health, Bethesda, Maryland 20205*

BRAY, BONNIE A. (507), *Department of Medicine, College of Physicians and Surgeons, New York, New York 10032*

BREEN, MOIRA (511, 561), *Research-in-Aging Laboratory, Veterans Administration Medical Center, North Chicago, Illinois 60064*

BRETON, MONIQUE (737), *Université de Paris VI, Faculté de Médecine Saint-Antoine, 27 rue Chaligny, 75571 Paris, France*

BROQUET, PIERRE (649), *Faculté de Médecine Lyon–Sud, Laboratoire de Biochimie Médicale, B.P. n° 12, 69600 Oullins, France*

BROWN, MICHAEL C. (517), *Department of Pathology, Tufts University School of Medicine, Boston, Massachusetts 02111*

BRUCKDORFER, KARL R. (747), *Department of Biochemistry, Queen Elizabeth College, Atkins Building, Campden Hill, London W8 7AH, England*

BUGGE, BIRGITTE (685), *Laboratory for Carbohydrate Research, Massachusetts General Hospital, Boston, Massachusetts 02114*

BUNDE, TERRY A. (105), *Department of Biochemistry, Baylor College of Medicine, Houston, Texas 77030*

BUSH, C. ALLEN (111), *Department of Chemistry, Illinois Institute of Technology, Chicago, Illinois 60616*

BUTKOWSKI, RALPH (257), *Department of Biochemistry, University of Kansas Medical Center, Kansas City, Kansas 66103*

CABEZAS, JOSE A. (867), *Departamento Interfacultativo de Bioquimica, Facultades de Ciencias y Farmacia, Universidad de Salamanca, Salamanca, Spain*

CACAN, RENÉ (1091), *Laboratoire de Chimie Biologique, Université des Sciences et Techniques de Lille I, B.P. n° 36, 59650 Villeneuve d'Ascq, France*

CALVO, PEDRO (867), *Departamento Interfacultativo de Bioquimica, Facultades de Ciencias y Farmacia, Universidad de Salamanca, Salamanca, Spain*

CAMPBELL, PATRICK (713), *Institute of Dental Research, University of Alabama in Birmingham, Birmingham, Alabama 35294*

CAPLAN, ARNOLD I. (189), *Department of Biology, Case Western Reserve University, Cleveland, Ohio 44106*

CARNE, LEONARD R. (1019), *MRC Clinical Research Center, Watford Road, Harrow, Middlesex HA1 3UJ, England*

CARTER, WILLIAM G. (1023), *Biochemical Oncology, Fred Hutchinson Cancer Research Center and University of Washington, Seattle, Washington 98104*

CARUBELLI, RAOUL (279), *Biomembrane Research Laboratory, Oklahoma Medical Research Foundation, Oklahoma City, Oklahoma 73104*

CASCIO, STEPHANIE (205), *Department of Biology, The Johns Hopkins University, Baltimore, Maryland 21218*

CASKEY, THOMAS C. (889), *Division of Medical Genetics, Baylor College of Medicine, Houston, Texas 77030*

CATERSON, BRUCE (115, 329), *Institute for Dental Research and School of Medicine, University of Alabama in Birmingham, Birmingham, Alabama 35294*

CHAKRABARTI, BIRESWAR (119), *Eye Research Institute of the Retina Foundation, Boston, Massachusetts 02114*

CHANTLER, ERIC (515), *Department of Obstetrics and Gynecology, Manchester University, Manchester M20 8LR, England*

CHAO, HELEN (959), *School of Basic Medical Sciences, Department of Biological Chemistry, University of Illinois at the Medical Center, Chicago, Illinois 60612*

CHENG, PIWAN (503), *Department of Pediatrics, Rainbow Babies and Children's Hospital, Case Western Reserve University, Cleveland, Ohio 44106*

CHIEN, JOW-LONG (209), *Department of Biochemistry and Delta Regional Primate Research Center, Tulane Medical Center, New Orleans, Louisiana 70112*

CHILDS, ROBERT (1095), *Clinical Research Centre, Division of Communicable Diseases, Watford Road, Harrow, Middlesex HA1 3UJ, England*

CHOU, TA-HSU (895), *Wayne State University School of Medicine, Department of Oncology, Harper-Grace Hospital, Detroit, Michigan 48202*

CHOW, PHILIP (959), *School of Basic Medical Sciences, Department of Biological Chemistry, University of Illinois at the Medical Center, 835 South Wolcott Avenue, Chicago, Illinois 60612*

CLEARY, EDWARD G. (125), *Department of Pathology, The University of Adelaide, Box 498, G.P.O., Adelaide, South Australia 5001*

CODINGTON, JOHN F. (517), *Laboratory for Carbohydrate Research, Massachusetts General Hospital, Boston, Massachusetts 02114*

COLLINS, JOAN M. (1059), *Department of Biochemistry, Memorial University of Newfoundland, St. John's, Newfoundland, Canada A1B 3X9*

COMERTON, AILEEN M. (723), *Department of Biochemistry, University of Ottawa, Ottawa, Ontario, Canada K1N 9A9*

CONCHA-SLEBE, IRINA (873), *Department of Chemistry, University of Notre Dame, Notre Dame, Indiana 46556*

COOPER, AMIEL G. (221, 517), *Department of Pathology, Tufts University School of Medicine, Boston, Massachusetts 02111*

CORFIELD, ANTHONY P. (131), *Biochemisches Institut, Christian-Albrechts-Universität, 2300 Kiel, West Germany*

CORLISS, DAVID A. (911), *Department of Engineering Biophysics, University Station, University of Alabama, Birmingham, Alabama 35294*

COURTIN, D. (937), *Laboratoire de Chimie Biologique, Université des Sciences et Techniques de Lille I, B.P. n° 36, 59650 Villeneuve d'Ascq, France*

CREMER-BARTELS, GERTRUD (653), *Eye Hospital of Westphalia, Wilhelms University, Münster, West Germany*

CROCKETT, CHRISTINE P. (217), *Department of Pharmacy, Downstate Medical Center, State University of New York, Brooklyn, New York 11203*

DAIN, JOEL A. (915), *Department of Biophysics, University of Rhode Island, Kingston, Rhode Island 02881*

DATEMA, ROELF (135), *Institut für Virologie, Fachbereich Veterinärmedizin, Justus Liebig–Universität Giessen, Frankfurter Strasse 107, D-6300 Giessen, West Germany*

DAVIDSON, EUGENE A. (95), *Department of Biological Chemistry, Milton S. Hershey Medical Center, Hershey, Pennsylvania 17033*

DAVIS, JAMES (503), *Department of Pediatrics, Rainbow Babies and Children's Hospital, Case Western Reserve University, Cleveland, Ohio 44106*

DAWSON, GLYN (877), *Departments of Pediatrics and Biochemistry, University of Chicago, Chicago, Illinois 60637*

DEBRAY, HENRI (1027), *Laboratoire de Chimie Biologique, Université des Sciences et Techniques de Lille I, B.P. n° 36, 59650 Villeneuve d'Ascq, France*

DEBRUYNE, ERIC (515), *I.C.P., Université Catholique de Louvain, Avenue Hippocrate 75, B-1200 Brussels, Belgium*

DEGAND, PIERRE (99), *Unité de Recherches n° 16 de l'I.N.S.E.R.M., Verdun, 59045 Lille, France*

DEJTER-JUSZYNSKI, MARTHA (1033), *Department of Biophysics, Weizmann Institute of Science, Rehovot, Israel*

DELERS, FRANCISCO (139), *Laboratoire de Chimie Biologique, U.E.R. Biomédicale des Saints-Pères, 45 rue des Saint-Pères, 75006 Paris, France*

DELMOTTE, FRANCIS (229, 881), *Centre National de la Recherche Scientifique, Centre de Biophysique Moléculaire et Laboratoire de Chimie Biologique, Université d'Orléans, 45045 Orléans, France*

DE LUCA, LUIGI M. (767), *National Cancer Institute, National Institutes of Health, Building 37, Room 2B-26, Bethesda, Maryland 20205*

DE PEDRO, MARÍA ANGELES (867), *Departamento Interfacultativo de Bioquimica, Facultades de Ciencias y Farmacía, Universidad de Salamanca, Salamanca, Spain*

DERAPPE, CHRISTIAN (899), *Centre de Recherches sur les Protéines, Faculté de Médecine Lariboisière Saint-Louis, Laboratoire de Biochimie, 45 rue des Saints-Pères, 75006 Paris, France*

DESAI, PARIMAL R. (659, 941), *Evanston Hospital, Immunological Research, Evanston, Illinois 60201*

DEUDON, ELISABETH (737), *Laboratoire de Biochimie, Université de Paris VI, Faculté de Médecine Saint-Antoine, 27 rue Chaligny, 75571 Paris, France*

DIELEMAN, BAUKJE (829), *Laboratorium voor Chemische Fysiologie, Vrije Universiteit, Amsterdam, The Netherlands*

DI FERRANTE, DANIELA T. (889), *Department of Biochemistry, Baylor College of Medicine, Houston, Texas 77030*

DI FERRANTE, NICOLA (889), *Department of Biochemistry, Baylor College of Medicine, Houston, Texas 77030*

DI GIANFILIPPO, F. (751), *Institute of Dental Research, School of Dentistry, and Departments of Medicine and Biochemistry, University of Alabama in Birmingham, Birmingham, Alabama 35294*

DISCHE, ZACHARIAS (653), *Institute of Ophthalmology, Columbia University, College of Physicians and Surgeons, New York, New York 10032*

DODEUR, MICHÉLE (1037), *Centre de Recherches sur les Protéines, Faculté de Médecine Lariboisière Saint-Louis, 45 rue des Saints-Pères, 75006 Paris, France*

DONNELLY, PATRICIA V. (889), *Department of Biochemistry, Baylor College of Medicine, Houston, Texas 77030*

DONOVAN, JOANE M. (915), *Department of Biochemistry and Biophysics, University of Rhode Island, Kingston, Rhode Island 02881*

DORFMAN, ALBERT (327, 823), *Departments of Pediatrics and Biochemistry, University of Chicago, Chicago, Illinois 60637*

DORLAND, LAMBERTUS (149), *Laboratory of Organic Chemistry, University of Utrecht, Croesestraat 79, Utrecht, The Netherlands*

DOWNS, FREDERICK (145, 233), *Department of Chemistry, Herbert Lehman College, City University of New York, Bronx, New York 10468*

EBERT, JAMES D. (xlix), *Marine Biological Laboratory, Woods Hole, Massachusetts 02543*

EBERT, WERNER (521), *Medizinische Poliklinik, Hospitalstrasse 3, D-6900 Heidelberg, West Germany*

EBNER, KURT E. (257), *Department of Biochemistry, University of Kansas Medical Center, Kansas City, Kansas 66103*

ECKHARDT, ALLEN E. (1043), *Department of Biological Chemistry, University of Michigan, Ann Arbor, Michigan 48109*

EGAMI, FUJIO (427), *Mitsubishi–Kasei, Institute of Life Sciences, 11 Minamiooya, Machida–shi, Tokyo 194, Japan*

EHRLICH, KARLHEINZ (771), *Biochemisches Institut, Christian Albrechts Universität, Olshausenstrasse 40–60, D-2300 Kiel, West Germany*

ELKINS, RONALD C. (279), *Biomembrane Research Laboratory, Oklahoma Medical Research Foundation, and Departments of Surgery and Biochemistry and Molecular Biology, University of Oklahoma Health Sciences Center, Oklahoma City, Oklahoma 73104*

EYLAR, EDWIN H. (261, 271), *Playfair Neuroscience Unit, 1 Spadina Crescent, Toronto, Ontario, Canada M5S 2J5*

FALKE, DIETRICH (1077), *Institut für Medizinische Mikrobiologie, Hochhaus am Augustusplatz, D-6500 Mainz, West Germany*

FEIZI, TEN (549, 1095), *Clinical Research Centre, Division of Communicable Diseases, Watford Road, Harrow, Middlesex HA1 3UJ, England*

FENYÖ, EVA MARIA (1085), *Department of Tumor Biology, Karolinska Institutet, S-104 01 Stockholm, Sweden*

FERWERDA, WŸNHOLT (681), *Laboratory for Medical Chemistry, Vrije Universiteit, van der Boechorststraat 7, Amsterdam, The Netherlands*

FIGUEROA, NORA (119), *Eye Research Institute of the Retina Foundation, Boston, Massachusetts 02114*

FIGURA, KURT, VON (951), *Westfälische Wilhelms–Universität, Physiologisch–Chemisches Institut, Waldeyerstrasse 15, 44 Münster (Westf.), West Germany*

FLASHNER, MICHAEL (955), *College of Environmental Science and Forestry, Syracuse Campus, State University of New York, Syracuse, New York 13210*

FLETCHER, MARY ANN (1047), *Division of Immunology, University of Miami, Miami, Florida 33152*

FLOWERS, HAROLD M. (1033), *Department of Biophysics, Weizmann Institute of Science, Rehovot, Israel*

FORSBERG, L. SCOTT (251), *Department of Biophysics, Pennsylvania State University, University Park, Pennsylvania 16802*

FORSTNER, GORDON (171), *Hospital for Sick Children, 555 University Avenue, Toronto, Ontario, Canada M5G 1X8*

FORSTNER, JANET (171), *Hospital for Sick Children, 555 University Avenue, Toronto, Ontario, Canada M5G 1X8*

FOURET, R. (317), *Laboratoire de Physique des Solides, Université des Sciences et Techniques de Lille I, B.P. n° 36, 59650 Villeneuve d'Ascq, France*

FOURNET, BERNARD (149, 317, 937, 945), *Laboratoire de Chimie Biologique, Université des Sciences et Techniques de Lille I, 59650 Villeneuve d'Ascq, France*

FOX, OWEN F. (279), *Departments of Surgery, and of Biochemistry and Molecular Biology, University of Oklahoma Health Sciences Center, Oklahoma City, Oklahoma 73104*

FRÉNOY, JEAN-PIERRE (567), *Centre de Recherches sur les Protéines, Faculté de Médecine Lariboisière Saint-Louis, 45 rue des Saints-Pères, 75006 Paris, France*

GALICKI, NINA I. (907), *Department of Biochemistry, New York Medical College, Valhalla, New York 10595*

GALLIGANI, LEONARDO (157), *Sezione Farmcologia, Recordati S.p.a., via Civitali 1, 20148 Milano, Italy*

GARG, HARI (161), *Shriner Burns Institute, Boston, Massachusetts 02114*

GATEAU, ODILE (663), *Laboratoire de Biochimie, Faculté de Médecine Lyon-Sud, B.P. n° 12, 69600 Oullins, France*

GEISEN, HANS PETER (521), *University of Heidelberg, Hospitalstrasse 3, D-6900 Heidelberg, West Germany*

GEUZE, HANS J. (805), *Department of Cell Biology, Center for Electron Microscopy, University of Utrecht College of Medicine, Nic. Beetsstraat 22, Utrecht, The Netherlands*

GIBBS, DOROTHY A. (885), *Division of Metabolic Diseases, MRC Clinical Research Center, Watford Road, Harrow, Middlesex HA1 3UJ, England*

GINSBERG, LEONARD C. (889), *Department of Biochemistry, Baylor College of Medicine, Houston, Texas 77030*

GLASGOW, LORRIE R. (247), *Duke University Medical Center, Durham, North Carolina 22710*

GLASS, GEORGE B. J. (295, 733), *Departments of Medicine and Biochemistry, Gastroenterology Research Laboratory, New York Medical College, New York, New York 10029*

GODELAINE, DANIÈLE (669, 843), *Laboratoire de Chimie Physiologique, Catholique Université de Louvain 75-39, Avenue Hippocrate 75, B-1200 Brussels, Belgium*

GOLDEN, JAMES F. (911), *Department of Engineering Biophysics, University Station, University of Alabama, Birmingham, Alabama 35294*

GOLDSTEIN, IRWIN J. (1043), *Department of Biological Chemistry, University of Michigan, Ann Arbor, Michigan 48109*

GOLOVTCHENKO–MATSUMOTO, ANNE MARIANNE (1065), *Division of Chemical Toxicology and Immunochemistry, Faculty of Pharmaceutical Sciences, University of Tokyo, Bunkyo-ku, Tokyo 113, Japan*

GOUSSAULT, YVES (1037), *Centre de Recherches sur les Protéines, Faculté de Médecine Lariboisière Saint-Louis, 45 rue des Saints-Pères, 75006 Paris, France*

GRANT, MICHAEL E. (795), *Department of Medical Biochemistry, University of Manchester Medical School, Manchester M13 9PT, England*

GRAVES, WILLIAM R. (1047), *Division of Immunology, University of Miami, School of Medicine, Miami, Florida 33152*

GREENWELL, PAMELA (719), *Division of Immunochemical Genetics, MRC Clinical Research Centre, Watford Road, Harrow, Middlesex HA1 3UJ, England*

GREGORY, JOHN D. (xxxix), *The Rockefeller University, New York, New York 10021*

GRIVET, JEAN-PHILIPPE (229), *Centre de Biophysique Moléculaire C.N.R.S. and Université d'Orléans, 45045 Orléans, France*

GUILMETTE, KATHLEEN (529), *Forsyth Dental Center, Boston, Massachusetts 02115*

GUIRE, PATRICK E. (1051), *6741 Tartan Curve, Eden Prairie, Minnesota 55344*

HABETS-WILLEMS, CHRISTINA (759), *Laboratory for Physiological Chemistry, State University of Utrecht, Vondellaan 24a, 3521 GG Utrecht, The Netherlands*

HADLER, NORTIN M. (237, 525), *Department of Medicine, University of North Carolina, School of Medicine, Chapel Hill, North Carolina 27514*

HAKOMORI, SEN-ITIROH (965, 1023), *Biochemical Oncology, Fred Hutchinson Cancer Research Center and University of Washington, Seattle, Washington 98104*

HANCOCK, LARRY W. (673), *Department of Biochemistry, University of Kentucky, College of Medicine, Lexington, Kentucky 40506*

HANDA, SHIZUO (949), *Department of Biochemistry, The University of Tokyo, Faculty of Medicine, Bunkyo-ku Tokyo 113, Japan*

HARDINGHAM, TIMOTHY E. (375), *Division of Biochemistry, Kennedy Institute of Rheumatology, Bute Gardens, London W6 7DW, England*

HARPAZ, NOAM (1033), *Department of Biophysics, Weizmann Institute of Science, Rehovot, Israel*

HARRISON, FREDERICK (1055), *Department of Biology, Western Carolina University, Cullowhee, North Carolina 28723*

HASCALL, VINCENT C. (189, 341), *Laboratory of Biochemistry, National Institute of Dental Research, National Institutes of Health, Bethesda, Maryland 20205*

HASILIK, ANDREJ (677), *National Institutes of Health, Building 10, Room 9N-238, Bethesda, Maryland 20205*

HAVERKAMP, JOHAN (149, 557, 771), *Laboratory for Organic Chemistry, University of Utrecht, Croesestraat 79, Utrecht, The Netherlands*

HAYASHI, SHIRO (893), *Department of Biochemistry, Fukushima Medical College, Fukushima 960, Japan*

HEANEY-KIERAS, JOY (165), *Department of Obstetrics and Gynecology, The New York Hospital–Cornell Medical Center, New York, New York 10021*

HEIJLMAN, JAN (681), *Laboratory for Medical Chemistry, Vrije Universiteit, van der Boechorststraat 7, Amsterdam, The Netherlands*

HEINEGÅRD, DICK K. (341), *Department of Physiological Chemistry, Box 750, S-22007 Lund 7, Sweden*

HERLANT-PEERS, MARIE-CLAIRE (945), *Laboratoire de Chimie Biologique, Université des Sciences et Techniques de Lille I, B.P. n° 36, 59650 Villeneuve d'Ascq, France*

HERP, ANTHONY A. (87), *Department of Biochemistry, New York Medical College, Valhalla, New York 10595*

HERSCOVICS, ANNETTE (685, 691), *Laboratory for Carbohydrate Research, Massachusetts General Hospital, Boston, Massachusetts 02114*

HILL, ROBERT L. (247, 641, 763), *Department of Biochemistry, Duke University Medical Center, Durham, North Carolina 27710*

HINEK, ALEKSANDER (725), *Department of Histology, Karolinska Institutet, S-104 01 Stockholm, Sweden*

HO, PEI-LEE (823), *Departments of Pediatrics and Biochemistry, University of Chicago, Chicago, Illinois 60637*

HOF, LISELOTTE (1055), *Department of Biochemistry, The Albany Medical College of Union University, Albany, New York 12208*

HOFLACK, BERNARD (1091), *Laboratoire de Chimie Biologique, Université des Sciences et Techniques de Lille I, B.P. n° 36, 59650 Villeneuve d'Ascq, France*

HOHM, CRAIG E. (955), *College of Environmental Science and Forestry, Syracuse Campus, State University of New York, Syracuse, New York 13210*

HOLROYDE, MICHAEL J. (247), *Duke University Medical Center, Durham, North Carolina 27710*

HOROWITZ, MARTIN I. (xxxvii, 907), *Biochemistry Department, New York Medical College, Valhalla, New York 10595*

HOUNSELL, ELIZABETH F. (167), *Clinical Research Center, Division of Communicable Diseases, Watford Road, Harrow, Middlesex HA1 3UJ, England*

HOWELL, DAVID S. (743), *Arthritis Division, University of Miami School of Medicine, Box 520875, Biscayne Annex, Miami, Florida 33152*

HSIEH, THOMAS C.-Y. (181, 673), *Department of Biochemistry, Albert B. Chandler Medical Center, University of Kentucky, Lexington, Kentucky 40506*

HUBBARD, S. CATHERINE (441), *Center for Cancer Research, E17-236A, Massachusetts Institute of Technology, Cambridge, Massachusetts 02139*

HUDSON, BILLY G. (257), *Department of Biochemistry, University of Kansas Medical Center, Kansas City, Kansas 66103*

HUGHES, R. COLIN (985), *National Institute of Medical Research, Mill Hill, London NW7 1AA, England*

HURST, ROBERT E. (911), *Department of Engineering Biophysics, University Station, University of Alabama, Birmingham, Alabama 35294*

INOMATA, MITSUSHI (701), *Department of Biochemistry, Tokyo Metropolitan Institute of Gerontology, Tokyo, Japan*

IRIMURA, TATSURO (1065), *Division of Chemical Toxicology and Immunochemistry, Faculty of Pharmaceutical Sciences, University of Tokyo, Bunkyo-ku, Tokyo 113, Japan*

ISHAQUE, ARMANA (271), *Playfair Neuroscience Unit, 1 Spadina Crescent, Toronto, Ontario, Canada M5S 2J5*

ISHIZUKA, INEO (701), *Department of Biochemistry, Teiko University School of Medicine, Kaga 2-11-1, Itabashi-ku, Tokyo 173, Japan*

IVATT, RAYMOND J. (705), *Department of Biology and Center for Cancer Research, Massachusetts Institute of Technology, Cambridge, Massachusetts 02139*

JABBAL, INDERJIT (171), *100 Rugby Road, West Bridgeford, Nottingham, England*

JACKSON, DAVID S. (795), *Department of Medical Biochemistry, University of Manchester Medical School, Manchester M13 9PT, England*

JACOBSSON, INGVAR (713), *Department of Medical Chemistry, Royal Veterinary College, Biomedical Center, S-751 23 Uppsala, Sweden*

JACQUET, MARIE-ANGE (1037), *Centre de Recherches sur les Protéines, Faculté de Médecine Lariboisière Saint-Louis, 45 rue des Saints-Pères, 75006 Paris, France*

JANSSON, LEIF (537), *Department of Clinical Chemistry, University Hospital, S-221 85 Lund, Sweden*

JEANLOZ, ROGER W., (xxxix, 161, 241, 517, 685, 691, 835, 839), *Laboratory for Carbohydrate Research, Massachusetts General Hospital, Boston, Massachusetts 02114*

JENSEN, JOHN (713), *Institute of Dental Research, University of Alabama in Birmingham, Birmingham, Alabama 35294*

JONES, CAROLYN J. P. (795), *Department of Rheumatology, University of Manchester Medical School, Manchester M13 9PT, England*

JONES, MICHAEL B. (167), *Clinical Research Center, Division of Communicable Diseases, Watford Road, Harrow, Middlesex HA1 3UJ, England*

JUNQUA, SIMONE (177), *Laboratoire de Biochimie du Tissu Conjonctif, Faculté de Médecine, Université de Paris XII, 6 rue du Général Sarrail, 94000 Créteil, France*

KANG, SARWAN S. (747), *Department of Biochemistry, Queen Elizabeth College, Atkins Building, Campden Hill, London W8 7AH, England*

KAPLAN, JERRY (817), *Department of Physiology, The University of Connecticut Health Center, Farmington, Connecticut 60632*

KASAI, NORIYUKI (533), *School of Dentistry, Hokkaido University 7W, 13N, Kita-ku, Sapporo 060, Japan*

KASHKET, SHELBY (529), *Forsyth Dental Center, Boston, Massachusetts 02115*

KATSURA, NOBUHIKO (533), *School of Dentistry, Hokkaido University 7W, 13N, Kita-ku, Sapporo 060, Japan*

KAUFMAN, BERNARD (275), *Departments of Biochemistry and Medicine, Duke University Medical Center, Durham, North Carolina 27710*

KAUL, KARAN (181), *Department of Biochemistry, Albert B. Chandler Medical Center, University of Kentucky, Lexington, Kentucky 40506*

KAYE, GORDON I. (653), *Department of Anatomy, The Albany Medical College of Union University, Albany, New York 12208*

KELLS, DAVID I. C. (171), *Department of Biochemistry, The Hospital for Sick Children, 555 University Avenue, Toronto, Ontario, Canada M56 1X8*

KERNES, STEWART M. (877), *Departments of Pediatrics and Biochemistry, University of Chicago, Chicago, Illinois 60637*

KESSEL, DAVID (895), *Wayne State University School of Medicine, Department of Oncology, Harper-Grace Hospital, Detroit, Michigan 48202*

KEWLEY, MICHAEL A. (493, 795), *Department of Medical Biochemistry, Medical School, University of Manchester, Oxford Road, Manchester M13 9PT, England*

KHAN, NASEEM (271), *Playfair Neuroscience Unit, 1 Spadina Crescent, Toronto, Ontario, M5S 2J5, Canada*

KHILANANI, PREM (895), *Wayne State University School of Medicine, Department of Oncology, Harper-Grace Hospital, Detroit, Michigan 48202*

KIANG, WEI-LAI (217), *Department of Pharmacology, Downstate Medical Center, State University of New York, Brooklyn, New York 11203*

KIEDA, CLAUDINE (881) *Centre de Biophysique Moléculaire and Laboratoire de Chimie Biologique, C.N.R.S. et Université d'Orléans, 45045 Orléans, France*

KIERAS, FRED J. (185), *The Rockefeller University, New York, New York 10021*

KIM, YOUNG S. (1081), *Veterans Administration Hospital, Building 12, Room 109, San Francisco, California 94121*

KIMURA, ATSUSHI (893), *Department of Biochemistry, Fukushima Medical College, Fukushima 960, Japan*

KIMURA, JAMES H. (189), *Laboratory of Biochemistry, National Institute of Dental Research, National Institutes of Health, Bethesda, Maryland 20205*

KIRSCH, KATHARINA (1073), *Gastrointestinal Unit, Massachusetts General Hospital, Boston, Massachusetts 02114*

KLEIN, GEORGE (517), *Department of Tumor Biology, Karolinska Institutet Medical School, S-104 01 Stockholm 60, Sweden*

KLENK, HANS-DIETER (283), *Institut für Virologie, Justus–Liebig Universität Giessen, D-6300 Giessen, West Germany*

KNEPPER, PAUL A. (511), *Research-in-Aging-Laboratory, Veterans Administration Medical Center, North Chicago, Illinois 60064*

KOBATA, AKIRA (193), *Department of Biochemistry, Kobe University, School of Medicine, Ikuta-ku, Kobe, Japan*

KORYTNYK, WALTER (809), *Department of Experimental Therapeutics, Roswell Park Memorial Institute, Buffalo, New York 14263*

KOŚCIELAK, JERZY (199), *Department of Biochemistry, Instytut Hematologii, Chocimska 5, Warsaw, Poland*

KOZDROWSKI, I. (637), *Medizinisch–Chemisches Institut, Universität Bern, CH-3000 Bern, Switzerland*

KRAMER, MEBIUS F. (799, 805), *Department of Cell Biology, Center for Electron Microscopy, University of Utrecht Medical School, Nic. Beetsstraat 22, Utrecht, The Netherlands*

KUHNS, WILLIAM J. (719), *Department of Pathology, School of Medicine, University of North Carolina, Chapel Hill, North Carolina 27514*

LABEDZ, RHONA I. (493), *Department of Medical Biochemistry, University of Manchester, Medical School, Oxford Road, Manchester M13 9PT, England*

LAINE, ROGER A. (181, 673), *Department of Biochemistry, Albert B. Chandler Medical Center, University of Kentucky, Lexington, Kentucky 40506*

LAMBLIN, GENEVIÈVE (99), *Unité de Recherches n° 16 de l'I.N.S.E.R.M., Place de Verdun, 59045 Lille, France*

LAYNE, DONALD S. (723), *Department of Biochemistry, University of Ottawa, Ottawa, Ontario, Canada K1N 9A9*

LECAT, DANIEL (899), *Centre de Recherches sur les Protéines, Faculté de Médecine Lariboisière Saint-Louis, Laboratoire de Biochimie, 45 rue des Saints-Pères, 75006 Paris, France*

LECOMTE, JACQUELINE (1095), *Clinical Research Center, Division of Communicable Diseases, Watford Road, Harrow, Middlesex HA1 3UJ, England*

LEE, NORA (517), *Department of Pathology, Tufts University School of Medicine, Boston, Massachusetts 02111*

LEE, REIKO T. (205), *Department of Biology, The Johns Hopkins University, Baltimore, Maryland 21218*

LEE, YUAN CHUAN (205), *Department of Biology, The Johns Hopkins University, Baltimore, Maryland 21218*

LEHLE, LUDWIG (779), *Fachbereich Biologie, Universität Regensburg, Universitätsstrasse, D-8400 Regensburg, West Germany*

LEMONNIER, MARGUERITE (899), *Centre de Recherches sur les Protéines, Faculté de Médecine Lariboisière Saint-Louis, Laboratoire de Biochimie, 45 rue des Saints-Pères, 75006 Paris, France*

LESTER, ROBERT L. (181), *Department of Biochemistry, Albert B. Chandler Medical Center, University of Kentucky, Lexington, Kentucky 40506*

LHERMITTE, MICHEL (541), *Unité de Recherches n° 16 de l'I.N.S.E.R.M. and Laboratoire d'Histologie, Faculté de Medecine, 59045 Lille, France*

LI, SU-CHEN (209, 903), *Department of Biochemistry, Delta Regional Primate Research Center, Tulane Medical Center, New Orleans, Louisiana 70112*

LI, YU-TEH (3, 209, 903), *Department of Biochemistry, Delta Regional Primate Research Center, Tulane Medical Center, New Orleans, Louisiana 70112*

LIAU, YUN-HAU (907), *Department of Biochemistry, New York Medical College, Valhalla, New York 10595*

LINDAHL, ULF (713), *Department of Medical Chemistry, Royal Veterinary College Biomedical Center, S-751 23 Uppsala, Sweden*

LINDBERG, BENGT (17), *Department of Organic Chemistry, University of Stockholm, S-106 91 Stockholm, Sweden*

LIS, HALINA (213), *Department of Biophysics, Weizmann Institute of Science, Rehovot, Israel*

LIU, THERESA (441), *Center for Cancer Research, E17-236A, Massachusetts Institute of Technology, Cambridge, Massachusetts 02139*

LO, TIMOTHY M. (1047), *Division of Immunology, University of Miami, Miami, Florida 33152*

LOEWENSTEIN, MICHAEL (545), *Department of Chemistry, Adelphi University, Garden City, New York 11530*

LOHMANDER, STEFAN (725), *Laboratory of Biochemistry, National Insti-*

tute of Dental Research, National Institutes of Health, Bethesda, Maryland 20205

LOMBART, CHRISTIAN (139), *Laboratoire de Chimie Biologique, U.E.R. Biomédicale des Saints-Pères, 45 rue des Saints-Pères, 75006 Paris, France*

LOUISOT, PIERRE (649, 663), *Laboratoire de Biochimie, Faculté de Médecine Lyon-Sud, B.P. n° 12, 69600 Oullins, France*

LOWMAN, CATHERINE (247), *Duke University Medical Center, Durham, North Carolina 27710*

LYNN, WILLIAM S. (91, 275, 923), *Department of Biochemistry, Box 3711, Duke University Medical Center, Durham, North Carolina 27710*

MADSEN, KJELL (725), *Department of Histology, Karolinska Institutet, S-104 01 Stockholm, Sweden*

MAGET-DANA, RÉGINE (229, 553), *Centre de Biophysique Moléculaire C.N.R.S.; Université d'Orleans, 45045 Orléans, France*

MARCHESSAULT, ROBERT H. (245), *Département de Chimie, Universite de Montréal, Montréal, Quebec, Canada H3C 3V1*

MARGOLIS, RENÉE K. (217), *Department of Pharmacology, Downstate Medical Center, State University of New York, Brooklyn, New York 11203*

MARGOLIS, RICHARD U. (217), *Department of Pharmacology, New York University, School of Medicine, New York, New York 10016*

MARSHALL, J. WAYNE (1059), *Department of Biochemistry, Memorial University of Newfoundland, St. John's, Newfoundland, Canada A1B 3X9*

MASON, G. S. (261), *Playfair Neuroscience Unit, 1 Spadina Crescent, Toronto, Ontario, Canada M5S 2J5*

MAZZUCA, MARC (541), *Unité de Recherches n° 16 de 1'I.N.S.E.R.M. and Laboratoire d'Histologie, Faculté de Médecine, 59045 Lille, France*

McARTHUR, JANET W. (241), *Department of Obstetrics and Gynecology, Massachusetts General Hospital, Boston, Massachusetts 02114*

MENDICINO, JOSEPH (753), *Department of Biochemistry, University of Georgia, Athens, Georgia 30602*

MENTER, JULIAN M. (911), *Department of Engineering Biophysics, University Station, University of Alabama, Birmingham, Alabama 35294*

MERSMANN, GUNTHER (951), *Westfälische Wilhelms–Universität, Physiologisch–Chemisches Institut, Waldeyerstrasse 15, 44 Münster (Westf.), West Germany*

MICHALSKI, JEAN-CLAUDE (945), *Laboratoire de Chimie Biologique, Université des Sciences et Techniques de Lille I, B.P. n° 36, 59650 Villeneuve d'Ascq, France*

MICHON, CATHERINE (899), *Centre de Recherches sur les Protéines, Faculté de Médecine Lariboisière Saint-Louis, Laboratoire de Biochimie, 45 rue des Saints-Pères, 75006 Paris, France*

MILLER, DOUGLAS (221), *Department of Physiology and Biophysics, CMDNJ–Rutgers Medical School, University Heights, Piscataway, New Jersey 08854*

MILLER, KATHRYN S. (783), *Department of Orthopedics and Rehabilitation, University of Virginia Medical Center, Charlottesville, Virginia 22903*
MILLER-PODRAZA, HALINA (199), *Department of Biochemistry, Instytut Hematologii, Chocimska 5, Warsaw, Poland*
MOCZAR, ELEMER (775), *Université de Paris Val-de-Marne, 6 rue du Général Sarrail, 94000 Créteil, France*
MONSIGNY, MICHEL (229, 553, 881), *Centre de Biophysique Moléculaire C.N.R.S. et Laboratoire de Chimie Biologique, Université d'Orléans, 45045 Orléans, France*
MONTGOMERY, REX (287), *Department of Biochemistry, College of Medicine, University of Iowa, Iowa City, Iowa 52242*
MONTREUIL, JEAN (35, 149, 317, 933, 937, 945, 1027, 1091), *Laboratoire de Chimie Biologique, Université des Sciences et Techniques de Lille I, B.P. n° 36, 59650 Villeneuve d'Ascq, France*
MOOKERJEA, SAILEN (1059), *Department of Biochemistry, Memorial University of Newfoundland, St. John's, Newfoundland, Canada A1B 3X9*
MOORE, JAMES D. (241), *Laboratory for Carbohydrate Research, Massachusetts General Hospital, Boston, Massachusetts 02114*
MORA, PETER T. (847), *Macromolecular Biology Section, National Cancer Institute, National Institutes of Health, Bethesda, Maryland 20205*
MORELIS, RENÉE (663), *Laboratoire de Biochimie, Faculté de Médecine Lyon-Sud, B.P. n° 12, 69600 Oullins, France*
MUIR, HELEN (375), *Kennedy Institute of Rheumatology, Bute Gardens, Hammersmith, London W6 7DW, England*
MULLER, FRANCISCO (743), *Arthritis Division, University of Miami, School of Medicine, Box 520875, Biscayne Annex, Miami, Florida 33152*
MURRAY, LOUANN W. (817), *Department of Biochemistry, The University of Connecticut Health Center, Farmington, Connecticut 06032*
MURTHY, SATYA M. (941), *Evanston Hospital of Immunological Research, Evanston, Illinois 60201*
MURTY, VARAHABHOTLA L. N. (145, 233), *Department of Biochemistry, New York Medical College, Valhalla, New York 10595*
MUTHIAH, PALANIAPPAN (125), *Department of Pathology, The University of Adelaide, Box 498, G.P.O., Adelaide, South Australia 5001*
MUTO, YASUTOSHI (949), *Department of Nutrition, School of Health Sciences, University of Tokyo, Faculty of Science, Bunkyo-ku, Tokyo 113, Japan*
NADER, FRANZ (521), *University of Heidelberg, Hospitalstrasse 3, D-6900 Heidelberg 1, West Germany*
NAKAMURA, NOBUTO (911), *Department of Medicine, The Center for Adult Disease, Osaka, 1-3-3 Nakamichi, Higashinari-ku, Osaka, Japan 537*
NAPIER, MARY A. (237, 525), *Department of Medicine, School of Medicine, University of North Carolina, Chapel Hill, North Carolina 27514*
NARASIMHAN, SAROJA (575), *Department of Biochemistry, University of*

Toronto and Research Institute, Hospital for Sick Children, Toronto, Canada M5G 1X8

NASIR-UD-DIN (241, 839), *Laboratory for Carbohydrate Research, Massachusetts General Hospital, Boston, Massachusetts 02114*

NATO, FARIDA (1037), *Centre de Recherches sur les Protéines, Faculté de Médecine Lariboisière Saint-Louis, 45 rue des Saints-Pères, 75006 Paris, France*

NORDÉN, NILS E. (537), *Department of Clinical Chemistry, University Hospital, S-221 85 Lund, Sweden*

NORDIN, JOHN H. (245), *Department of Biochemistry, University of Massachusetts, Amherst, Massachusetts 01003*

NOTANI, KEN-ICHI (533), *School of Dentistry, Hokkaido University 7M, 13N, Kita-ku, Sapporo 060, Japan*

OBRENOVITCH, ANGÈLE (553), *Centre de Biophysique Moléculaire, C.N.R.S., 45045 Orléans, France*

O'BRIEN, PAUL J. (729), *Laboratory of Vision Research, National Eye Institute, National Institutes of Health, Bethesda, Maryland 20205*

OGAMO, AKIRA (903), *Delta Regional Primate Research Center, Covington, Louisiana 70433*

OGURI, KAYOKO (851), *Mitsubishi-Kasei Institute of Life Sciences, 11 Minamiooya, Machida, Tokyo 194, Japan*

OLIVER, RODERICK T. D. (719), *St. Bartholomew's Hospital, London E.C.1, England*

OSAWA, TOSHIAKI (1065), *Division of Chemical Toxicology and Immunochemistry, Faculty of Pharmaceutical Sciences, University of Tokyo, Bunkyo-ku Tokyo 113, Japan*

OSDOBY, PHILIP (189), *Department of Biology, Case Western Reserve University, Cleveland, Ohio 44106*

OWEN, ALBERT J. (915), *Department of Biochemistry and Biophysics, University of Rhode Island, Kingston, Rhode Island 02881*

PAL, SUBHASH (393), *New York University School of Medicine, New York, New York 10016*

PARK, JOHN W. (119), *Eye Research Institute of the Retina Foundation, Boston, Massachusetts 02114*

PARKER, TERENCE L. (131, 917), *Arbeitsgruppe für Zellchemie, Ruhr-Universität, D-4630 Bochum, West Germany*

PATKOWSKA, MARGARET (733), *Gastroenterology Research Laboratory, Department of Medicine and Biochemistry, New York Medical College, New York, New York 10029*

PAULSON, JAMES C. (247, 763), *Department of Biochemistry, Duke University Medical Center, Durham, North Carolina 27710*

PAZUR, JOHN H. (251), *108 Althouse Laboratory, Pennsylvania State University, University Park, Pennsylvania 16802*

PEREZ-GONZALEZ, MARIA-NIEVES (649), *Laboratoire de Biochimie Médicale, Université de Lyon, Faculté de Médecine Lyon-Sud, B. P. n° 12, 69600 Oullins, France*

PERRET, VERA (87), *Department of Biochemistry, New York Medical College, Valhalla, New York 10595*

PETERSON, CHRISTINE (145), *Department of Chemistry, Herbert Lehman College, City University of New York, Bronx, New York 10468*

PICARD, JACQUES (549, 737), *Laboratoire de Biochimie, Université de Paris VI, Faculté de Médecine Saint-Antoine, 27 rue Chaligny, 75571 Paris, France*

PIGMAN, WARD* (87, 145, 233), *Department of Biochemistry, New York Medical College, Valhalla, New York 10595*

PITA, JULIO C. (743), *Arthritis Division, University of Miami School of Medicine, Box 520875, Biscayne Annex, Miami, Florida 33152*

PORTER, CARL W. (809), *Department of Experimental Therapeutics, Roswell Park Memorial Institute, Buffalo, New York 14263*

POWELL, SUSAN (299), *Shriner Burns Institute, Boston, Massachusetts 02114*

PRASAD, RAJANI (257), *Department of Biochemistry, University of Kansas Medical Center, Kansas City, Kansas 66103*

PRESPER, KATHLEEN A. (873), *Department of Chemistry, University of Notre Dame, Notre Dame, Indiana 16556*

PRICE, ROBERT G. (305, 747), *Department of Biochemistry, Atkins Building, Queen Elizabeth College, Campden Hill, London W8 7AH, England*

PRIEELS, JEAN-PAUL (641), *Duke University Medical Center, Durham, North Carolina 27710*

QUARONI, ANDREA (1073), *Gastrointestinal Unit, Massachusetts General Hospital, Boston, Massachusetts 02114*

QUINTARELLI, GIULIANO (313, 751), *Regina Elena Institute for Cancer Research, Viale R. Elena 291, Rome, Italy*

RABIN, PETER (261), *Playfair Neuroscience Unit, 1 Spadina Crescent, Toronto, Ontario, Canada M5S 2J5*

RAO, A. KALYAN (753), *Department of Biochemistry, University of Georgia, Athens, Georgia 30602*

RATNAM, SAMUEL (1059), *Department of Biochemistry, Memorial University of Newfoundland, St. John's, Newfoundland, Canada A1B 3X9*

REARICK, JAMES I. (763), *Department of Biochemistry, Duke University Medical Center, Durham, North Carolina 27710*

REGLERO, ANGEL (867), *Facultades de Ciencias y Farmacia, Universidad de Salamanca, Salamanca, Spain*

REINHOLD, VERNON N. (241, 265, 839), *Department of Biological Chemistry, Harvard Medical School, Boston, Massachusetts 02115*

REINKING, ARNOLD (759), *Laboratory for Physiological Chemistry, State University of Utrecht, Vondellaan 24 A, 3521 GG Utrecht, The Netherlands*

*Deceased

REUVERS, FRANS (759), *Laboratory for Physiological Chemistry, State University of Utrecht, Vondellaan 24 A, 3521 GG Utrecht, The Netherlands*

REVOL, JEAN-FRANÇOIS (245), *Département de Chimie Université de Montréal, Montréal, Québec, Canada H3C 3V1*

RITCHIE, R. GEORGE S. (83), *Department of Chemistry, York University, Downsview, Toronto, Ontario, Canada M3J 1P3*

ROBBINS, PHILLIPS W. (441), *Center for Cancer Research, E-17-233, Massachusetts Institute of Technology, Cambridge, Massachusetts 02139*

ROBERT, LADISLAS (177), *Laboratoire de Biochimie du Tissu Conjonctif, Faculté de Médecine, Université Paris XII, 6 rue du Général Sarrail, 94000 Créteil, France*

ROBERTS, ANNE E. (885), *Division of Inherited Metabolic Diseases, MRC Clinical Research Center, Watford Road, Middlesex HA1 3UJ, England*

ROCHE, ANNIE-CLAUDE (229, 541, 553), *Centre de Biophysique Moléculaire du C.N.R.S., 45045 Orléans, France*

RODÉN, LENNART (313, 713), *Department of Medicine and Biochemistry, School of Dentistry, Institute of Dental Research, University of Alabama in Birmingham, Birmingham, Alabama 35294*

ROELCKE, DIETER (521), *Institute of Immunology and Serology, University of Heidelberg, Hospitalstrasse 3, D-6900 Heidelberg 1, West Germany*

ROOMI, WAHEED (271), *Playfair Neuroscience Unit, 1 Spadina Crescent, Toronto, Ontario, Canada M5S 2J5*

ROSE, MARY CALLAGHAN (275), *Department of Biochemistry, Duke University, Durham, North Carolina 27705*

ROSENBERG, LAWRENCE (393), *Orthopedic and Connective Tissue Research Laboratories, Montefiore Hospital and Medical Center, Bronx, New York 10467*

ROSTAD, STEVEN (691), *Laboratory for Carbohydrate Research, Massachusetts General Hospital, Boston, Massachusetts 02114*

ROUSSEL, PHILIPPE (99, 541), *Unité des Protéines de l'I.N.S.E.R.M. n° 16 Place de Verdun, 59045 Lille, France*

ROUX, MICHEL (245), *Département de Chimie, Université de Montréal, Montréal, Québec, Canada H3C 3V1*

ROWLAND, FREDERICK N. (817), *Department of Biochemistry, The University of Connecticut Health Center, Farmington, Connecticut 06032*

SACHDEV, GOVERDHAN P. (279), *Oklahoma Medical Foundation, Oklahoma City, Oklahoma 73104*

SADLER, J. EVAN (763), *Department of Biochemistry, Duke University Medical Center, Durham, North Carolina 27710*

SAHU, SAURA (923), *Department of Biochemistry, Duke University, Durham, North Carolina 27710*

SANDER, MICHAEL (557, 927), *Arbeitsgruppe für Zellchemie, Ruhr-Universität, D-4630 Bochum, West Germany*

SASAK, WLODZIMIERZ (767), *National Cancer Institute, National Institutes of Health, Bethesda, Maryland 20205*

SATO, MAYUMI (949), *Department of Nutrition, School of Health Sciences, University of Tokyo, Faculty of Science, Bunkyo-ku, Tokyo 113, Japan*

SAWICKA, THÉRÈSE (933), *Laboratoire de Chimie Biologique, Université des Sciences at Techniques de Lille I, B.P. n° 36, 59650 Villeneuve d'Ascq, France*

SCANLON, EDWARD F. (941), *Evanston Hospital for Immunological Research, Evanston, Illinois 60201*

SCHACHTER, HARRY (575), *Department of Biochemistry, Hospital for Sick Children, 555 University Avenue, Toronto, Ontario, Canada M5G 1X8*

SCHAFER, DOROTHY (955), *College of Environmental Science and Forestry, Syracuse Campus, State University of New York, Syracuse, New York 13210*

SCHAUER, ROLAND (131, 557, 597, 771, 775, 917, 927), *Biochemisches Institut, Universität Kiel, Olshausenstr. 40-60, D-2300 Kiel, West Germany*

SCHIPHORST, WIETSKE E. C. M. (637, 829), *Laboratorium voor Chemische Fysiologie, Vrije Universiteit, Amsterdam, The Netherlands*

SCHMID, KARL (149), *Department of Biochemistry, Boston University School of Medicine, Boston, Massachusetts 02118*

SCHMIDT, MICHAEL F. G. (283, 779), *Institute für Virologie der Justus-Liebig Universität Giessen, Frankfurter Strasse 107, D-6300 Giessen, West Germany*

SCHNEIDER, DIETMAR H. (1077), *Institute for Medical Microbiology, Hochhaus am Augustusplatz, D-6500 Mainz, West Germany*

SCHROEDER, TERRY (279), *Biomembrane Research Laboratory, University of Oklahoma Health Sciences Center, Oklahoma City, Oklahoma 73104*

SCHUT, BERNARD L. (149), *Laboratory for Organic Chemistry, University of Utrecht, Croesestraat 79, Utrecht, The Netherlands*

SCHUTZBACH, JOHN S. (751, 791), *Department of Microbiology, University of Alabama, Birmingham, Alabama 35294*

SCHWARTZ, EDITH R. (783), *Department of Orthopedics, Tufts University School of Medicine, Boston, Massachusetts 02111*

SCHWARTZ, NANCY B. (787), *Department of Pediatrics, University of Chicago, Box 413, Chicago, Illinois 60637*

SCHWARZ, RALPH T. (283, 779), *Institut für Virologie der Justus-Liebig Universität Giessen, Frankfurter Strasse 107, D-6300 Giessen, West Germany*

SEAR, CHRISTOPHER, H. J. (795), *Beecham Pharmaceuticals Research Division, Great Burgh, Yew Tree Bottom Road, Epsom, Surrey KT18 5XQ, England*

SEYMOUR, FRED R. (105), *Department of Biochemistry, Baylor College of Medicine, Houston, Texas 77030*

SHAN, ALICE (1051), *Midwest Research Institute, Kansas City, Missouri 64110*

SHARON, NATHAN (213, 459, 1033), *Department of Biophysics, The Weizmann Institute of Science, Rehovot, Israel*

SHEPHERD, VIRGINIA (287), *Department of Biochemistry, University of Iowa College of Medicine, Iowa City, Iowa 52242*

SHIONO, MASAKI (533), *School of Dentistry, Hokkaido University 7W, 13N, Kita-ku, Sapporo 060, Japan*

SIDDIQUI, BADER (1081), *Veterans Administration Hospital, San Francisco, California 94121*

SIX, PIERRE (933), *Laboratoire de Chimie Biologique, Université des Sciences et Techniques de Lille I, B.P. n° 36, Villeneuve d'Ascq 59650, France*

SLOMIANY, AMALIA (291, 295, 733), *Gastroenterology Research Laboratory, Department of Medicine and Biochemistry, New York Medical College, New York, New York 10029*

SLOMIANY, BRONISLAW (291, 295, 733), *Gastroenterology Research Laboratory, Department of Medicine and Biochemistry, New York Medical College, New York, New York 10029*

SOSSO, MARIA CRISTINA (157), *Institute of Biological Chemistry, Faculty of Sciences, University of Pavia, 27100 Pavia, Italy*

SOTMAN, STUART (299), *Shriner Burns Institute, Boston, Massachusetts 02114*

SPEE-BRAND, RITA (799), *Department of Histology and Cell Biology, Medical School, University of Utrecht, Utrecht, The Netherlands*

SPELLACY, ELIZABETH (885), *Division of Inherited Metabolic Diseases, MRC Clinical Research Center, Watford Road, Middlesex HA1, 3UJ, England*

SPEZIALE, PIETRO (157), *Institute of Biological Chemistry, Faculty of Sciences, University of Pavia, 27100 Pavia, Italy*

SPIK, GENEVIÈVE (149, 317, 933, 937), *Laboratoire de Chimie Biologique, Université des Sciences et Techniques de Lille I, B.P. n° 36, 59650 Villeneuve d'Ascq, France*

SPIRO, MARY JANE (613), *Elliott P. Joslin Research Laboratory, Boston, Massachusetts 02215*

SPIRO, ROBERT G. (613) *Elliott P. Joslin Research Laboratory, Boston, Massachusetts 02215*

SPRINGER, GEORG F. (563, 659, 941), *Immunochemical Research, Evanston Hospital, Evanston, Illinois 60201*

STARCHER, BARRY (751), *Pulmonary Division, WUMC, 499 South Euclid, St. Louis, Missouri 63110*

STERN, ROBERT C. (503), *Department of Pediatrics, Rainbow Babies and Children's Hospital, Case Western Reserve University, Cleveland, Ohio 44106*

STIRLING, JOHN L. (937), *Department of Biochemistry, Queen Elizabeth College, Atkins Building, Campden Hill, London W8 7AH, England*

STRECKER, GÉRARD (149, 317, 937, 945, 951), *Laboratoire de Chimie Biologique, Université des Sciences et Techniques de Lille I, B.P. n° 36, 59650 Villeneuve d'Ascq, France*

STROUS, GER, J. A. M. (799, 805), *State University Utrecht, Laboratory of Histology and Cell Biology, Nic. Beetsstraat 22, Utrecht, The Netherlands*

SUFRIN, JANICE R. (809), *Department of Experimental Therapeutics, Roswell Park Memorial Institute, Buffalo, New York 14263*

SUZUKI, AKEMI (949), *Department of Biochemistry, Faculty of Medicine, The University of Tokyo, Bunkyo-ku, Tokyo 113, Japan*

SWANN, DAVID A. (299), *Shriner Burns Institute, Boston, Massachusetts 02114*

SWANSON, MELVIN (1051), *Midwest Research Institute, Kansas City, Missouri 64110*

SWEET, M. BARRY E. (813, 819), *Orthopaedic Research Laboratories, University of Witwatersrand, Johannesburg, South Africa*

SWIFT, HEWSON (165), *Department of Biology, University of Chicago, Chicago, Illinois 60637*

TAI, TADASHI (193), *Department of Biochemistry, Kobe University, School of Medicine, Ikuta-ku, Kobe, Japan*

TAKETOMI, TAMOTSU (309), *Department of Biochemistry, Institute of Adaptation Medicine, Shinshu University, Matsumoto 390, Japan*

TAKITA, HORIYUKI (533), *School of Dentistry, Hokkaido University 7W, 13N, Kita-ku, Sapporo 060, Japan*

TANENBAUM, STUART W. (955), *College of Environmental Science and Forestry, Syracuse Campus, State University of New York, Syracuse, New York 13210*

TANG, LIN-HENG (393), *Orthopedic and Connective Tissue Research Laboratories, Montefiore Hospital and Medical Center, Bronx, New York 10467*

TANNER, WIDMAR (677), *Universität Regensburg, Fachbereich Biologie, D-8400 Regensburg, West Germany*

TANZER, MARVIN L. (817), *Department of Biochemistry, The University of Connecticut Health Center, Farmington, Connecticut 06032*

TAYLOR, SARAH A. (305, 747), *Department of Biochemistry, Queen Elizabeth College, Atkins Building, Campden Hill, London W8 7AH, England*

TCHILIAN, MARIE JOSÉ (899), *Laboratoire de Biochimie, Centre de Recherches sur les Protéines, Faculté de Médecine, 45 rue des Saints-Pères, 75006 Paris, France*

THONAR, EUGENE J.-M. A. (813, 819), *Orthopaedic Research Laboratories, University of the Witwatersrand, Johannesburg, South Africa*

TROY, FREDERIC A. (1085), *Department of Biological Chemistry, University of California School of Medicine, Davis, California 95616*

TSUJI, TSUTOMU (1065), *Division of Chemical Toxicology and Immunochemistry, Faculty of Pharmaceutical Sciences, University of Tokyo, Bunkyo-ku, Tokyo 113, Japan*

TSURUMI, KOICHI (893), *Department of Biochemistry, Fukushima Medical College, Fukushima 960, Japan*

TURCO, SAM (441), *E17-236, Center for Cancer Research, Massachusetts Institute of Technology, Cambridge, Massachusetts 02139*

UEMURA, KEIICHI (309), *Department of Biochemistry, Institute of Adaptation Medicine, Shinshu University, Matsumoto 390, Japan*

UENO, KUNIHIRO (701), *Department of Biochemistry, Tokyo Metropolitan Institute of Gerontology, Tokyo, Japan*

ULLRICH, KURT (951), *Westfälische Wilhelms-Universität, Physiologisch-Chemisches Institut, Waldeyerstrasse 15, 44 Münster (Westf.), West Germany*

UPHOLT, WILLIAM B. (823), *Departments of Pediatrics and Biochemistry, University of Chicago, Chicago, Illinois 60637*

VALENT, BARBARA S. (415), *Department of Chemistry, University of Colorado, Boulder, Colorado 80309*

VAN DEN EIJNDEN, DIRK H. (637, 829), *Department of Medical Chemistry, Vrije Universiteit, Amsterdam, The Netherlands*

VAN HUIS, GERARD A. (799, 805), *Department of Cell Biology, Center for Electron Microscopy, College of Medicine, University of Utrecht, Nic. Beesstraat 22, Utrecht, The Netherlands*

VEH, RÜDIGER W. (557, 917, 927), *Arbeitsgruppe für Zellchemie, Ruhr-Universität, D-4630 Bochum, West Germany*

VERBERT, ANDRÉ (1091), *Université des Sciences et Techniques de Lille I, Laboratoire de Chimie Biologique, B.P. n° 36, 59650 Villeneuve d'Ascq, France*

VERTEL, BARBARA M. (823), *Department of Pediatrics and Biochemistry, University of Chicago, Illinois 60637*

VITELLO, LIDIA B. (511), *Research-in-Aging Laboratory, Veterans Administration Medical Center, North Chicago, Illinois 60064*

VLIEGENTHART, JOHANNES F. G. (35, 149), *Laboratory for Organic Chemistry, University of Utrecht, Croesestraat 79, Utrecht, The Netherlands*

VOCATURO, AMINO (313, 751), *Institute of Dental Research, Departments of Medicine and Biochemistry, University of Alabama in Birmingham, University Station, Birmingham, Alabama 35294*

VORDERBRUEGGE, WILLIAM F. (1011), *Departments of Internal Medicine and Biological Chemistry, Simpson Memorial Institute, The University of Michigan, Ann Arbor, Michigan 48109*

WALKER-NASIR, EVELYNE (835), *Laboratory for Carbohydrate Research, Massachusetts General Hospital, Boston, Massachusetts 02114*

WAN, CHIN CHIN (209), *Department of Biochemistry and Delta Regional Primate Research Center, Tulane Medical Center, New Orleans, Louisiana 70112*

WANG, PHILIP (955), *College of Environmental Science and Forestry, Syracuse Campus, State University of New York, Syracuse, New York 13210*

WARIN, VINCENT (317), *Laboratoire de Physique des Solides, Université des Science et Techniques de Lille I, B.P. n° 36, 59650 Villeneuve d'Ascq, France*

WARREN, CHRISTOPHER D. (839), *Laboratory for Carbohydrate Research, Massachusetts General Hospital, Boston, Massachusetts 02114*

WATKINS, WINIFRED M. (719, 1019), *Division of Immunochemical Genetics, MRC Clinical Research Centre, Watford Road, Harrow, Middlesex HA1 3UJ, England*

WATTS, R. W. E. (885), *Division of Inherited Metabolic Diseases, MRC Clinical Research Center, Watford Road, Middlesex HA1 3UJ, England*

WEBER, ERNST (951), *Westfälische Wilhelms-Universität, Physiologisch-Chemisches Institut, Waldeyerstrasse 15, 44 Münster (Westf.), West Germany*

WEBER, PETER (1055), *Department of Biochemistry, The Albany Medical College of Union University, Albany, New York 12208*

WEICKER, HELMUT (521), *University of Heidelberg, Hospitalstrasse 3, D-6900 Heidelberg, West Germany*

WEINSTEIN, HYMAN (511, 561), *Research-in-Aging Laboratory, Veterans Administration Medical Center, North Chicago, Illinois 60064*

WEISER, MILTON M. (637, 1073), *Gastrointestinal Unit, Massachusetts General Hospital, Boston, Massachusetts 02114*

WEISSMANN, BERNARD (959), *Department of Biological Chemistry, School of Basic Medical Sciences, University of Illinois at the Medical Center, Chicago, Illinois 60612*

WELTEN-VERSTEGEN, G. (759), *Laboratory for Physiological Chemistry, State University of Utrecht, Vondellaan 24 A, 3521 GG Utrecht, The Netherlands*

WEMBER, MARGARET (771, 775), *Biochemisches Institut, Universität Kiel, D-2300 Kiel, West Germany*

WEN, GARY (279), *Biomembrane Research Laboratory, University of Oklahoma Health Sciences Center, Oklahoma City, Oklahoma 73104*

WEST, SEYMOUR S. (911), *Department of Engineering Biophysics, University Station, University of Alabama, Birmingham, Alabama 35294*

WIBO, MAURICE (669, 843), *Laboratoire de Chimie Physiologique, Université Catholique de Louvain 75-39, Avenue Hippocrate 75, B-1200 Brussels, Belgium*

WILLIAMSON, DENIS G. (723), *Department of Biochemistry, University of Ottawa, Ottawa, Ontario, Canada K1N 9A9*

WILSON, JAMES R. (575), *Department of Biochemistry, University of Toronto and Research Institute, Hospital for Sick Children, Toronto, Canada*

WINTER, WILLIAM T. (321), *Department of Chemistry, Polytechnic Institute of New York, Brooklyn, New York 11201*

WINTERBOURNE, DAVID J. (847), *Macromolecular Biology Section, National Cancer Institute, National Institutes of Health, Bethesda, Maryland 20205*

WIRTH, DYANN (441), *E17-236A, Center for Cancer Research, Massachusetts Institute of Technology, Cambridge, Masachusetts 02139*

WOOD, EDWIN (1095), *Clinical Research Center, Division of Communicable Diseases, Watford Road, Harrow, Middlesex HA1 3UJ, England*

WRIGHT, JOHN A. (167), *Clinical Research Center, Division of Communicable Diseases, Watford Road, Harrow, Middlesex HA1 3UJ, England*

YAMAGATA, TATSUYA (427, 851), *Mitsubishi-Kasei Institute of Life Sciences, 11 Minamiooya, Machida, Tokyo 194, Japan*

YAMAKAWA, TAMIO (701, 949), *Department of Biochemistry, Faculty of Medicine, The University of Tokyo, Bunkyo-ku, Tokyo 113, Japan*

YAMASHITA, KATSUKO (193), *Department of Biochemistry, Kobe University, School of Medicine, Ikuta-ku, Kobe, Japan*

YANG, HUNG-JU (563), *Evanston Hospital, Evanston, Illinois 60201*

YANO, SHINGO (733), *Gastroenterology Research Laboratory, Department of Medicine and Biochemistry, New York Medical College, New York, New York 10029*

YASUMOTO, SHIGERU (851), *Mitsubishi-Kasei Institute of Life Sciences, 11 Minamiooya, Machida, Tokyo 194, Japan*

YEUNG, KWOKAN K. (915), *Department of Biochemistry and Biophysics, University of Rhode Island, Kingston, Rhode Island 02881*

YU, ROBERT K. (79), *Department of Neurology, Yale University School of Medicine, New Haven, Connecticut 06510*

YUDKIN, JOHN (747), *Department of Biochemistry, Queen Elizabeth College, Atkins Building, Campden Hill, London W8 7AH, England*

ZDEBSKA, EWA (199), *Department of Biochemistry, Instytut Hematologii, Chocimska 5, Warsaw, Poland*

ZENTZ, CHRISTIAN (567), *Centre de Recherches sur les Protéines, Faculté de Médecine Lariboisière Saint-Louis, 45 rue des Saints-Pères, 75006 Paris, France*

TRIBUTE

It is with deep sorrow that we note the passing of Professor Ward Pigman. He was stricken by a heart attack during our recent meeting at Woods Hole, Massachusetts. Ward was highly regarded for the texts that he wrote, *Advances in Carbohydrate Chemistry,* which he initiated, and for his efforts as a councillor to the American Chemical Society and as a Chairman of its Division of Carbohydrate Chemistry. His research, which began with studies of mutarotation and the behavior of simple sugars, di- and trisaccharides (with H. S. Isbell), progressed successfully to investigations with his students and collaborators into the action of almond emulsin, role of saliva in dental carries, remineralization of teeth, mechanisms of degradation of hyaluronic acid, purification and structure of submaxillary mucins, the alkaline catalyzed β-elimination reaction, to name but a few. Equal to, or perhaps surpassing his strictly scientific contributions, were his efforts toward advancing the field of glycoconjugate research by improving communication between participants in this field. Ward organized, together with Sophie Jakowska, the 1963 *Conference on Mucous Secretions,* convened under the auspices of the New York Academy of Sciences. This conference was widely heralded as marking a turning point toward sophisticated research on glycoproteins and acid mucopolysaccharides. In 1965, he organized (in collaboration with E. A. Balazs and R. W. Jeanloz) the *First International Symposium on Glycoconjugates* at Swampscott, Massachusetts. He was known to all of us as a founding father and past president of the Society of Complex Carbohydrates, and a vital member who worked to foster international cooperation among the various individuals and societies interested in glycoconjugates. The recent *Fourth International Symposium on Glycoconjugates* represented to him the ideal assembly and a culmination of many years of work and hope. It is altogether fitting and appropriate that the participants in the symposium voted unanimously to dedicate this volume of the proceedings in his memory. We shall sorely miss Ward Pigman, our friend and colleague.

Martin I. Horowitz

PREFACE

These volumes are a record of the Fourth International Symposium on Glycoconjugates held from September 27 to October 1, 1977, at Woods Hole, Massachusetts.

International meetings on glycoconjugates are now being held every two years, a frequency that is well justified by advances in the field. This meeting was held in spectacularly beautiful weather and surroundings. In addition, the meeting provided a welcome opportunity for a special symposium in honor of Martin Mathews for his many years of devotion and constructive research in the field of connective tissue biochemistry.

A cloud unfortunately appeared, however, when news came of the death, during the meeting, of Ward Pigman, an outstanding personality in the area of carbohydrate chemistry, a prime motive force in the organization of the Society for Complex Carbohydrates, and one of the organizers of the First International Symposium on Glycoconjugates. Our shock and sense of loss are well documented on a later page by his colleague Martin Horowitz, and these reports are therefore dedicated to the memory of Ward Pigman.

The field of complex carbohydrates, or glycoconjugates, is one of growing interest and activity because of the recognition of the functional importance of these compounds at the surface of cells and in the extracellular matrices, and because of the development of new methods of studying their structure and their biosynthesis. These aspects are here represented by many examples. The field encompasses a rather wide range of compounds: polysaccharides, glycoproteins, glycolipids, proteoglycans, and all their varieties in plants, animals, and microorganisms. The structures all have in common, however, the glycosidic linkage, and new knowledge in one part of the field is often of value in others. The participants in this meeting were thus linked by a common interest in complex biological molecules containing carbohydrates.

The value of a symposium report lies in its ability to give the reader a feeling for the state of knowledge in a wide area. With this aim, each section is headed by the presentation of one or more invited speakers in order to set the stage for the shorter and more specific contributions that follow. The divisions are, of course, somewhat arbitrary, and the editors accept the responsibility for the assignment of papers to the divisions.

The organizing committee of the symposium expresses its appreciation and thanks to the following organizations for financial support that made possible many of the arrangements. Major support from the National Institute of Arthritis, Metabolism, and Digestive Disease, U.S.P.H.S. (Grants AM-03564, -20089, and -20945), the Diabetes Trust Fund, the University of Alabama in Birmingham, and the International Union of Biochemistry (BBA

Fellowships) was supplemented by contributions from Hoffman-LaRoche, Inc., Abbott Laboratories, and the Ortho Research Foundation.

The program committee (L. Rodén, chairman) is grateful to those who served as session chairmen: A. Dorfman, R. L. Hill, B. Lindberg, Y.-T. Li, W. T. J. Morgan, R. Schauer, N. Sharon, and W. M. Watkins. The editors have been ably assisted in the editing of the manuscripts by Dorothy A. Jeanloz, and in their preparation by Andrea Morris, Phyllis Tura, and Nalayini Fernando.

John D. Gregory
Roger W. Jeanloz

CONTENTS OF VOLUME II

BIOSYNTHESIS AND REGULATION

Asterisks denote invited speakers

GLYCOCONJUGATES OF CELL MEMBRANES

WELCOME

There is no better way to greet you—and to set the stage, for what we confidently expect to be an outstanding symposium—than with the words of the distinguished embryologist E. G. Conklin, "There are many marine laboratories in the world but there is only one Woods Hole."

I welcome you to the unique scientific community that is Woods Hole. This small village is a world center for marine science. Its reputation may seem out of proportion with its size, yet it contains two federal laboratories—the Northeast Fisheries Center of the National Marine Fisheries Service and the Branch of the Atlantic-Gulf of Mexico Geology of the U.S. Geological Survey; and two private nonprofit institutions—the Marine Biological Laboratory, which I am privileged to serve, and our younger sister, the Woods Hole Oceanographic Institution. I mention the National Marine Fisheries Service first since it has the distinction of being the oldest laboratory in the village. This international fisheries center has gone under several names, prior to 1904, as the "United States Fish Commission," and from 1904 until recently as the "Bureau of Fisheries." The work of the Commission was established in Woods Hole in temporary quarters in 1871 and its first definitive laboratory was erected in 1885.

The Marine Biological Laboratory is just entering its ninetieth year and in 1980 the Woods Hole Oceanographic Institution will celebrate its fiftieth birthday.

Taken together, these four institutions occupy more than 30 buildings, use 13 research vessels and collecting boats, employ more than 1000 persons, and offer instruction to approximately 400 students annually. I hope that during your stay many of you will have an opportunity to at least sample some of the scientific offerings in this remarkable concatenation of institutions whose interests range from the ocean floor to the biology of salt marshes to marine biomedicine. In fact, in the latter field the glycoconjugates are not unknown in Woods Hole.

There are at least three key ingredients in the making of a successful symposium. First, the subject must be timely, indeed, it must be forward-looking. Over a century and a half ago, Thomas Jefferson wrote in a letter to John Adams, "I like the dreams of the future better than the history of the past." So it is with scientists. Any symposium must, to some degree, deal with past and already cold events, but the architects of this symposium have fashioned a program focusing on today's and, hopefully, tomorrow's ideas.

The second ingredient is one over which the organizers and the hosts have relatively little control. It is the audience. An audience at an international symposium must have at least a slight bias toward youth. You will understand if I define a "youth" as anyone under the age of 55. More importantly the audience should have come prepared to work. I understand that participants in the earlier symposia of this group have done just that. I hope that this symposium will be no exception.

Finally, a successful symposium requires an attractive setting and congenial hosts. To that end we will do our best. Thank you for coming.

James D. Ebert
President
Marine Biological Laboratory
Woods Hole, Massachusetts

Structure of Complex Carbohydrates

GLYCOCONJUGATE RESEARCH / Volume I
Proceedings of the Fourth International Symposium on Glycoconjugates

Enzymatic Methods for Structural Analysis of Complex Carbohydrates

Yu-Teh Li

Recent investigations in various fields have revealed that complex carbohydrates play many important and intriguing biological roles. To appreciate fully the biological significance of glycoconjugates, the exact structure of carbohydrate chains in these macromolecules must be determined. The structural analysis of a complex carbohydrate chain involves the determination of the anomeric configuration, the sequential arrangement of saccharide units, and the linkage between different sugar units.

Four years ago, at the Second International Symposium on Glycoconjugates in Lille (France), both Dr. Egami and I reported the usefulness of applying *exo*-glycosidases for the structural analysis of glycoconjugates. Today I am happy to report to you that, in addition to *exo*-glycosidases, several *endo*-glycosidases are now available for structural analysis. The purpose of this presentation is to give a current review on the use of glycosidases for the structural analysis of complex carbohydrates.

At the center of Scheme 1 is the schematic illustration of a complex carbohydrate chain. I use A, B, C, D, and E to designate different sugar units. The arrows indicate the four possible positions at the pyranose ring that may be involved in forming glycosidic linkages. As we all know, the structural analysis of complex carbohydrates involves the determination of the anomeric configuration and of the sequential arrangement of saccharide units, as well as that of the linkage-points between sugar units. The stepwise release of monosaccharide units from the nonreducing terminal by specific *exo*-glycosidases can provide us with information concerning the anomeric configuration and the sequential arrangement of sugar units.

ISBN 0-12-301301-1

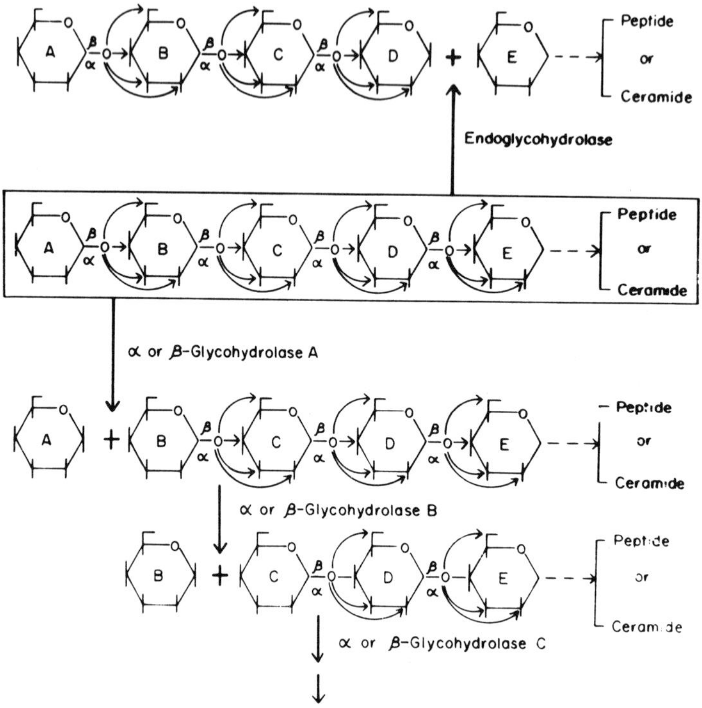

Scheme 1.

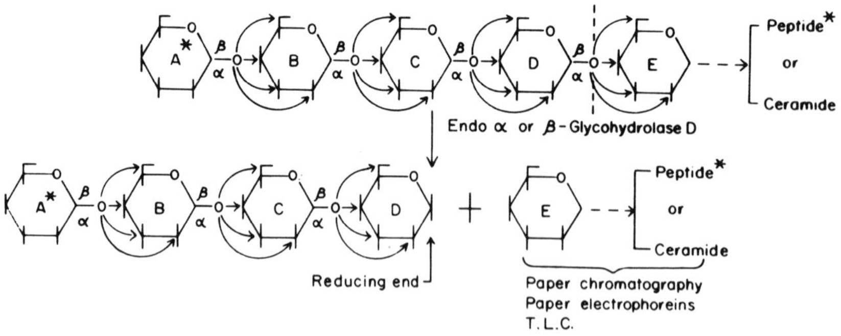

Scheme 2.

One can also use a specific *endo*-glycosidase to cleave the internal glycosidic linkage to produce one or several oligosaccharides (Scheme 2). In some cases, an *endo*-glycosidase can tell us a great deal about the structure of a saccharide chain.

There are about ten *exo*-glycosidases available for the structural analysis of complex carbohydrates, namely, neuraminidase, α- and β-D-galactosidases, α- and β-D-mannosidases, β-*N*-acetyl-D-hexosaminidase, α-*N*-acetyl-D-galactosaminidase, α-L-fucosidase, β-D-xylosidase, and α-D-glucuronidase. Their sources, methods of isolation, specificities, and properties have been reviewed in articles or monographs (1-4). There are three kinds of *endo*-glycosidases available at the present time: *endo*-β-*N*-acetyl-D-glucosaminidase, *endo*-β-D-galactosidase, and *endo*-α-*N*-acetyl-D-galactosaminidase. *Endo*-β-*N*-acetylglucosaminidase has been isolated from *Diplococcus pneumoniae* (5), *Streptococcus plicatus* (6), *Clostridium perfringens* (7,8), hen oviduct (9), and fig latex (10). *Endo*-β-galactosidase has been isolated from *Escherichia freundii* (11,12), *Coccobacillus* (13), and *D. pneumoniae* (14). *Endo*-α-*N*-acetylgalactosaminidase has been isolated from *Cl. perfringens* (15) and *D. pneumoniae* (16, 17).

For the structural analysis of complex carbohydrates with glycosidases, usually 0.02-0.1 μmol of the substrate is required. When a glycolipid is the substrate, it is necessary to add a detergent, such as sodium taurodeoxycholate, or a protein activator in the reaction mixture. The concentration of the detergent is usually between 100 and 200 μg per 100 μl of the reaction mixture and that of the buffer between 10 and 50 mM at the optimum pH of *exo*-glycosidases, which is usually between pH 4 to 5. For *endo*-glycosidases, the optimum pH is between pH 5 and 6. The total volume of the incubation mixture is usually between 100 and 500 μl, and the incubation is carried out at 37°C for 2-16 h. In some cases, the rate of the hydrolysis is greatly affected by the ionic strength of the buffer. For example, the ionic strength of 50 mM sodium citrate buffer is too high for liver β-*N*-acetylhexosaminidase A to carry out the hydrolysis of GM_2-ganglioside.

The monosaccharide units generated by *exo*-glycosidases can be conveniently determined by an automated sugar or gas-liquid chromatograph. However, it is also very important to analyze the remaining saccharide chain. When a glycolipid is used as substrate, one can take advantage of the ceramide moiety by extracting the incubation mixture with 4-5 vol. of chloroform--methanol (2:1) and detect the lipid product by t.l.c. The compounds produced by *endo*-glycosidases are generally more difficult to analyze than those of *exo*-glycosidases. Since *endo*-glycosidases cleaves the internal glycosidic linkages, one can detect the increment of the reducing power in the reaction mixture (Scheme 2). One can also tag the peptide moiety and detect the newly formed radioactive glycopeptide by paper chromatography or paper electrophoresis. When a glycolipid is used as substrate, one can detect the newly formed glycolipid by t.l.c. Above all, it is essential to determine the nature of the oligosaccharide produced.

After obtaining a positive as well as a negative result, some consideration should be given to carefully interpret the results. If no hydrolysis was detected, before making a negative conclusion, one should carefully consider the following: (a) The specificity of the glycosidase, as it is well known that the same enzymes isolated from different sources may differ considerably in their substrate specificity. Most of the intact glycoproteins are poor substrates for glycosidases, except for neuraminidase. It is, therefore, very difficult to prepare a so-called "carbohydrate-free glycoprotein" by use of glycosidases. (b) When a glycolipid is used as the substrate, one should carefully consider the amount and the nature of the detergent or of the activator to be included in the reaction mixture. (c) One should also consider the substrate concentration and the ionic strength of the buffer. In some cases, enzymic activity is greatly affected by the substrate concentration and the ionic strength. (d) It is very important to realize that *O*-acetylated or *O*-methylated sugar residues are often resistant to glycosidase. For example, 4-*O*-acetyl-*N*-acetylneuraminic acid is resistant to clostridial neuraminidase. (e) Since all the glycosidases available so far isolated are pyranosidases in nature, furanoside compounds will be resistant to the pyranosidases. If hydrolysis is observed, we should carefully consider the contamination by other glycosidases and the multiple specificity of the enzymes before making a positive conclusion. Contamination by other glycosidases is probably the most serious shortcoming in the use of enzymes as an analytical tool.

Specific examples to explain the specificity, the activator requirement, and the multiple specificity of glycosidases are given in the following lines. Fig. 1 shows the specificity of β-galactosidases isolated from *E. coli*, jack bean, and human liver toward *N*-acetyllactosamine and the (1→3) and (1→6) isomers. Among the three compounds, the (1→6)-linked is the best substrate for *E. coli* β-galactosidase, followed by the (1→3)- and (1→4)-linked compounds. In the case of jack bean β-galactosidase, the best substrate is the (1→6)-linked isomer followed by the (1→4)-linked. For some unknown reason, jack bean β-galactosidase hydrolyzes the (1→3)-linked isomer with great difficulty. In contrast to jack bean β-galactosidase, the β-galactosidase isolated from human liver hydrolyzes most rapidly the (1 3)-linked isomer. This enzyme hydrolyzes the (1 4)-linked compound at a slow rate. The (1→6)-linked isomer is very resistant to β-galactosidase of human liver. Although *E. coli* β-galactosidase is very active in hydrolyzing such oligosaccharides as lactose and the *N*-acetyllactosamine isomers, this enzyme can barely hydrolyze the β-D-linked galactosyl residues in glycoproteins and glycolipids.

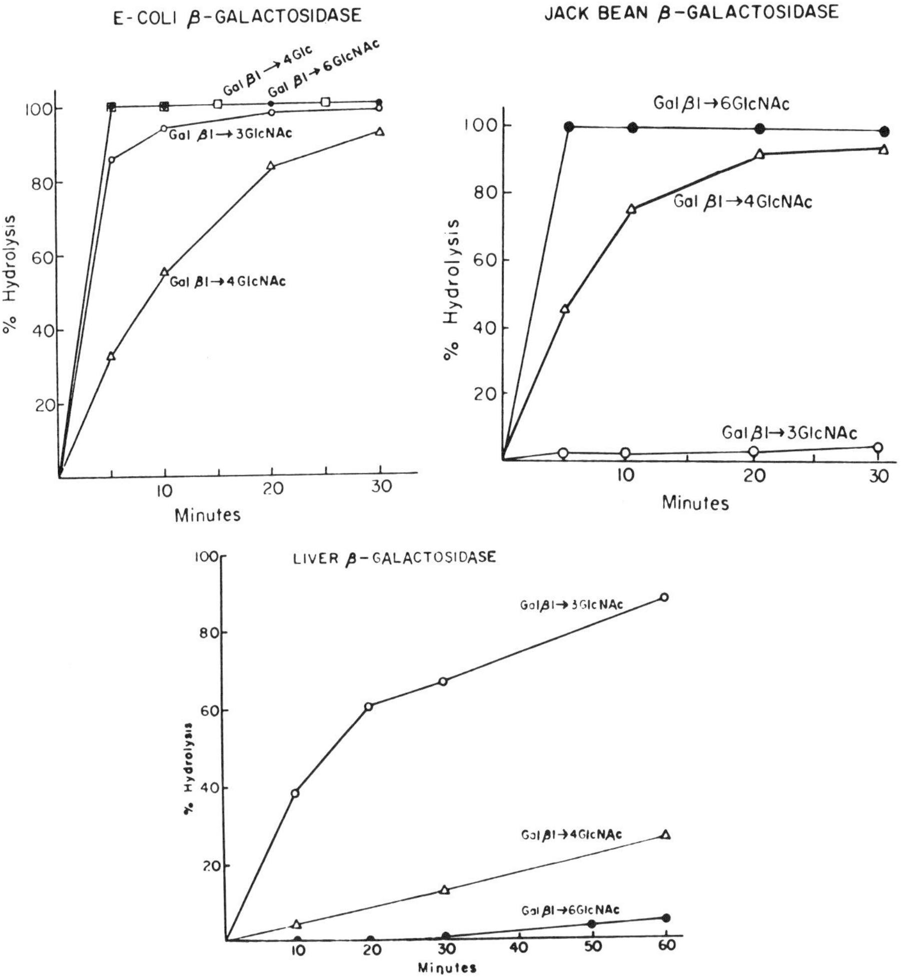

Fig. 1. Specificity of β-galactosidases of E. coli, *jack bean, and human liver.*

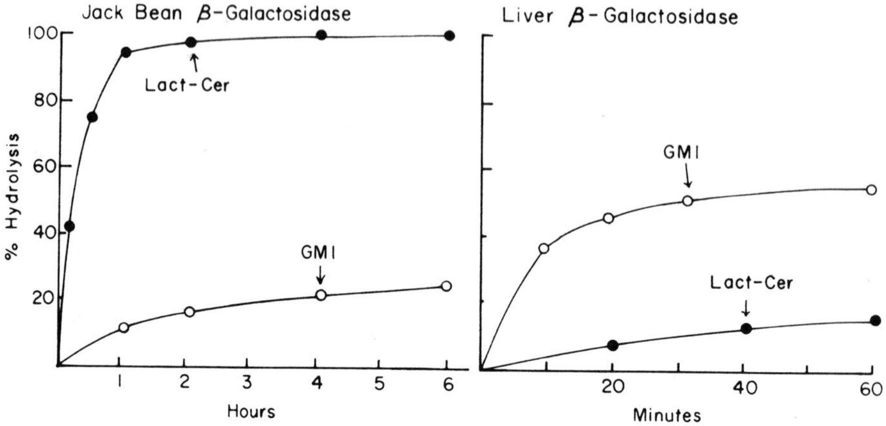

Figure 2. Action of jack bean and human liver β-galactosidases on lactosylceramide and GM_1 ganglioside.

Fig. 2 shows that lactosylceramide is more susceptible to jack bean β-galactosidase than GM_1 ganglioside, whereas the β-galactosidase isolated from human liver hydrolyzes GM_1 much faster than it hydrolyzes lactosylceramide.

In addition to the linkage specificity, we also need to consider the multiple specificity of glycosidases; this is to say that some glycosidases may be able to recognize more than one kind of sugar residue. The best example is the dual specificity of β-*N*-acetylhexosaminidase. As shown in the upper half of Scheme 3, β-*N*-acetylhexosaminidase can hydrolyze β-linked 2-acetamido-2-deoxy-D-glucopyranosyl as well as -D-galactopyranosyl residues. Apparently the enzyme does not require absolute stereospecificity toward the configuration of the hydroxyl group at C-4 of the pyranose ring. As shown in the lower half of Scheme 3, several laboratories have recently reported that α-*N*-acetylgalactosaminidase can hydrolyze 2-acetamido-2-deoxy-α-D-galactopyranosyl residue as well as α-D-galactosyl residue (18-20).

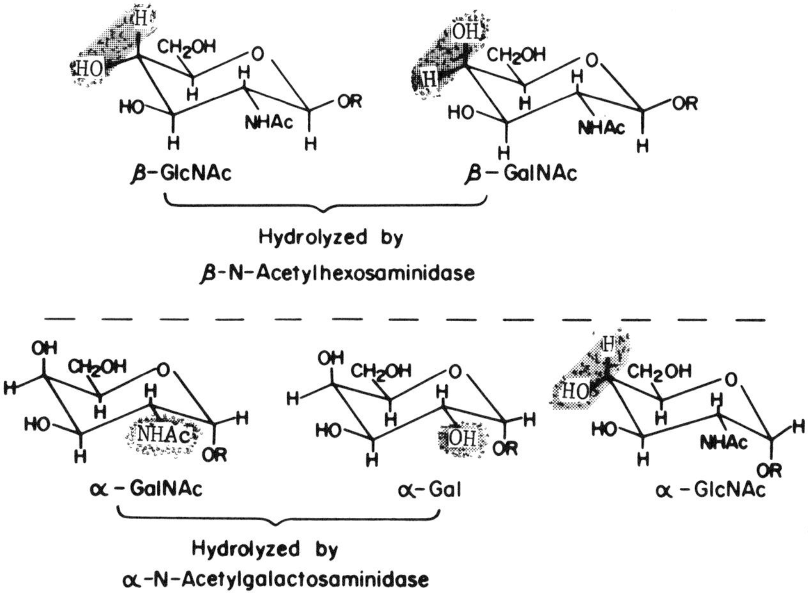

Scheme 3. Specificity of β-N-*acetylhexosaminidase and* α-N-*acetylgalactosaminidase.*

This may indicate that the acetamido group at C-2 is not involved in forming the enzyme-substrate complex. I would like to point out that although α-D-galactopyranosides can be hydrolyzed by α-*N*-acetylgalactosaminidase, α-galactosidase, on the other hand, has not been reported to hydrolyze 2-acetamido-2-deoxy-α-D-galactopyranosides, and 2-acetamido-2-deoxy-α-D-glucopyranosides are not hydrolyzed by α-*N*-acetylgalactosaminidase. Fig. 3 shows the sequential hydrolysis of the Forssman hapten α-D-GalNAc*p*-(1→3)-β-D-GalNAc*p*-(1→3)-α-D-Gal*p*-(1→4)-β-D-Gal*p*-(1→4)-β-D-Glc*p*-Cer, by α-*N*-acetylgalactosaminidase and other exoglycosidases. At both ends of Fig. 3 are standards for Forssman hapten, globoside, ceramide trihexoside, lactosylceramide, and glucosylceramide. Lane 1 is a control with Forssman hapten and no enzyme. Lane 2 is an incubation mixture of Forssman hapten with α-*N*-acetylgalactosaminidase isolated from limpet. As expected, this glycolipid was converted into globoside. Lane 3 is an incubation mixture of Forssman hapten with both α-*N*-acetylgalactosaminidase and β-*N*-acetylglucosaminidase. From the structure of the Forssman hapten, we expected to obtain ceramide trihexoside from this treatment. However, to our surprise, we detected the production of lactosylceramide instead. The reason for the conversion of the Forssman hapten into lactosylceramide by α-*N*-acetylgalactosaminidase and β-*N*-acetylhexosaminidase is due to the ability of the α-*N*-acetylgalactosaminidase to cleave the terminal 2-acetamido-2-deoxy-α-D-galactosyl, as well as the α-D-galactosyl residue exposed by β-*N*-acetylhexosaminidase.

Some glycoconjugates are susceptible to only those glycosidases that are isolated from mammalian tissues. Recently, it has been reported that the hydrolysis of sphingoglycolipids by mammalian exoglycosidases requires the presence of a protein activator. An activator for glucocerebrosidase was reported by Ho and O'Brien (21), and Peters and Glew (22). Fischer and Jatzkewitz reported an activator that stimulates the hydrolysis of cerebroside sulfatase (23). We have isolated an activator that can stimulate the enzymic hydrolysis of GM_1, GM_2, and ceramide trihexoside (24,25). Recently, Hechtman reported the presence of an activator that stimulates the enzymic hydrolysis of GM_2 ganglioside (26). The following points concerning the role of the activator remain to be clarified: (a) Does the activator act upon the enzyme or the substrate? (b) Does one activator serve many glycosidases or does each glycosidase require a specific activator? (c) Does the activator require a detergent for its activity? (d) What is its physiological function?

$\text{GalNAc} \xrightarrow{\alpha} \text{GalNAc} \xrightarrow{\beta} \text{Gal} \xrightarrow{\alpha} \text{Gal} \xrightarrow{\beta} \text{Glc} \longrightarrow \text{Cer}$

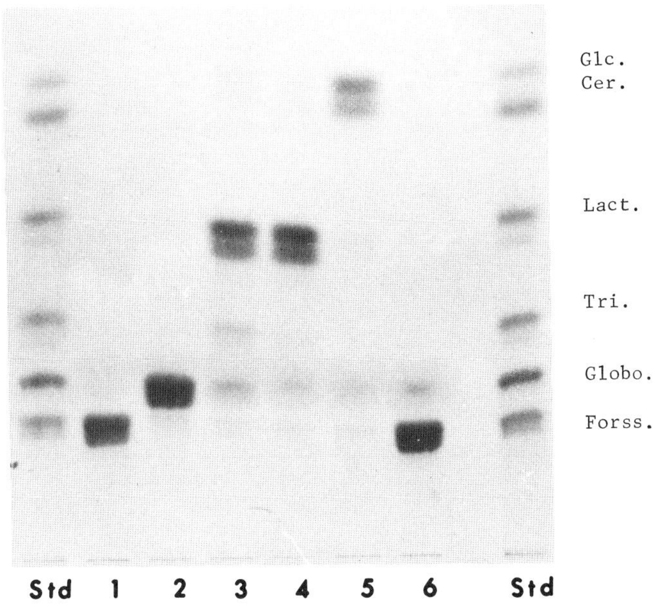

Fig. 3. Sequential hydrolysis of the Forssman hapten by N-*acetylhexosaminidases. See text for explanations.*

Next follows a brief overview of endoglycosidases. Scheme 3 shows the specificity of two *endo*-β-*N*-acetylglucosaminidases toward glycopeptides. The structures are taken from Prof. Montreuil's recent review (27) to represent the two types of saccharide chains in the plasma-type glycoproteins. The upper half of Scheme 4 shows the acidic or *N*-acetyllactosamine type which contains *N*-acetylglucosamine, galactose, and sialic acid or fucose at the peripheral position. α_1-Acid glycoprotein is an example for this type of saccharide chain. The structure in the lower half is called neutral or oligomannoside-type chain, it contains no galactose nor sialic acid. Ovalbumin contains this type of sugar chain. The *endo*-glycosidases isolated from fig latex (*endo*-glycosidase I), *D. pneumoniae*, and *Cl. perfringens* (*endo*-glycosidase I) cleave the chitobiosyl linkage of the inner core of the acidic type glycoprotein, provided that the sialic acid, galactose, and *N*-acetyl glucosamine residues at the peripheral position have been removed. The *endo*-glycosidase from *Streptococcus griseus, Cl. perfringens* (*endo*-glycosidase II), and fig latex (*endo*-glycosidase II), on the other hand, hydrolyze the chitobiosyl linkage in the neutral-type sugar chain.

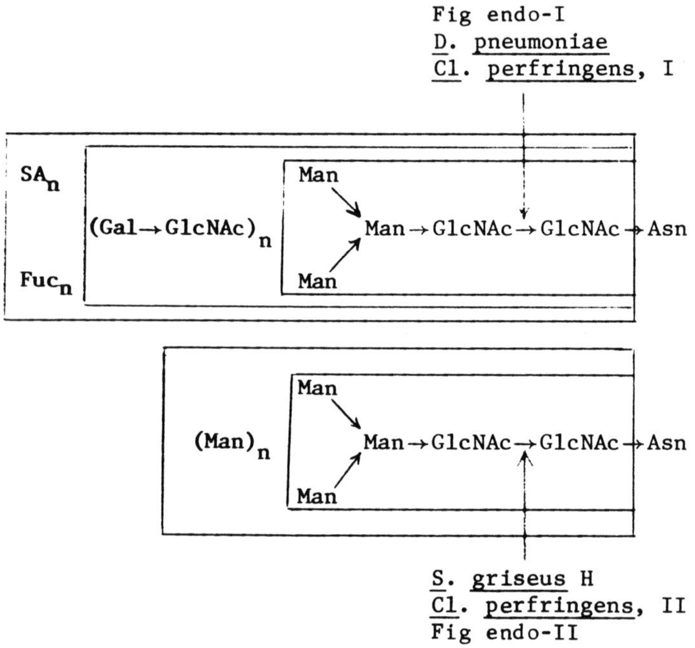

Scheme 4.

Scheme 5 shows the specificity of *endo*-β-galactosidase and *endo*-α-*N*-acetylgalactosaminidase. There are two types of *endo*-β-galactosidase. The *endo*-β-galactosidase isolated from *Escherichia freundii* hydrolyzes the internal β-D-galactosyl residue in keratan sulfate. We found that this enzyme can also cleave off the galactosyl residue (1→4)-linked to glucose, instead of to *N*-acetylglucosamine. The *endo*-β-galactosidase from *D. pneumoniae* was found to cleave the internal galactosyl unit in the type II chain of the blood-group A or B substances. The *endo*-α-*N*-acetylgalactosaminidase cleaves off the internal 2-acetamido-2-deoxy-D-galactosyl residue in human erythrocyte glycoprotein and porcine submaxillary glycoprotein.

E. freundii endo-β-galactosidase:

(1→3)-[β-Gal-(1↕4)-β-GlcNAc]$_n$→ – → Protein   Keratan sulfate

D. pneumoniae endo-β-galactosidase:

```
α-GalNAc-(1→3)-β-Gal-(1↕4)-Glc   or   (GlcNAc)
or  -Gal            2
                    ↑
                  α-Fuc
```

Cl. perfringens and *D. pneumoniae endo*-α-galactosaminidase:

β-Gal-(1→3)-α-GalNAc-(1↕3)Ser or (Thr)-Peptide

Scheme 5. Specificities of endo-β-*galactosidases and* endo-α-N-*acetylgalactosaminidase.*

Finally, I would like to give an example of the use of *endo*-β-galactosidase to study the structure of a glycolipid isolated from bovine erythrocytes. The upper part of Scheme 6 shows the basic structure of keratan sulfate, which contains the repeating unit β-D-galactosyl-(1→4)-2-acetamido-2-deoxy-D-glucosyl. The *endo*-β-galactosidase of *E. freundii* hydrolyzes this internal D-galactosyl unit. We found that the sphingo-glycolipids isolated from bovine erythrocytes also contain the same repeating unit found in keratan sulfate. This repeating unit is attached to lactosylceramide through a β-(1→3)-linkage. In addition to cleaving the internal β-D-Gal-(1→4)-GlcNAc linkage of the glycolipid, *endo*-β-galactosidase also cleaves the linkage between galactose and glucose units. We applied this enzyme to the study of the structure of a hexaglycosyl-ceramide. Incubation of hexaglycosylceramide with *endo*-β-galactosidase resulted in the production of trihexosylceramide and glucosylceramide, which means that the *endo*-β-galactosidase cleaved the internal β-D-galactosyl-(1→4)-D-glucose linkage to produce glucosylceramide. From this result, we were able to propose the structure of the hexaglycosylceramide as Gal→GlcNAc→Gal→GlcNAc→Gal→Glc→Cer, based on the specificity of the *endo*-β-galactosidase. Later we used *exo*-glycosidases and per-methylation analysis to prove that the complete structure of this glycolipid is β-Gal-(1→4)-β-GlcNAc-(1→3)-β-Gal-(1→4)-β-GlcNAc-(1→3)-β-Gal-(1→4)-Glc-Cer. Although the domain of *endo*-glycosidases as a whole is still in its infancy, the potential usefulness of these enzymes is very obvious. In conclusion, I would like to state that glycosidases will continue to be indispensable for the structural analysis of glycoconjugates.

→3-[β-Gal-(1↕4)-β-GlcNAc]$_n$→ – → Protein

→3-[β-Gal-(1↕4)-β-GlcNAc]-(1→3)-β-Gal-(1↕4)-Glc-Cer
ca. 1-3
Lac-Cer

Scheme 6. Action of E. freundii endo-β-*galactosidase on keratan sulfate (upper part) and bovine erythrocyte sphingolipids (lower part).*

ACKNOWLEDGMENTS

This work was supported by grants NS 09626 and RR 00164 from the National Institutes of Health and grant PCM 76-16881 from the National Science Foundation.

REFERENCES

1. Spiro, R. G., *Adv. Protein Chem. 27,* 349 (1973).
2. Lee, Y. C., *in* "Methods in Enzymology" (V. Ginsberg, ed.), vol. 28, p. 699, Academic Press, New York (1972).
3. Li, Y. T., *Int. Colloq. C.N.R.S. 221,* 283 (1973).
4. Li, S. -C., and Li, Y. -T., *in* "Handbook Biochemistry" Molecular Biology" (G. D. Fasman, ed.), 3rd ed., CRC Press (1976).
5. Koide, N., and Muramatsu, *J. Biol. Chem. 249,* 4897 (1974).
6. Tarentino, A. L., and Maley, F., *J. Biol. Chem. 249,* 811 (1974).
7. Ito, S., Muramatsu, T., and Kobata, A., *Arch. Biochem. Biophys. 171,* 78 (1975).
8. Chien, S. F., Yevich, S. J., Li, S. -C., and Li, Y. -T., *Biochem. Biophys. Res. Commun. 65,* 683 (1975).
9. Tarentino, A. L., and Maley, F., *J. Biol. Chem. 251,* 6537 (1976).
10. Chein, S. F., Weinburg, K., Li, S. -C., and Li, Y. -T., *Biochem. Biophys. Res. Commun. 76,* 317 (1977).

11. Kitamikado, M., and Ueno, R., *Nippon Suisan Gakkaishi 36,* 1175 (1970).
12. Fukuda, M. N., and Matsumura, G., *J. Biol. Chem. 251,* 6218 (1976).
13. Hirano, S., and Meyer, K., *Conn. Tissue Res. 2,* 1 (1973).
14. Takasaki, S., and Kobata, A., *J. Biol. Chem. 251,* 3603 (1976).
15. Huang, C. C., and Aminoff, D., *J. Biol. Chem. 247,* 6737 (1972).
16. Bhavanandan, V. P., Umemoto, J., and Davidson, E. A., *Biochem. Biophys. Res. Commun. 70,* 738 (1976).
17. Endo, Y., and Kobata, A., *J. Biochem. (Tokyo) 80,* 1 (1976).
18. Schram, A. W., Hamers, M. N., and Tager, J. M., *Biochim. Biophys. Acta 482,* 138 (1977).
19. Dean, K. J., Sung, S. J., and Sweeley, C. C., *Biochem. Biophys. Res. Commun. 77,* 1411 (1977).
20. Uda, Y., Li, S. -C., Li, Y. -T., and McKibbin, J. M., *J. Biol. Chem. 252,* 5194 (1977).
21. Ho, M. W., and O'Brien, J. S., *Proc. Natl. Acad. Sci. U.S.A. 68,* 2810 (1971).
22. Peters, S. P., Coyle, P., Coffe, C. J., and Glew, R. H., *J. Biol. Chem. 252,* 563 (1977).
23. Fischer, G., and Jatzkewitz, H., *Hoppe Seyler's Z. Phys. Chem. 356,* 605 (1975).
24. Li, Y. -T., Mazzotta, M. Y., Wan, C. C., Orth, R., and Li, S. -C., *J. Biol. Chem. 248,* 7512 (1973).
25. Li, S. -C., and Li, Y. -T., *J. Biol. Chem. 251,* 1159 (1976).
26. Hechtman, P., *Can. J. Biochem. 55,* 315 (1977).
27. Montreuil, J., *Pure Applied Chem. 42,* 431 (1975).

GLYCOCONJUGATE RESEARCH / Volume I
Proceedings of the Fourth International Symposium on Glycoconjugates

Methods for the Structural Elucidation of Complex Carbohydrates

Bengt Lindberg

Until recently, it was not very easy to determine the structure of complex carbohydrates and it is probably safe to say that most structures published more than 10 years ago should be reinvestigated. There are, however, notable exceptions, such as the *Pneumococcus* type 3 capsular polysaccharide (Scheme 1), published in 1941 by Reeves and Goebel (1), the more complicated type 8 polysaccharide (Scheme 2), published by Jones and Perry (2) in 1957, and the human milk oligosaccharides (Scheme 3), the structures of which were determined by Kuhn's and Montreuil's groups in the fifties. There are, of course, other examples, but these are in my opinion outstanding.

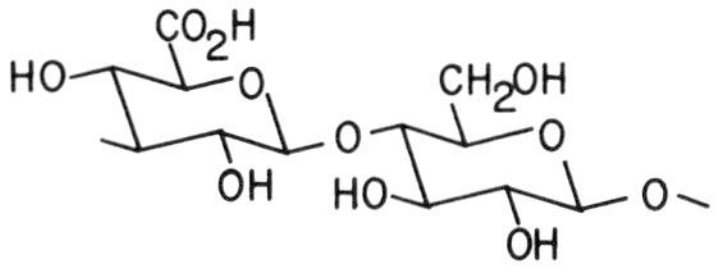

Scheme 1. Disaccharide repeating unit of the Pneumococcus *type 3 capsular polysaccharide.*

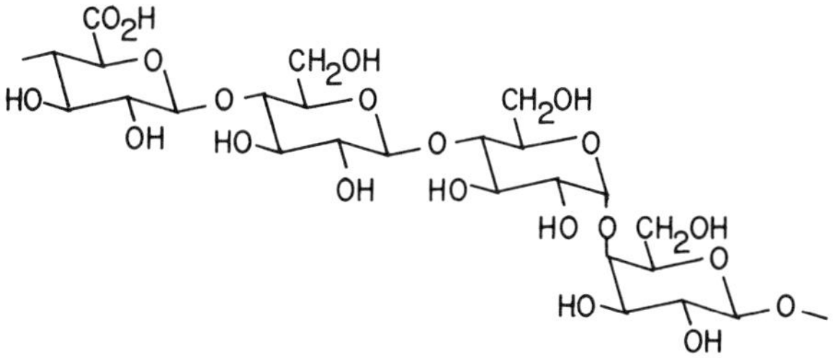

Scheme 2. Tetrasaccharide repeating unit of the Pneumococcus *type 8 capsular polysaccharide.*

ISBN 0-12-301301-1

```
2'-Fucosyl-lactose            α-Fuc-(1→2)-β-Gal-(1→4)-Glc
Lacto-difucotetraose          α-Fuc-(1→2)-β-Gal-(1→4)-Glc
                                                       3
                                                       ↑
                                                       1
                                                    α-Fuc
Lacto-N-tetraose              β-Gal-(1→3)-β-GlcNAc-(1→3)-β-Gal-(1→4)-Glc
Lacto-N-neotetraose           β-Gal-(1→4)-β-GlcNAc-(1→3)-β-Gal-(1→4)-Glc
Lacto-N-fucopentaose I
               α-Fuc-(1→2)-β-Gal-(1→3)-β-GlcNAc-(1→3)-β-Gal-(1→4)-Glc
Lacto-N-fucopentaose II       β-Gal-(1→3)-β-GlcNAc-(1→3)-β-Gal-(1→4)-Glc
                                           4
                                           ↑
                                           1
                                        α-Fuc
Lacto-N-difucohexaose I
               α-Fuc-(1→2)-β-Gal-(1→3)-β-GlcNAc-(1→3)-β-Gal-(1→4)-Glc
                                           4
                                           ↑
                                           1
                                        α-Fuc
Lacto-N-difucohexaose II      β-Gal-(1→3)-β-GlcNAc-(1→3)-β-Gal-(1→4)-Glc
                                           4                         3
                                           ↑                         ↑
                                           1                         1
                                        α-Fuc                     α-Fuc
```

Scheme 3. Oligosaccharides from human milk.

The improvements during recent years are due to new methods, and for the organic chemist gas-liquid chromatography-mass spectrometry (g.l.c.-m.s.) and nuclear magnetic resonance (n.m.r.) have involved the main changes. I shall briefly discuss some new methods in structural carbohydrate chemistry and illustrate them with examples from our current work.

In sugar and methylation analyses, the monomers are generally analyzed as their alditol acetates, by g.l.c.-m.s. Some groups use acetylated aldononitriles for the same purpose. These have the same advantage as the alditol acetates in that they are easy to prepare (Scheme 4) and that each sugar gives a single derivative. They are also well separated on g.l.c. (Fig. 1). We have found that they give good response on g.l.c. with an electron-capture (e.c.) detector. It should therefore be possible to scale down sugar and methylation analyses from the mg scale, which we are using now, to the μg scale. For sugar analysis this is simple. For methylation analysis, however, we need relative retention times for some hundred derivatives, and this will take time. The mass spectra of these derivatives are readily interpreted (Scheme 5) (3). One further advantage is that the response on e.c. is mainly due to the nitrile group, and for this reason all derivatives give approximately the same molar response.

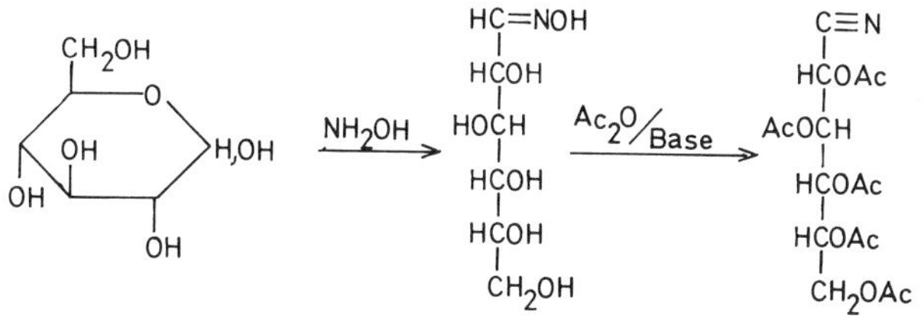

Scheme 4. Synthesis of acetylated aldononitriles.

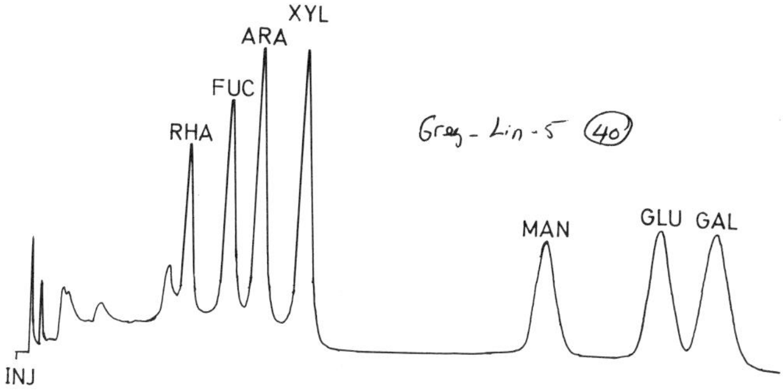

Fig. 1. Separation of acetylated aldononitriles from some common sugars by g.l.c. on DV-225.

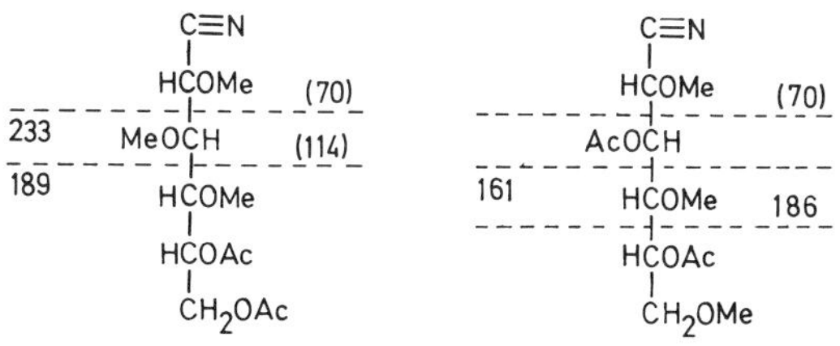

Scheme 5. Fragmentation of acetylated aldononitriles from two methylated sugars on electron impact.

There are examples of horrible mistakes in the literature when sugars have been identified by chromatographic methods only. The identification should be confirmed, *e.g.* by m.s., but this is not sufficient as several sugars occur both as D and L forms in Nature. These may even occur together in the same biological material. We have developed a new method for distinguishing between D and L sugars, which is fast, simple, and requires only small amounts of material (4). The sugar is treated with an optically active alcohol and an acidic catalyst. The glycosides formed from the D and L form are diastereoisomers and separable, *e.g.*, by g.l.c. of their acetates (Fig. 2). It is not necessary to have both forms as reference materials, as the chromatography picture given by one of the forms may be estimated from the pictures given by the other form after reaction with *rac* and optically active alcohol, respectively. The method is illustrated with an analysis of hydrolyzed *Helix pomatia* galactan (Fig. 3), known to contain both D- and L-galactose (5).

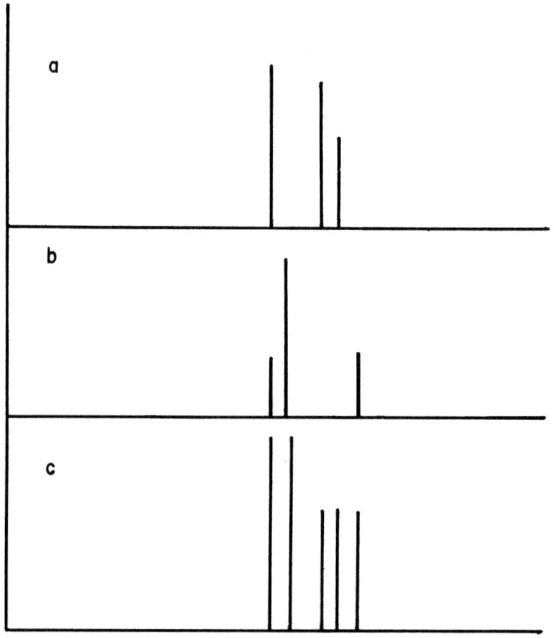

Fig. 2. Separation of acetylated 2-octyl arabinosides by g.l.c.: (a) (+)-2-octyl D-arabinoside; (b) (+)-2-octyl L-arabinoside; and (c) rac*-2-octyl D-arabinoside.*

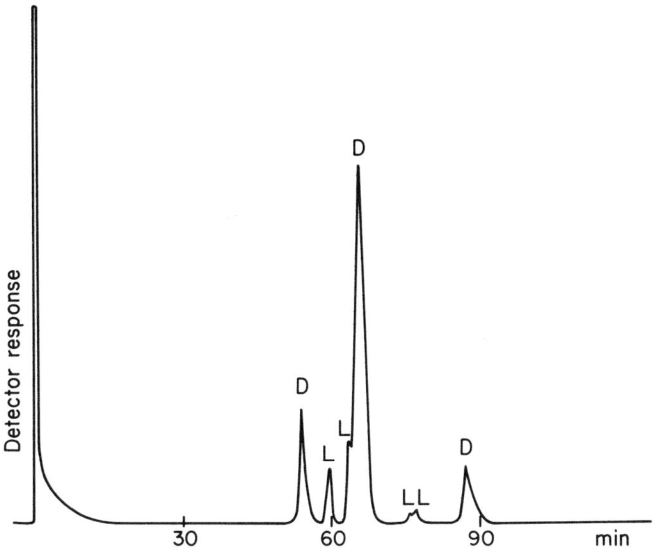

Fig. 3. Analysis of the hydrolyzate of Helix pomatia *galactan as acetylated (+)-2-octyl glycosides.*

Specific degradations may be used in order to determine the sequence of sugar residues in complex carbohydrates. It is often convenient to follow these degradations by methylation analysis. The first specific degradations used in carbohydrate chemistry were those devised by Barry (6) and Smith (7). It is suitable, therefore, to start with an example in which a Smith degradation gives the final piece of information needed to establish the complete structure.

Studies of the capsular polysaccharide from *Rhizobium meliloti* demonstrated that it contains the sugar residues listed in Scheme 6, and that all these are β-linked (8). It was not possible to decide, however, if the polysaccharide was composed of octasaccharide repeating units or if it had a less regular structure. This problem has now been solved by subjecting the polysaccharide to a number of specific degradations (9). The starting point was the terminal D-glucopyranosyl group with pyruvic acid linked in acetal form to O-4 and O-6. The polysaccharide was methylated and carboxyl-reduced, and the acetal group was removed by mild acidic hydrolysis. In the resulting product all positions, except O-4 and O-6 in the terminal residue, are therefore methylated. The terminal residue was eliminated by oxidizing the alcohol functions to carbonyl functions, β-elimination by treatment with base, and mild acidic hydrolysis (Scheme 7).

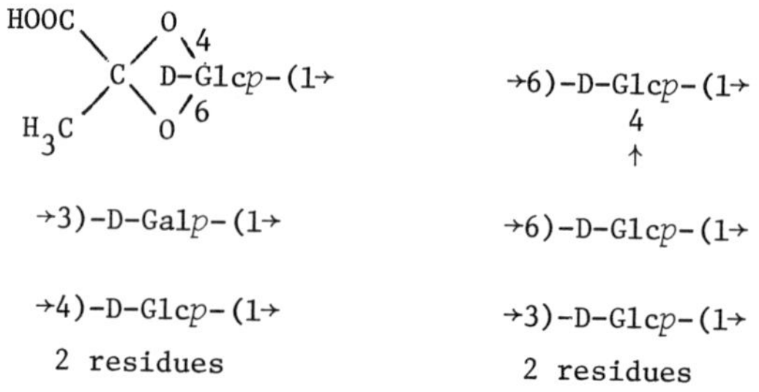

Scheme 6. Structural units of the Rh. meliloti *polysaccharide.*

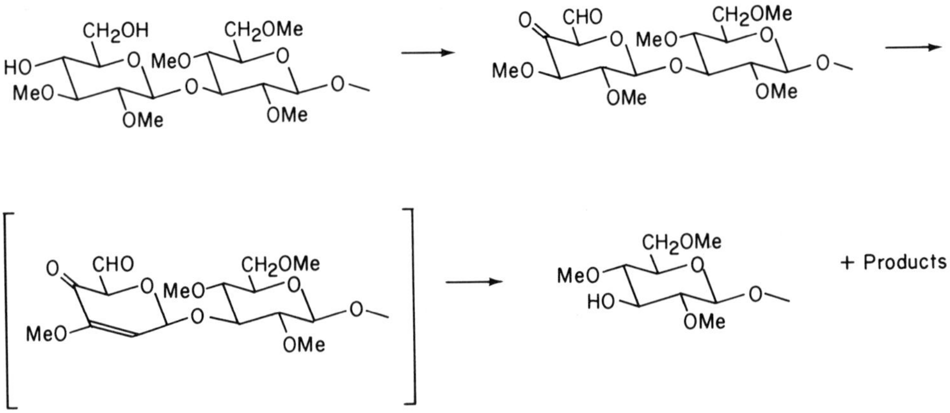

Scheme 7. First degradation of the Rh. meliloti *polysaccharide.*

Part of the product was trideuteriomethylated and analysis of a hydrolyzate, by g.l.c.-m.s., revealed that a new terminal residue having a free hydroxyl group at C-3, had been formed by the degradation. The linkage between the terminal and the next-terminal residue in the polysaccharide was thereby established. The hydroxyl group exposed by the first degradation was oxidized, and the oxidized residue was eliminated as just described, and this procedure was repeated twice (Schemes 8, 9, 10). After these four consecutive degradations, the branching point had been reached, and the partial structure given in Scheme 11 was established. This contains five of the eight residues illustrated in Scheme 6, the remaining being one D-galactopyranosyl residue linked through O-3 and two D-glucopyranosyl residues linked through O-4.

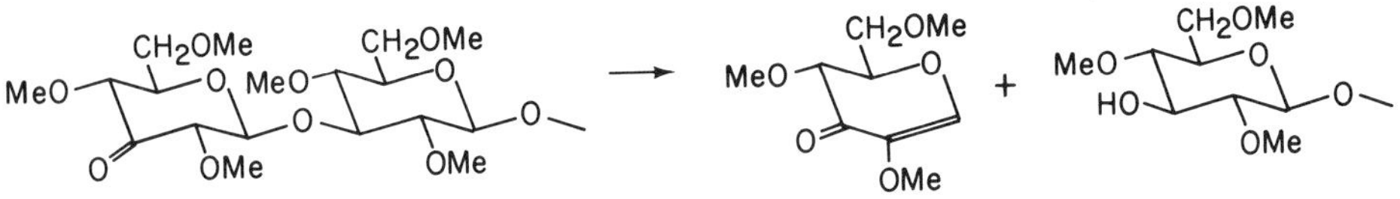

Scheme 8. Second degradation of the Rh. meliloti *poly-saccharide.*

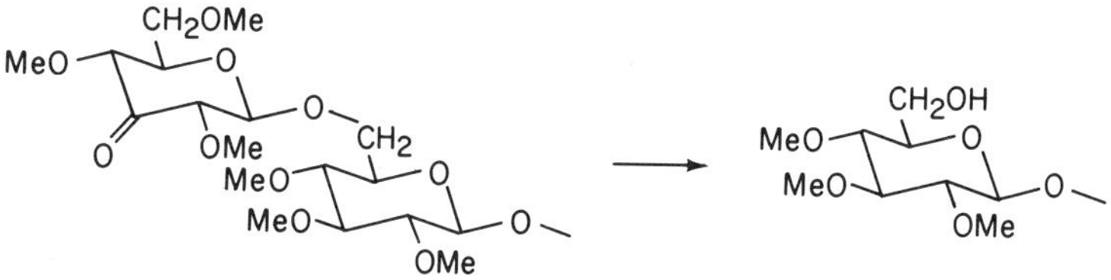

Scheme 9. Third degradation of the Rh. meliloti *poly-saccharide.*

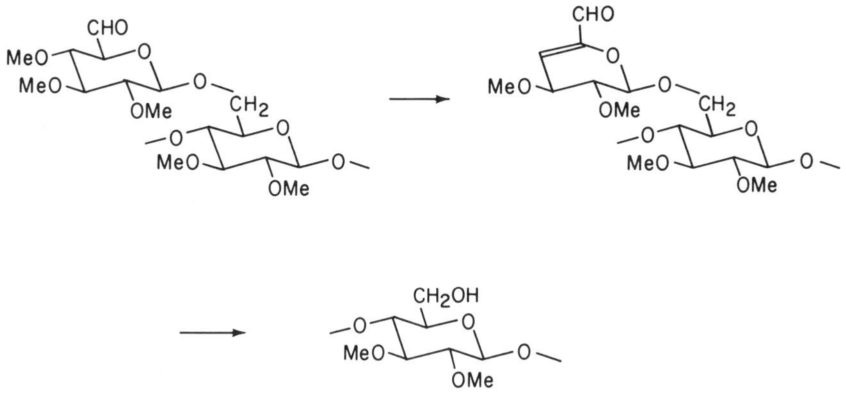

Scheme 10. Fourth degradation of the Rh. meliloti *poly-saccharide.*

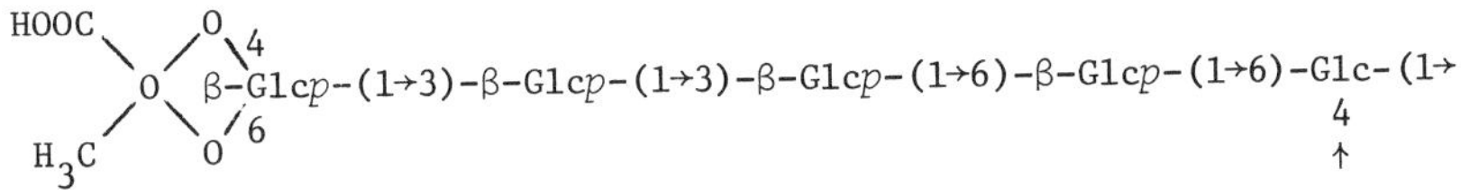

Scheme 11. Side chain in the Rh. meliloti *polysaccharide.*

A possible sequence, which proved to be correct, is given in Scheme 12. The polyalcohol obtained on periodate oxidation-borohydride reduction of this structure (Scheme 13, R = H) should give an erythritol galactoside (Scheme 14) on mild acid hydrolysis. The same D-galactoside should, however, be formed if the D-galactose residue was linked to O-4 of one of the chain D-glucose residues. If the polyalcohol is methylated before the Smith hydrolysis (Scheme 13, R = CH_3) and the product then ethylated, more detailed information is obtained. Now the product (Scheme 15) contains ethoxyl groups in the positions exposed during the mild hydrolysis, and analysis by g.l.c.-m.s. revealed that the D-galactose residue was, indeed, linked to O-4 of the branching D-glucose residue. Identification of another product, a trisaccharide (Scheme 16), confirmed a sequence of three sugar residues in the side chain. By use of five specific degradations, it was therefore possible to determine the structure of the octasaccharide repeating unit (Scheme 17) of the *Rhizobium meliloti* capsular polysaccharide.

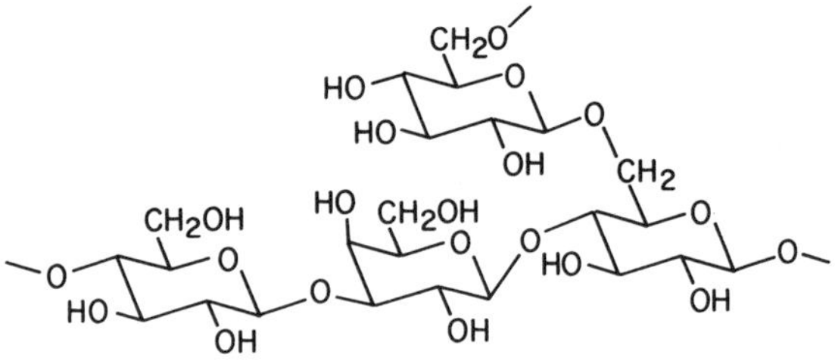

Scheme 12. Structure around the D-galactopyranosyl residue in the Rh. meliloti *polysaccharide.*

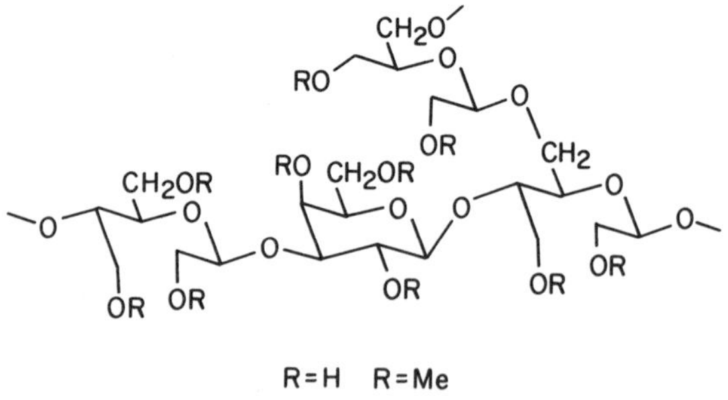

Scheme 13. Polyalcohol obtained by periodate oxidation-borohydride reduction of the Rh. meliloti *polysaccharide.*

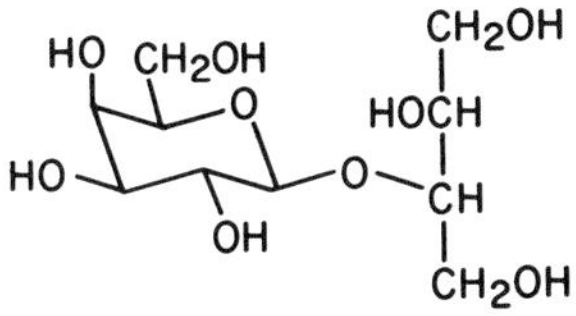

Scheme 14. Erythritol galactoside obtained on Smith degradation of the Rh. meliloti *polysaccharide.*

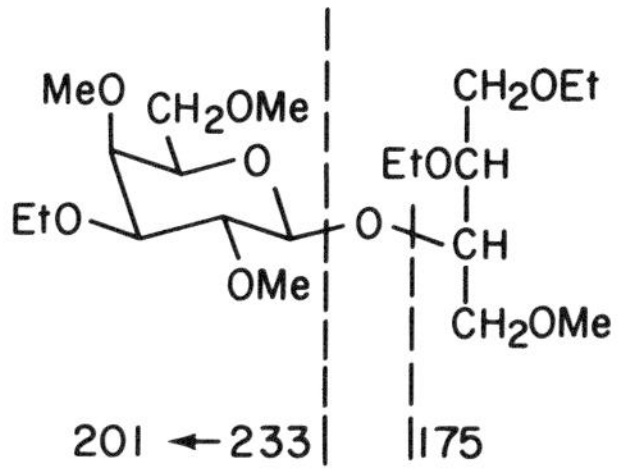

Scheme 15. Etherified erythritol galactoside obtained on modified Smith degradation of the Rh. meliloti *polysaccharide.*

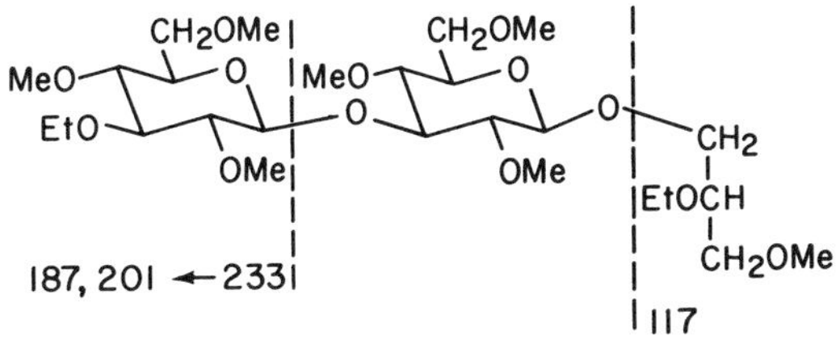

Scheme 16. Trisaccharide obtained on modified Smith degradation of the Rh. meliloti *polysaccharide.*

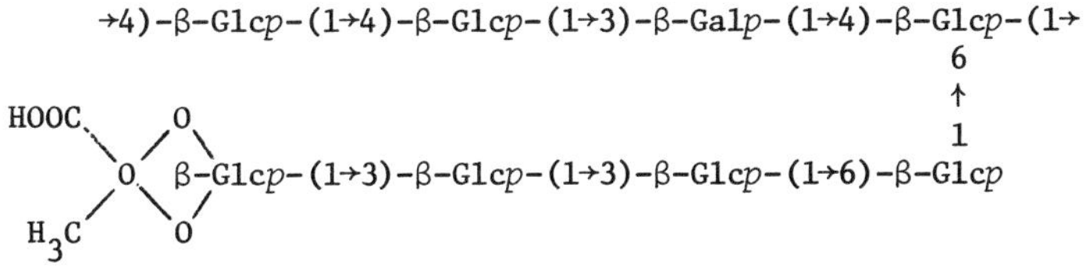

Scheme 17. Octasaccharide repeating unit of the Rh. meliloti *polysaccharide.*

Smith and Lewis (10) observed that the Smith degradation is sometimes complicated by acetal migration, with the formation of cyclic acetals of glycolaldehyde, as shown in Scheme 18. Another advantage of the modified Smith degradation, in which the carbohydrate is methylated before the acidic hydrolysis, is that such acetal migration is excluded.

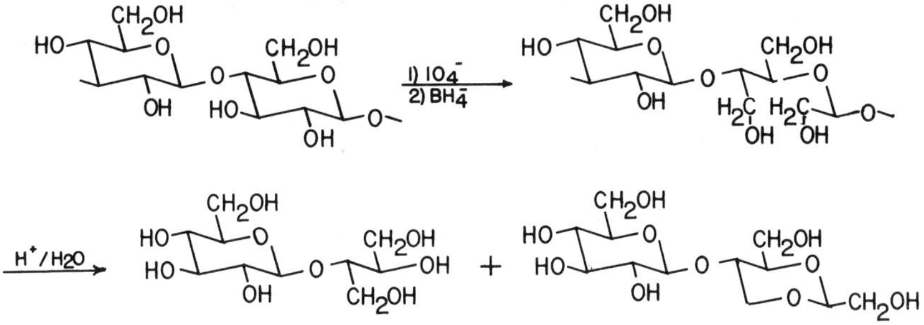

Scheme 18. Acetal migration during a Smith degradation.

Uronic acids offer starting points for specific degradations, a possibility that has been investigated by Aspinall's group and ours, and used in numerous structural studies. The polymer is completely methylated and esterified by a Hakomori methylation. As indicated in Scheme 19, all substituents, either methoxyl groups or sugar residues, are released on treatment with base followed by mild acidic hydrolysis, I shall give a simple example of this degradation, in which the uronic acid residue was first created (11). Methylation analysis of *Pneumococcus* type 14 polysaccharide revealed that it contained a terminal D-galactopyranosyl group which, according to other evidence, should be β-linked. This residue was transformed into a D-galactopyranosyluronic acid residue by oxidation, first with galactose oxidase, and then with hypoiodite (Scheme 20). The product was then subjected to a uronic acid degradation, trideuteromethylation and hydrolysis. The 3-*O*-methyl derivative of D-glucosamine observed in the methylation analysis of the original polysaccharide was now replaced by a 3,4-di-*O*-methyl derivative, with a trideuteromethyl group at O-4. The β-D-galactopyranosyl group is consequently linked to that position, as indicated in the structure (Scheme 21) of the tetrasaccharide repeating unit.

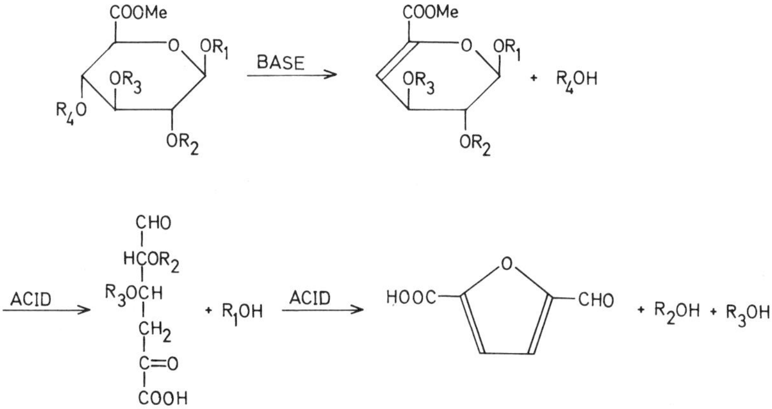

Scheme 19. Uronic acid degradation.

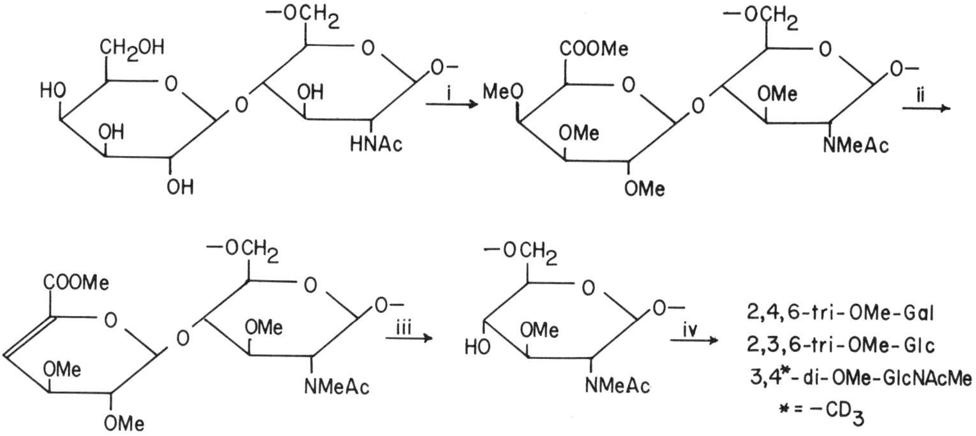

Scheme 20. Specific degradation of the Pneumococcus *type 14 capsular polysaccharide.*

→6)-β-D-GlcNAc-(1→3)-β-D-Gal-(1→4)-D-Glc-(1→
4
↑
1
β-D-Gal

Scheme 21. Structure of the Pneumococcus *type 14 capsular polysaccharide.*

Angyal and James (12) demonstrated that acetylated aldopyranosides in which the aglycon is equatorially disposed in the most stable chair form are oxidized to esters of 5-ketoaldonic acid on treatment with chromium trioxide in acetic acid (Scheme 22). The anomers having axially disposed aglycons are oxidized only slowly. We have used this method to determine the anomeric configuration of sugar residues in oligo- and poly-saccharides (13). In favourable situations, it may also be used to determine the sequences of sugar residues. In the *Klebsiella* type 37 polysaccharide (Scheme 23), one of the two D-glucopyranosyl residues is α-linked, all the other residues being β-linked (14). Residue A is D-glucopyranosyluronic acid, etherified with L-lactic acid at O-4 (Scheme 24) (15). On oxidation of the acetylated polysaccharide with chromium trioxide, only the α-D-glucopyranosyl residue should remain intact (Scheme 25). Reduction of the product with borodeuteride yielded an α-D-glucopyranoside of galactitol (and some of the corresponding L-altritol derivative). Because of the deuterium labeling, it was easy to demonstrate, by g.l.c.-m.s. of the methylated product, that the α-D-glucopyranosyl group was linked to O-4 of D-galactitol. The α-linked D-glucopyranosyl residue in the polysaccharide was thereby located.

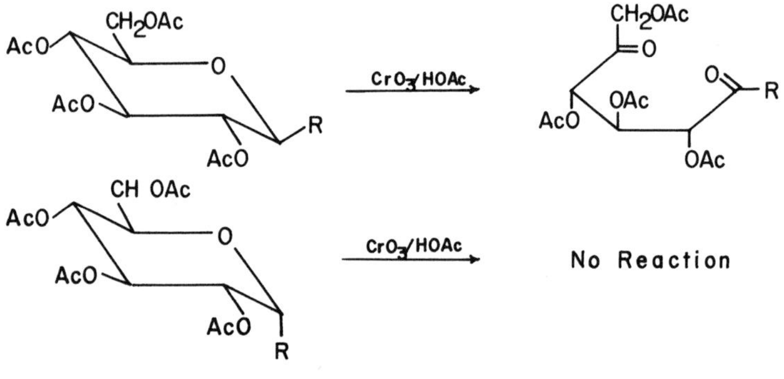

Scheme 22. Oxidation of acetylated glycopyranosides with chromium trioxide in acetic acid.

```
→3)-D-Galp-(1→4)-D-Glcp-(1→
         4
         ↑
         1
A-(1→6)-D-Glcp
```

Scheme 23. Structure of the Klebsiella *type 37 polysaccharide.*

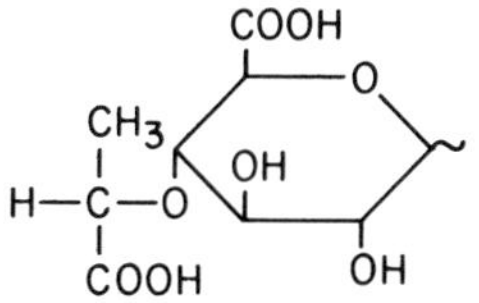

Scheme 24. Acidic sugar in Klebsiella *type 37 polysaccharide.*

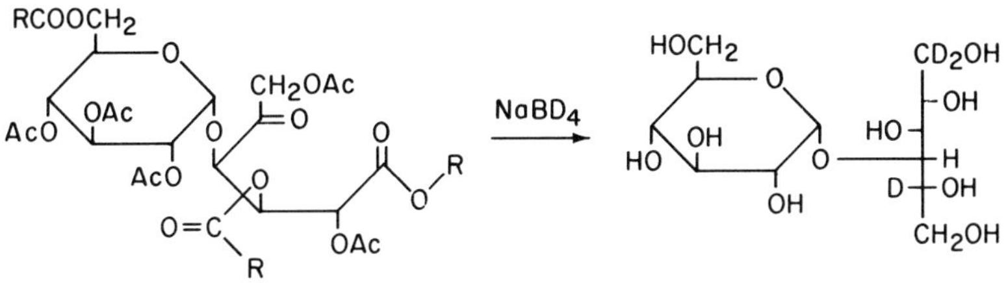

Scheme 25. Specific degradation of the Klebsiella *type 37 polysaccharide.*

Many complex carbohydrates contain aminodeoxy sugars, which are generally *N*-acetylated. After *N*-deacetylation, they offer suitable starting points for specific degradation. We recently devised a method for *N*-deacetylation by which the material is treated with sodium hydroxide--sodium thiophenolate in aqueous dimethylsulfoxide (16). From the ^{1}H-n.m.r. spectra of O-antigen from *Shigella flexneri* before (Fig. 4) and after (Fig. 5) *N*-deacetylation, it is evident that the reaction is complete. The protons of *N*-acetyl and *O*-acetyl groups, which are also removed by the treatment, appear at δ 2.1 and 2.2, respectively. Deamination of the *N*-deacetylated material was applied in structural studies of the *Shigella flexneri* variant Y O-antigen (17), which has the repeating unit given in Scheme 26. Methylation analysis of original and deaminated-reduced product (Scheme 27) demonstrated that D-glucosamine is linked to O-2 of an L-rhamnosyl residue in the original polysaccharide. The 2,5-anhydro-D-mannitol residue formed by the degradation is symmetrical, but, by performing the reduction with borodeuteride, the 2,5-anhydro-1,4,6-tri-*O*-methyl-D-mannitol formed in the methylation analysis could be distinguished from its 1,3,6-isomer by m.s. (Scheme 28). The identification of the former compound established that the D-glucosamine residue is linked through O-3 in the polysaccharide.

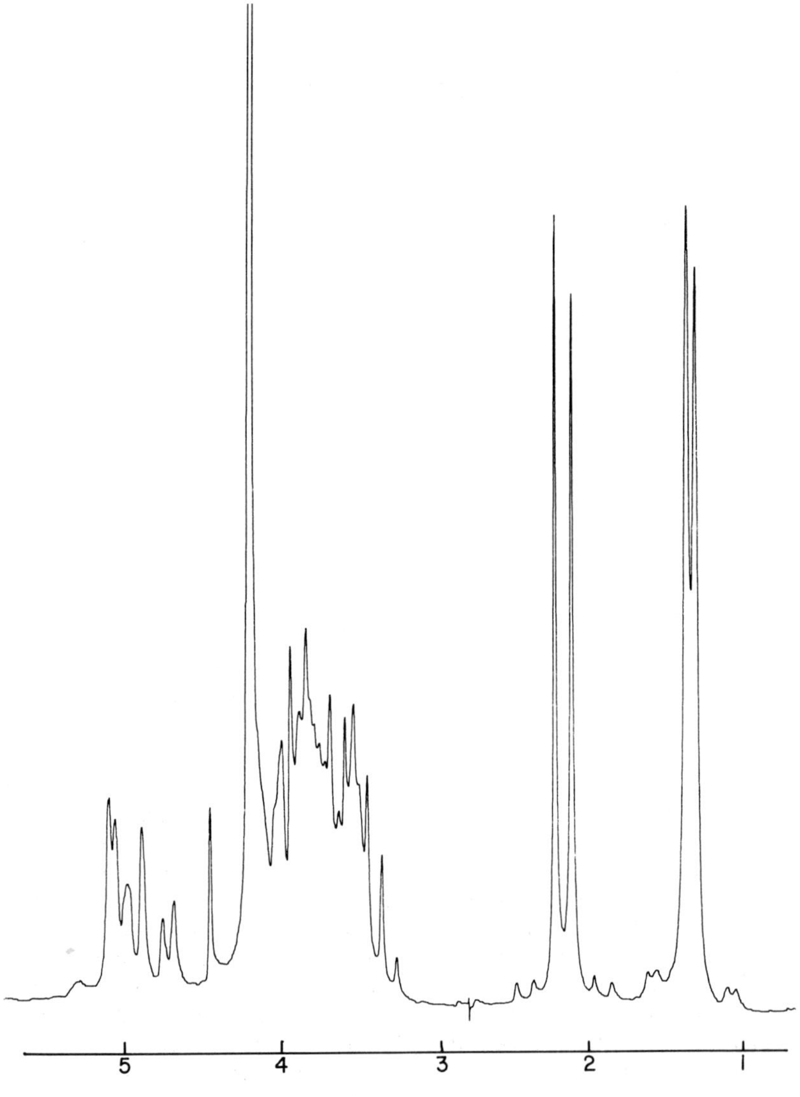

Fig. 4. ^{1}*H-N.m.r. spectrum of* Shigella flexneri *O-antigen.*

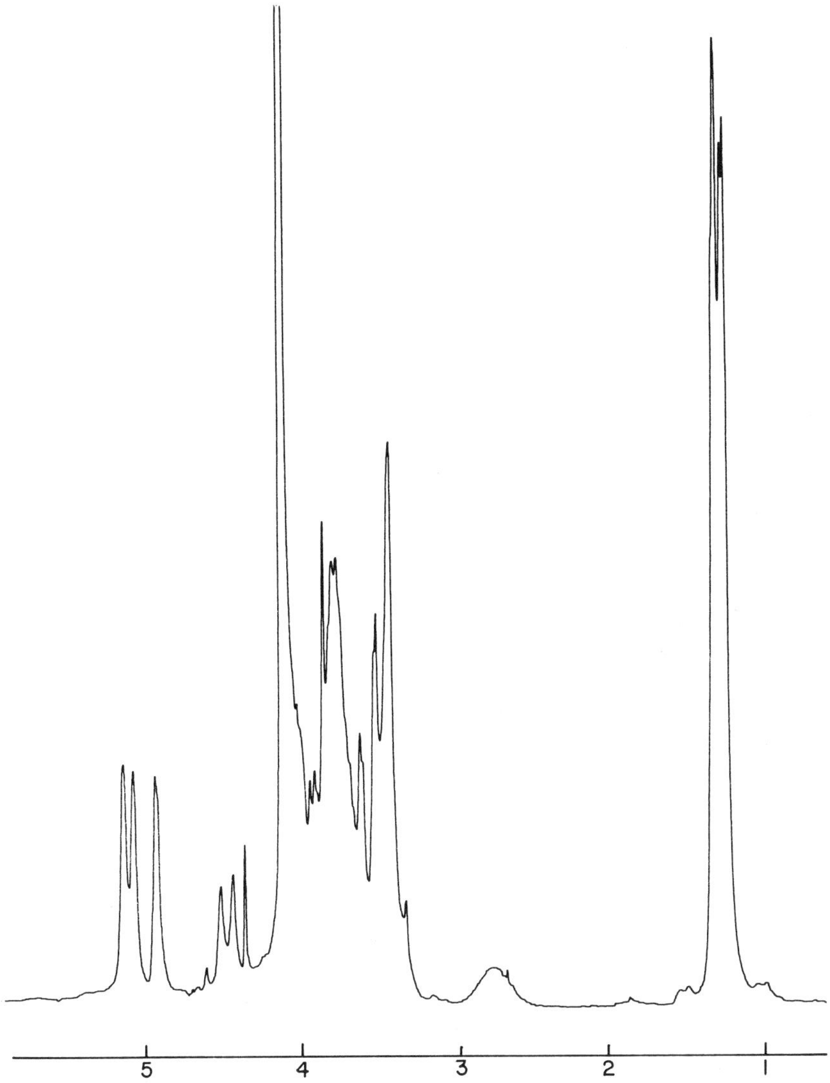

Fig. 5. ^{1}*H-N.m.r. spectrum of deacetylated* Shigella flexneri *O-antigen.*

→3)-β-D-GlcNAc*p*-(1→2)-α-L-Rha*p*-(1→2)-α-L-Rha*p*-(1→3)-α-L-Rha*p*-(1→

Scheme 26. Tetrasaccharide repeating unit of the Shigella flexneri *O-antigen.*

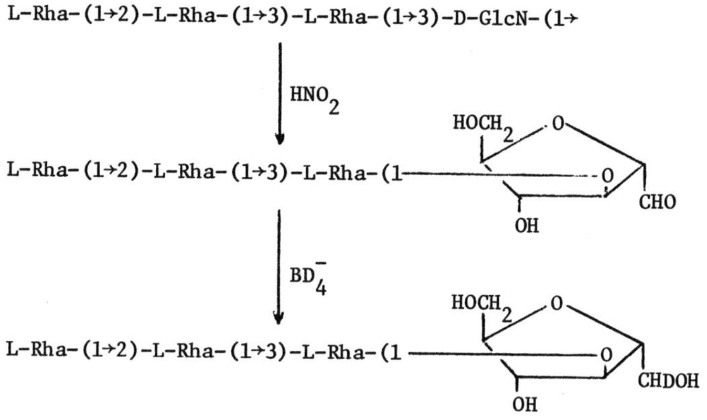

2,4-di-OMe-Rha	2,3,4-tri-OMe-Rha
3,4-di-OMe-Rha	1,4,6-tri-OMe-2,5-Anhydrorhamnitol

Scheme 27. Deamination of N-*deacetylated* Shigella flexneri *O-antigen.*

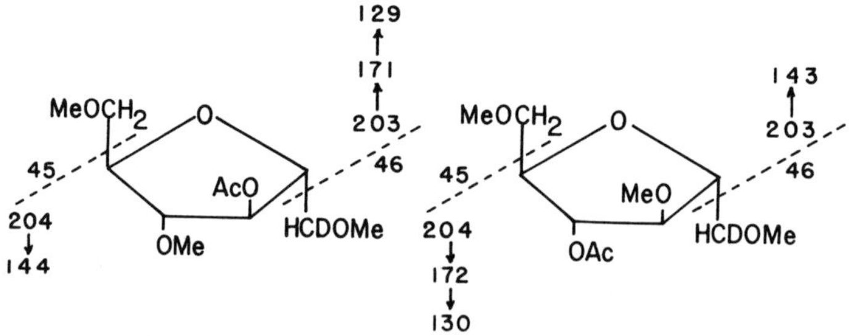

Scheme 28. Fragmentation of 2,5-anhydro-D-[1-^{2}H]mannitol derivatives on electron impact.

In order to determine the sequence of the two remaining L-rhamnose residues, the 2,5-anhydro-D-mannose derivative formed by deamination was treated with base, which eliminated the trisaccharide residue linked to its 3-position (Scheme 29). Reduction of this trisaccharide and methylation analysis gave the desired information. Optical rotations and ^{1}H-n.m.r. spectra of the oligosaccharides formed by these degradations showed that all L-rhamnose residues are α-linked.

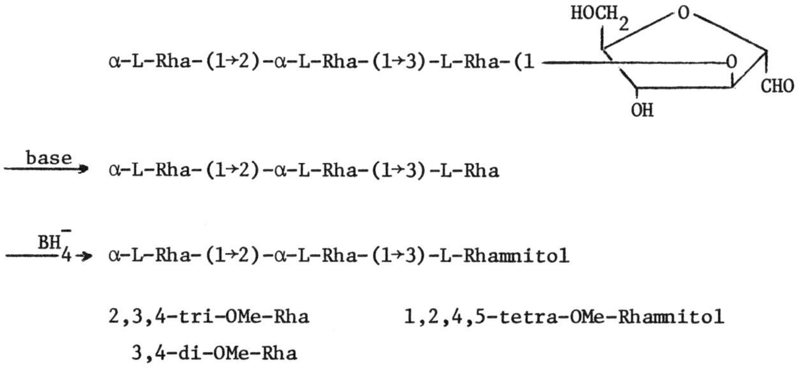

Scheme 29. Alkaline β-elimination applied to the tetrasaccharide obtained by N-*deacetylation-deamination of* Shigella flexneri *O-antigen.*

ACKNOWLEDGMENT

It is a pleasant duty to thank the members of our research group, listed in the bibliography, for their contributions to theoretical and practical aspects of this work.

REFERENCES

1. Reeves, R. E., and Goebel, W. F., *J. Biol. Chem. 139*, 511 (1941).
2. Jones, J. K. N., and Perry, M. B., *J. Am. Chem. Soc. 79*, 2787 (1957).
3. Dmitriev, B. A., Backinowsky, L., V., Chizhov, O. S., Zolotarev, B. M., and Kochetkov, N. K., *Carbohydr. Res. 19*, 432 (1971).
4. Leontein, K., Lindberg, B., and Lönngren, J., *Carbohydr. Res. 62*, 359 (1978).
5. Bell, D. J., and Baldwin, E., *J. Chem. Soc. 125*, (1941).

6. Barry, V. C., *Nature (London) 152*, 538 (1943).
7. Goldstein, I. J., Hay, G. W., Lewis, B. A., and Smith, F., *Abstr. Papers Am. Chem. Soc. Meet. 135*, 3D (1959).
8. Björndal, H., Erbing, C., Lindberg, B., Fåhraeus, G., and Ljunggren, H., *Acta Chem. Scand. 25*, 1281 (1971).
9. Jansson, P. -E., Kenne, L., Lindberg, B., Ljunggren, H., Lönngren, J., Rudén, U., and Svensson, S., *J. Am. Chem. Soc. 99*, 3812 (1977).
10. Lewis, B. A., and Smith, F., *Abstr. Papers Am. Chem. Soc. Meet. 144*, 8D (1963).
11. Lindberg, B., Lönngren, J., and Powell, D. A., *Carbohydr. Res. 58*, 177 (1977).
12. Angyal, S. J., and James, K., *Aust. J. Chem. 23*, 1209 (1970).
13. Hoffman, J., Lindberg, B., and Svensson, S., *Acta Chem. Scand. 26*, 66. (1972).
14. Lindberg, B., Lindqvist, B., Lönngren, J., and Nimmich, W., *Carbohydr. Res. 58*, 443 (1977).
15. Lindberg, B., Lindqvist, B., Lönngren, J., and Nimmich, W., *Carbohydr. Res. 49*, 411 (1976).
16. Erbing, C., Granath, K., Kenne, L., and Lindberg, B., *Carbohydr. Res. 47*, C5 (1975).
17. Kenne, L., Lindberg, B., Petersson, K., and Romanowska, E., *Carbohydr. Res. 56*, 363 (1977).

GLYCOCONJUGATE RESEARCH / Volume I
Proceedings of the Fourth International Symposium on Glycoconjugates

Primary Structure and Conformation of Glycans *N*-Glycosically Linked to Peptide Chains

Jean Montreuil and Johannes F. G. Vliegenthart

For a long time, our knowledge of the structure of glycoprotein glycans was limited to that of some linear glycans, such as the acidic mucopolysaccharides, and to that of some simple glycans like those of submaxillary mucins. Our ignorance was essentially due to the lack of precise and sensitive methods for structural investigation of complex glycans. But, in the past few years, important results have been accumulated because the far reaching biological importance of glycoconjugates, in general, and of glycoproteins, in particular, has become understood after the following discoveries: (a) Glycoconjugates are cell-surface antigens and their structure and function are modified in virus transformed cells and in cancerous cells. (b) They play an important role in intracellular adhesion and recognition, and in cell-contact inhibition. (c) They are receptor sites for viruses, proteins, and hormones. (d) Glycans protect the protein moiety against proteolytic attack. (e) The carbohydrate moiety may influence the conformation of the peptide chain. (f) Glycan groups permit the exit of proteins outside of the cell according to Eylar's hypothesis (1), which leads to the concept of Winterburn and Phelps (2), that glycoproteins are synthesized by cells for cells. (g) According to Ashwell (3), the sugar component regulates the catabolism of circulating proteins by different tissues, and the lifetime of proteins and cells. (h) A pathology of glycoproteins that is due to a lack of lysosomal glycosidases now exists and the term "glycoproteinosis" has been coined.

ISBN 0-12-301301-1

Thus, in the past few years, the complete primary structure of numerous glycans has been firmly established, and this rapid advance of our knowledge is due to the improvement of chemical, physical, and enzymic methods. The results obtained in the past four years entirely confirm, from a comparative biochemical point of view, the concepts that were developed in 1974 at the VIIth Symposium on Carbohydrate Chemistry held in Bratislava (4). In fact, glycan structures may be classified into families, within each of which glycan structures are very similar and present common oligosaccharide structures, even if they originate from animals, plants, microorganisms, or viruses. Consequently, a series of classes may be established: Glycans are conjugated to the peptide chains through two types of primary covalent linkages: (a) *O*-Glycosyl linkages (i) between L-serine or L-threonine and *N*-acetyl-D-galactosamine, D-mannose, or D-xylose, and (ii) between L-hydroxylysine or L-hydroxyproline and D-galactose. (b) *N*-Glycosyl linkage between L-asparagine and *N*-acetyl-D-glucosamine, which is until now the only one to have been characterized. Two types of glycans are present: (a) Linear glycans such as the acidic mucopolysaccharides; these structures are relatively simple because they result from the polymerization of a disaccharide unit; they have been known for several years. (b) Branched glycans, which present a more complex structure; some show a single branch, such as the human α_1-acid glycoprotein; others show a highly branched structure, such as the ovomucoid from hen egg white, which has 7 branching points (5).

β-Gal-(1 → 4)-β-Glc-(1→)-Ceramide A

β-Gal-(1 → 3)-α-GalNAc-(1→)-Ser (Thr) B

β-Gal-(1 → 3)-β-Gal-(1 → 4)-β-Xyl-(1→)-Ser C

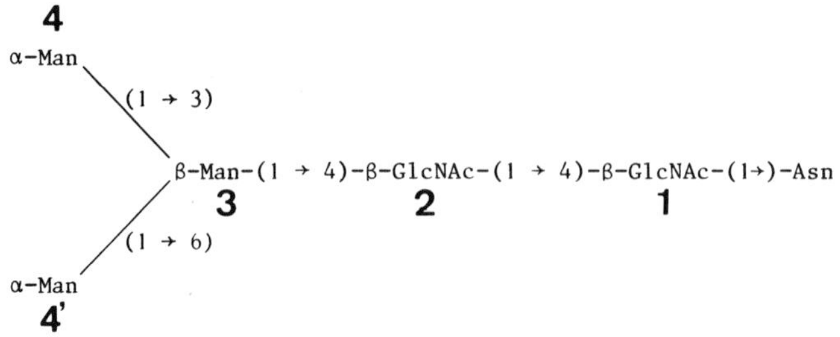

D

Scheme 1. Oligosaccharide cores of glycoconjugates.

The knowledge of the structure of glycoconjugates, in general, and of glycoproteins, in particular, leads to the establishment of the core concept: All glycan structures derive from substitutions at a common and nonspecific oligosaccharide core linked to the noncarbohydrate component (see Scheme 1): core A is that of glycolipids; core B is that found in numerous glycoproteins, such as MN and blood-group antigens, kappa caseins, antarctic fish antifreezing glycoprotein, IgA immunoglobulins from human serum and milk, and glycophorin; core C is that found at the reducing end of acidic mucopolysaccharides; and core D is that present in almost all *N*-glucosically conjugated glycans, which results from a mannotriose residue β-linked to a di-*N*-acetylchitobiose residue, linked itself to an asparagine residue.

In *N*-glycosically conjugated glycans, the existence of two types of structures has been established. In the first one (Scheme 2A), the pentasaccharide core is substituted only by D-mannose residues, the term "oligomannosidic type" was proposed (4). The glycans illustrated in Schemes 3-5 belong to this type. In the second type (see Scheme 2B), the same pentasaccharide core is substituted by a variable number of *N*-acetyllactosamine, and sialic acid, or L-fucose (or both) residues. These structures belong to the "*N*-acetyllactosaminic type" the substituting and substituted *N*-acetyllactosamine residues were called "antennae" (4).

Until now, the types of substitution of the pentasaccharide core (Scheme 1) that have been demonstrated are as follows: (a) "biantennary structures" C-2 of mannose-4 and -4' residues (see Schemes 6 and 7A); (b) "triantennary structures" C-2 and C-4 of mannose-4, and C-2 of mannose-4' residue; or C-2 of mannose 4, and C-2 and C-6 of mannose-4' residue (see Schemes 7B, 8, and 9); (c) "tetraantennary structures" C-2 and C-4 of mannose-4, and C-2 and C-6 of mannose-4' residue (see Scheme 7C); (d) C-4 of mannose-3 residue substituted by an *N*-acetylglucosamine residue (Schemes 6 and 10); (e) C-6 of the first *N*-acetylglucosamine of the di-*N*-acetylchitobiose residue substituted by a fucose residue (Schemes 6, 8, 9, and 10); (f) C-3 of the *N*-acetylglucosamine residues of the *N*-acetyllactosamine residues substituted by a fucose residue (Schemes 6 and 7).

However, the concept of the existence of common cores and of similar structures in glycans must not be accepted as a dogma, because dogmas are dangerous and because some "non-orthodox" structures have been described. They generally differ from "orthodox" structures by the existence of only one *N*-acetylglucosamine residue at the "reducing terminus" and of "iso-*N*-acetylactosamine"-like structures (see Scheme 11). On the other hand, oligosialosyl sequences have been characterized, such as α-NeuAc-(2→8)-α-NeuAc-(2→6)-β-Gal in glycoproteins of various tissues (14).

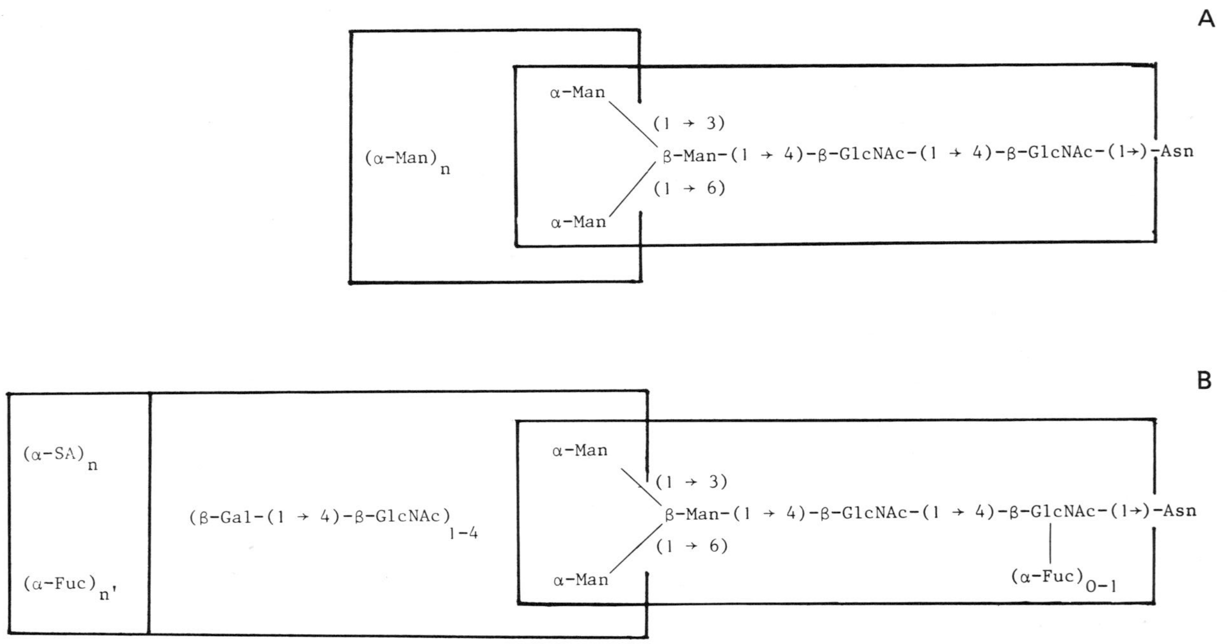

Scheme 2. General structural scheme of glycans of oligomannoside type (A) and of N-acetyllactosamine type (B).

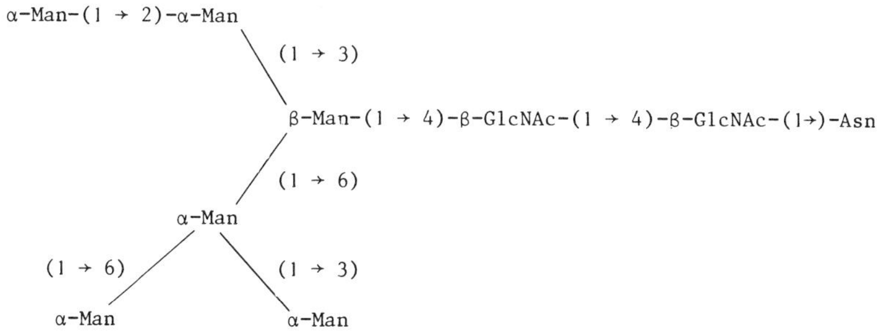

Scheme 3. Structure of ovalbumin (6) and of Taka-amylase A (7) glycans.

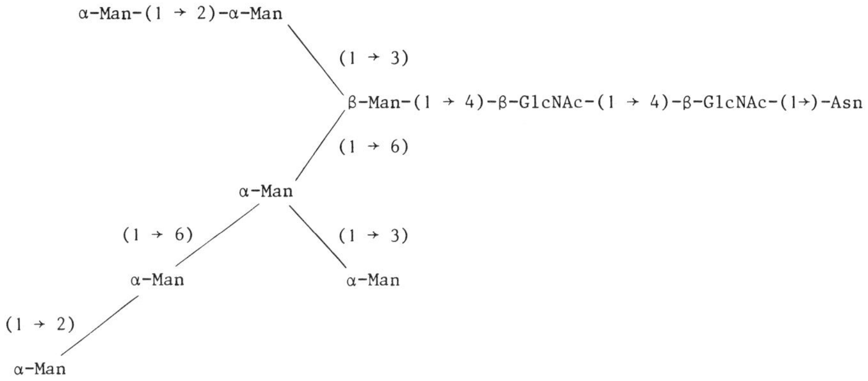

Scheme 4. Structure of Sindbis virus S-4 glycan.

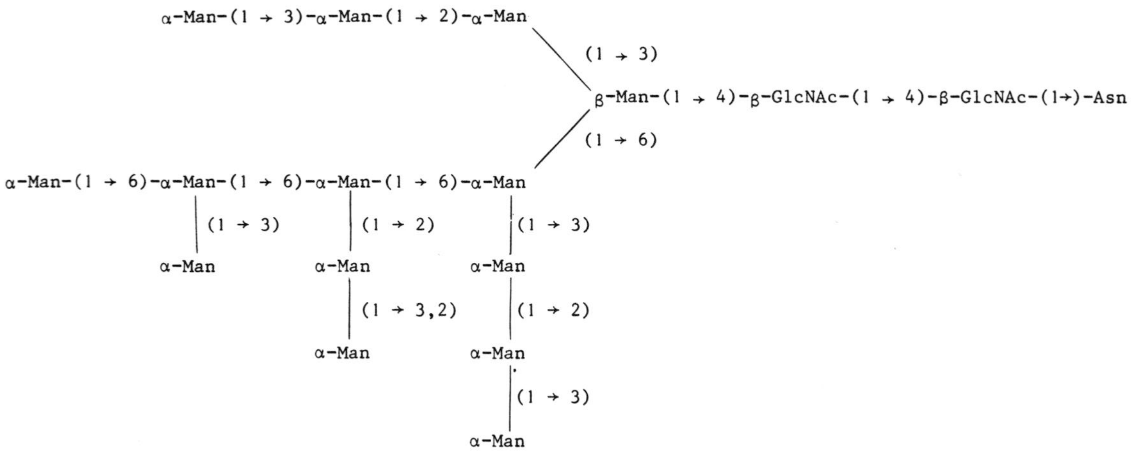

Scheme 5. Core of S. cerevisiae *X-2180-1A wild type mannan (8).*

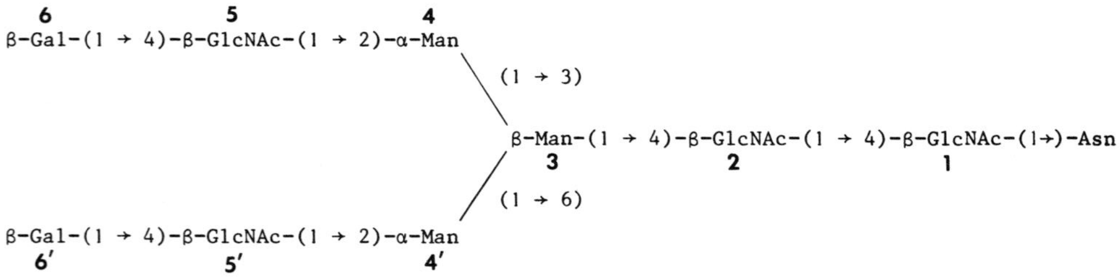

Scheme 6. Structure of biantennary glycans.

Most of the structures just described have been slowly and patiently determined by combining the use of chemical (partial acetolysis and hydrolysis, hydrazinolysis-nitrous deamination, chromic and periodate oxidation, and methylation) with enzymatic (hydrolysis by *exo-* and *endo-*glycosidases) well-defined methods. Recently, our laboratories have entered into a collaboration for exploring the conformation of human serotransferrin glycan (15) by high-resolution n.m.r. spectroscopy, by use of reference spectra obtained by enzymatic degradation of the transferrin glycan and from a large collection of oligosaccharides prepared by partial hydrolysis and acetolysis of ovomucoid (16), or isolated from urine of patients with lysosomal diseases (see later, refs. 17-26, and Strecker *et al.*, in this volume). No information concerning the conformation of the glycan molecule was obtained, as we expected, but the unexpected and exciting results obtained concerning the primary structure of glycans allow us to propose a method for determining the complete primary sequence of monosaccharides of the *N*-acetyllactosamine-type glycans on the basis of only methylation and 360-MHz ^{1}H-n.m.r. spectroscopy.

360-MHZ ^{1}H-N.M.R. SPECTROSCOPY OF CARBOHYDRATE CHAINS OF GLYCOPROTEINS

In recent years, high-resolution n.m.r. spectroscopy has become an extremely valuable technique in the study of biopolymers. In particular, for the investigation of structures, conformations, and intermolecular interactions in the fields of protein and nucleic acid chemistry, a vast amount of significant results has been described (27-31). However, the application of this technique to glycoconjugates (glycolipids and glycoproteins) is still rather limited. Several high-resolution ^{1}H-n.m.r. data are available for derivatives of mono-, oligo-, and polysaccharides; especially peracetyl (32-34), pertrimethylsilyl (35), and permethyl derivatives (36) have been studied. These spectra give valuable information on the configuration of the glycoside linkages, and on the type and conformation of the (constituting) monosaccharides.

A smaller number of papers deal with high-resolution ^{1}H-n.m.r. spectroscopy of non-derivatized carbohydrates in solution of deuterium oxide. For applications in the biochemical field, the latter system has several advantages, namely: (a) A chemical modification of the compounds can be omitted, (b) deuterium oxide is a good solvent for a wide range of carbohydrates and glycoconjugates, and (c) intramolecular hydrogen bonds are preserved, thus providing more insight into the complete structure of the compound in aqueous solution. To understand the biochemical role of the carbohydrate part of glycoconjugates, a detailed knowledge of the primary and

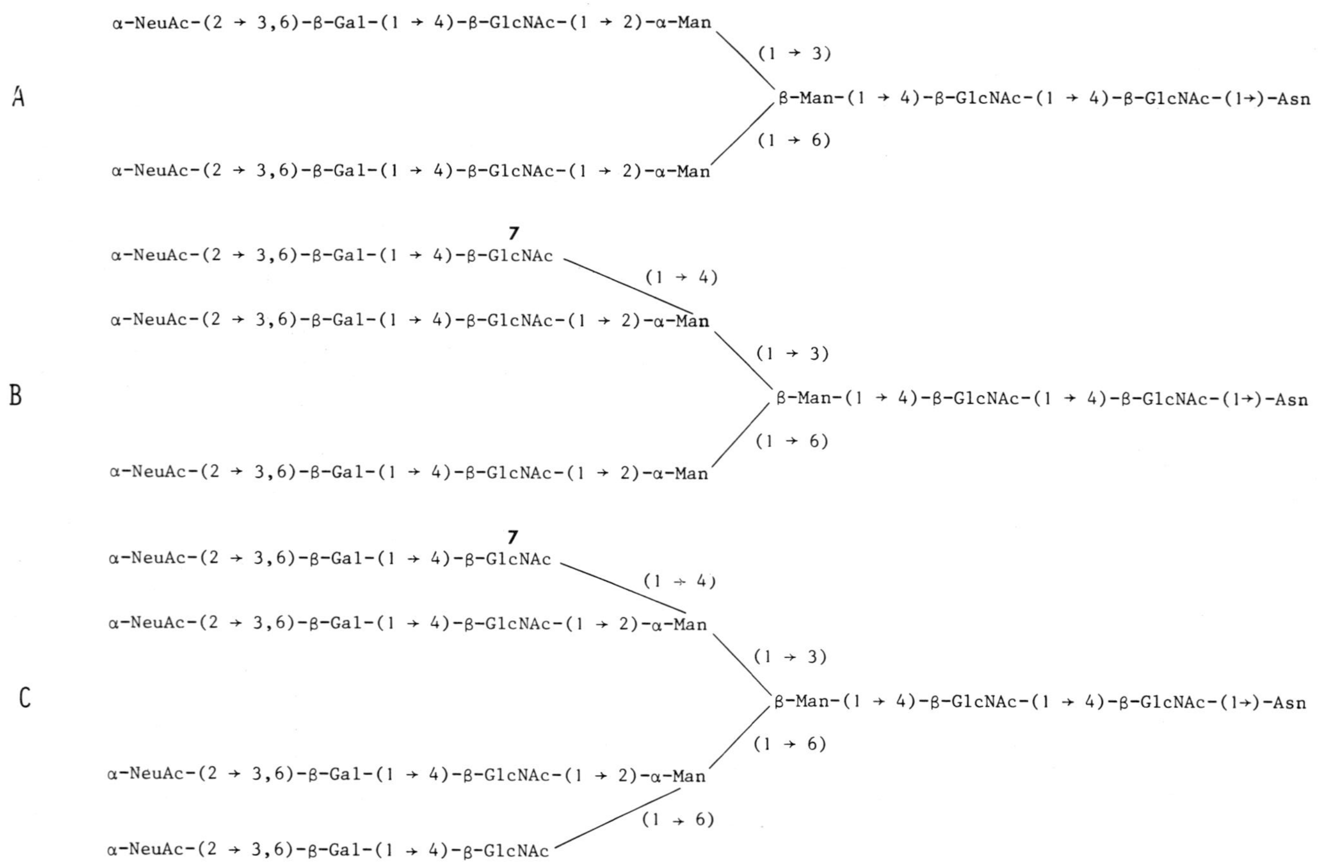

A
α-NeuAc-(2 → 3,6)-β-Gal-(1 → 4)-β-GlcNAc-(1 → 2)-α-Man
(1 → 3)
β-Man-(1 → 4)-β-GlcNAc-(1 → 4)-β-GlcNAc-(1→)-Asn
(1 → 6)
α-NeuAc-(2 → 3,6)-β-Gal-(1 → 4)-β-GlcNAc-(1 → 2)-α-Man
B
7
α-NeuAc-(2 → 3,6)-β-Gal-(1 → 4)-β-GlcNAc
(1 → 4)
α-NeuAc-(2 → 3,6)-β-Gal-(1 → 4)-β-GlcNAc-(1 → 2)-α-Man
(1 → 3)
β-Man-(1 → 4)-β-GlcNAc-(1 → 4)-β-GlcNAc-(1→)-Asn
(1 → 6)
α-NeuAc-(2 → 3,6)-β-Gal-(1 → 4)-β-GlcNAc-(1 → 2)-α-Man
C
7
α-NeuAc-(2 → 3,6)-β-Gal-(1 → 4)-β-GlcNAc
(1 → 4)
α-NeuAc-(2 → 3,6)-β-Gal-(1 → 4)-β-GlcNAc-(1 → 2)-α-Man
(1 → 3)
β-Man-(1 → 4)-β-GlcNAc-(1 → 4)-β-GlcNAc-(1→)-Asn
(1 → 6)
α-NeuAc-(2 → 3,6)-β-Gal-(1 → 4)-β-GlcNAc-(1 → 2)-α-Man
(1 → 6)
α-NeuAc-(2 → 3,6)-β-Gal-(1 → 4)-β-GlcNAc

Scheme 7. *(See following page.)*

Scheme 7. Structure of bi-, tri-, and tetraantennary glycans of α_1-acid glycoprotein (B. Fournet et al., in this volume). GP-II-6: structure A; GP-II-5, GP-II-7, and GP-V-5: structure B; GP-III-6 and GP-V-4: structure C. In GP-III-5, GP-V-2, and GP-V-3 glycopeptides an additional α-Fuc-(1→3) residue is linked to the GlcNAc residue (7) of structure C.

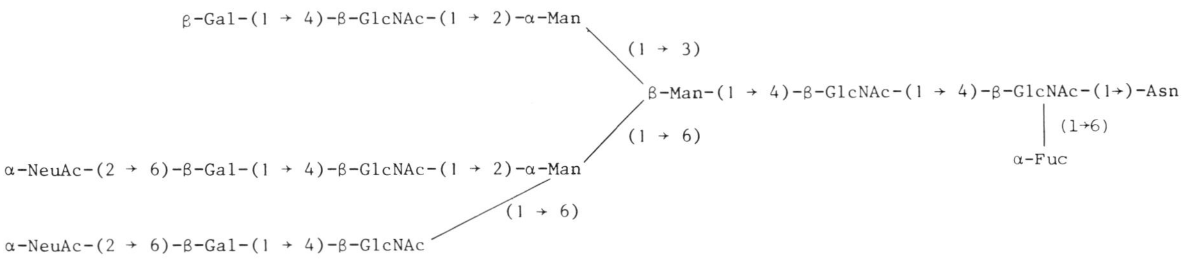

Scheme 8. Structure of unit-B type glycopeptide Gp-3 of porcine thyroglobulin (9).

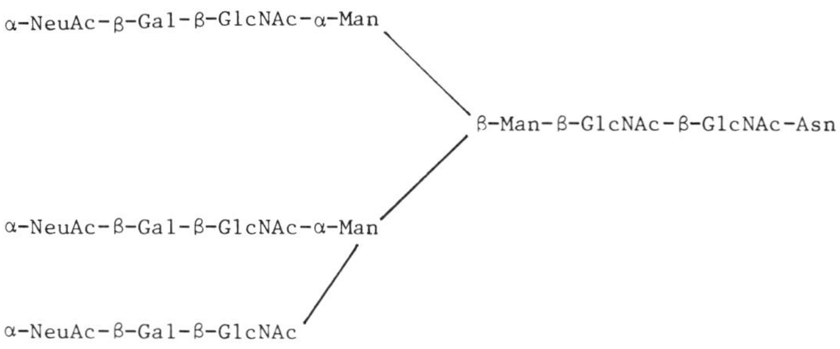

Scheme 9. Structure of type A glycopeptide of Semliki Forest virus (10).

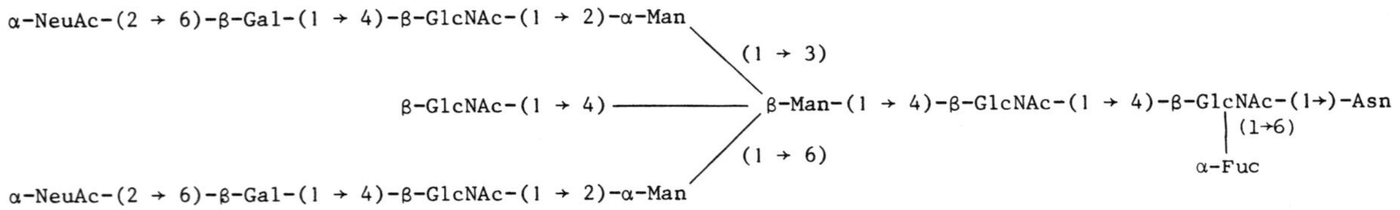

Scheme 10. Structure of human serum IgG glycopeptide (11,12).

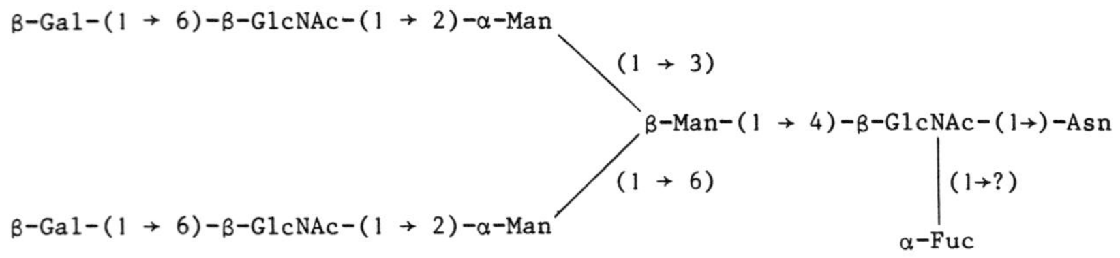

Scheme 11. Structure of human myeloma IgM glycan (13).

preferably also of the three-dimensional structure, in aqueous solution, of these moieties is indispensable. For this purpose, n.m.r. spectroscopy is in principle one of the most powerful methods.

This report deals with the 360-MHz ^{1}H-n.m.r. spectroscopy of complex carbohydrate chains, coupled *via* an *N*-glycosyl linkage to an asparagine residue in glycoproteins. For reference purposes, the spectral data of some partial (oligosaccharide and glycopeptide) structures have been included.

The 360-MHz ^{1}H-n.m.r. spectrum of the general structural element β-GlcNAc-(1→4)-Asn is shown in Fig. 1. The resonances of the anomeric proton (H-1), the CH_2-3 residue group of the asparagine residue, and the *N*-acetyl protons of the GlcNAc residue are found clearly separated from the bulk of the non-anomeric carbohydrate protons (3.5-4.0 p.p.m.). The low-field position of the (axially oriented) anomeric proton (5.09 p.p.m.) and the rather large coupling constant $J_{1,2}$ (9.8 Hz) are due to the linkage of the amide nitrogen atom to the anomeric carbon atom (37). The two methylene-3 protons of the asparagine residue are not equivalent; their chemical shifts (2.93-2.87 p.p.m.) and the chemical shift of the methine-2 proton (3.99 p.p.m.) are in accordance with those found for the zwitterionic form of free asparagine (28,30,31). The bulk of the non-anomeric proton signals at 3.5-4.0 p.p.m. could be completely assigned, as indicated in Fig. 2, by use of specific proton-decoupling and spectrum simulation. The $^4C_1$ (D) chair conformation of the pyranosyl ring of the GlcNAc residue could be deduced from the proton coupling-constants by use of an adapted Karplus equation (38). Another structural element frequently occurring in glycoproteins is α-Fuc-(1→6)-β-GlcNAc-(1→4)-Asn, the spectrum of which is given in Fig. 3 (37). Characteristic for the fucose residue are the resonances of the anomeric proton (4.90 p.p.m., $J_{1,2}$ 3.75 Hz) and the methyl group (1.21 p.p.m.). The coupling constant of 3.75 Hz is indicative of an α-L glycosidic bond of the fucose residue. The spectrum could be completely interpreted and the ring conformation of the sugar residues determined. The attachment of fucose to C-6 of a GlcNAc residue gives rise to changes in the chemical shifts for H-4,-5, and -6 of the GlcNAc residue as compared to the shifts of β-GlcNAc-(1→4)-Asn. Also a change in the geminal coupling constant of H-6 and -6' of the GlcNAc residue (12.7 to 11.4 Hz) occurs.

The (1→6) linkage of the fucose residue was unambiguously proven by ^{13}C-n.m.r. spectroscopy (39). Only two carbon atoms of the GlcNAc residue of the glycopeptide Fuc→GlcNAc→Asn show significant shifts with respect to the resonances of the GlcNAc→Asn residue, namely a downfield shift of 6.7 p.p.m. for C-6 and an upfield shift of 0.8 p.p.m. for C-5. These shifts point directly to glycosylation of the OH-6 of the GlcNAc residue (40,41).

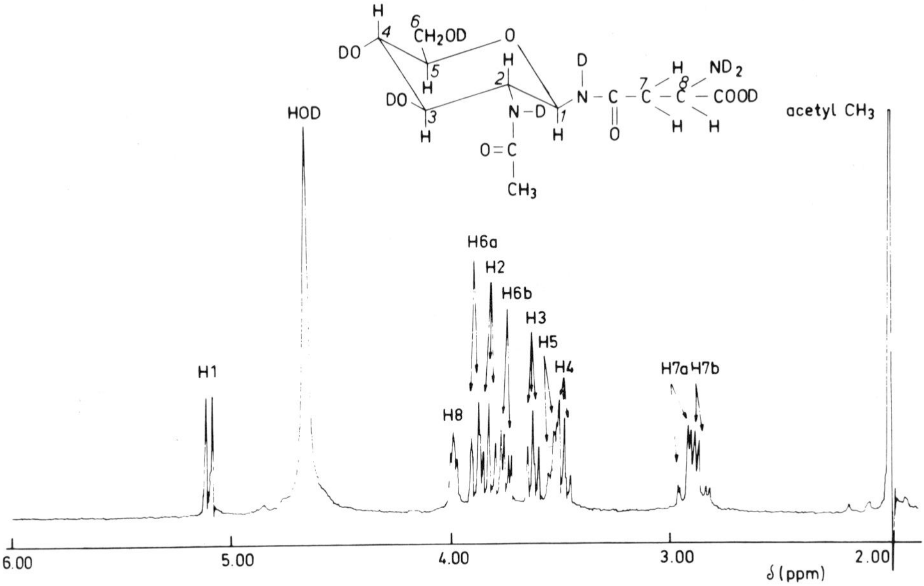

Fig 1. 360-MHz ^{1}H-n.m.r. spectrum of β-GlcNAc-(1→4)-Asn in D_2O.

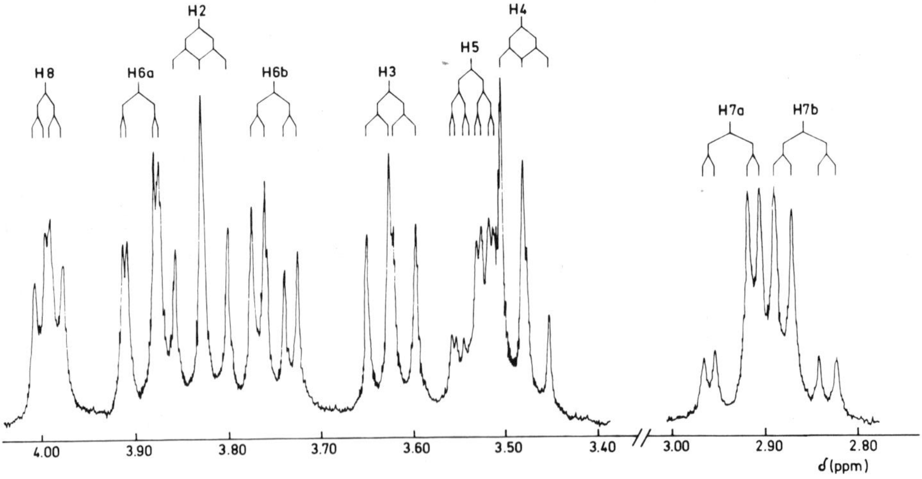

Fig. 2. 360-MHz ^{1}H-n.m.r. spectrum of the non-anomeric protons and β-methylene protons of β-GlcNAc-(1→4)-Asn.

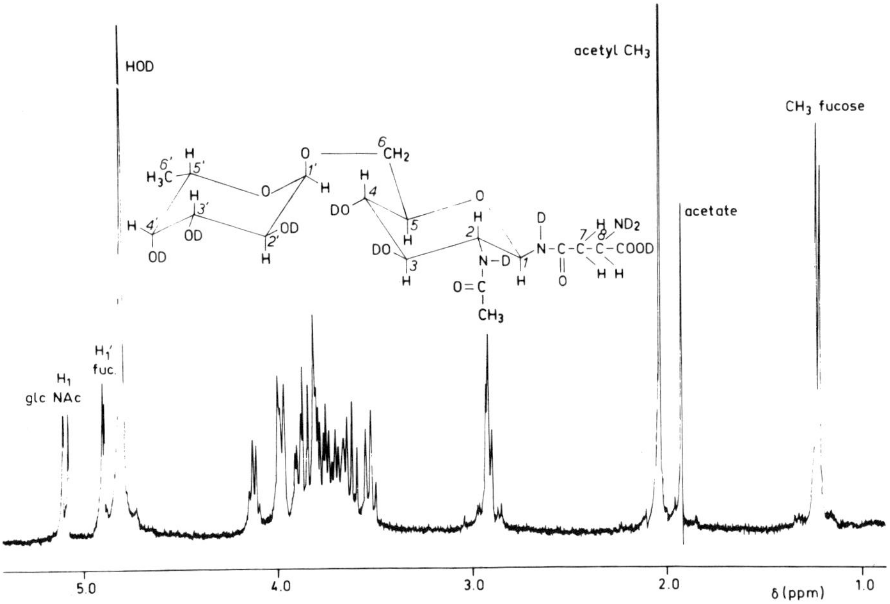

Fig. 3. 360-MHz ^1H-n.m.r. spectrum of α-Fuc-(1→6)-β-GlcNAc-(1→4)-Asn in D_2O.

An example of a complex Asn-glycopeptide is the asialoglycopeptide obtained from human serotransferrin. The 360-MHz ^1H-n.m.r. spectrum and the structure are shown in Fig. 4 (42). To illustrate the high resolution in this spectrum, the 90-MHz ^1H-n.m.r. spectrum is presented in Fig. 5. Comparison of both spectra makes clear that the low-resolution instrument is unsuitable for unraveling structural details of complex carbohydrate chains. The resonances shown in Fig. 4 can roughly be divided into signals from the following groups of protons: (a) Anomeric protons; at ambient temperature one anomeric-proton resonance is hidden under the HOD line. It can be visualized by recording the spectra at other temperatures. (b) H-2 protons of mannose residues. (c) Bulk of the non-anomeric protons. (d) The methylene-3 group of the asparagine residue. (e) The methyl group protons of the acetamido group. So far, it is impossible to interpret the bulk of the non-anomeric protons. Many structural details of this compound are, however, reflected in the chemical shifts of the anomeric protons and the H-2 of the mannose residues.

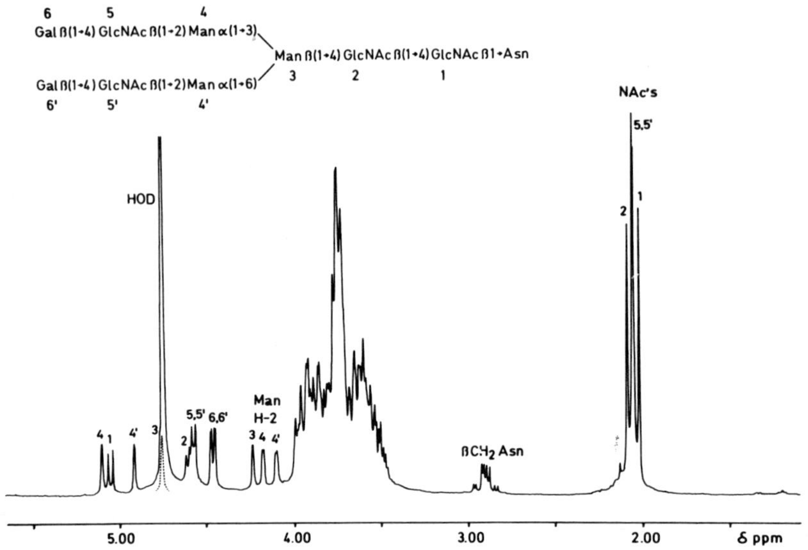

Fig. 4. 360-MHz ^{1}*H-n.m.r. spectrum of the asialo-glycan-Asn isolated from human serotransferrin.*

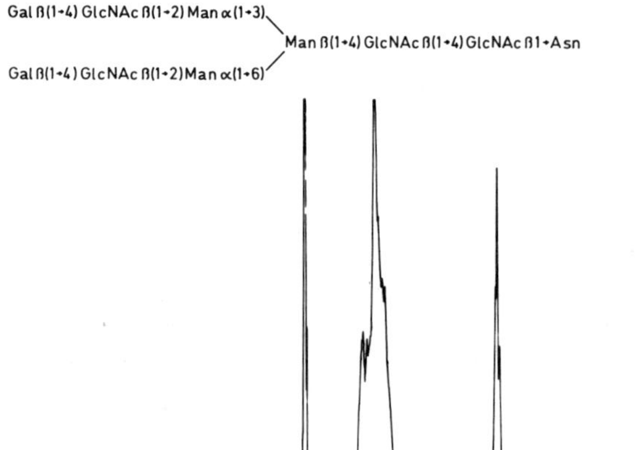

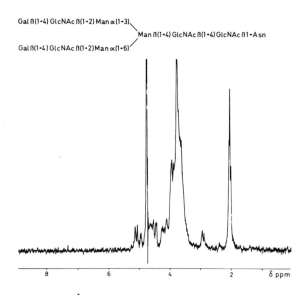

Fig. 5. 90-MHz ^{1}*H-n.m.r. spectrum of the asialo-glycan-Asn isolated from human serotransferrin.*

Integration showed that nine anomeric protons are present, which is in accordance with the proposed number of monosaccharide units. In fact, integration of peak areas is an accurate method for the determination of the number of constituting monosaccharides. An additional control for the number of amino sugar residues may be obtained from the integration of the NAc signals. For assignment of the anomeric protons, the spectra of a large series of partial structures has been recorded. A few relevant representatives of this group of compounds are given in Fig. 6 together with a schematic presentation of some spectral data.

The anomeric signal of the GlcNAc residue (*1*) can easily be recognized; among others, the large coupling constant is typical. Comparison of the structures *B*, *E*, and *F* yields the anomeric signal of the GlcNAc residue (*2*) ($J_{1,2}$ *ca.* 7 Hz, H-1β). The anomeric signals of the mannose residues follow from correlation of the spectra of *B* and *C*, taking into account that an H-1 bound to a C   in α-D-glycosidic linkage resonates at a field lower than that of an H-1 bound to a β-D-linked C. Comparison of the spectra of *D*, *E*, and *F* makes clear that the anomeric protons of the GlcNAc residues (*5*) and (*5'*) resonate both at *ca.* 4.57 p.p.m. The anomeric protons of the residue (*6*) and (*6'*) are also indistinguishable and resonate both at *ca.* 4.47 p.p.m., as is evident from comparison of the spectra of *E* and *F*. By selective irradiation, the signals of the H-2 of the mannose residues could be assigned as indicated. The *N*-acetyl resonances could be interpreted on the basis of a set of reference compounds. The expanded region of the interpretable parts of the spectrum of the asialoglycan→Asn glycopeptide of human serotransferrin is depicted in Fig. 7.

A few conclusions can be drawn: (a) The chemical shifts of the various anomeric protons in the intact glycopeptide and the partial structures thereof occur at characteristic positions. (b) The primary structure and the type of glycosidic linkages are reflected by the chemical shifts and the coupling constants of the anomeric protons of the various monomers. (c) The total n.m.r. spectrum can be used as a fingerprint, *e.g.*, on this basis the occurrence of the "asialoglycan→Asn" part as a structural element in rabbit serotransferrin, human lactotransferrin, orosomucoid, IgA, and IgG could be demonstrated. (d) The mannotriose branching-core, surrounded by GlcNAc residues, can be recognized on the basis of the pattern of the mannose H-2 resonances.

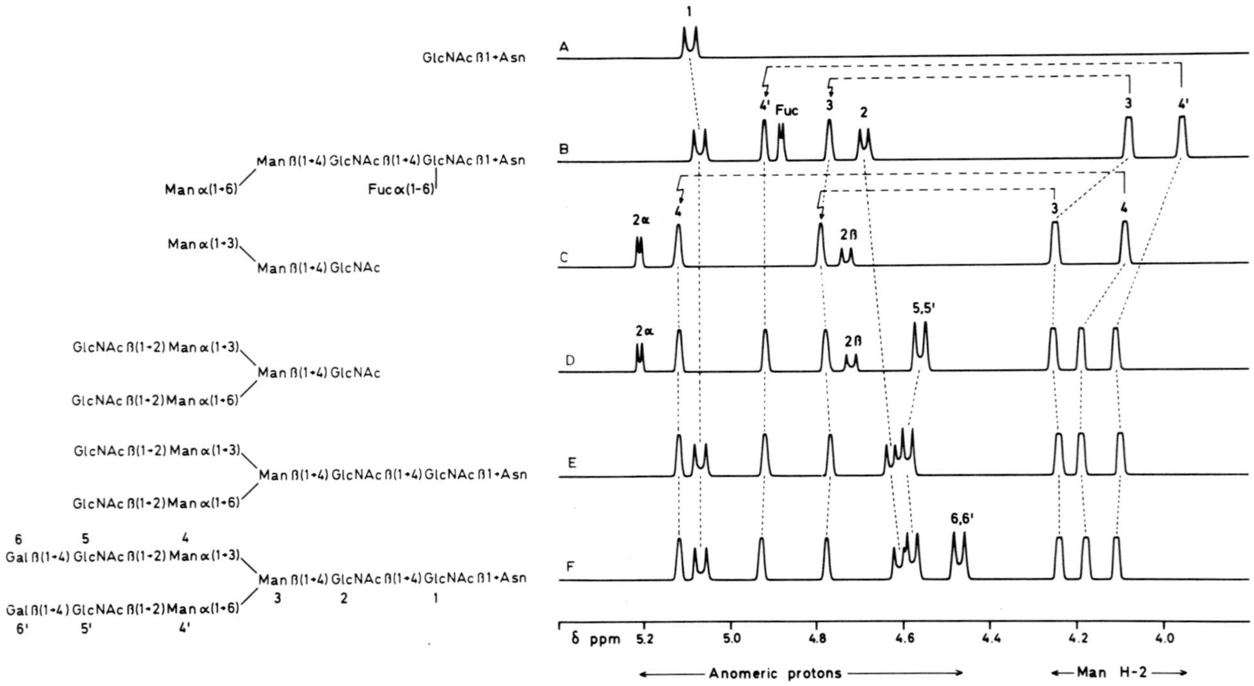

Fig. 6. Schematic representation of 360-MHz ^{1}H-n.m.r. data of the asialo-glycan-Asn and some reference compounds.

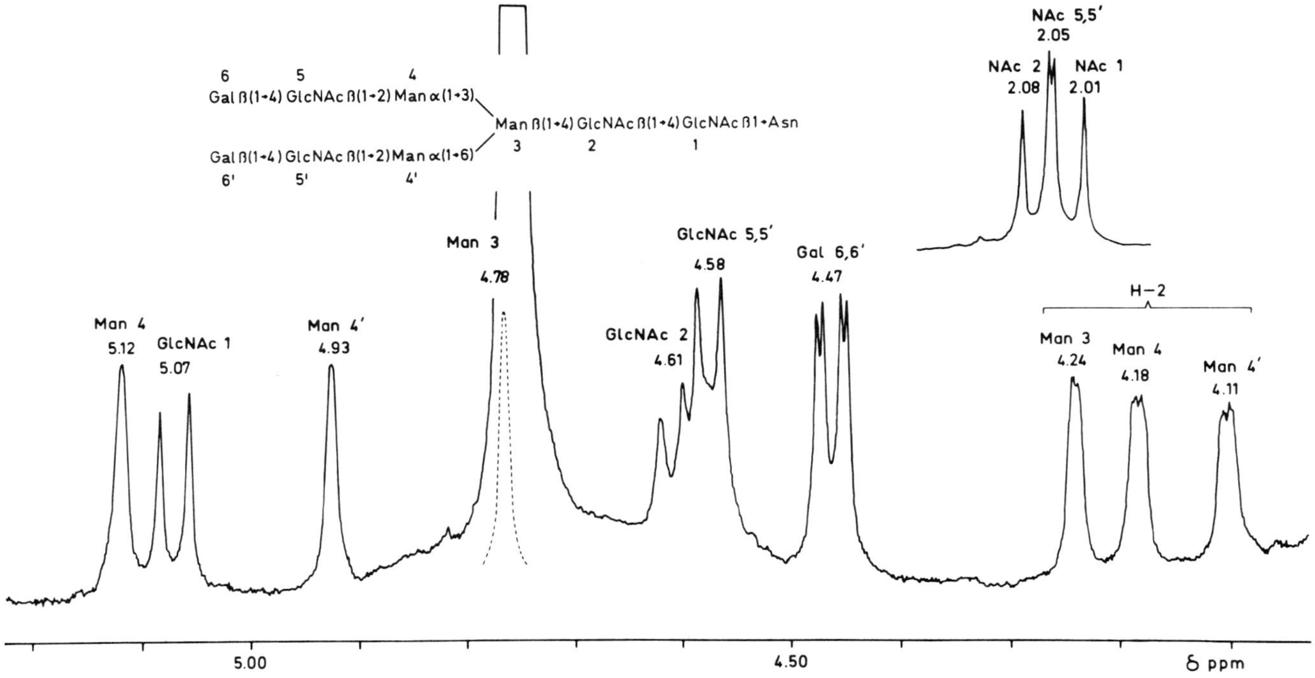

Fig. 7. Expanded regions of the anomeric protons, the mannose H-2 protons, and the N-*acetyl protons of the 360-MHz* ^{1}*H-n.m.r. spectrum of the asialo-glycan-Asn.*

The occurrence of sialic acid residues in terminal position of the biantennary structure has a remarkable effect on the spectrum, depending on the type of glycosidic linkage. First, the (2→6)-linked sialic acid residue occurring in the glycan of human serotransferrin will be considered. In Fig. 8, the 360-MHz ^{1}H-n.m.r. spectrum of this glycan Asn-Lys chain is presented. The significant additional signals in this spectrum are those of the H-3*eq* and H-3*ax* protons of sialic acid. These are typical resonances for sialic acid, which do not coincide with the bulk of the non-anomeric protons. Their chemical shifts may be used to assign the position of the glycosidic linkage as will be illustrated later. The introduction of a sialic acid residue at C-6 of the galactose residues gives rise to a few small, but characteristic shift increments for some anomeric protons, as indicated in Fig. 9. Downfield shifts are observed for the anomeric protons of the Man (*4*) and (*4'*) residues as well as of the GlcNAc (*5*) and (*5'*) residues, whereas the anomeric protons of the Gal (*6*) and (*6'*) residues show upfield shifts. These shift increments may be used to determine the position of sialic acid in a biantennary structure having only one sialic acid residue.

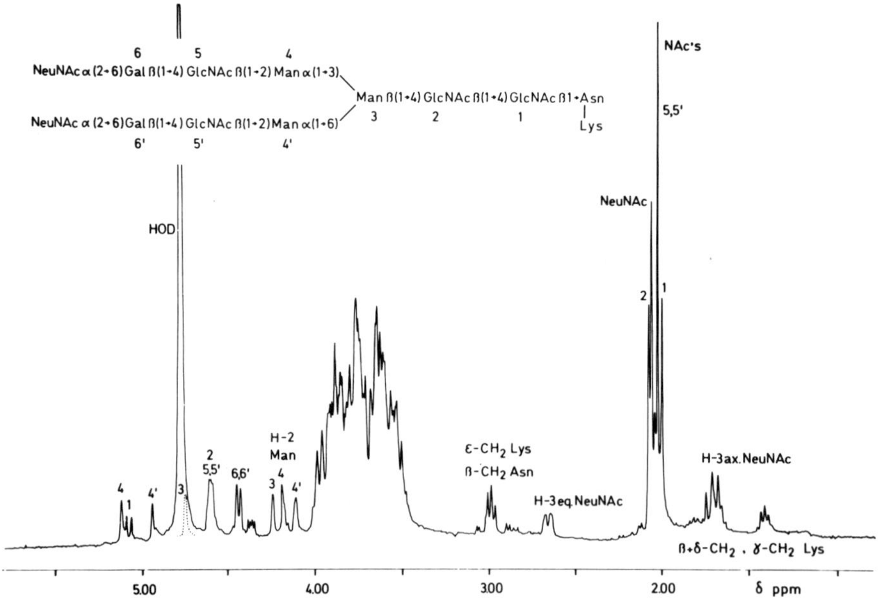

Fig. 8. 360-MHz ^{1}H-n.m.r. spectrum of the glycan-Asn-Lys isolated from human serotransferrin.

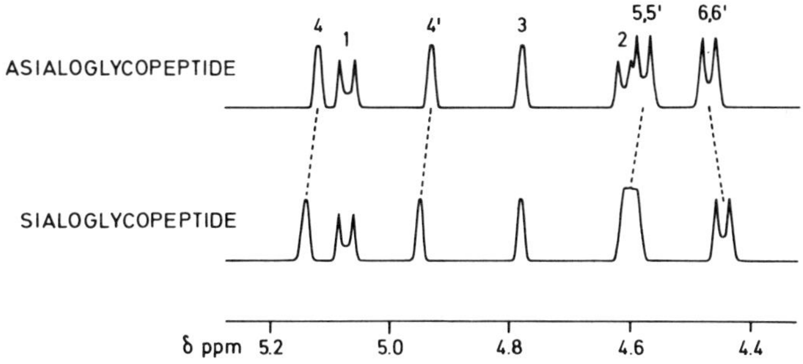

Fig. 9. Comparison of the anomeric regions of the asialo glycopeptide and the sialo glycopeptide.

In Table I, the 360-MHz ^{1}H-n.m.r. spectral data of asialo-glycan→Asn and the oligosaccharide V, which has the sialic acid residue attached to the upper branch, are compared. The shift increments of the anomeric protons of neither the Gal (*6*) and (*6'*) residue, nor of the GlcNAc (*5*) and (*5'*) residues provide structural information, since it is not known which signals stem from the upper or the lower branch. However, it is easy to distinguish the resonances of the anomeric protons of the Man (*4*) and (*4'*) residues as only the anomeric proton of the Man (*4*) residue undergoes a shift when a sialic acid residue is linked to the upper branch (see Table I). This resonance is therefore indicative of the position of the sialic acid residue. On the basis of these data and those of a set of reference compounds, the following conclusions may be drawn: (a) The attachment of a sialic acid residue to one branch in a biantennary structure has no effect on the chemical shifts of the anomeric protons in the other branch. (b) The (2→6)-linked sialic acid residue has a long distance effect on the mannose residue that occurs in the same branch. (c) The chemical shifts of the mannose (*4*) and (*4'*) anomeric protons may be used to establish the position of a (2→6)-linked sialic acid residue. Sialic acid residues may also occur in a (2→3)-glycosidic linkage to D-galactose residues. Such a sialic acid residue may easily be recognized from the chemical shifts of the H-3*eq* and -*ax* protons, as indicated in Table II. The only anomeric proton that undergoes a significant shift upon attachment of a sialic acid residue to C-3 of the galactose residue is that of the galactose residue itself. In contrast to the (2→6)-linked sialic acid residue, it is not possible to distinguish whether the (2→3)-linked sialic acid residue is present in the upper or the lower branch because the anomeric protons of Gal (*6*) and (*6'*) residues have the same chemical shift, and the chemical shifts of the mannose (*4*) and (*4'*) anomeric protons are unaffected.

Table I. Changes in the Chemical Shift of Anomeric Protons Due to the Presence of Sialic Acid in the Upper Branch of the Biantennary Structure

Oligosaccharide V

```
                        6           5              4
α-NeuNAc-(2→6)-β-Gal-(1→4)-β-GlcNAc-(1→2)-α-Man-(1→3)
                                                      \
                                                       β-Man-(1→4)-GlcNAc
                                                      /  3           2
               β-Gal-(1→4)-β-GlcNAc-(1→2)-α-Man-(1→6)
                 6'          5'             4'
```

Structures	Chemical shifts (δ) of H-1 of residues					
	4	5	6	4'	5'	6'
Asialoglycan→Asn	5.119	4.581	4.470	4.926	4.581	4.470
Oligosaccharide V	5.131	4.603	4.447	4.929	4.583	4.468

*Table II. Comparison of the Chemical Shifts of H-3*eq. *and H-3*ax. *of Sialic Acid Residue (2→3)- or (2→6)-linked to the Galactose Residue of* N-*Acetyllactosamine*

Structures	*Chemical shifts (δ)* H-3eq	H-3ax
α-*NeuNAc-(2→6)-Gal*	*2.67*	*1.72*
α-*NeuNAc-(2→3)-Gal*	*2.76*	*1.80*

In a biantennary structure having one (2→3)-linked and one (2→6)-linked sialic acid residue, the occurrence and position of both types may easily be deduced from the 360-MHz ^{1}H-n.m.r. spectrum. As shown in Fig. 10, two sets of H-3 protons characteristic for the two types of sialic acid residues are present. The signal of the anomeric proton of the mannose (*4*) residue is shifted from 5.12 to 5.14 p.p.m., thus indicating the presence of a (2→6)-linked sialic acid residue in the upper branch. Consequently, the (2→3)-linked sialic acid residue is attached to the lower branch. In more complex structures, for example in a triantennary structure, it is often possible to draw useful conclusions on the position of sialic acid residues. As shown in Fig. 11, the 360-MHz ^{1}H-n.m.r. spectrum of a triantennary structure having 3 sialic acid residues, the integration of the H-3-*eq* and -*ax* signals of the sialic acid residue indicates that the ratio of (2→3)- to (2→6)-linked sialic acid residues is 1:2. The resonance position of the anomeric protons of the mannose (*4*) and (*4'*) residues indicate the occurrence of a (2→6)-linked sialic acid residue in the corresponding branches. Therefore, it must be concluded that the (2→3)-linked sialic acid residue is located in the additional branch. Characteristic for the triantennary structure are the chemical shifts of the H-2 protons of the mannose residues, as will be discussed later.

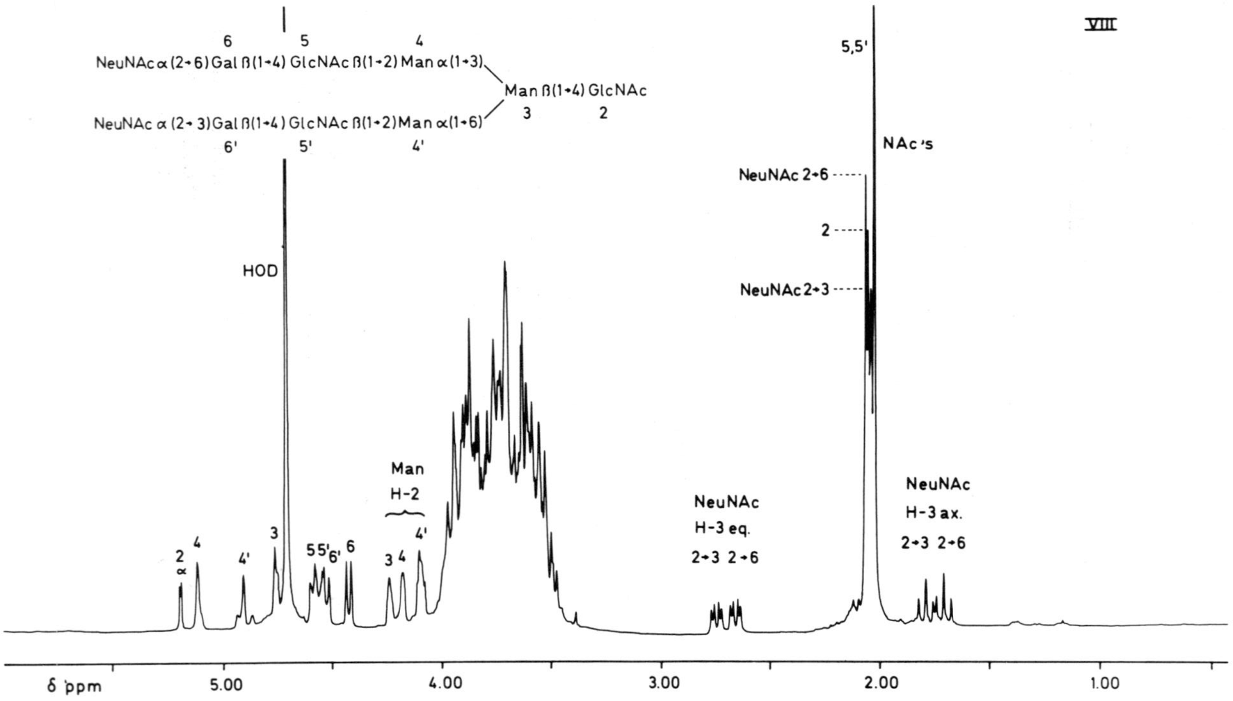

Fig. 10. *360-MHz ^{1}H-n.m.r. spectrum of oligosaccharide VIII, isolated from urine of a sialidosis patient, having two different linkages of the sialic acid residues.*

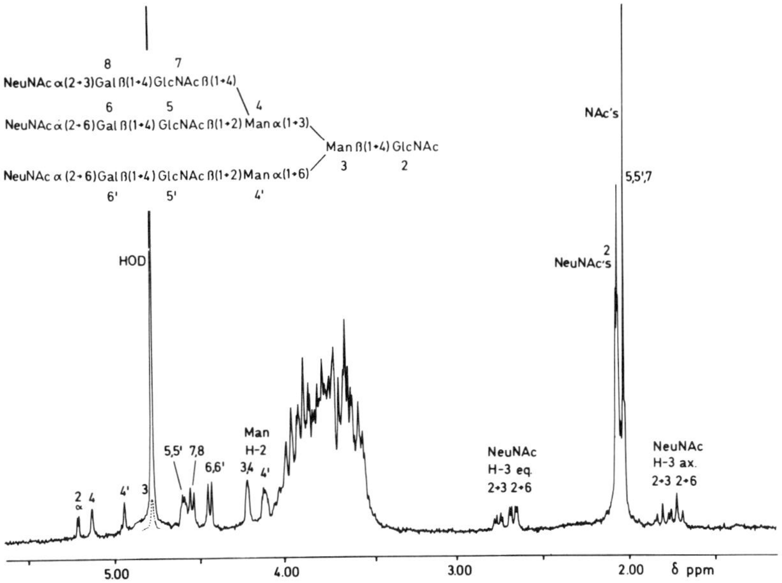

Fig. 11. 360-MHz ^{1}H-n.m.r. spectrum of an urinary oligosaccharide having a triaantennary structure.

Tetraantennary structures occur in α_1-acid glycoprotein. The 360-MHz ^{1}H-n.m.r. spectrum of such a structure in the asialo form is shown in Fig. 12. The quantitative composition of this glycopeptide can easily be derived from integration of the peak areas: The 5 *N*-acetyl signals account for 6 *N*-acetyl groups. The methyl signals (2 doublets) indicate the presence of 1 fucose residue, besides 1 threonine residue. Furthermore, 3 mannose and 4 galactose residues are present. Characteristic for this structure is the extension of the mannotrioside core with two additional branches. This is reflected, in the ^{1}H-n.m.r. spectrum, in the chemical shifts of the H-1 and H-2 protons of the mannose residues. These chemical shifts are summarized in Table III for the bi-, tri-, and tetra-antennary structures. On the basis of these data, it is possible to deduce the substitution pattern of the mannotrioside core, provided that only the structures reported here are taken into consideration. It is to be expected that other structures give rise to other spectral data.

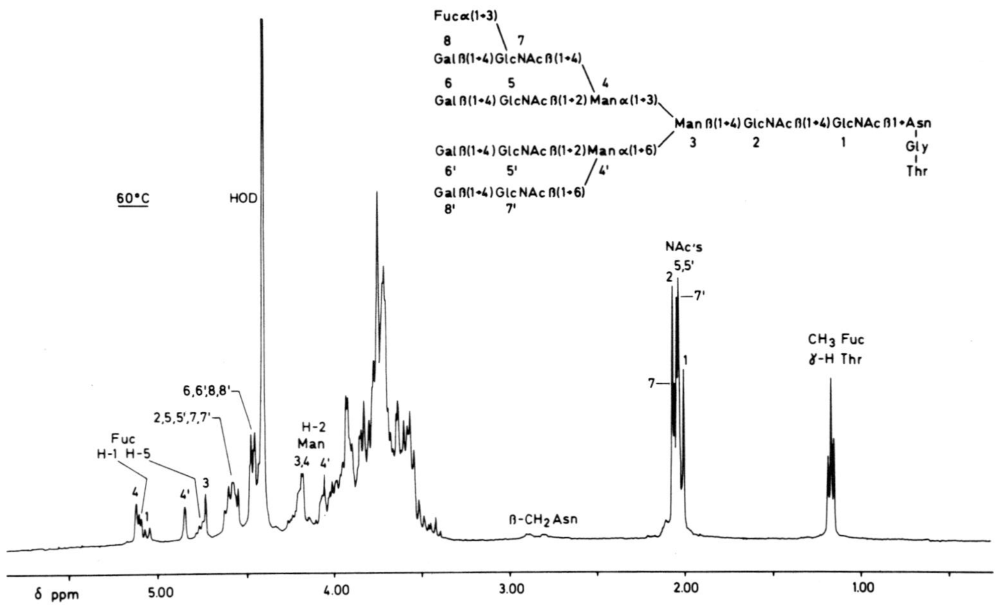

Fig. 12. 360-MHz ^{1}H-n.m.r. spectrum of a glycopeptide, isolated from α_1-acid glycoprotein, having a tetraantennary structure.

Table III. Chemical Shift Data of the Mannose H-1 and H-2 Protons in Bi-, Tri-, and Tetra-antennary Structures

Structures	Chemical shift (δ) of H-1 Mannose residue 3	Chemical shift (δ) of H-1 Mannose residue 4	Chemical shift (δ) of H-1 Mannose residue 4'	Chemical shift (δ) of H-2 Mannose residue 3	Chemical shift (δ) of H-2 Mannose residue 4	Chemical shift (δ) of H-2 Mannose residue 4'
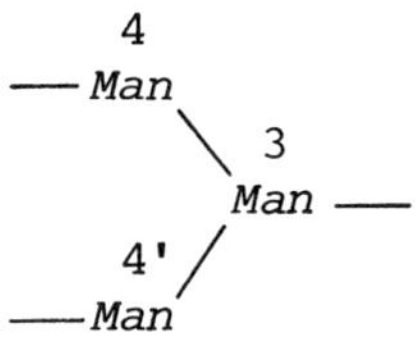	4.77	5.12	4.93	4.24	4.18	4.11
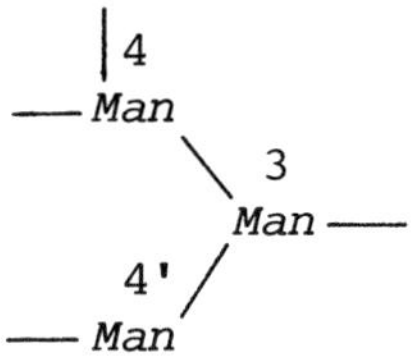	4.76	5.12	4.93	4.21	4.21	4.11
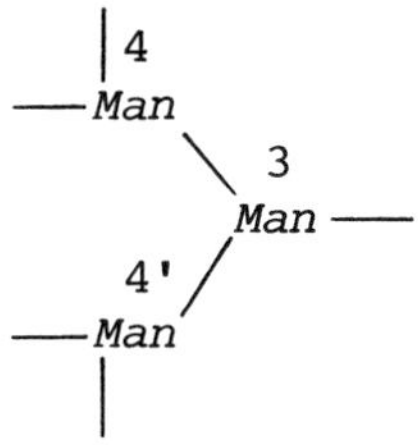	4.77	5.12	4.86	4.22	4.22	4.09

The position of the L-fucose residue is an interesting feature of the tetraantennary structure illustrated in Fig. 12. The unusual chemical shift of the H-5 proton of the fucose residue suggests a location of this residue in vicinal position to the galactose residue, both being substituents of an *N*-acetylglucosamine residue. A similar abnormal behavior of the signal of the H-5 proton of a fucose residue has been reported for the trisaccharide α-Fuc-(1→4)-[β-Gal-(1→3)]-GlcNAc having Lewis[a] blood-group activity (43,44). A small but significant difference exists, however, between the chemical shifts of the H-5 in the tetraantennary structure and that of the Lewis[a] structure. Furthermore, the tetraanntennary structure is completely devoid of Lewis activity, which excludes the occurrence of such a structural element in α_1-acid glycoprotein. Interestingly, the attachment of a fucose residue to the tetraantennary structure has a typical effect on the chemical shift of one of the *N*-acetyl groups. By comparison of the spectra of reference compounds, as illustrated in Fig. 13, this *N*-acetyl group was identified as being part of the GlcNAc (7) residue. Therefore, it was concluded that the fucose residue is located in the most upper branch, as indicated in Fig. 12.

In summary, high-resolution ^{1}H-n.m.r. spectroscopy is an extremely powerful tool for the elucidation of carbohydrate structures of glycopeptides. As it is a nondestructive technique, it can easily be incorporated in the usual analysis routes, provided that sufficient material of high purity is available.

PREDICTION OF GLYCAN STRUCTURES

All the described structures are in good agreement with the present concepts on the biosynthesis of glycoproteins. It is known that glycoprotein biosynthesis begins in the rough endoplasmic reticulum, where a part of the *N*-acetylglucosamine residues and the totality of the mannose residues are conjugated, and terminates in the smooth reticulum where the remaining *N*-acetylglucosamine residues and the totality of galactose, sialic acid, and fucose residues are conjugated.

On the other hand, the study of oligosaccharides, isolated from the urine of patients with lysozomal diseases, has furnished some interesting information on the catabolism of glycoproteins. These diseases result from a deficit of *exo*-glycosidases, which leads to an accumulation reaching often several grams per liter of oligosaccharides in tissues and urines. Schemes 12-16 illustrate the structures of oligosaccharides extracted from urines of patients with deficits in galactosidase (GM_1 gangliosidosis type I), mannosidase (mannosidosis), *N*-acetylglucosaminidase (Sandhoff's disease), fucosidase (fucosidosis), and neuraminidase (mucolipidosis I, mucolipidosis II or I-Cell disease, mucolipidosis III, and mucolipidosis De P. and N). All these structures possess an *N*-acetylglucosamine residue at the reducing terminal position, to which a β-D-mannose residue is linked at C-4.

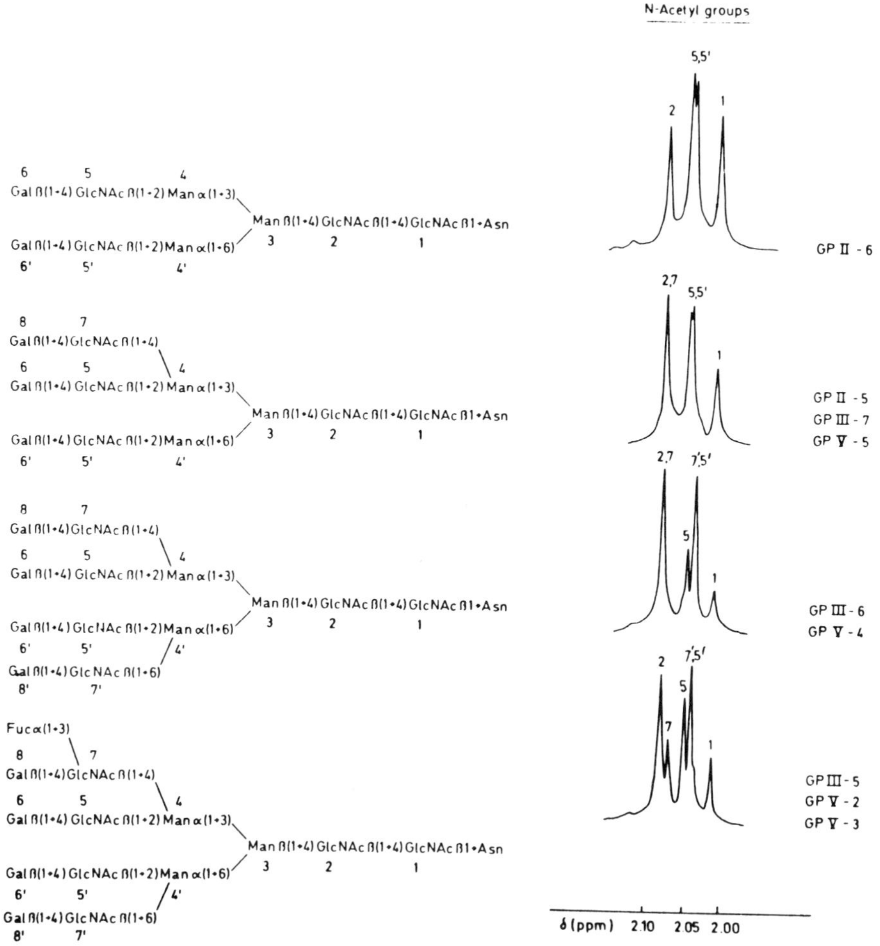

Fig. 13. Influence of the fucose residue on the N-*acetyl protons in the tetraantennary structure.*

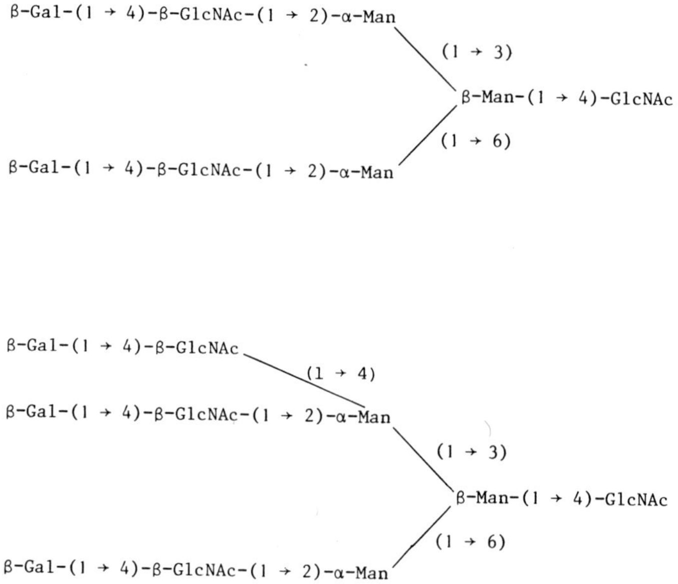

Scheme 12. Structure of oligosaccharides isolated from the liver of GM_1-gangliosidosis type I patients (45).

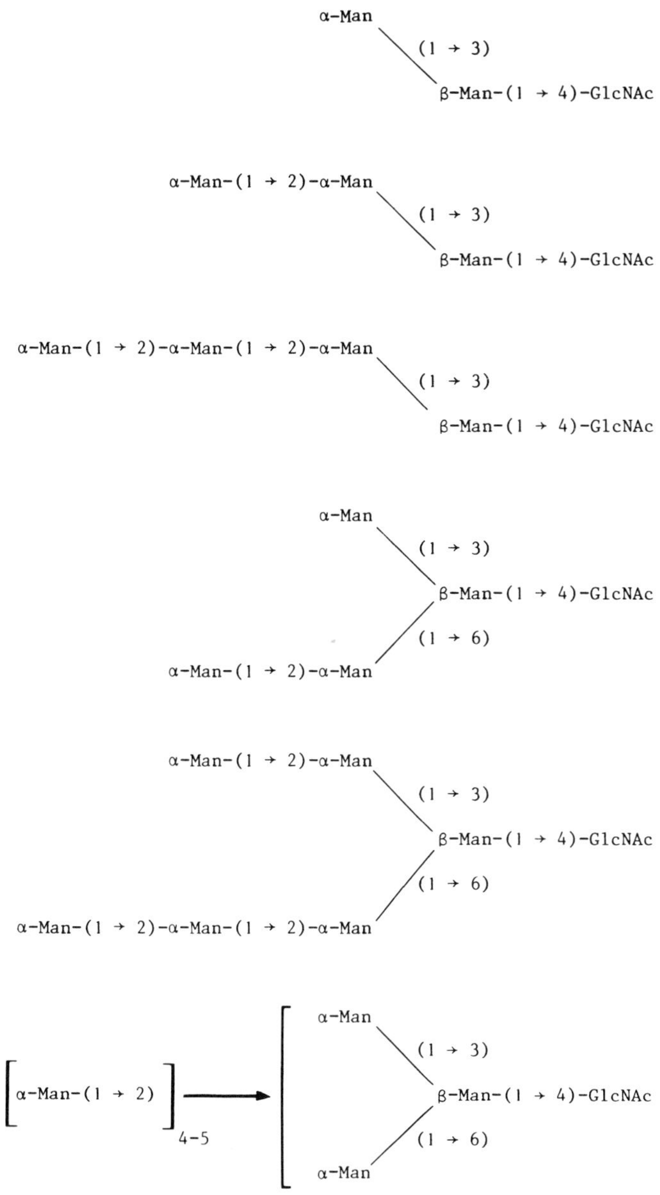

Scheme 13. Structure of oligosaccharides isolated from urine of mannosidosis patients (oligosaccharides 1 to 3, ref. 18; oligosaccharides 5 to 8, ref. 20).

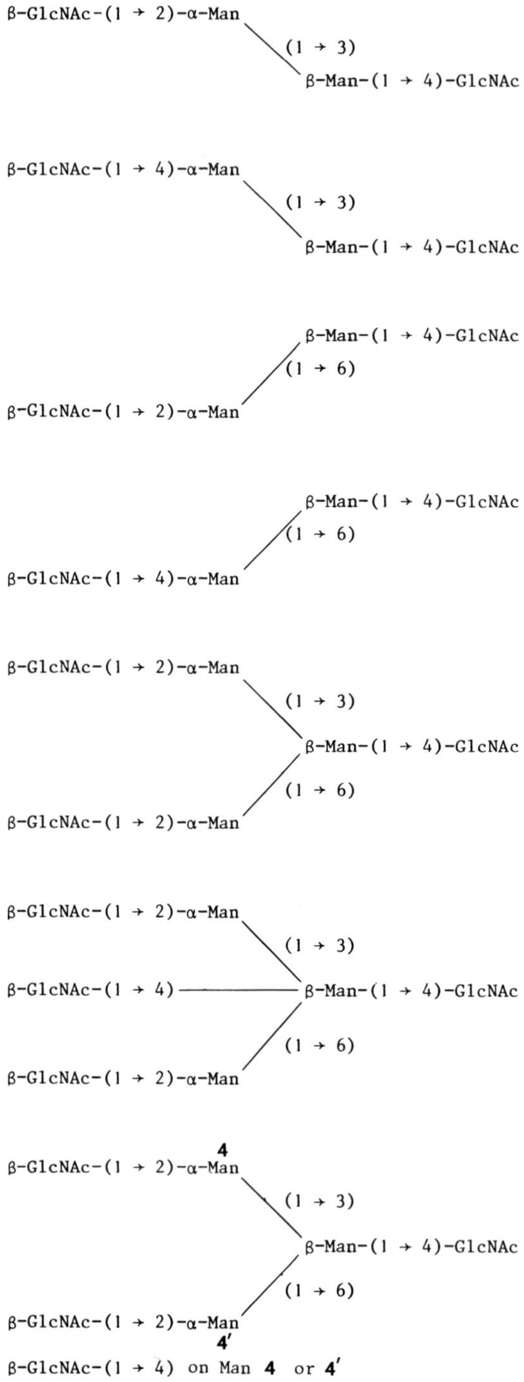

Scheme 14. Structure of oligosaccharides isolated from urine of Sandhoff's disease patient (23).

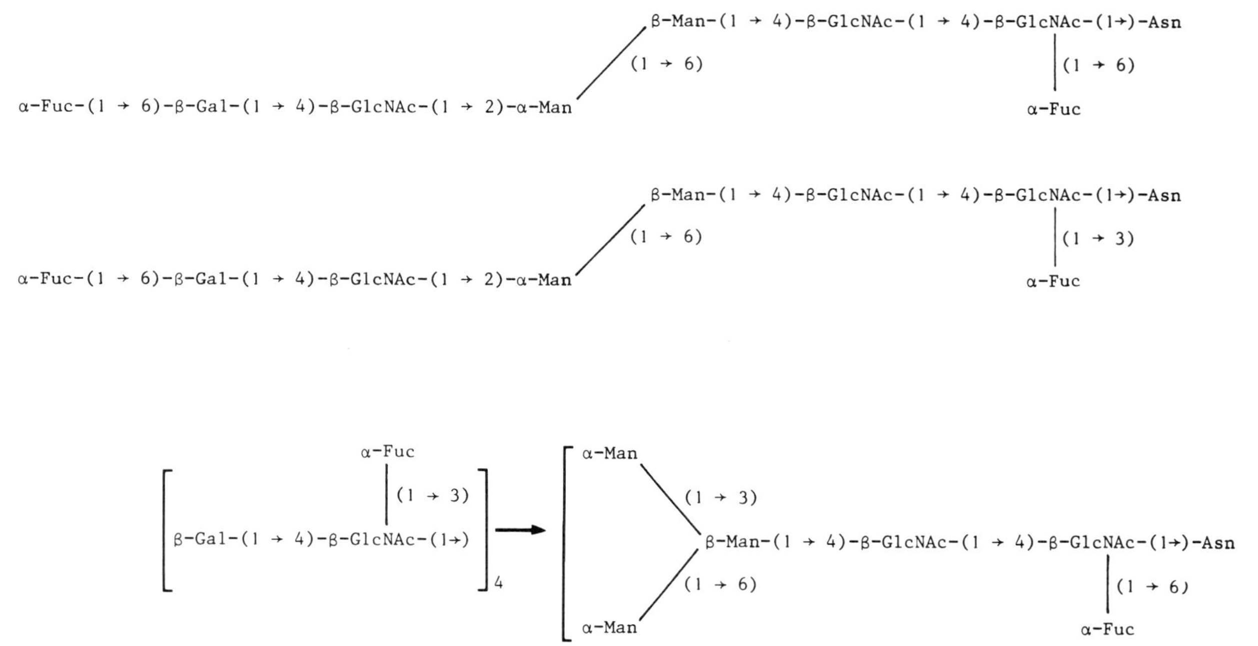

Scheme 15a. Structure of oligosaccharides isolated from the urine of a fucosidosis patient (21).

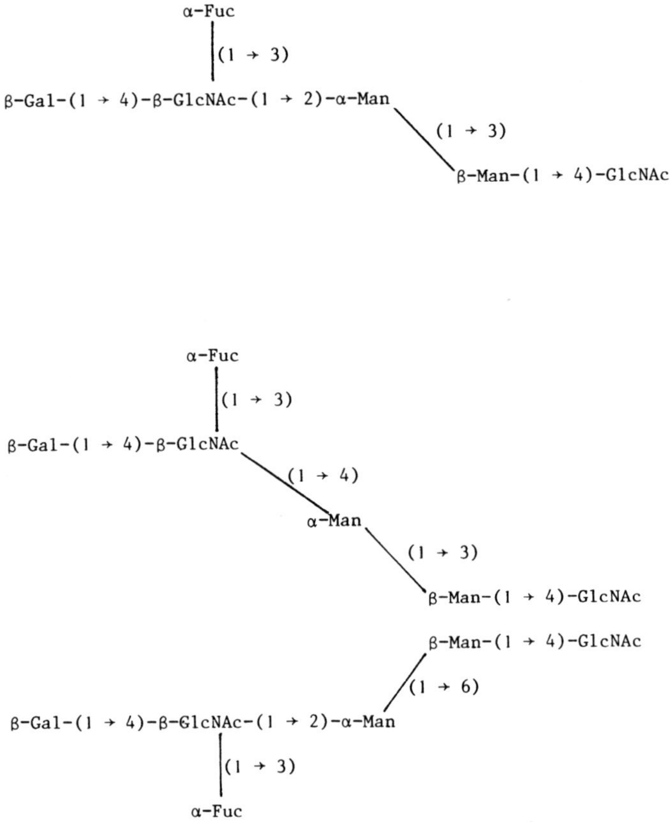

Scheme 15b. Structure of oligosaccharides isolated from the urine of a fucosidosis patient (21).

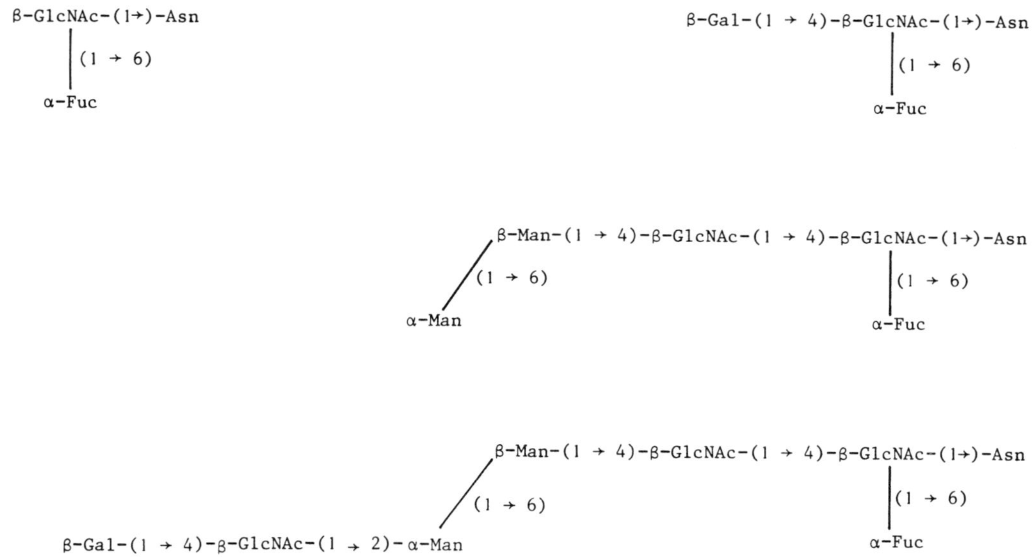

Scheme 15c. Structure of oligosaccharides isolated from the urine of a fucosidosis patient (21).

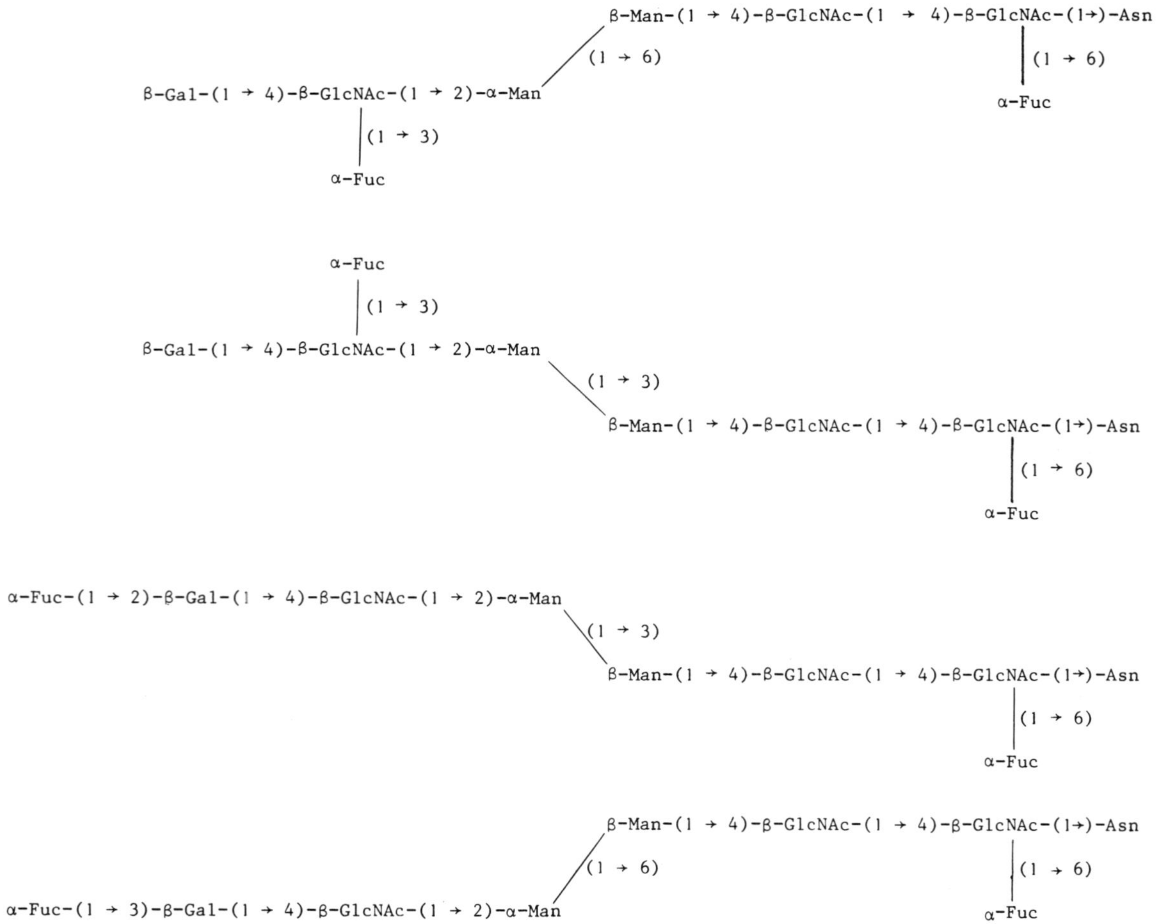

Scheme 15d. Structure of oligosaccharides isolated from the urine of a fucosidosis patient (21).

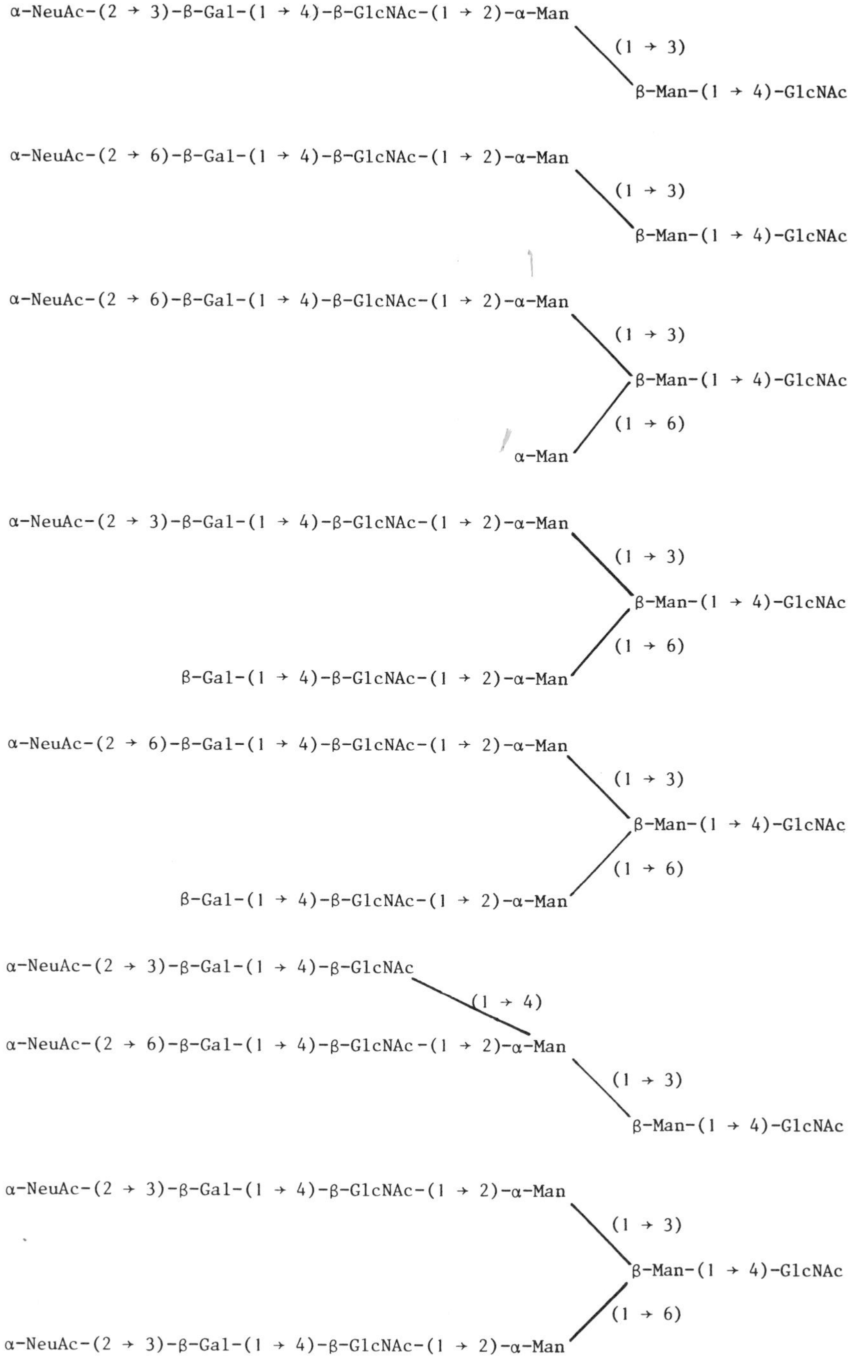

Scheme 16a. Structure of oligosaccharides isolated from the urine of a sialidosis patient (17,22,25,26).

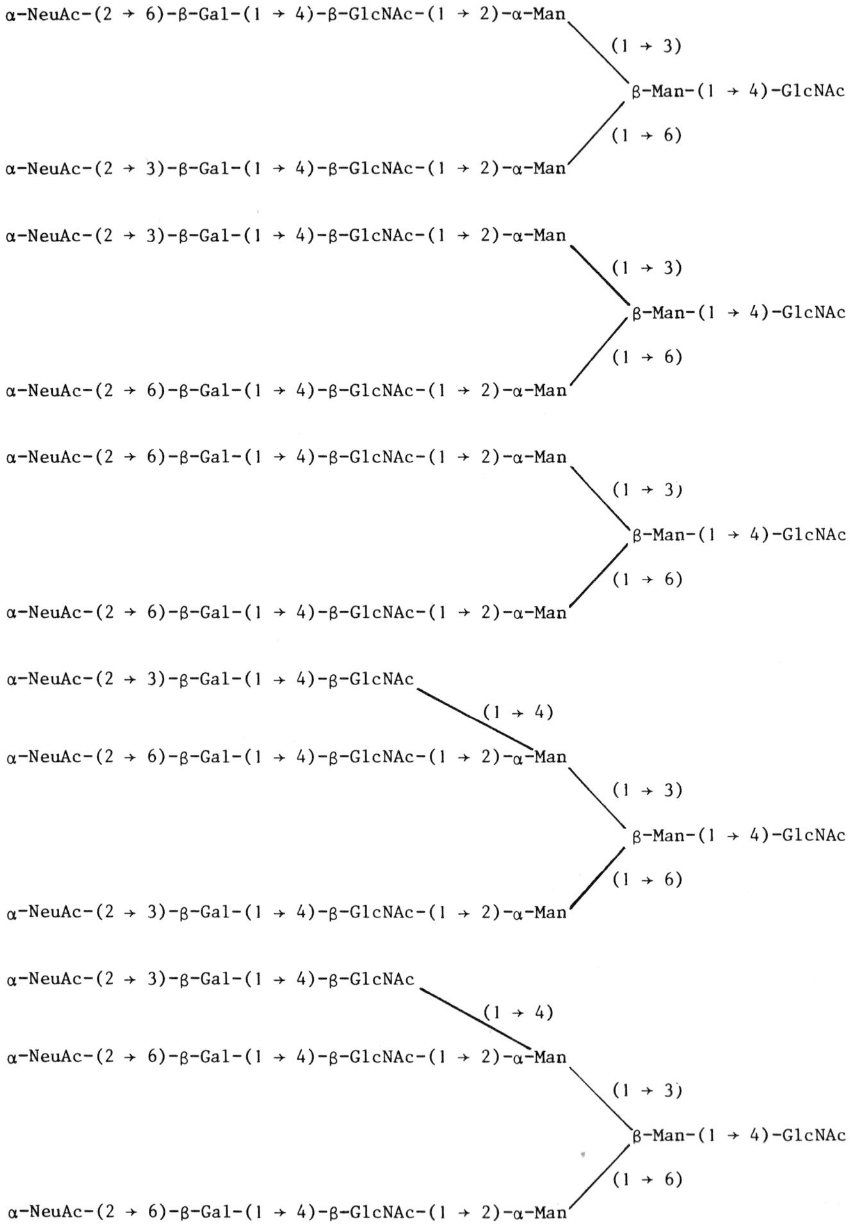

Scheme 16b. Structure of oligosaccharides isolated from the urine of a sialidosis patient (7,22,25,26).

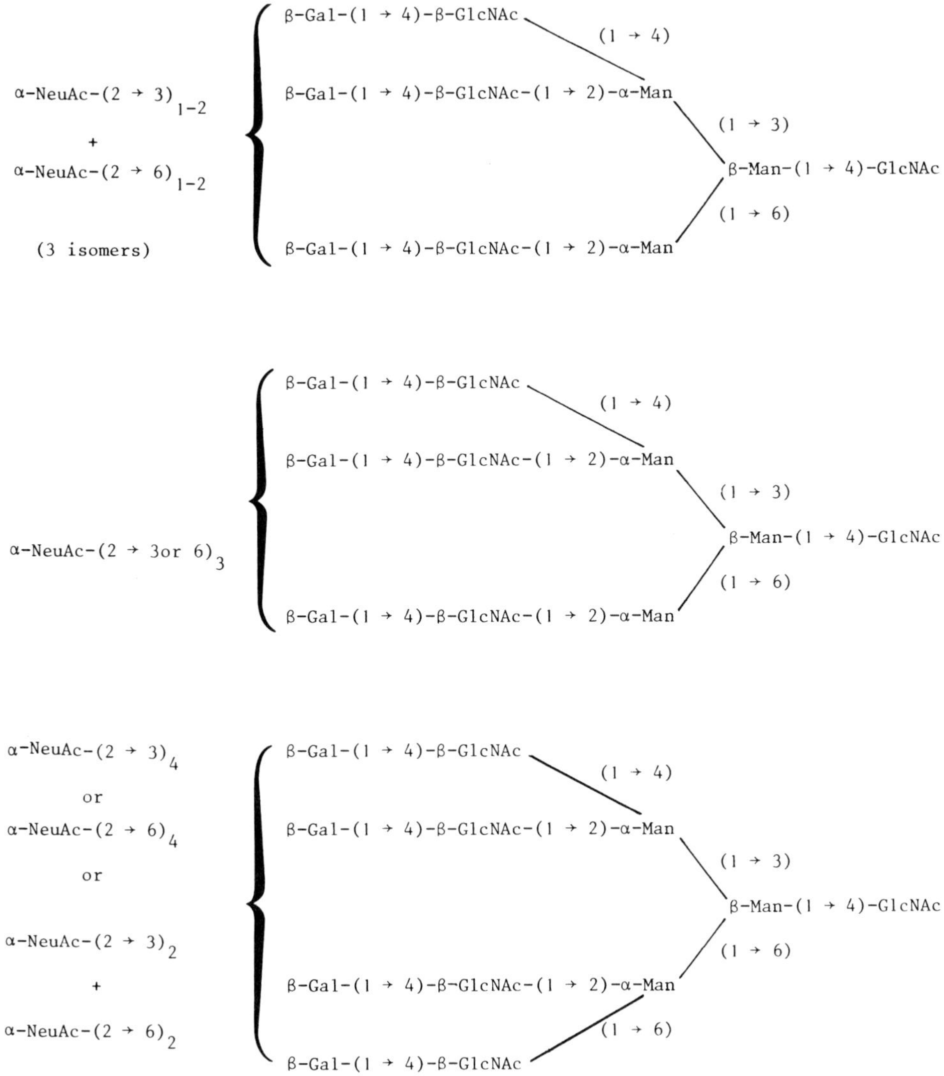

Scheme 16c. Structure of oligosaccharides isolated from the urine of a sialidosis patient (7,22,25,26).

Fragments of glycan structures, as they exist in numerous glycoproteins, are present. For example, the structures of the oligosaccharides of Sandhoff's disease may be considered as parts of the structures of serotransferrin (Fig. 8), of α_1-acid glycoprotein (Scheme 7) and of IgG (Scheme 10) glycans, respectively, and the oligosaccharides shown in Scheme 16 as parts of the structures of serotransferrin glycan or of α_1-acid glycoprotein *A*, *B*, or *C* glycans (Scheme 7).

On the basis of these observations, we have proposed the 3 following hypotheses: (a) The oligosaccharides from urines or tissues of lysozomal diseases originate from glycoprotein glycans, and we proposed to call this type of disease "glycoproteinoses" (4,19,25). (b) Consequently, we have reconstituted the first step of glycan catabolism, postulating that it begins with the action of *endo*-β-*N*-acetylglucosaminidases that cleave the di-*N*-acetylchitobiose residue (4,19,25). Thus, oligosaccharides having an *N*-acetylglucosamine residue in terminal reducing position would be liberated. *Endo*-β-*N*-acetylglucosaminidases acting on glycans of the oligomannoside type have been already characterized (46,47), but enzyme splitting sialo- and fuco-glycans of the *N*-acetyllactosamine type remain to be discovered. (c) Correlatively, we must postulate that the structures corresponding to the oligosaccharides of mannosidosis, sialidoses, and fucosidosis pre-exist in glycoprotein structures, even if these structures have not yet been characterized. For example, by adding the sequence β-GlcNAc-(1→4)-Asn to the tenth oligosaccharide of Scheme 16, we are able to reconstitute the complete glycan of serotransferrin. By adding the same sequence (Scheme 17) to all the oligosaccharides, accumulation of which is due to a lack in "terminal" exoglucosidases (fucosidases, neuraminidases, and mannosidases), we are able to reconstitute unknown but foreseen glycan structures. That such glycans have not yet been characterized might be due to the fact that they exist in too low quantities, probably in the cytoplasm or the cell membrane, or both. As the products of the action of *endo*-β-*N*-acetylglucosaminidases are being protected because of the lack of *exo*-glycosidases, they accumulate in the cells, and then in the urine.

PREDICTION OF GLYCAN CONFORMATION

Is it possible, on the basis of our knowledge of glycan primary structure, to imagine the spatial conformation of glycans, and is this conformation compatible with the biological role of glycoproteins, in particular with their role of recognition signals?

Oligosaccharides of urines of mannosidosis, sialidosis and fucosidosis (Structures of figures 29, 31 and 32)	-β-GlcNAC-(1 → 4)-β-GlcNAc-(1→)-Asn (1 → 6) $(\alpha\text{-Fuc})_{0-1}$

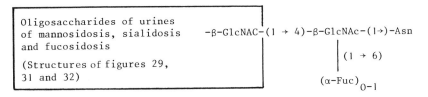

Scheme 17. General scheme of foreseeable glycan structures.

Fig. 14A illustrates the disposition in space of the biantennary glycan of human serotransferrin and shows that one can distinguish two parts in this structure. The first one is a compact zone, linked to the protein, and consisting of the pentasaccharide core itself. This part, common to numerous glycoprotein molecules, is non-specific and invariable. This was the reason to call it the invariant (inv) part (*4*), as in the case of the invariable peptide component of immunoglobulins. On the other hand, the trisaccharide fragment Man→GlcNAc→GlcNAc is practically flat (see Fig. 15). Moreover, this core is rigid because hydrogen bonds stabilize the structure.

To the compact and rigid core are attached the antennae. This part of the molecule is variable (var) and directs the specificity of the glycans. These antennae could occupy two positions in the space. The first one leads to the "Y conformation" (Fig. 14A), and the second one to the "T conformation" (Fig. 14B). The results from X-ray diffractometry obtained in collaboration with R. Fouret's group in our University are in favor of the second conformation. In fact, the analysis of the crystalline trisaccharide α-Man-(1→3)-β-Man-(1→4)-GlcNAc shows that the α-mannose residue is in a position perpendicular to that of the β-Man-GlcNAc residue.

The last information given by the construction of molecular models concerns the conformation of antennae which, after the formation of hydrogen bonds, appears helical (Fig. 15).

In conclusion, I would like to go back 52 years to present from Levene's book "Hexosamine and Mucoproteins" (48) the first structures of chondroitin sulfate and of fibrin "hyaloidin" (Schemes 18 and 19) so that we may contemplate the route that has been travelled. These structures are of course erroneous, however we should not smile but remember the pionners' works that during one century have lead us to the present knowledge of glycoprotein structure.

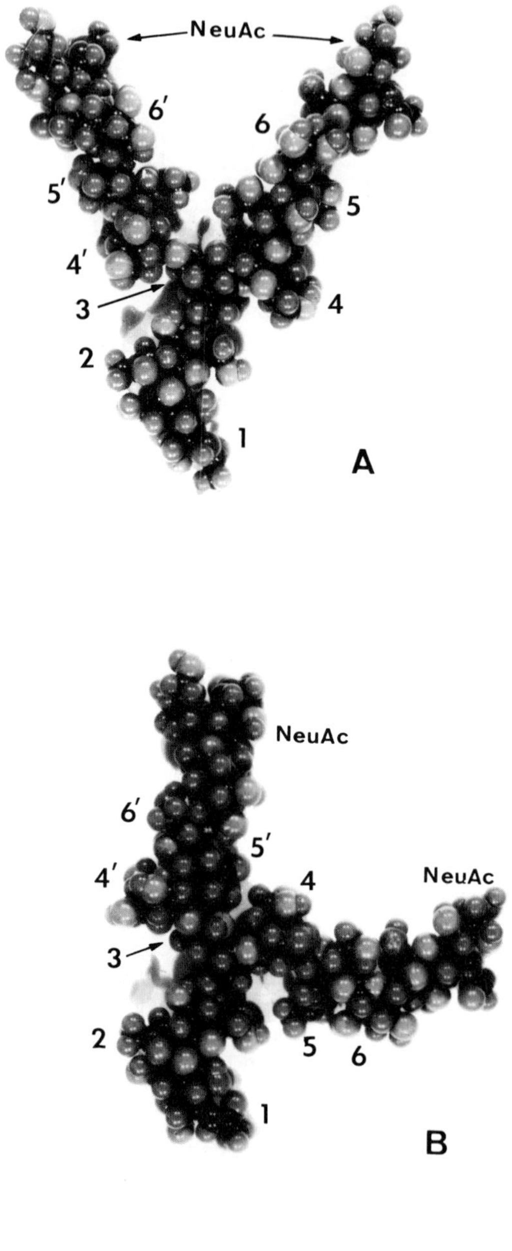

Fig. 14. Molecular model of human serotransferrin biantennary glycan: A,Y conformation; B,T conformation. The numbers correspond to those of Scheme 6.

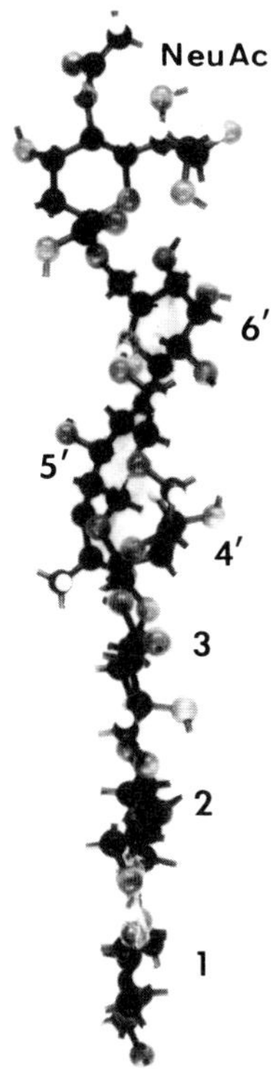

Fig. 15. Lateral view of the molecular model of N-*acetyl-lactosamine type. The numbers correspond to those of Scheme 6.*

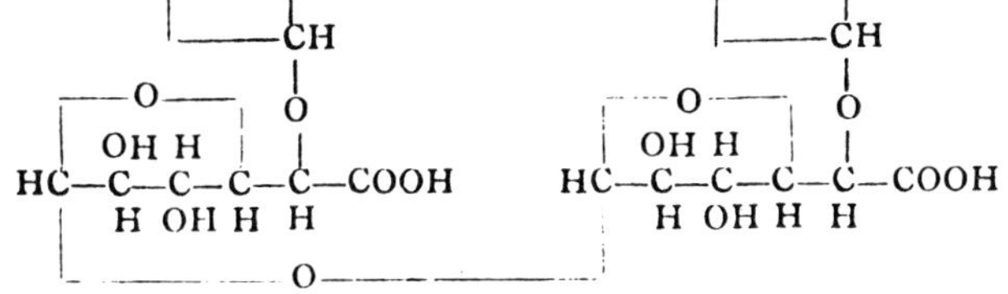

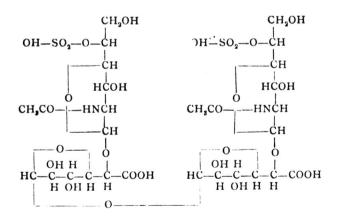

Scheme 18. Schmiedeberg's structure of chondroitin sulfuric acid (in Levene, ref. 48).

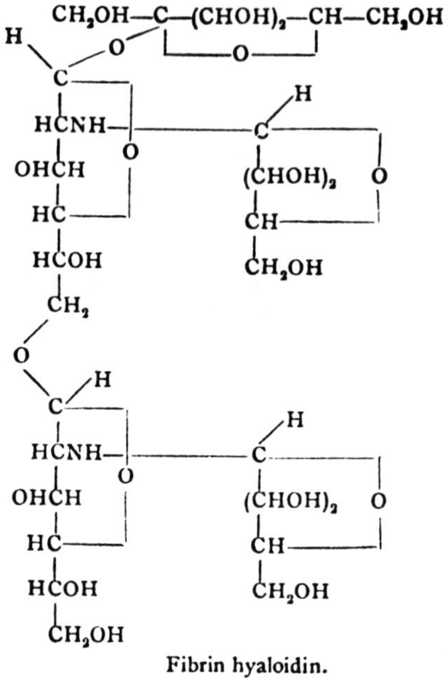

Scheme 19. Schmiedeberg's structure of fibrin hyalosidin (in Levene, ref. 48).

ACKNOWLEDGMENTS

The authors are much indebted to Drs. L. Dorland, B. Fournet, J. Haverkamp, B. L. Schut, G. Spik and G. Strecker for their enthusiastic contributions. These investigations were supported by The Netherlands Foundation for Chemical Research (SON) with financial aid from The Netherlands Organization for the Advancement of Pure Research (ZWO), the Centre National de la Recherche Scientifique (Laboratoire Associé n° 217), and the Université des Sciences et Techniques de Lille.

REFERENCES

1. Eylar, H., *J. Theor. Biol. 10*, 89 (1965).
2. Winterburn, P.-J., and Phelps, C.F., *Nature (London) 236*, 147 (1972).

3. Ashwell, G., and Morell, A. G., *in* "Glycoproteins of Blood Cells and Plasma" (G. A. Jamieson and T. J. Greenwalt, eds.), p. 173, J. B. Lippincott Co., Philadelphia (1971).
4. Montreuil, J., *Pure Appl. Chem. 42,* 431 (1975).
5. Bayard, B., and Montreuil, J., *C.N.R.S. Int. Symp. 221,* 209 (1974).
6. Tai, T., Yamashita, K., Ogata-Arakawa, M., Koide, N., Muramatsu, T., Shintaro, I., Inoue, Y., and Kobata, A., *J. Biol. Chem. 250,* 8569 (1975).
7. Yamaguchi, H., Ikenaka, T., and Matsushima, Y., *J. Biochem. (Tokyo) 70,* 587 (1971).
8. Nakajima, T., and Ballou, C. E., *J. Biol. Chem. 249,* 7685 (1974).
9. Kondo, T., Fukuda, M., and Osawa, T., *Carbohydr. Res. 58,* 405 (1977).
10. Pesonen, M., and Renkonen, O., *Biochim. Biophys. Acta 455,* 510 (1976).
11. Baenziger, J., and Kornfeld, S., *J. Biol. Chem. 249,* 7260; 7270 (1974).
12. Kearns, D. R., and Shulman, R. G., *Accounts Chem. Res. 7,* 33 (1974).
13. Miller, F., *Immunochemistry 9,* 217 (1972).
14. Finne, J., Krusius, T., and Rauvala, H., *Biochem. Biophys. Res. Commun. 74,* 405 (1977).
15. Spik, G., Bayard, B., Fournet, B., Strecker, G., Bouquelet, S., and Montreuil, J., *FEBS Lett. 50,* 296 (1975).
16. Bayard, B., Fournet, B., Bouquelet, S., Strecker, G., Spik, G., and Montreuil, J., *Carbohydr. Res. 24,* 445 (1972).
17. Michalski, J.-C., Strecker, G., and Fournet, B., *FEBS Lett. 79,* 101 (1977).
18. Norden, N.E., Lundblad, A., Svensson, S., Ockerman, P. A., and Autio, S., *J. Biol. Chem. 248,* 6210 (1973); *Biochemistry 13,* 871 (1974).
19. Strecker, G., *in* "J.-P. Farriaux, Les oligosaccharidoses" (Crouan et Roques, eds.), p. 13, Lille (1977).
20. Strecker, G., Fournet, B., Bouquelet, S., Montreuil, J., Dhondt, J. L., and Farriaux, J.-P., *Biochimie 58,* 579 (1976).
21. Strecker, G., Fournet, X., Spik, G., Montreuil, J., Durand, P., and Tondeur, M., *C.R. Acad. Sci. Ser. D. 284,* 84 (1977); and unplublished results.
22. Strecker, G., Hondi-Assah, T., Fournet, B., Spik, G., Montreuil, J., Maroteaux, P., Durand, P., and Farriaux, J.-P., *Biochim. Biophys. Acta 444,* 349 (1976).
23. Strecker, G., Herlant-Peers, M-C., Fournet, B., and Montreuil, J.; Dorland, L., Haverkamp, J., and Vliegenthart, J.F.G.; Farriaux J.-P., *Eur. J. Biochem. 81,* 165 (1977).
24. Strecker, G., and Lemaire-Poitau, A., *J. Chromatog. (Biochem. Applic.) 1,* 553 (1977).

25. Strecker, G., and Montreuil, J., *Biochimie*, in press.
26. Strecker, G., Peers, M.-C., Michalski, J.-C., Hondi-Assah, T., Fournet, B., Spik, G., Montreuil, J., Farriaux, J.-P., Maroteaux, P., and Durand, P., *Eur. J. Biochem. 75*, 391 (1977) and unpublished results.
27. Bovey, F. A., "High Resolution NMR of Macromolecules", Academic Press, New York, 1972.
28. Dwek, R. A., "Nuclear Magnetic Resonance (NMR) in Biochemistry, Applications to Enzyme Systems" Clarendon Press, Oxford (1973).
29. Kearns, D. R., and Shulman, R. G., *Acc. Chem. Res. 7*, 33 (1974).
30. Roberts, G. C. K., and Jardetzky, O., *Adv. Protein Chem. 24*, 447 (1970).
31. Wüthrich, K., "NMR in Biological Research: Peptides and Proteins", North-Holland Pub. Co., Amsterdam (1976).
32. Durette, P. L., and Horton, D., *Adv. Carbohydr. Chem. Biochem. 26*, 49 (1971).
33. Hall, L. D., *Adv. Carbohydr. Chem. 19*, 51 (1964).
34. Inch, T. D., *Ann. Rev. NMR Spectroscop. 2*, 35 (1969).
35. Streefkert, D. G., Thesis, University of Utrecht, 1973.
36. Haverkamp, J., Thesis, University of Utrecht, 1974.
37. Dorland, L., Schut, B. L., Vliegenthart, J. F. C., Strecker, G., Fournet, B., Spik, G., and Montreuil, J., *Eur. J. Biochem. 73*, 93 (1977).
38. Streefkerk, D. G., De Bie, M. J.A., and Vliegenthart, J. F. G., *Tetrahedron 29*, 833 (1973).
39. Dorland, L., Haverkamp, J., Vliegenthart, J. F. G., Strecker, G., Fournet, B., Spik, G., and Montreuil, J., to be published.
40. Usui, T., Yamaoka, N., Matjuda, K., Tuzimura, K., Sugiyama, H., and Seto, S., *J. Chem. Soc. Perkin I*, 2425 (1973).
41. Colson, P., and King R. R., *Carbohydr. Res. 47*, 1 (1976).
42. Dorland, L., Haverkamp, J., Schut, B. L., Vliegenthart, J. F. C., Spik, G., Strecker, G., Fournet, B., and Montreuil, J., *FEBS Lett. 77*, 15 (1977).
43. Lemieux, R. U., Bundle, D. R., and Baker, D. A., *J. Am. Chem. Soc. 97*, 4076 (1975).
44. Lemieux, R. U., and Driguez, H., *J. Am. Chem. Soc. 97*, 4063 (1975).
45. Wolfe, L. S., Senior, R. G., and Ng Ying Kin, N. M. K., *Fed. Proc. 32*, 484 (1973); *J. Biol. Chem. 249*, 1828 (1974).
46. Muramatsu, T., *J. Biol. Chem. 246*, 5535 (1971).
47. Tarentino, A. L., Plummer, T. H., Jr., and Maley, F. J., *J. Biol. Chem. 248*, 5547;(1973); Tarentino, A. L., and Maley, F., *J. Biol. Chem. 249*, 811 (1974).
48. Levene, P. A., "Hexosamine and Mucoproteins", Longmans Green, and Co., London (1925).

GLYCOCONJUGATE RESEARCH / Volume I
Proceedings of the Fourth International Symposium on Glycoconjugates

Isolation and Structural Study of a Novel Fucose-Containing Disialoganglioside from Human Brain

Susumu Ando and Robert K. Yu

Fucose-containing gangliosides (fucogangliosides) have been found in several animal organs (1-4). A monofuco-monosialoganglioside was first found in bovine liver by Wiegandt (1), who obtained the oligosaccharide moiety by ozonolysis and established the structure as a G_{M1} (NeuGl) derivative containing a fucose residue attached to C-2 of the terminal galactose residue. Suzuki *et al.* (2) isolated a similar fucoganglioside from boar testes as the intact lipid and showed that it contained *N*-acetylneuraminic acid in place of *N*-glycolylneuraminic acid. This latter ganglioside was also isolated from bovine brain by Ghidoni *et al.* (3). Ohashi and Yamakawa (4) recently reported another type of fucoganglioside obtained from pig adipose tissue. The ganglioside has one residue of fucose attached to the terminal galactose residue of the G_{M1} structure through an α-L-(1→3) linkage.

We have recently isolated a novel monofuco-disialoganglioside from human and bovine brains. The ganglioside turned out to be a G_{D1b} derivative, which contains a fucose residue linked to the terminal galactose residue.

Total gangliosides were prepared from human whole brain by a modified method of Ledeen *et al.* (5). Gangliosides were separated according to their acidity by DEAE-Sephadex column chromatography, as reported previously (6). The novel fucoganglioside was recovered in the fractions between the major mono- and di-sialo peaks (Fig. 1).

ISBN 0-12-301301-1

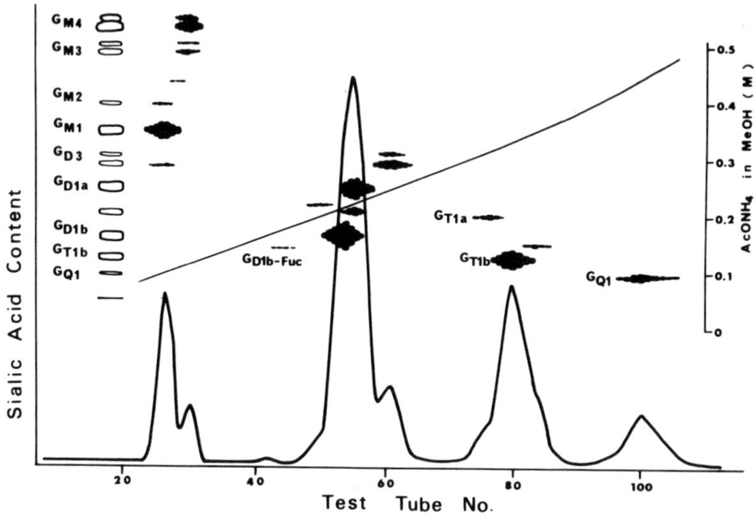

Fig. 1. Separation of gangliosides on a DEAE-Sephadex column. Gangliosides were eluted with a continuous gradient elution of ammonium acetate in methanol.

The ganglioside was further purified by chromatography on an Iatrobeads column. The yield was about 0.05% of the total gangliosides. The ganglioside migrated between G_{D1b} and G_{T1b} on a thin-layer plate (Silica gel 60, Merck) developed with chloroform-methanol-water (55:45:10, v/v, containing 0.02% $CaCl_2$) and between G_{T1b} and G_{Q1b} on a similar plate with chloroform methanol-2.5 N ammonia (60:35:8, v/v).

The sugar composition of the fucoganglioside was determined by g.l.c. (7). It was found to have glucose, galactose, *N*-acetylgalactosamine, *N*-acetylneuraminic acid, and fucose in the molar ratio of 1.00:2.00:1.14:1.90:0.92. Neuraminidase (*Cl. perfringens*, Sigma) treatment of the ganglioside gave only one glycolipid product, chromatographically similar to the fuco-G_{M1} isolated from boar testes (2), as shown in Fig. 2.

The ganglioside was partially degraded to G_{D1b} by α-fucosidase (beef kidney, Bohringer, Mannheim). A nonspecific hydrolysis of the glycolipid by heating at 100°C for 5 min with 0.01 N HCl yielded G_{D1b}, fuco-G_{M1}, and G_{M1}. When the fucoganglioside was subjected to Smith degradation, it produced approximately one equiv. each of 1,2-dihydroxypropane, glycerol, and glyceraldehyde. The results established the terminal structure as Fuc-(1→2)-Gal-. The data are therefore in accord with a structure having a G_{D1b} backbone with one residue of fucose joined to the terminal galactose residue through an α-L-(1→2) linkage: α-Fuc-(1→2)-β-Gal-(1→3)-β-GalNAc-(1→4)-[(α-NeuAc)-(2→8)-α-NeuAc-(2→3)]-β-Gal-(1→4)-β-Glc-(1→1)-Cer.

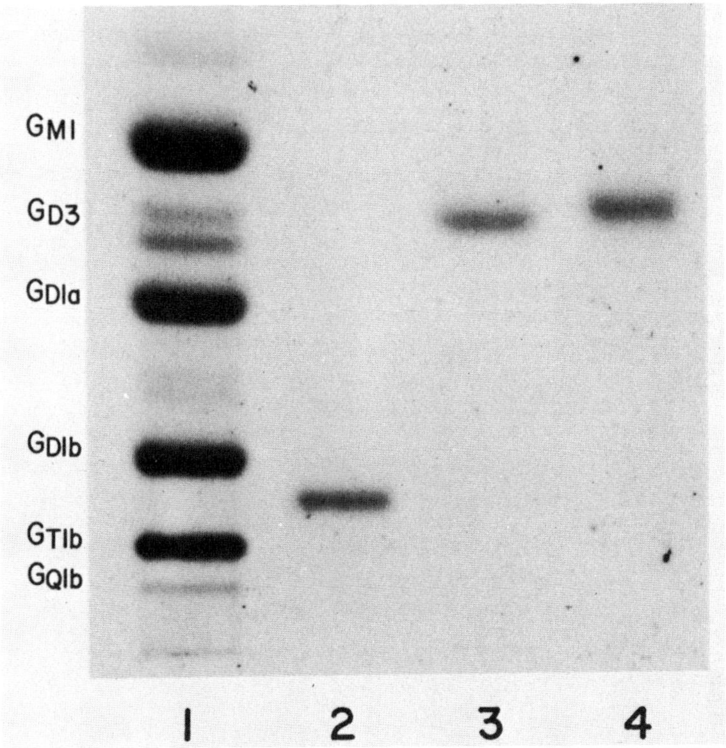

Fig. 2. Neuraminidase treatment experiment. Lanes: (1) Total gangliosides from human white matter; (2) Fuc-G_{D1b} before treatment; (3) Fuc-G_{D1b} after treatment; and (4) Fuc-G_{M1} from boar testis (ref. 2). The plate was developed with chloroform-methanol-water (55:45:10, v/v, containing 0.02% $CaCl_2$).

ACKNOWLEDGMENT

This work was supported by a USPHS grant NS-11853, and grants from The Kroc Foundation and the Klingenstein Fund.

REFERENCES

1. Wiegandt, H., *Hoppe-Seyler's Z. Physiol. Chem. 354,* 1049 (1973).
2. Suzuki, A., Ishizuka, I., and Yamakawa, T., *J. Biochem. 78,* 947 (1975).
3. Ghidoni, R., Sonnino, S., Tettamanti, G., Wiegandt, H., and Zambotti, V., *J. Neurochem. 27,* 511 (1976).
4. Ohashi, M., and Yamakawa, T., *J. Biochem. 81,* 1675 (1977).
5. Ledeen, R. W., Yu, R. K., and Eng, L. F., *J. Neurochem. 21,* 829 (1973).
6. Momoi, T., Ando, S., and Nagai, Y., *Biochim. Biophys. Acta 441,* 488 (1976).
7. Ando, S., and Yamakawa, T., *J. Biochem. (Tokyo) 70,* 335 (1971).

Deamination of Methylated Amino-Oligosaccharide Chains from Mucins

Gerald O. Aspinall and R. George S. Ritchie

N-Deacetylation followed by nitrous acid deamination is a highly specific technique for the controlled depolymerization of oligosaccharide chains containing *N*-acetylhexosamine units, but account must also be taken of side-reactions leading to the formation of 2-*C*-formylpentofuranosides (1). In applying the procedure to the corresponding methylated oligosaccharides it is necessary to use dimethyl sulfate and sodium hydroxide, rather than more efficient reagents, in order to achieve *O*-methylation without accompanying *N*-methylation, and rather drastic conditions are required to effect *N*-deacetylation. Simultaneous derivative formation, from 2,5-anhydrohexose units and sugar residues carrying aglyconic hydroxyl groups, may be performed by (A) reduction with sodium borodeuteride followed by trideuteriomethylation, or (B) oximation followed by treatment with acetic anhydride to give aldononitriles and sugar acetates. An important consideration is the formation of sufficiently volatile and thermally stable derivatives of mono- and di-saccharides (and possibly tri-saccharides) for analysis by g.l.c.-mass spectrometry.

N-Deacetylation followed by deamination of per-*O*-methylated lacto-*N*-tetraitol [β-D-Gal*p*-(1→3)-β-D-GlcNAc*p*-(1→3)-β-D-Gal*p*-(1→4)-D-Galol] affords two major products, 2,5-anhydro-4,6-di-*O*-methyl-3-*O*-(2,3,4,6-tetra-*O*-methyl-β-D-galactopyranosyl)-D-mannose and 1,2,3,5,6-penta-*O*-methyl-4-*O*-(2,4,6-tri-*O*-methyl-β-D-galactopyranosyl)-D-glucitol, whose structures were established by the formation of derivatives by procedures (A) and (B). The additional formation of 2,3,4,6-tetra-*O*-methyl-D-galactose as a product from the side-reaction involving an alternative ring contraction (1) provides confirmatory evidence for the presence of a 3-*O*-substituted hexosamine residue in the tetrasaccharide.

ISBN 0-12-301301-1

Hog gastric mucin contains a mixed population of oligosaccharide side-chains which may be liberated as oligosaccharide alditols by reductive elimination on treatment of the glycoprotein with sodium borohydride-sodium hydroxide. Six such penta- and hexa-saccharides have been individually characterized by Kochetkov *et al.* (2) and shown to contain a common tetrasaccharide core with additional fucose or *N*-acetylglucosamine end groups (or both) attached (Scheme 1). We have confirmed by methylation analysis the presence of the same structural units in the mixed population of oligosaccharides derived from commercial hog gastric mucin. Similar structural units are also present in the oligosaccharides liberated from the glycoprotein of the rat intestinal goblet-cell (3).

[or α-L-Fuc*p*-(1→2)]
α-D-GlcNAc*p*-(1→4)-β-D-Gal*p*-(1→3)-D-GalolNAc
6
↑
1
α-L-Fuc*p*-(1→2)-β-D-Gal*p*-(1→4)-β-D-GlcNAc*p*
[or β-D-GlcNAc*p*-(1→4)]

Scheme 1.

In order to assess the complex mixture of compounds derived from the hog glycoprotein, we have examined the products formed on deamination of the hog oligosaccharide alditols after *N*-deacetylation. The products were reduced with sodium borodeuteride, methylated, and examined by g.l.c.-mass spectrometry. Methylated derivatives of the following alditols have been characterized: 2,5-anhydromannitol, galactitol, 2,5-anhydro-4-(or 3-)*O*-galactopyranosylmannitol, *O*-fucopyranosyl (1→2)-*O*-galactopyranosyl-(1→[4 or 3])-2,5-anhydromannitol, and (from the galactosaminitol terminus) *O*-galactopyranosyl-(1→3)-2-deoxyhexitol and an *O*-galactopyranosyl-(1→3[or 4])-hexitol. With the exception of the galactitol, whose origin is as yet not clear, the formation of these compounds is consistent with their derivation from oligosaccharides of the general type shown in Scheme 1. Further structural information is now being obtained from the characterization of the corresponding fragments formed on deamination of the methylated oligosaccharide alditols, followed by reduction with sodium borodeuteride, and re-alkylation with trideuteriomethyl iodide.

ACKNOWLEDGMENTS

We acknowledge financial support from the National Research Council of Canada and the Atkinson Charitable Foundation.

REFERENCES

1. Erbing, C., Lindberg, B., and Svensson, S., *Acta Chem. Scand.* *27*, 3692 (1973).
2. Kochetkov, N. K., Serevitskaya, V. A., and Arbatsky, N. P., *Eur. J. Biochem.* *67*, 129 (1976).
3. Forstner, J. F., Jabbal, I., and Forstner, G. G., *Can. J. Biochem.* *51*, 1154 (1973).

GLYCOCONJUGATE RESEARCH/Volume I
Proceedings of the Fourth International Symposium on Glycoconjugates

Nucleotide-Activated Peptides from Rat Colonic Cells

Jiri T. Beranek, Ward Pigman, Anthony A. Herp, and Vera Perret

Earlier work has shown that nucleotide-activated peptides are present in rat liver, *E. coli* homogenates (1), and porcine submaxillary glands (2). The occurrence of repetitive glycopeptide sequences in bovine submaxillary glycoproteins (3) raised the question whether such activated peptides are precursors of large secretory glycoproteins. In this study, we report on the partial chemical characterization of activated peptides isolated from a single cell suspension of rat colonic secretory cells. Such cells produce high mol. wt. mucous glycoproteins (4).

Single cell suspensions were prepared from 57.6 g of rat colonic epithelium by the procedure of Perret *et al.* (5), which avoids the use of enzymes. The cells were washed twice in saline and in water at 4°C, homogenized, sonicated, and centrifuged for 30 min at 11 000*g*. The supernatant solution was collected and ultrafiltered *in vacuo* (15 mm Hg), by means of a dialysis tubing (12 000 mol. wt. filtration limit). The ultrafilterable fraction was resolved on a Sephadex G-100 column. The water eluents were monitored spectrophotometrically at 260 nm (nucleotides) and 280 nm (peptides), and pooled as indicated in Fig. 1. Fraction D was pooled, based on the yellow color of the effluent in tubes No. 44 through 58. Reaction with ninhydrin before and after hydrolysis (6 N HCl, 22 h, 110°C)

ISBN 0-12-301301-1

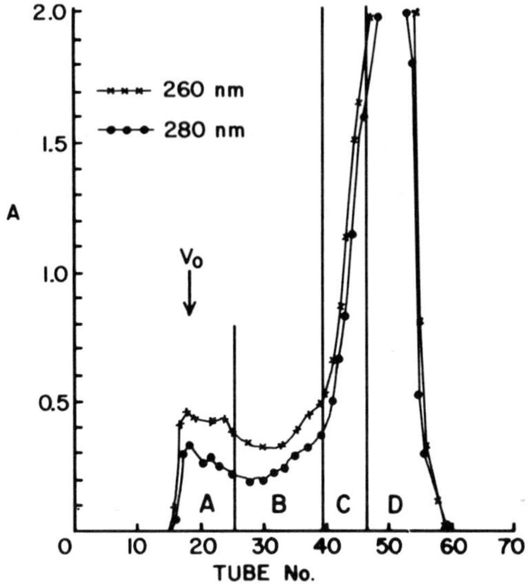

Fig. 1. Gel-filtration chromatography of the ultrafiltrate from rat colonic mucosal cells on Sephadex G-100 (2 x 100 cm); 5-ml effluents were collected at a flow rate of 12 ml/h. Fractions were pooled as indicated.

gave an average peptide size of about 20 amino acids for fraction A, 11 for B, 5 for C, and 2 for D. The dry weight of the preparation of single cells was 5.12 g, that of the ultrafiltrate 0.76 g. Fractions A, B, C, and D represented, respectively, 0.5, 3.6, 14.4, and 54.2% of the dry weight of the ultrafiltrate. Fraction A was too small for further studies. Fraction B contained 0.53; fraction C, 1; and fraction D, 7.3 µmol of ribose/ml. Deoxyribose was negative for all three fractions. Free phosphates were found in fractions C and D only. After elimination of free phosphates by precipitation with calcium chloride and potassium oxalate, the fractions were hydrolyzed (1 N HCl, 1 h, 100°C) and the amount of acid-labile phosphates was determined. Fraction B contained 0.56 ; fraction C, 1.25; and fraction D, 6.9 µmol of phosphates/ml. Thus, for each of the 3 fractions, the ratio of ribose to phosphates was close to one.

None of the three fractions showed migration of u.v.-absorbing spots on paper chromatography when tested for nucleotides; however, after acid hydrolysis (1 N HCl, 1 h, 100°C), adenine and guanine migrating spots were identified in each fraction.

The amino acid composition of the hydrolyzed fractions (6 N HCl, 22 h, 110°C) revealed a flagrant similarity between fractions B and C (in mol/100 mol: aliphatic amino acids, 35; carboxylic amino acids, 27; hydroxyamino acids, 13; basic amino acids, 13; imino acids, 4; and sulfur-containing amino acids, 3). Fraction D differed from fractions B and C particularly by its high amounts of imino (13 mol/100 mol) and aromatic (9.5 mol/100 mol) amino acids, and by a low proportion of basic amino acids (5 mol/100 mol).

These results seem to confirm the presence of nucleotide-activated peptides, which may have important physiological implications in rat colonic cells.

ACKNOWLEDGMENT

This research was supported by NCI grant 5R26 CA17168.

REFERENCES

1. Herp, A., Liska, M., Payza, N., Pigman, W., and Vittek, J., *FEBS Lett. 6*, 321 (1970).
2. Moschera, J., Mound, R., Payza, N., Pigman, W., and Weiss, M., *FEBS Lett. 6*, 326 (1970).
3. Pigman, W., Moschera, J., Weiss, M., and Tettamanti, G., *Eur. J. Biochem. 32*, 148 (1973).
4. Murty, V. L. N., Downs, F., and Pigman, W., *Carbohydr. Res. 61*, 139 (1978).
5. Perret, V., Lev, R., and Pigman, W., *Gut. 18*, 382 (1977).

GLYCOCONJUGATE RESEARCH/Volume I
Proceedings of the Fourth International Symposium on Glycoconjugates

Structural Relationship between Two Glycoproteins Isolated from Alveoli of Patients with Alveolar Proteinosis

Sambhunath N. Bhattacharyya and William S. Lynn

In alveolar proteinosis (1), a chronic lung disease, the alveoli and terminal bronchioles of the lung are filled with excessive amounts of amorphous material. This material can be removed by pulmonary lavage. On analysis, the lavage material was found to contain approximately 50% of lipid and the rest was protein. Analysis of the protein components by dodecyl sodium sulfate (SDS)-acrylamide gel electrophoresis revealed three major bands, two of which (mol. wt. 62 000 and 36 000) were found to contain carbohydrates (2). The two glycoproteins were separated and isolated by gel filtration (3,6) on a Bio-Gel column. The purified glycoproteins were found to be homogeneous by SDS-acrylamide gel electrophoresis and contained only one *N*-terminal amino acid residue, *i.e.*, threonine, in both glycoproteins. The chemical compositions (3,4) of the two glycoproteins indicated the presence of hydroxyproline, of a high content of glycine, and of 3 residues of methionine, 3 residues of sialic acid, 4 residues of mannose, 4 residues of galactose, 1 residue of fucose, and 6 residues of glucosamine, as shown in Table I.

Cyanogen bromide (CNBr) treatment of the 62 000 mol. wt. glycoprotein resulted in four peptides of mol. wts. 18 000, 12 000, 5 000, and 27 000, three of the peptides containing hydroxyproline and two of the peptides containing carbohydrates. CNBr treatment of the 36 000 mol. wt. glycoprotein resulted in four peptides of mol. wts. 18 000, 12 000, 5 000, and 1 000, the chemical composition of the first three peptides being identical with those derived from the 62 000 mol. wt. glycoprotein. The 18 000 and 12 000 mol. wt. peptides were glycopeptides. Threonine was the *N*-terminal residue in both glycoproteins and in the 18 000 mol. wt. CNBr fragment, indicating that this peptide is the *N*-terminal peptide of the native glycoprotein molecules. The 1 000 and 27 000 mol. wt. peptides were C-terminal extension

ISBN 0-12-301301-1

Table I. Composition of Glycoprotein of Mol. Wt. 62 000 and 36 000

	Glycoprotein	
Component[a]	62 000 mol. wt.	36 000 mol. wt.
Hydroxyproline	5	5
Glycine	72	50
Sialic acid	3	3
Mannose	4	4
Galactose	4	4
Fucose	1	1
Glucosamine	6	6

[a] *In residue/peptide*

peptides of the 36 000 and 62 000 mol. wt. glycoproteins.

The structure of the oligosaccharide moieties of the two glycoproteins has been established (5) and an *N*-acetylglucosamine residue was found to be attached to an asparagine residue of the glycoprotein molecule. The proposed structure has many similarities to the structure of oligosaccharide units of many glycoproteins (6). However, the most striking difference between the proposed oligosaccharide structures of the alveolar glycoproteins and those of other animal glycoproteins is the absence of chiotobiose units in the core of the oligosaccharide structure of the alveolar glycoproteins. Glycoproteins containing β-GlcNAc→Asn structure in the core have been found to occur in IgE (7) and secretory glycoproteins isolated from human urine (8,9). It seems from this observation that there is (are) biosynthetic pathway(s) other than the dolichol pathway by which the glycoproteins are synthesized.

Another interesting feature of these glycoproteins is the presence of a Gly-Pro-Hyp-Gly sequence in the peptide chain. This indicates that collagen-like regions are present in the peptide chain (3,4). Limited trypsin digestion of the 62 000 mol. wt. glycoprotein resulted in an insoluble precipitate containing the 36 000 mol. wt. glycoprotein and a soluble fraction containing the 26 000 mol. wt. peptide, the former being a glycoprotein found in the lavage material. The 26 000 mol. wt.

Thr - Asx - Gly - Tyr - Ala - Phe - Thr - Gly - Pro - Hyp - Gly - Ser -
18 000

CHO | CNBr ↓ | CHO | CNBr ↓ | CNBr ↓

— Asn ——— Met ———— Asn ———— Met ——— Met - Ser - Glx - Gly -
12 000 | 5 000

Trypsin ↓

— Pro - Ala - Asx - Leu - Gly - Arg ↓ - Ala - Ser - Pro - Tyr - Gly -
1 000 | 26 000

Scheme 1. *Structural relationship between 26 000 mol. wt. peptide and 62 000 mol. wt. glycoprotein.*

peptide was found to be present in the lavage supernatant solution. Trypsin attacked the Arg-Ala bond in the peptide chain of the 62 000 mol. wt. glycoprotein, yielding the two peptides as just mentioned. Thus, the 36 000 and 26 000 mol. wt. peptides are proteolytic products of the 62 000 mol. wt. glycoprotein molecule and the 26 000 mol. wt. peptide is the C-terminal extension peptide of the 62 000 mol. wt. glycoprotein. The structural relationship between these two glycoproteins is illustrated in Scheme 1.

REFERENCES

1. Rosen, S.H., Castelman, B., and Liebow, A.A., *N. Engl. J. Med. 258,* 1123 (1958).
2. Passero, M.A., Tye, R.W., Kilburn, K.H., and Lynn, W.S., *Proc. Natl. Acad. Sci. U.S.A. 70,* 973 (1973).
3. Bhattacharyya, S.N., Sahu, S., and Lynn, W.S., *Biochim. Biophys. Acta 427,* 91 (1976).
4. Bhattacharyya, S.N., and Lynn, W.S., *Biochim. Biophys. Acta 494,* 150 (1977).
5. Bhattacharyya, S.N., and Lynn, W.S., *J. Biol. Chem. 252,* 1172 (1977).
6. Sharon, N., *Annu. Rev. Biochem. 35,* 485 (1966).
7. Baenziger, J., Kornfeld, S., and Kochwa, S., *J. Biol. Chem. 249,* 1889 (1974).
8. Sugahara, K., Funakoshi, S., Funakoshi, I., Aula, P., and Yamashina, I., *J. Biochem. (Tokyo) 78,* 673 (1975).
9. Sugahara, K., Funakoshi, S., Funakoshi, I., Aula, P., and Yamashina, I., *J. Biochem. (Tokyo) 80,* 195 (1976).

Sialoglycopeptides and Glycosaminoglycans Produced by Cultured Human Melanoma Cells and Melanocytes

Veerasingham P. Bhavanandan, John R. Banks, and Eugene A. Davidson

The glycoconjugates produced by cultured human melanoma cells (HM7) and by human fetal uveal melanocytes (FM6), which are tumorigenic and nontumorigenic, respectively, in athymic mice, were investigated (1,2). The cells were cultured in the presence of D-[^{3}H]glucosamine and [$^{35}SO_4^{2-}$], and the glycoconjugates prepared by Pronase digestion of the chloroform-methanol extracted cells and of the dialyzed media. The glycosaminoglycans in the digests were precipitated with cetylpyridinium chloride (CPC) and fractionated by stepwise elution of the precipitate with an NaCl solution of increasing molarity. Whereas about 44% of the ^{3}H-labeled glycoconjugates shed into the media by melanocytes were glycosaminoglycans, this class accounted for only 14% in the case of the melanoma culture.

The glycosaminoglycans were further fractionated by gel filtration on controlled-pore, glass-bead columns or on a column of DEAE-Sepharose employing a gradient elution with LiCl. The purified components were characterized by identification of their hexosamines, susceptibility to hyaluronidase (leech, testicular), chondroitinases, and heparanase, and by electrophoresis on cellulose acetate strips in different buffer systems. Chondroitin sulfate, tentatively identified as the 4-sulfate, was the major product of both cell lines, which also produced heparan sulfate and hyaluronic acid. However, the HM7 cells produced significantly lower proportions of hyaluronic acid in contrast to the FM6 cells. This ability to synthesize increased proportions of sulfated polysaccharides (chondroitin sulfate and heparan sulfate) by malignant cells, compared to normal cells, appears to be a general phenomenon (3,4).

In contrast to BI6 mouse melanoma cell cultures (3,5), significant quantities of sialoglycopeptides were not precipitated by CPC. The CPC-soluble glycopeptides were isolated from the

ISBN 0-12-301301-1

CPC-supernatant by removal of CPC with KCNS, followed by dialysis. A significant difference between the glycopeptides from HM7 and FM6 cells was in the proportion specifically bound to a WGA-Sepharose 4B column and eluted with 0.1 M *N*-acetylglucosamine. Of the glycopeptides isolated from spent media of HM7 and FM6 cells, 6.2% and 2.3%, respectively, interacted with a wheat germ agglutinin affinity column. The difference was much more striking in the case of HM7 and FM6 cell-associated glycopeptides, where 25.1% and 3.9%, respectively, bound to the affinity column. The glycopeptides with specificity for WGA were isolated from HM7 spent media (HMM class I) and cells (HMC class I) and compared to those which do not bind WGA (HMM class II and HMC class II). The distribution of ^{3}H-activity in *N*-acetylneuraminic acid and hexosamines in these glycopeptides and in the glycopeptides produced by FM6 cells is shown in Table 1.

Table I. Distribution of ^{3}H Activity in Sialic Acid and Hexosamines of the Sialoglycopeptides from HM7 and FMG Cells

	Components		
Sialoglycopeptides	*Acid*	*GlcN*	*GalN*
HMM Class I	*27.4*	*30.5*	*42.1*
HMM Class II	*22.9*	*72.5*	*4.6*
HMC Class I	*31.0*	*32.6*	*36.4*
HMC Class II	*19.6*	*68.3*	*12.1*
FMM	*34.0*	*54.8*	*11.2*
FMC	*23.0*	*73.2*	*3.8*

A noticable difference is the higher percent of galactosamine and sialic acid in the class I glycopeptides compared to class II. The higher sialic acid content of the class I glycopeptides was also reflected by their elution from DEAE-sepharose column at a higher concentration of pyridine acetate (0.5 M), in contrast to the class II glycopeptides, which emerged between 0.2 and 0.5 M. Further, the elution profiles of class I and class II glycopeptides on columns of controlled-pore glass-bead and Sephadex G-50 indicate that the former were larger in size (about 10 000) than the latter (below 4000). The galactosamine-rich class I glycopeptides were found to have mainly *O*-glycosically-linked carbohydrate chains. This conclusion is based on treatment with alkaline borohydride (1.0 M $NaBH_4$-0.1 M NaOH, 37°C, 72 h), followed by chromatography on Bio-Gel P-4 and P-6 columns. The significance of the ability of the malignant HM7 cells to produce an increased proportion of *O*-glycosically-linked glycopeptides in contrast to the normal fetal cells (FM6), is not clear, but one may reasonably expect this to reflect changes in surface properties of these cells.

ACKNOWLEDGMENT

We are grateful to Ms. Anne Katlic and Mr. Jeffrey Kemper for their able assistance in this work, to Dr. J.W. Kreider for the HM7 cell line, and to Drs. B.C. Giovanella and K.J. McCormick for the FM6 cell line. This work was supported by USPHS Grants CA17686 and CA 15483.

REFERENCES

1. Banks, J., Kreider, J.W., Bhavanandan, V.P., and Davidson, E.A., *Cancer Res. 36,* 424 (1976).
2. Giovanella, B.C., Stehlin, J.S., Santomaria, C., Yim, S.O., Morgan, A.C., Williams, L.J., Leibovitz, A., Fialkow, P.J., and Mumford, D.M., *J. Natl. Cancer Inst. 56,* 1131 (1976).
3. Satoh, C., Banks, J., Horst, P., Kreider, J., and Davidson, E.A., *Biochemistry 13,* 1233 (1974).
4. Goggins, J.F., Johnson, G.S., and Pastan, I., *J. Biol. Chem. 247,* 5759 (1972).
5. Bhavanandan, V.P., and Davidson, E.A., *Biochem. Biophys. Res. Commun. 70,* 139 (1976).

GLYCOCONJUGATE RESEARCH / Volume I
Proceedings of the Fourth International Symposium on Glycoconjugates

Purification and Structural Studies of Proline-Rich Glycoprotein of Human Parotid Saliva

Arnold Boersma, Geneviève Lamblin, Philippe Roussel, and Pierre Degand

Previous studies have demonstrated the presence in human parotid saliva of proline-rich proteins and glycoprotein (1,5). Interest in these components is related to their possible role in mineralization processes of enamel, secretion mechanism of serous cells, and α-amylase activity in parotid saliva (6). A study of the carbohydrate moiety of the proline-rich glycoprotein was carried out.

MATERIAL AND METHODS

The lyophilized parotid saliva (100 mg) was applied to a column of Sephadex G-200 equilibrated with 0.01 M sodium phosphate and 0.2 M sodium chloride buffer, pH 8.0. The orcinol-reacting fraction was dialyzed against distilled water and lyophilized. The proline-rich glycoprotein (300 mg, fraction F_2 of the Sephadex G-200 column) was digested by papain and Pronase, and then purified by gel-filtration chromatography as previously described (see Figs. 1 and 2) (7).

The glycopeptides were oxidized with 0.01 M sodium metaperiodate, pH 4.0 at 4°C. After oxidation for 9 h, the material was treated with sodium borohydride and then hydrolyzed with sulfuric acid, pH 1.0, for 24 h at room temperature. After filtration on a Sephadex G-15 column, the carbohydrate content was determined. The glycopeptides were treated first with β-galactosidase and β-*N*-acetylhexosaminidase (jack bean), and then with a human liver *endo*-β-*N*-acetylglucosaminidase (8). After reduction with sodium borotritide and hydrolysis with 2 N hydrochloric acid for 2 h at 100°C (8), paper chromatography on Whatman No. 3 was performed to identify, in the presence of standards, glucosaminitol as the reducing end of the oligosaccharides.

ISBN 0-12-301301-1

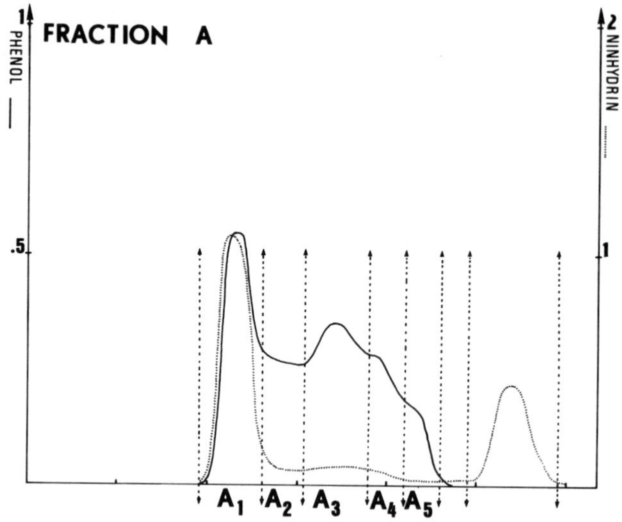

Fig. 1. Diagram of elution, from a Bio-Gel P-2 column, of the oligosaccharides obtained after treatment with liver endo-β-N-*acetylglucosaminidase of Fraction A previously treated with* exo-*glycosidases.*

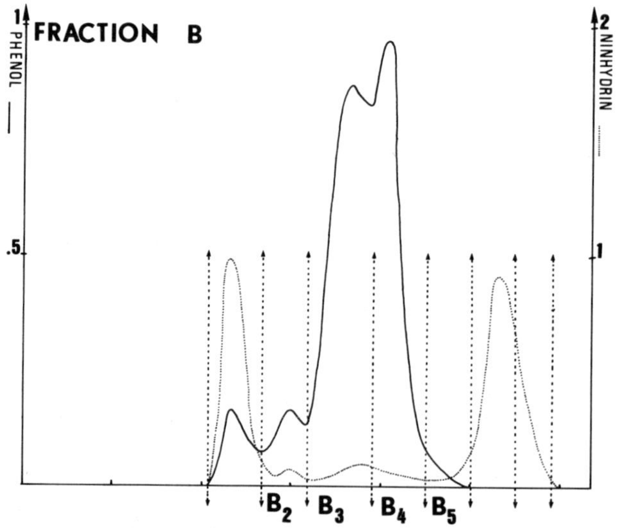

Fig. 2. Diagram of elution, from a Bio-Gel P-2 column, of the oligosaccharides obtained after treatment with liver endo-β-N-*acetylglucosaminidase of Fraction B previously treated with* exo-*glycosidases.*

RESULTS AND DISCUSSION

The amino acid and carbohydrate composition of the native glycoprotein and of the glycopeptides obtained by papain and Pronase digestion is presented in Table I.

Table I. Chemical Composition of the Parotid Glycoprotein and Treatment of the Glycopeptides Obtained by Papain-Pronase

Components	*Native glycoprotein (F_2, Sephadex G-200)*	*Glycopeptides*
	(residues/100 amino acid residues)	
Asp	5.4	14.5
Thr	1.0	2.7
Ser	5.4	19.5
Glu	20.5	25.5
Pro	35.6	1.8
Gly	16.5	24.0
Ala	0.9	2.3
Cys	0	0
Val	1.0	1.7
Met		0
Ile	0.5	0.8
Leu	1.0	1.0
Tyr	0.4	
Phe	0.5	
Lys	5.0	2.4
His	1.1	1.2
Arg	5.1	2.4
	(mmol/g of dry weight preparation)	
Fucose	520	980 (3.2)[a]
Mannose	440	910 (3)
Galactose	380	760 (2.5)
N-Acetylglucosamine	520	1160 (3.8)

[a] *In parentheses, values expressed in relation to mannose taken as 3*

Proline, glycine, and glutamic acid (or glutamine) were the amino acid residues present in the greatest concentrations. The carbohydrate units were fucose, galactose, mannose, and *N*-acetylglucosamine. Minor variations were observed from one donor to another.

As previously demonstrated, the glycopeptides were homogenous by amino acid sequence determination, and show the asparaginyl-*N*-acetylglucosamine-type linkage. After treatment with jack bean β-galactosidase and β-*N*-acetylhexosaminidase, the glycopeptides were applied to a Bio-gel P-2 column. Two fractions (A and B) were separated and then submitted to liver *endo*-β-*N*-acetylglucosaminidase. After filtration on Bio-Gel P-2, all fractions located by ninhydrin and phenol-sulfuric acid reactions were analyzed for their chemical composition. The results are presented in Table II.

Table II. Carbohydrate Composition of the Glycopeptides ($FA-P_2$ and $FB-P_2$) and of the Fractions Obtained After Endo-β-N-*acetylglucosaminidase*[a].

Components	Fraction A					
	$FA-P_2$	A_1	A_2	A_3	A_4	A_5
Fucose	457	171	195	67		
Mannose	806	467	878	1 194	1 616	1 256
Galactose	361	128	178	139		
N-*Acetylglucosamine*	989	452	810	995	1 000	765
Man/GlcNAc	0.81	1.0	1.1	1.2	1.6	1.6

Components	Fraction B				
	$FB-P_2$	B_2	B_3	B_4	B_5
Fucose	524	134	110	55	55
Mannose	1 211	861	1 744	1 722	1 750
Galactose	228	283	111		
N-*Acetylglucosamine*	1 113	715	1 086	1 217	457
Man/GlcNAc	1.1	1.2	1.6	1.4	1.6

[a] *µmol/g*

The carbohydrate composition of the same preparation of glycopeptides, after periodate oxidation, showed 3 mol of sugar/ mol of aspartic acid with a molar ratio of mannose to *N*-acetylglucosamine of 1:2.

The following comments can be made: (a) Combining *exo*-glycosidases and human liver *endo*-glycosidase, we were able to obtain a series of oligosaccharides. (b) The chemical composition of the oligosaccharides and of the periodate-treated glycopeptides suggests a serum-type carbohydrate unit for the proline-rich glycoprotein. (c) By use of the same protocole on proline-rich glycoproteins purified from the parotid saliva of various donors, it was possible to observe differences in the carbohydrate composition (slight differences in fucose and galactose in relation to mannose and *N*-acetylglucosamine).

Further structural investigations, especially methylation, have to be carried out to determine the presence of a heteropolysaccharide unit in this secreted parotid glycoprotein.

ACKNOWLEDGMENT

This work was supported by Contrat DGRST n°75-7-0915 and Contrat de Recherches UER III des Sciences Médicales de Lille.

REFERENCES

1. Mandel, I.D., Thompson, R.N., and Ellison, S.A., *Arch. Oral. Biol. 10,* 499 (1965).
2. Levine, M.J., Weill, J.C., and Ellison, S.A., *Biochim. Biophys. Acta 188,* 165 (1969).
3. Andjic, J., Bonte, M., and Havez, R., *C.R. Acad. Sci., Ser. D. 270,* 564 (1970).
4. Bennick, A., and Connell, G.E., *Biochem. J. 123,* 455 (1971)
5. Oppenheim, F.G., Hay, D.L., and Franzblau, C., *Biochemistry 10,* 4233 (1971).
6. Degand, P., Aubert, J.P., Boersma, A., Richet, C., Loucheux-Lefebvre, M.H., and Biserte, G., *FEBS Lett. 63,* 137 (1976).
7. Degand, P., Boersma, A., Roussel, Ph., Richet, C., and Biserte, G., *FEBS Lett. 54,* 189 (1975).
8. Boersma, A., Lamblin, G., Roussel, Ph., Degand, P., and Biserte, G., *C.R. Acad. Sci., Ser. D. 281,* 1269 (1975).

GLYCOCONJUGATE RESEARCH / Volume I
Proceedings of the Fourth International Symposium on Glycoconjugates

Sulfated Mucins from Marine Prosobranch Snails

Terry A. Bunde, Fred R. Seymour, and Stephen H. Bishop

Hypobranchial gland mucin from *Buccinum undatum* is composed of a glycoprotein and a β-D-(1→4)-glucan 6-sulfate (1). By contrast, hypobranchial gland mucus from *Busycon canaliculatum* appeared to be composed of a glycoprotein and a sulfated hexosamine polysaccharide (2). The differences in composition were surprising considering their similar physical and functional properties, and the close evolutionary relationship of the two snails.

The snails were collected in the Cape Code, MA region. The mucins were extracted from excised glands and fractionated by use of hot 40% phenol (3). Chromatography on DEAE-cellulose removed the residual glycoprotein from the sulfated glycan polymers. The sulfated polysaccharide migrated as a single band on cellulose acetate strip-electrophoresis in 0.1 N HCl (Alcian Blue staining), and as a species with one negative charge per monomer unit. The amino acid and wet chemical analyses (methods in ref. 4) (Tables I and II) indicated that both species contained similar nonsulfated glycoprotein and sulfated polysaccharide components. After hydrolysis of both sulfated polysaccharides (2 N HCl at 100°C for 14 h) and formation of the peracetylated aldononitrile derivative (PAAN) (5), g.l.c.-m.s. revealed the PAAN derivative of glucose as the single major component. Neither sulfated glucans contained hexosamine, but both contained a small amount of galactose (<3%). The two sulfated polysaccharide polymers were desulfated (6). The resulting polysaccharides and microcrystalline cellulose (as control) were subjected to Hakomori (7) and Kuhn (8) methylations. The permethylated products were hydrolyzed, converted to the PAAN derivatives, and analyzed by g.l.c.-m.s. (9).

ISBN 0-12-301301-1

Table I. Overall Percent Composition of Polymers from Snail Mucins

Component	B. undatum		B. canaliculatum	
	Glyco-protein	Poly-saccharide	Glyco-protein	Poly-saccharide
Neutral sugar	8.4	45	8	44
Hexosamine	7.6	1[a]	14	3[a]
Uronic acid	0	0	0	0
Sialic acid	0	0	0	0
Sulfate	0	51	0	50
Phosphate	0	0	0	0
Protein	84	3	78	3

[a] No hexosamine found by ion-exchange chromatography

Table II. Amino Acid Composition of the Glycoprotein Components

Residue	B. undatum (residues/1000 residues)	B. canaliculatum (residues/1000 residues)
Lys	73	71
His	26	27
Arg	58	53
Asp	110	104
Thr	47	54
Ser	34	51
Glu	131	119
Pro	51	56
Gly	85	79
Ala	80	80
Cys	8	8
Val	71	68
Met	21	25
Ile	42	46
Leu	90	86
Tyr	32	31
Phe	41	41

The only observed compound in each polymer was the 4,5-di-*O*-acetyl-2,3,6-tri-*O*-methyl-D-glucononitrile derivative, indicating a linear glucan composed of (1→4) linkages (DP>80).

The purified sulfated glycan from *Buccinum* (35 mg) was dissolved in D_2O (2 ml) and the ^{13}C-n.m.r. spectrum recorded with a Varian XL-100-15 spectrometer under conditions (90°C) previously described using Fourier-transform (FT) data analysis (10). The ^{13}C-n.m.r. spectrum (Fig. 1A) indicates only β anomeric glucan resonances (A). The resonance at 80.3 p.p.m. (B) is diagnostic for (1→4)-glucan linkages (10), thereby confirming a linear β-D-(1→4)-linked structure. However, the expected C-6 resonance at 62 p.p.m. is very weak (F), with a strong resonance at 68.5 p.p.m. (E) indicating a downfield shift of approximately 6 p.p.m. This $\Delta\delta$ of 6 p.p.m. is entirely consistent with extensive, though not complete, C-6 sulfation, as the sulfate group is known to result in downfield $\Delta\delta$ of attached carbon positions (11). The resonances for the C-2, C-3, and C-5 positions are found in the usual region of the spectrum (C-D).

The FT i.r. spectrum of the sulfated *Busycon* polymer was recorded on a Nicolet 7199 i.r. spectrometer on 3-mm KBr pellets (Fig. 1B). Two prominent ester sulfate resonances are present at 1220 (general sulfate ester) and 810 cm^{-1} [glucan 6-sulfate (12)]. These resonances are greatly enhanced over those seen with a conventional scanning instrument. The FT i.r. difference spectrum for the *Buccinum* sulfated polysaccharide and microcrystalline cellulose with approximately 1:1 glycan subtraction (Fig. 1C) enhance the resonance (1220 and 810 cm^{-1}) due to sulfate ester substitution. Additional diagnostic sulfate ester peaks in the 850-1220 cm^{-1} region appear in the difference spectrum.

Our data confirm the *Buccinum* mucin structure predicted by Hunt and Jevons (1), and demonstrate the essential identity of the mucins from *Busycon* and *Buccinum*. The data do not confirm the premise of the sulfated hexosamine polymer, previously reported for *Busycon canaliculatum* (2).

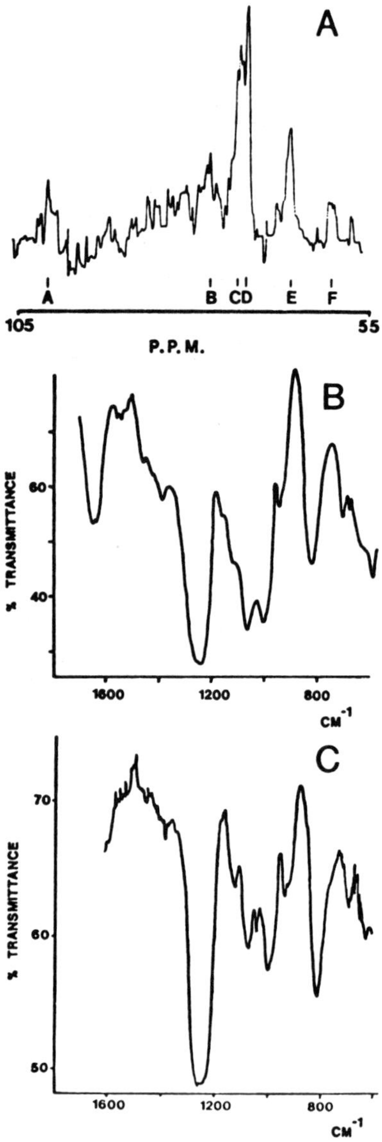

Fig. 1. ^{13}*C-N.m.r. spectrum of* Buccinum undatum *sulfated glycan (A), and FT i.r. spectra of* Busycon canaliculatum *(B) and* B. undatum *(C) sulfated glycans*

ACKNOWLEDGMENTS

Supported by Grants in Aid to Stephen H. Bishop from the NSF, NIH, and the Robert A. Welch Foundation. We thank Dr. R. Julian (Nicolet Instr. Corp.) for recording the FT i.r. spectra.

REFERENCES

1. Hunt, S., and Jevons, F. R., *Biochem. J. 97*, 701 (1965).
2. Shashoua, V. E., and Kwart, H., *J. Am. Chem. Soc. 81*, 2899 (1958).
3. Hunt, S., and Jevons, F. R., *Biochem. J. 98*, 522 (1966).
4. Hurley, J. C., Bunde, T. A., Dell, J. C., Kirkpatrick, D. S., and Bishop, S. H., *Comp. Biochem. Physiol. B 57*, 233 (1977).
5. Seymour, F. R., Chen, E. C. M., and Bishop, S. H., *Carbohydr. Res. 68*, 113 (1979).
6. Kantor, T. G., and Schubert, M., *J. Am. Chem. Soc. 79*, 152 (1957).
7. Hakomori, S., *J. Biochem. (Tokyo) 55*, 205 (1969).
8. Hall, D. H., and Adams, G. A., *Can. J. Chem. 37*, 1012 (1959).
9. Seymour, F. R., Slodki, M. E., Plattner, R. D., and Jeanes, A., *Carbohydr. Res. 53*, 153 (1977).
10. Seymour, F. R., Knapp, R. D., and Bishop, S. H., *Carbohydr. Res. 51*, 179 (1976).
11. Perlin, A. S., *Fed. Proc. 36*, 106 (1977).
12. Turvey, J. R., *Adv. Carbohydr. Chem. 20*, 183 (1965).

GLYCOCONJUGATE RESEARCH / Volume I
Proceedings of the Fourth International Symposium on Glycoconjugates

Far Ultraviolet Circular Dichroism of Oligosaccharides

C. Allen Bush

N-Acetylamino sugars possess the amide chromophore, whose optical properties are familiar to biochemists as a result of their importance in polypeptide and protein CD. The CD spectra of acetamido sugars exhibit bands near 209 nm due to the amide n-π* transition, the same transition that gives rise to large negative bands near 222 nm in α-helical polypeptides and in proteins (1). GlcNAc and GalNAc have similar negative n-π* bands, whereas ManNAc has a positive band as a result of the opposite configuration at C-2. The n-π* bands of GlcNAc and GalNAc are not very sensitive to the anomeric configurations of the sugar (2). The insensitivity of the 209-nm band to the anomeric configuration is also illustrated by the spectrum of the methyl α- and β-glycosides of GlcNAc and GalNAc (2). The amide π-π* transition, on the other hand, gives rise to CD bands in the 185-192-nm region whose wavelength and magnitude depend on anomeric configuration (3).

It has been possible to apply some of the theoretical techniques used in peptide CD to the calculation of the n-π* CD in acetamido sugars. As in the case of peptides, the magnetically allowed n-π* transition is made optically active through the one-electron mechanism (4). The electron paths are asymmetrically bent by the gradient of the electrostatic field of the surrounding groups, especially by the static dipole moments of OH-1 and more importantly by OH-3 (5). The calculations of Yeh and Bush (5) show that rotation of the dihedral angle about the C-3--O bond may cause changes in both the sign and magnitude of the rotational strength. That such significant changes in the spectrum could be brought about by such an apparently trivial geometric modification as the movement of a single hydrogen atom suggested that changes in the hydrogen-bonding properties of the solvent might change the experimentally observed CD at 209 nm.

ISBN 0-12-301301-1

Therefore, we have concluded that the sign reversal of CD of GlcNAc and GalNAc on transfer from aqueous solution to hexafluoro-2-propanol (HFIP) results from a change of the OH-3 orientation between the two solvents (6). On the other hand, the CD of methyl 2-acetamido-2-deoxy-3-*O*-methyl-β-D-glucopyranoside (I), in contrast to acetamido sugars having a free OH-3, remains negative in both HFIP and in water (6). Similarly the CD of lacto-*N*-tetraose (Fig. 1), in which GlcNAc has substituents in the same positions as in I, also shows a negative CD at 209 nm in both HFIP and in water. These results may be contrasted with the CD of β-GlcNAc-(1→6)-GlcNAc, which shows a positive CD band at 209 nm in HFIP solution (6). Chitobiose [β-(1→4)-linked] shows a negative band at 209 nm, an effect we ascribe to formation in HFIP of a hydrogen bond between OH-3 of the reducing residue with the ring O atom of the nonreducing terminal residue (6).

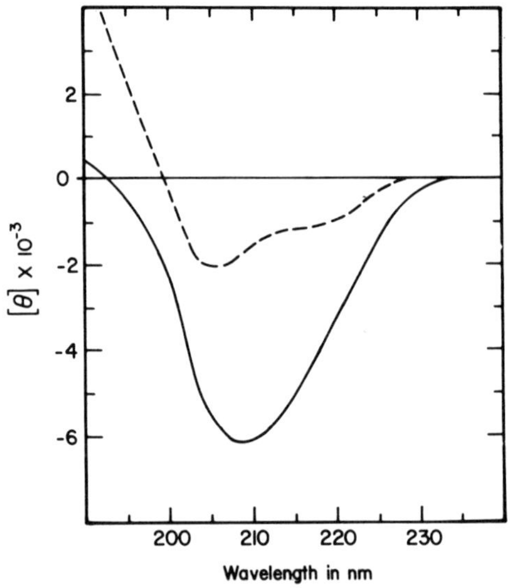

*Figure 1. Circular dichroism of lacto-*N*-tetraose [β-Gal-(1→3)-β-GlcNAc-(1→3)-β-Gal-(1→4)-Glc], in water (——) and in hexafluoro-2-propanol (----) solution.*

In 2,3-diacetamido-2,3-dideoxy-D-glucose (7), the two amide functions are sufficiently close to promote mutual coupling effects in the CD, much as in the case of regular polypeptides. This exciton coupling splits the 189 nm π-π* transition into a strong positive CD band at 198 nm and a negative band of *ca.* equal magnitude near 180 nm. Such paired bands yield the so-called conservative Cotton effect seen in dinucleosides (8).

2-Acetamido-1-*N*-(L-aspart-4-oyl)-2-deoxy-D-glucosylamine (II) may be viewed as an equatorial, vicinal 1,2-diacetamidodideoxy sugar for interpretation of its CD spectrum (2). We have used a theoretical method for calculating the CD of the n-π* and π-π* exciton bands for vicinal diacetamidodideoxy sugars based on a model introduced by Schellman and coworkers for dipeptides (9). Our results show a strong dependence of the calculated n-π* rotational strength and of the exciton bands on the dihedral angles about the C--N bond joining the amide group to the pyranose ring. For II, our calculations agree with the experimental CD, in both the n-π* and π-π* region, for the conformation that places both amide protons *trans* to their respective C--H protons, in agreement with the conformation found in the crystalline state (10). The conformation of 2,3-diacetamido-2,3-dideoxy-D-glucose, for which our calculations agree with the experimental data, places the amide protons *cis* to their respective pyranose-ring C--H protons.

ACKNOWLEDGMENT

This research was supported by NIH grant AI-11014 and by NSF grant CHE 76-16783.

REFERENCES

1. Kabat, E.A., Lloyd, K.O., and Beychok, S., *Biochemistry 8*, 747 (1969).
2. Coduti, P.L., Gordon, E.C., and Bush, C.A., *Anal. Biochem. 78*, 9 (1977).
3. Bush, C.A., *Proc. Jerusalem Symp. 10* (1977).
4. Stigter, D., and Schellman, J.A., *J. Chem. Phys. 51*, 3397 (1969).
5. Yeh, C.Y., and Bush, C.A., *J. Phys. Chem. 78*, 1829 (1974).
6. Dickinson, H.R., Coduti, P.L., and Bush, C.A., *Carbohydr. Res. 56*, 249 (1977).
7. Keilich, G., Roppel, J., and Mayer, H., *Carbohydr. Res. 51*, 129 (1976).
8. Bush, C.A., and Brahms, J., *J. Chem. Phys. 46*, 79 (1967).
9. Bayley, P.M., Neilsen, E.B., and Shellman, J.A., *J. Phys. Chem. 73*, 228 (1969).
10. Delbaere, L.T.J., *Biochem. J. 143*, 197 (1974).

The Link Proteins as Specific Components of Cartilage Proteoglycan Aggregates

Bruce Caterson and John Baker

Cartilage proteoglycan aggregates isolated according to the procedure of Sajdera and Hascall (1) may be considered "artificial", *i.e.*, they have been dissociated during isolation in 4 M guanidine and then allowed to reassociate upon lowering the guanidine concentration to 0.4 M. As there are many different proteins present in the cartilage extract, it may be considered possible that the link proteins are artificially incorporated into such reformed aggregates. The experiments outlined below have been designed to determine whether the link proteins are specific components of cartilage proteoglycan aggregates *in vivo*.

EXPERIMENT 1

Proteoglycan aggregates from bovine nasal cartilage were subjected to equilibrium density gradient centrifugation under conditions chosen to minimize the chances of extraneous proteins nonspecifically associating (entangling) with the proteoglycan aggregates, *i.e.*, centrifugations were performed at low proteoglycan concentrations (1 mg/ml) and the proteoglycan aggregates (A1) were either (*a*) rerun associatively or (*b*) redissociated in 4 M guanidine, allowed to reassociate, and then rerun associatively. Recentrifugation at dilute concentration by either procedure (*a*) or (*b*) caused the release of some extraneous protein and hexosamine-containing material, there being more released by procedure (*a*). Analyses of these fractions (*i.e.*, fractions other than A1) by polyacrylamide-gel electrophoresis is SDS (SDS-PAGE) showed that the link proteins were not present. Examination of the aggregate fractions (A1) isolated by both procedures

ISBN 0-12-301301-1

indicated that the link proteins were the predominant proteins present, strongly suggesting that they were specific components of the proteoglycan aggregates. Nonetheless, because of the prior use of a dissociative extraction step in the A1 preparation, we sought to find an independent method for the isolation of the proteoglycan aggregates under non-dissociative conditions.

EXPERIMENT 2

Proteoglycans were sequentially extracted from rat chondrosarcomas under associative conditions (0.5 M guanidine) followed by dissociative extraction in 4 M guanidine; 43% of the tissue proteoglycans were extracted in 0.5 M guanidine and an additional 50% was released by raising the guanidine concentration to 4 M. The aggregate fraction (A1) from each extract was recovered by equilibrium density gradient centrifugation under associative conditions. Over 90% of the extracted hexuronate was found in the A1 fraction from both the 0.5 M and 4 M extracts. Samples from each A1 fraction were chromatographed on Sepharose CL-2B to determine whether proteoglycan aggregates were present (Fig. 1). Both the 0.5 M

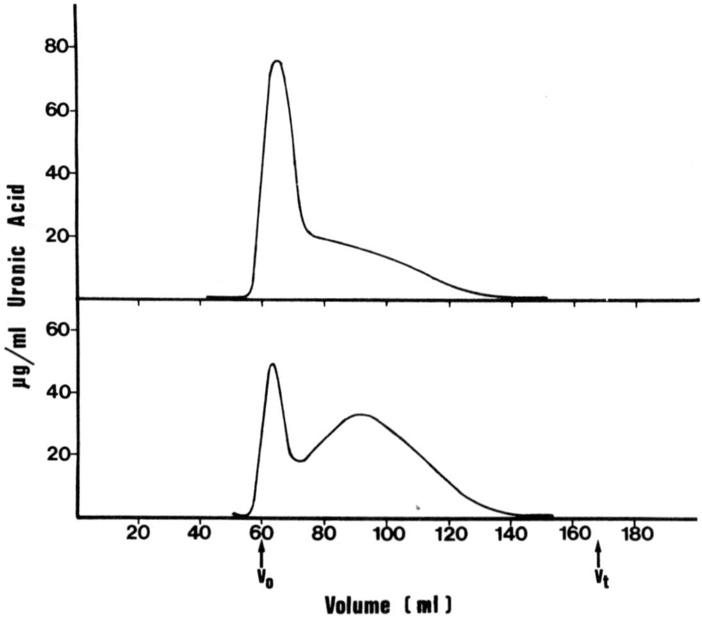

Fig. 1. Sepharose CL-2B chromatography of A1 fractions: upper, 0.5 M A1; lower, 4.0 M A1.

A1 and 4 M A1 fraction show hexuronic acid in the void volume, indicating the presence of proteoglycan aggregates. There is clearly a greater yield of aggregate from the 0.5 M guanidine preparation. The occurrence of the material in the void volume of the A1 fraction from the 0.5 M extract indicates that proteoglycan aggregates can occur *in vivo*. However the question still remains as to whether the link proteins are specifically associated with these structures. Fig. 2 illustrates the separation by SDS-PAGE of the fluorescamine-labeled A1 preparations from the 0.5 M and 4 M guanidine extracts. Note that only one link protein is present (link protein 2), which occurs in both aggregate preparations.

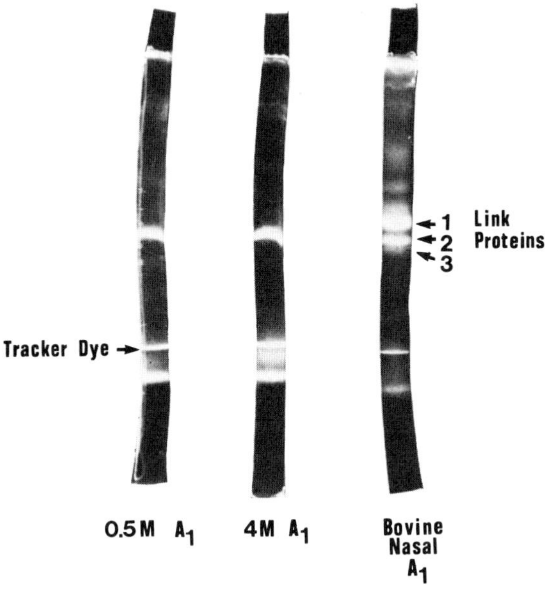

Fig. 2. SDS-PAGE of A1 fractions

Rat chondrosarcoma has proven a useful tissue to study the nature of proteoglycan complexes extracted under mild associative conditions and has indicated that macromolecular aggregate structures consistent with the model proposed by Heinegård and Hascall (2) do exist *in vivo*. In addition, the results presented here reveal that the link proteins are truly part of proteoglycan aggregates and not artifacts introduced by reaggregation after their dissociation in 4 M guanidine.

ACKNOWLEDGMENTS

This work was supported by NIH Grants DE 02670 and HL 11310. The expert technical assistance of Mr. J. D. Ford is greatly appreciated. Mr. Allen Lenoir performed much of this work as a summer project for Birmingham Southern College.

REFERENCES

1. Sajdera, S. W., and Hascall, V. C., *J. Biol. Chem. 244*, 77 (1969).
2. Heinegård, D., and Hascall, V. C., *J. Biol. Chem. 249*, 4250 (1974).

GLYCOCONJUGATE RESEARCH/Volume I
Proceedings of the Fourth International Symposium on Glycoconjugates

Can Hyaluronic Acid Exist in Solution as a Helix?

Bireswar Chakrabarti, Nora Figueroa, and John W. Park

Earlier physico-chemical studies of hyaluronate in low concentration in dilute salt solution (1) have characterized the molecule as a "random coil with some stiffness." The detailed three-dimensional structure remained unknown until the early 70's when several investigators almost simultaneously reported the existence of various ordered conformations, including a double helix, in oriented fibers (2, 3) and in solution (4-6). Some of these workers, after further X-ray diffraction studies, rejected the double helical model and proposed, instead, several forms of aggregated single helices for hyaluronate (7). The possibility that the polymer can also adopt such a helical structure in solution is at present debatable. This report summarizes the results of our recent investigations and presents ideas about the helical structure of the molecule in solution.

Fig. 1. shows how the circular dichroism (CD) spectra of hyaluronate film (8), hyaluronic acid in aqueous-organic solvent at pH 2.6, and Cu(II)-hyaluronate complex (pH 6.8) differ from those of acid and neutral forms of hyaluronate. In all three spectra the appearance of a strong negative band in the $\pi \rightarrow \pi^*$ amide transition region and the apparent loss of $n \rightarrow \pi^*$ CD minima are evident. In addition, hyaluronic acid in mixed solvents shows a weak positive band at 226 nm, and the Cu(II) complex displays a positive CD peak at 250 nm. In

ISBN 0-12-301301-1

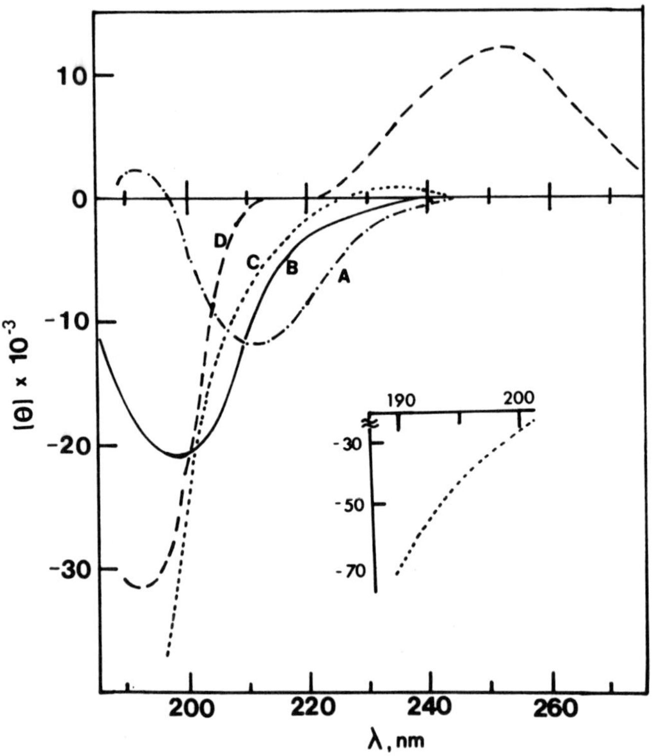

Fig. 1. CD spectra of hyaluronic acid in aqueous solution at pH 3.0 (A), hyaluronate film (B), hyaluronic acid in 20% ethanol at pH 3.0 (C), and Cu(II)-hyaluronate complex at pH 6.6 (D). Hyaluronate above pH 5 exhibits the same spectrum as A except for slight increase in ellipticity near 230 nm and absence of positive band near 190 nm.

aqueous-organic solvents, hyaluronic acid undergoes a major conformational change (9a,b), and the transition is cooperative with respect to temperature and solvent composition (Fig. 2) as well as with respect to pH. The conformational transition seems to be associated with intramolecular hydrogen bonding between a protonated carboxyl group and acetamido oxygen; supporting evidence for this is the fact that methyl

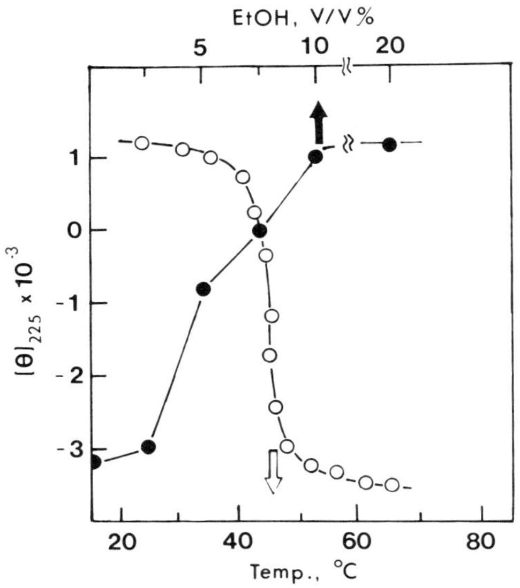

Fig. 2. CD ellipticity at 225 nm of hyaluronic acid in 10% ethanol, pH 2.6, at various temperatures (○), and at pH 2.6 in various ethanol concentrations (●) displaying conformational transition.

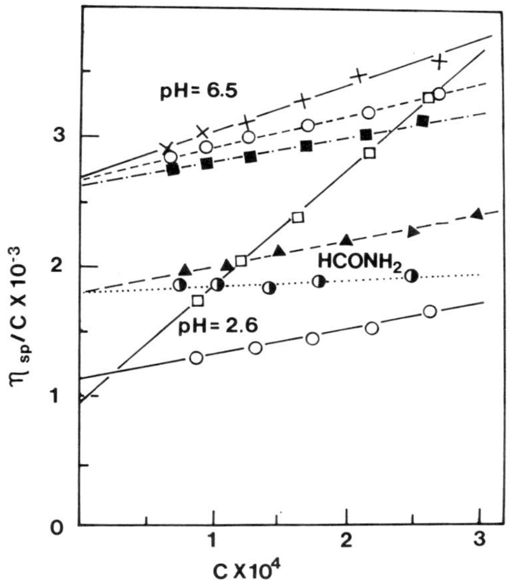

Fig. 3. Intrinsic viscosity of hyaluronic acid in various solvent environments at pH 2.5 and 6.5, in aqueous solution (■,◑,○) and in 10% ethanol (x, ▲,□). The open circles at pH 6.5, ▲, and ◑ represent viscosity values of solutions containing 1% $HCONH_2$ (ref. 9b).

hyaluronate fails to show such CD changes (9c). The hydrogen bonding restricts the rotation of the acetamido group, giving rise to a strong dichroism in the $\pi \rightarrow \pi^*$ amide transition region. A similar effect can be observed in hyaluronate film where intramolecular hydrogen bonding between acetamido and carboxyl oxygen has been proposed (8), and also when Cu(II) is chelated to carboxyl and acetamido groups intramolecularly (10). The 4-fold helical model of Guss *et al.* (7), which allows such interaction, is thus proposed for hyaluronate in film, in acidic mixed solvent, and in Cu(II)-hyaluronate complex. In the 4-fold helical model (7), there is also hydrogen bonding between the C-4 hydroxyl group of the glucosamine residues and the ring oxygen of the glucuronic acid moiety. The fact that chondroitin, which differs from hyaluronic acid only in the C-4 hydroxyl group configuration, does not show any CD change in acidic aqueous-organic solvent (9c) further supports our conclusion.

Viscosity data (Fig. 3) suggest that hydrogen bonding is not significantly involved in the overall geometry of the molecule at neutral pH. In acidic solution, the effect of formamide indicates that hyaluronic acid is in a compact form, thus favoring intramolecular hydrogen bonding, which probably involves hydroxyl groups and ring oxygen; this conformation could be similar to the 3-fold helical structure (7). In mixed solvent, the molecule appears to be in even more compact form than in acidic solution. We believe that all three conformations, at higher concentration of the polymer, aggregate through intermolecular interaction. The sharp decrease in optical rotation with urea (6), the CD change in acidic solution (4), and the gel-like state of the molecule in mixed solvent (9a) and when complexed with Cu(II) (10) are the properties of the aggregated state. The exact nature of the conformation of this aggregate requires further investigation.

ACKNOWLEDGMENTS

This study was supported by PHS grant EY 01760-02 and RCDA 1K04 EY 00070-01.

REFERENCES

1. Laurent, T. C., in "Chemistry and Molecular Biology of the Intercellular Matrix" (Balazs, E. A., ed.), Vol. 2, pp. 703-732, Academic Press, New York (1970).
2. Atkins, E. D. T., and Sheehan, J. K., *Science 179*, 562 (1973).
3. Dea, I. C. M., Moorhouse, R., Rees, D. A., Arnott, S. A., Guss, J. M., and Balazs, E. A., *Science 179*, 560 (1973).
4. Chakrabarti, B., and Balazs, E. A., *J. Mol. Biol.*, *78*, 135 (1973).
5. Chakrabarti, B., and Balazs, E. A., *Biochem. Biophys. Res. Commun.*, *52*, 1170 (1973).
6. Hirano, S., and Kondo-Ikeda, S., *Biopolymers*, *13*, 1357 (1974).
7. Guss, J. M., Hukins, D. W. L., Smith, R. J. C., Winter, W. T., Arnott, S., Moorhouse, R., and Rees, D. A., *J. Mol. Biol.*, *95*, 359 (1975).
8. Buffington, L. A., Pysh, E. S., Chakrabarti, B., and Balazs, E. A., *J. Amer. Chem. Soc.*, *99*, 1730 (1977).
9. Park, J. W., and Chakrabarti, B., (a) *Biopolymers 16*, 2807 (1977); (b) *Biopolymers 17*, 1323 (1978); (c) *Biochim. Biophys. Acta 541*, 263 (1978).
10. Figueroa, N., and Chakrabarti, B., *Biopolymers 17*, 2415 (1978).

Proteoglycans of Human Aorta

Edward G. Cleary and Palaniappan Muthiah

There have been few reported studies on aortic proteoglycans (PGs). Even so, there is evidence suggesting that aortic proteoglycans are relatively small molecules with a molecular mass less than 100 000 daltons (1,2). In none of these studies has account been taken of the high content of proteolytic enzymes in aorta homogenates. We have examined the extractability and behaviour of aortic proteoglycans using conventional and dissociative extraction methods, followed by equilibrium density gradient centrifugation (E.D.G.C.) and column chromatography, all in the presence of enzyme inhibitors.

MATERIALS AND METHODS

Non-atherosclerotic aortas from accident victims, aged 15-30, were obtained at autopsy within 18 h of death. Our studies of glycosaminoglycan (GAG) composition confirmed that there is considerable variation in these components from one location to another, especially in regard to hyaluronic acid and chondroitin sulphate (3). For this reason, the examinations of PGs have been confined to segments of ascending thoracic aorta. Segments have been pooled from 8-9 aortas of similar age and sex, and then crushed in liquid nitrogen to a fine powder. Portions have then been extracted both directly and sequentially at 4°C with 15 volumes of 0.15 M NaCl, 0.4 M guanidinium chloride (GuaHCl), and 4 M GuaHCl in the presence of Tris buffered to pH 7.4. Portions of the 4 M GuaHCl (direct) extracts have been concentrated and subjected to gel chromatography at 4°C on agarose columns in the extracting buffer, and under associative and dissociative conditions. Other portions were examined by E.D.G.C. as described by Hascal and Sajdera (4). Throughout the examinations, buffers have included such proteolytic enzyme

ISBN 0-12-301301-1

inhibitors as 75 mM EDTA, 10 mM *N*-ethylmaleimide, and 2 mM phenylmethylsulfonyl fluoride.

RESULTS AND DISCUSSION

Table I shows that a significant proportion of aortic PGs are extractable under associative conditions and with buffered saline solution.

Table I. Human Aortic Proteoglycans Extractability at pH 7.4

Extractant	*Uronic acid extracts (%)*		
	1st	*2nd*	*3rd*
0.15 M NaCl			*23*
0.4 M GuaHCl		*52*	*16*
4 M GuaHCl	*75*	*28*	*42*
Total	*75*	*80*	*81*

When portions of the 4 M GuaHCl extract were subjected to E.D.G.C. under associative conditions, uronic acid-containing material was spread throughout the gradient, indicating a wide range of flotation densities, and thus a complicated mixture of PGs. This was true even at initial densities as low as 1.38 g/ml. The lower flotation density reflects a considerably higher protein-to-GAG ratio in aortic PGs than in those from cartilage.

On gel chromatography under associative conditions, PG material was excluded from Sepharose 4B (Fig. 1), and a portion of this was also excluded from Sepharose 2B, indicating large molecular size. When the excluded material was chromatographed under dissociative conditions on Sepharose 4B (Fig. 2), an excluded PG and a uronic acid-negative protein peak, with apparent molecular size *ca*. 130 000 daltons, were obtained. A protein of similar size and amino acid analysis was obtained by gel chromatography of dissociative extracts from bovine aorta and bovine nasal cartilage. The amino acid composition of these materials is very similar to that reported for link protein by Hascall and Sajdera (5).

When the column eluates from dissociative gel chromatography were recombined, dialysed to associative conditions, and rechromatographed under these conditions (Fig. 3), a single, excluded proteoglycan peak was again obtained.

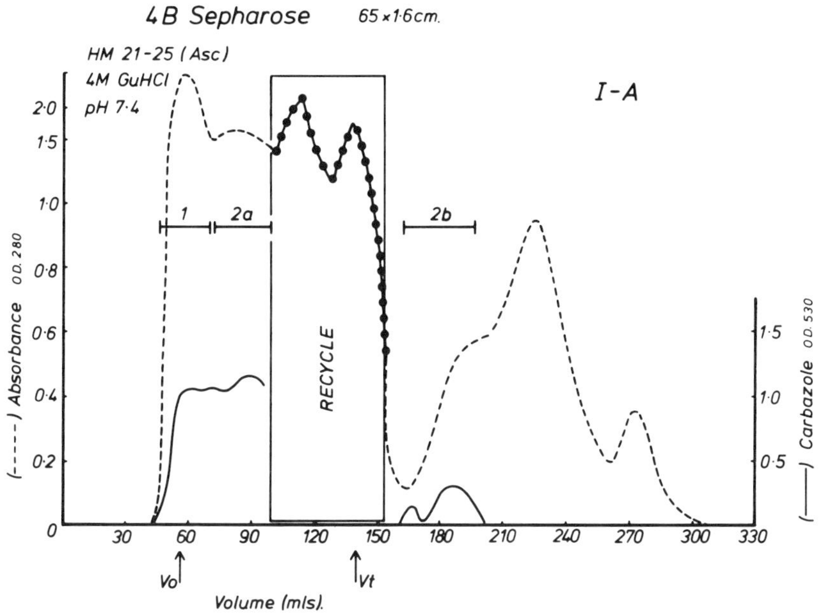

Fig. 1. Gel chromatography of 4 M GuaHCl extracts (direct) of ascending aorta on Sepharose 4B (Pharmacia) columns. Uronic acid assays made with 0.1 ml per fraction. Extract dialysed vs. *0.4 M GuaHCl (associative conditions) with recycling as shown. Eluant: 0.4 M GuaHCl. Peak 1 (excluded) from Fig. 1 concentrated and dialysed to dissociative conditions. Eluant: 4 M GuaHCl. Equivalent portions of peaks PG1, PG2, and P3 from Fig. 2 were mixed together, dialysed for 36 h* vs. *0.4 M GuaHCl, and re-chromatographed under associative conditions. Eluant: 0.4 M GuaHCl.*

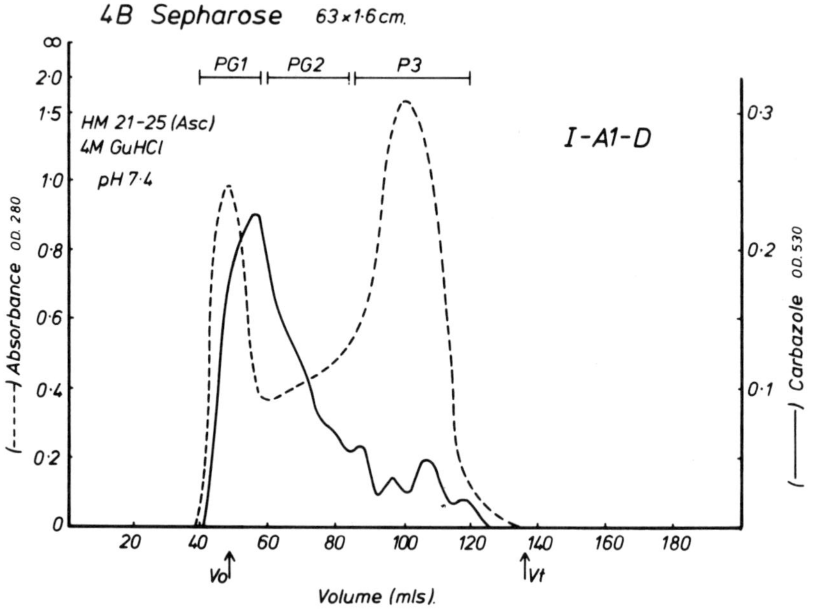

Fig. 2. Gel chromatography of 4 M GuaHCl extracts (direct) of ascending aorta on Sepharose 4B (Pharmacia) columns. Uronic acid assays made with 0.1 ml per fraction. Extract dialysed vs. 0.4 M GuaHCl (associative conditions) with recycling as shown. Eluant: 0.4 M GuaHCl. Peak 1 (excluded) from Fig. 1 concentrated and dialysed to dissociative conditions. Eluant: 4 M GuaHCl. Equivalent portions of peaks PG1, PG2, and P3 from Fig. 2 were mixed together, dialysed for 36 h vs. 0.4 M GuaHCl, and re-chromatographed under associative conditions. Eluant: 0.4 M GuaHCl.

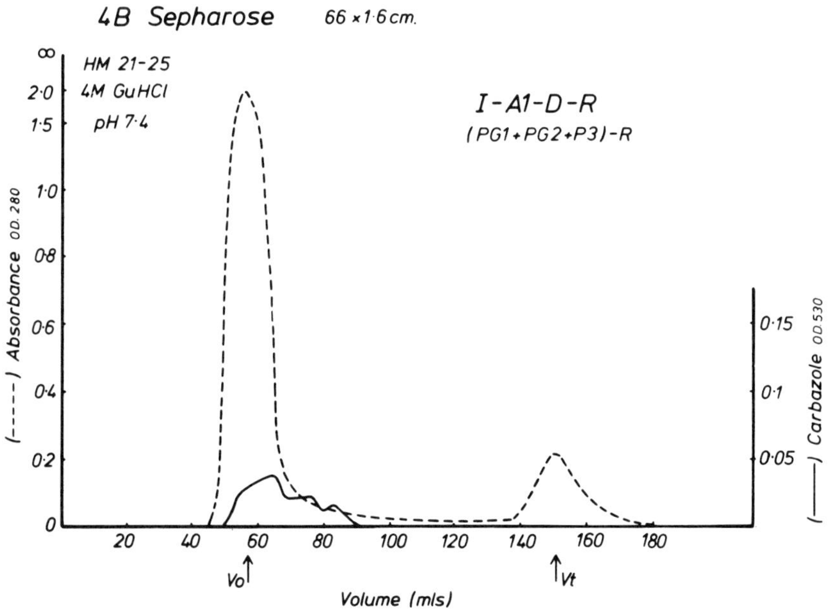

Fig. 3. Gel chromatography of 4 M GuaHCl extracts (direct) of ascending aorta on Sepharose 4B (Pharmacia) columns. Uronic acid assays made with 0.1 ml per fraction. Extract dialysed vs. *0.4 M GuaHCl (associative conditions) with recycling as shown. Eluant: 0.4 M GuaHCl. Peak 1 (excluded) from Fig. 1 concentrated and dialysed to dissociative conditions. Eluant: 4 M GuaHCl. Equivalent portions of peaks PG1, PG2, and P3 from Fig. 2 were mixed together, dialysed for 36 h* vs. *0.4 M GuaHCl, and re-chromatographed under associative conditions. Eluant: 0.4 M GuaHCl.*

We have thus demonstrated a form of aggregation-disaggregation behaviour of large aortic proteoglycan molecules kept in the presence of inhibitors of proteolytic enzymes. Viscosity studies under associative and dissociative conditions support this conclusion, but our initial studies in the analytical ultracentrifuge have failed to demonstrate this phenomenon, suggesting that the complexes are not as stable as those formed by cartilage PGs.

REFERENCES

1. Franek, M.D., and Dunstone, J.R., *Biochim. Biophys. Acta 165*, 555 (1968).
2. Ehrlich, K.C., Radhakrishnamurthy, B., and Berenson, G.S., *Arch. Biochem. Biophys. 171*, 361 (1975).
3. Muthiah, P., and Cleary, E.G., *Clin. Exp. Pharmacol. Physiol. 3*, 270 (1976).
4. Hascall, V.C., and Sajdera, S.W., *J. Biol. Chem. 244*, 2384 (1969).
5. Hascall, V.C., and Sajdera, S.W., *J. Biol. Chem. 245*, 4920 (1970).

GLYCOCONJUGATE RESEARCH / Volume I
Proceedings of the Fourth International Symposium on Glycoconjugates

Quantitation and Uses of Immobilized Sialic Acid-Containing Ligands

Anthony P. Corfield, Terence L. Parker, and Roland Schauer

Eight sialic acid-containing ligands were covalently coupled to a variety of insoluble matrices (Table I). The matrices were activated with either CNBr or IO_4^-, and the ligands labeled in the sialic acid moiety (1), or assayed spectrophotometrically after coupling to 2-(4-aminophenyl)ethylamine as spacer arm, in the case of the two reducing oligosaccharides. Hydrazido supports were prepared by both CNBr and IO_4^- methods and activated with HNO_2, as described by Junowicz and Charm (2).

Binding was carried out by micro-incubations of a 0.5-ml matrix suspension with 0.5 ml of ligand in binding buffer. Standard incubation mixtures contained 1 mg/ml of ligand in 0.1 M $NaHCO_3$, pH 8.0, for CNBr-activated and hydrazido matrices (20 h), or 0.1 M acetate, pH 6.0, containing 10 mM $NaBH_3CN$ for IO_4^--activated supports (12 h). The volume of support per test was quantitatively determined.

Binding was quantitatively determined by direct measurement of matrix-bead radioactivity after washing with 100-fold volumes of H_2O, 2 M NaCl, and H_2O. These results were correlated with spectrophotometric and radioactive measurements of the supernatants and wash solutions, and with sialic acid estimation after mild acid hydrolysis of the immobilized ligands.

Subsequent blocking of the remaining activated groups by ethanolamine or Tris was quantitatively determined with radioactive ethanolamine.

ISBN 0-12-301301-1

Table I. Range of Binding of Glycoproteins, Glycopeptides, and Oligosaccharide Measured with Periodate-(P) and Cyanogen Bromide-Activated (C) Supports[a]

Compounds		Sephadex G25	Sepharose 4B
Fetuin	P	0.288-15.1	0.036- 32.1
	C	0.02 - 2.0	1.8 -201.0
α_1-Acid glycoprotein	P	0.060- 0.310	0.020- 9.2
	C	0.02 - 1.5	2.4 -153.0
Bovine submandibular mucin (BSM)	P	0.007	0.039
	C	0.01 - 1.7	0.60 - 96.0
Brain glycopeptides	P	0.014	0.048
	C	0.01 - 0.3	0.29 - 48.0
α_1-Acid glycoprotein-glycopeptides	P	0.034	0.029
	C	0.02 - 0.2	0.12 - 36.0
BSM-glycopeptides	P	0.09 - 2.5	0.15 - 8.3
	C	0.02 - 0.3	0.18 - 45.0
Neuraminyllactose	P	0.112	0.045- 0.230
	C	0.23	2.8 - 11.2
BSM-disaccharide	P	0.069	0.028- 0.342
	C	0.17	2.1 - 19.9
Ethanolamine	P	(0.065- 0.42)	(0.025- 0.68)
	C	(1.7 -26.4)	(3.0 -264)

[a] Determined as mg/g of dry weight of support, or in parentheses, μmol/g of dry weight of support

[b] Hydrazido supports were prepared by coupling adipic acid dihydrazide to activated supports. Sephadex G-25: P, 132.4 μmol/g; C, 56 μmol/g. Sepharose 4B: P, 182.2 μmol/g; C, 306 μmol/g. Cellulose: P, 142.7 μmol/g; C, 54 μmol/g

Cellulose MN 2100	Hydrazido-[b] Sephadex G-25	Hydrazido-[b] Sepharose 4B	Hydrazido-[b] cellulose Mn 2100	CPG Glycophase
0.138-11.9	0.172	1.440	0.507	0.04-0.79
0.21 -25.4	0.03	2.52	0.15	
0.030- 1.3	0.103	0.630	0.475	0.06-0.34
0.35 -22.1	0.02	6.00	0.11	
0.013	0.069	1.86	0.076	0.01-0.07
0.11 -17.8	0.01	1.32	0.04	
0.008	0.218	0.450	0.049	0.01-0.95
0.07- 4.9	0.01	0.72	0.01	
0.031	0.184	0.510	0.050	0.01-0.02
0.05- 4.9	0.02	0.52	0.01	
0.29- 2.1	0.126	0.705	0.036	0.01-0.03
0.05- 5.2	0.01	0.50	0.01	
0.018				
0.36				
0.011				
0.49				
(0.012- 0.73)	(0.041)	(0.009)	(0.024)	(1.5-6.1)
(0.20 -10.8)	(6.2)	(109.5)	(10.8)	

The method described is rapid and accurate, and allows measurement of binding capacities for ligands under different conditions. The results obtained (Table I) show that each ligand has individual binding characteristics for each matrix. The variation of ligand concentration, activation conditions, or binding time can be used to obtain a range of accurately determined, immobilized ligand.

Periodate-activated matrices yield more stable derivatives, as compared to the CNBr-activated supports, but in general show lower binding capacities. The value of these immobilized ligands for affinity chromatography of sialic acid-binding proteins and enzymes involved in sialic acid metabolism, for enzyme assays with radioactive ligand or substrate, and for molecular interactions and ion-binding are currently under investigation.

REFERENCES

1. Veh, R. W., Corfield, A. P., Sander, M., and Schauer, R., *Biochim. Biophys. Acta 486*, 145 (1977).
2. Junowicz, E., and Charm, S. E., *Biochim. Biophys. Acta 428*, 157 (1976).

Polyanion-Polycation Interaction in Hyphal Walls from *Mucor mucedo*

Roelf Datema

Hyphal walls from *Mucor* contain *N*-nonacetylated glucosaminyl residues (1) and are, therefore, susceptible to nitrous acid degradation. Treatment of walls with HNO_2 resulted, among others, in the solubilization of an acidic polysaccharide containing all of the neutral sugars and uronic acids present in the wall. Monomer analysis of the glycuronan revealed that fucose, mannose, galactose, and glucuronic acid occurred in a 5:1:1:6 molar ratio (2). The polysaccharide was further characterized by partial acid hydrolysis, electrophoresis, and gel filtration (2).

Suspending hyphal walls in 6 M LiCl (4.4 nmol/mg wall) resulted in the solubilization of an acidic polysaccharide undistinguishable from the HNO_2-solubilized polysaccharide (composition, acid susceptibility, size, and electrophoretic mobility), while no GlcN-containing material was solubilized (2). As both destruction of GlcN residues (HNO_2) and interference with ionic bonds (LiCl) solubilized the glycuronan, it is proposed that, in the wall, the insoluble GlcN-containing polymers bind ionically the diffusable component, the glycuronan.

Heating a solution of the glycuronan in 1 N HCl (1 h, 100°C) resulted in the formation of crystalline polyglucuronic acid (mucoric acid) containing 60% of the glucuronic acid (GlcUA) present in the glycuronan (2). Apparently, these residues form homopolymeric segments in the glycuronan. The supernatant of the hydrolysate was analyzed by Bio-Gel P-2 chromatography. Man, Gal, and the remaining GlcUA residues were present as oligosaccharides of mol. wt. 1000-1200; Fuc was almost exclusively present as monosaccharide (3). In contrast, when mucoran, an alkali-soluble glycuronan from walls of *M. rouxii*, was partially acid hydrolyzed, formation of mucoric acid and monomeric Fuc was not observed (4).

ISBN 0-12-301301-1

To study the nature of the polycationic components, and the effects of HNO_2 upon them, the polymers were specifically labeled by growing mycelium in [^{14}C]-GlcNAc-containing medium (5). After HNO_2 treatment (pH 3, 1.5 h, 25°C), low-mol. wt. products were separated on a Bio-Gel P-2 column. Anhydromannose (Aman), the reaction product of GlcN residues with HNO_2, was the main product: 49% of the radioactive products was solubilized by HNO_2. In addition, free GlcNAc (9%), a disaccharide composed of GlcNAc and Aman (15%), a trisaccharide composed of 2 GlcNAc and 1 Aman residues (6%), and higher oligosaccharides composed of 3-5 GlcNAc and 1 Aman residues (10%) were detected. The Aman residue was present at the reducing end of the oligosaccharides. The remaining 10% was made up by an unidentified component excluded from the gel, (see 3 and 5). The nitrous acid-soluble products were present in similar relative amounts in walls labeled for varying periods of time (3), and point to the general occurrence of GlcNAc-GlcN copolymers in these walls. For example, the sequence GlcNAc→GlcNAc→GlcN gives rise to the above mentioned trisaccharide when walls are treated with HNO_2. Approximately 30% of the GlcNAc residues in the wall occur in this heteropolymeric form.

Not all of the nitrous acid-insoluble GlcNAc residues occur as chitin. When the hexosamine-containing, LiCl-insoluble cell-wall fraction was treated with Pronase, a part (40%) of the nitrous acid-resistant, GlcNAc-containing polymers became nitrous acid-soluble. Scheme 1 shows a possible explanation for this result: Amino sugar residues are bound to amino acids *via* the *N*-acetyl groups.

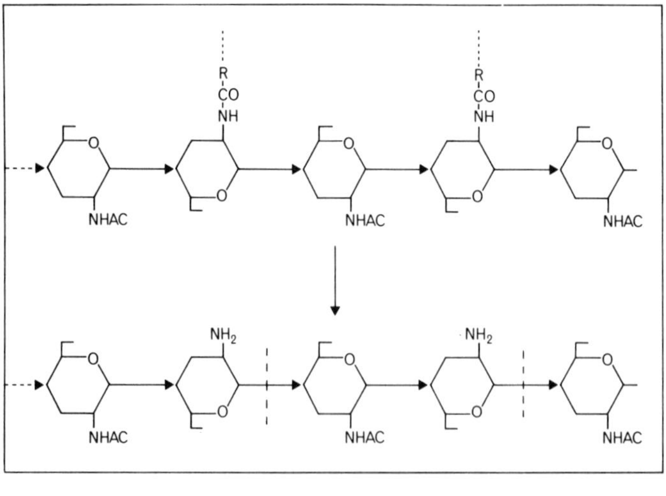

Scheme 1.

Pronase treatment generates primary amino groups, and the molecule becomes susceptible to nitrous acid deamination. As a consequence, glycosidic bonds can be split (vertical dotted lines). Amino acids are present and comprise 10% of the LiCl-insoluble wall fraction. A total amino acid analysis has been reported (3).

Chitin, defined as poly(GlcNAc), was present in material resisting successive treatments with Pronase and HNO_2; or alkali, acid, and HNO_2. This conclusion was based on X-ray diffraction, i.r. spectroscopy, and analysis of 6 N HCl hydrolysates. In these hydrolysates, we could not find Aman, indicating the absence of binding of chitin to GlcN-containing polymers *via* GlcNAc residues. This chitin fraction made up 4% of the cell-wall dry weight and *ca.* 30% of cell-wall GlcNAc residues.

ACKNOWLEDGMENTS

This study was supported by the Foundation for Fundamental Biological Research (BION), which is subsidized by the Netherlands Organization for the Advancement of Pure Research (ZWO). Discussions with Drs. H. van den Ende, J. H. Sietsma, and J. G. H. Wessels are appreciated. This study was performed at the Department of Developmental Plant Biology, State University of Groningen, the Netherlands.

REFERENCES

1. Rosenberg, R.F., *in* "The Filamentous Fungi" (J.E. Smith and D.R. Berry, eds.) vol. 2, pp. 328, E. Arnold, London (1976).
2. Datema, R., van den Ende, H., and Wessels, J.G.H., *Eur. J. Biochem. 80,* 611 (1977).
3. Datema, R., *Ph. D. Thesis,* Groningen (1977).
4. Bartnicki-Garcia, S., and Lindberg, B., *Carbohydr. Res. 23,* 7585 (1972).
5. Datema, R., Wessels, J.G.H., and van den Ende, H., *Eur. J. Biochem. 80,* 621 (1977).

Isolation and Partial Characterization of a Peptide from Bovine Cervical Mucin

Francisco Delers and Christian Lombart

Structural studies of the cervical mucin seem extremely important to understand its cyclic changes during the ovulatory cycle and its physiological function. In this work, we undertook to purify the mucin from the bovine estrous mucus in order to study its subunit composition in the presence of denaturing agents. First, we elaborated a mild purification procedure in order to obtain an undegraded material (see Scheme 1).

1 vol + 1 vol of solution 2M NaCl
↓
gentle agitation
↓
centrifugation 3000g 15 min.
↙ ↘
pellet | supernatant
↓
dialysis acetic buffer pH 3.9
↓
centrifugation 3000g 15 min.
↙ ↘
pellet | supernatant
↓
chromatography on Sepharose 4 B

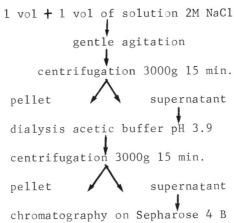

Scheme 1. Preparation of bovine cervical mucin from estrous mucus.

The final stage of purification gave a pure mucin, free of contaminants as judged by immunoelectrophoresis against various antisera: anti-bovine serum, anti-lactoferrin, anti-albumin, and anti-IgA. Moreover, 0.1% SDS-(6%) acrylamide gel electro-

ISBN 0-12-301301-1

phoresis did not reveal any contaminant after staining either with the periodic acid-Schiff reagent or Coomassie Blue. Under those conditions, the mucin itself did not penetrate the gel. The glycoprotein was analyzed for amino acid and carbohydrate content. The data (see Table I) are in good agreement with those of Bushana-Rao *et al.* (1) and Meyer *et al.* (2).

Table I. Amino Acid and Carbohydrate Compositions of Bovine Cervical Mucin

(Residues/1000 residues)

Amino Acid		*Carbohydrate*[b]	
Lysine	*30*		
Histidine	*20*		
Arginine	*38*	*Fucose*	*108*
Aspartic acid	*58*	*Galactose*	*346*
Threonine[a]	*162*	*Sialic acid*	*114*
Serine[a]	*87*	N-*Acetylgalactosamine*	*171*
Glutamic acid	*81*	N-*Acetylglucosamine*	*251*
Proline	*73*		
Glycine	*114*		
Alanine	*73*		
Cysteine	*20*		
Valine	*100*		
Methionine	*38*		
Isoleucine	*18*		
Leucine	*60*		
Tyrosine	*18*		
Phenylalanine	*22*		

[a] *Values obtained by extrapolation*
[b] *Values obtained by gas chromatography according to the procedure of Chambers and Clamp (3)*

The glycoprotein was then subjected to reduction and alkylation according to the following procedure: a sample was incubated under nitrogen in 0.02 M phosphate buffer, pH 8, containing 6 M guanidine and 0.13 M dithiothreitol (DTT) during 20 h. After reduction, iodoacetamide was added in a molar concentration slightly higher than that of DTT. The sample was then exhaustively dialyzed against 0.02 M phosphate buffer, pH 6.8, containing 0.1%SDS, and filtered through a Sepharose 4B C.L. column. Two well-separated peaks were obtained: the first eluted

(P_1) in the void volume contained practically all the carbohydrates of the original material, and the second retarded peak contained a peptide fraction (P_2) devoid of carbohydrate. After being pooled, the two Fractions P_1 and P_2 were dialyzed against water and lyophilized; after removal of SDS with acetone, the amino acid compositions of Fractions P_1 and P_2 were determined and are reported in Table II.

Table II. Amino Acid Composition of Fractions P_1 and P_2

(Residues/1000 residues)	Fractions	
	P_1	P_2
Lysine	40	88
Histidine	16	28
Arginine	50	63
Aspartic acid	100	113
Threonine	157[a]	48
Serine	120[a]	76
Glutamic acid	55	137
Proline	37	91
Glycine	136	100
Alanine	75	102
CM-Cysteine	18	0,5
Valine	48	39
Methionine	41	2
Isoleucine		24
Leucine	55	34
Tyrosine	b	10
Phenylalanine		28

[a] Values corrected for destruction
[b] Not determined

P_1 shows an amino acid composition very similar to that of the original mucin and, in comparison, P_2 possesses larger proportions of the basic amino acids and also of glutamic and aspartic acids. Since larger proportions of NH_4^+ were released by acid hydrolysis, one may conclude that aspartic and glutamic residues exist in their amide form in the original mucin. When P_1 and P_2 were applied onto a polyacrylamide gradient-gel system (4-28%)

in 0.1%SDS at neutral pH, P_1 showed an electrophoretic pattern similar to that of the reduced and alkylated mucin. Multiple bands probably correspond to different aggregates of the same subunit (Fig. 1).

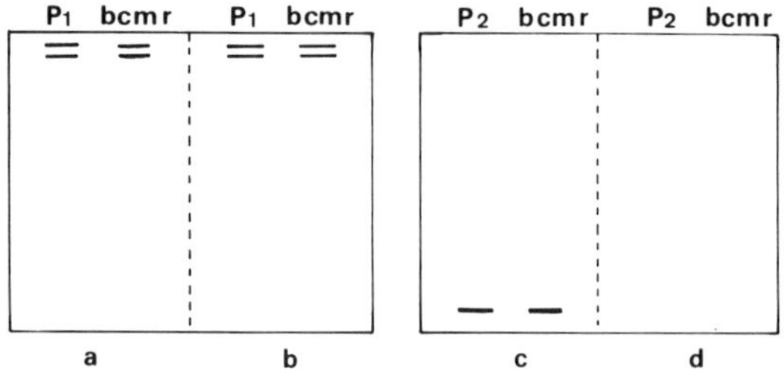

Fig. 1. Acrylamide gradient (4-8%) gel electrophoresis of fractions P_1 and P_2.

(a), (b): pattern obtained in 0.1%SDS-phosphate buffer (pH 7.2)
(c), (d): pattern obtained in acid buffer (pH 4.37)

(a), (b): staining with Amido Black
(b), (e): staining with periodic acid-Schiff reagent (PAS)

bcmr : reduced and alkylated mucin

On the contrary, P_2 did not move into the gel at neutral pH, but only migrated as a single band towards the cathode in an acidic buffer system (β-alanine-acetic acid, pH 4.37) (4). This electrophoretic behavior of P_2 clearly demonstrates its basic character (Fig. 1).

These results suggest that the mucin molecule is formed of two different types of subunits: one is represented by high mol. wt. glycopeptides, and the other, the "naked" peptide, probably corresponds to the cross-linking factor that has been postulated by several authors to account for the highly organized structure of the estrous gel (5,6). Evidently, further analysis of this peptide is needed to verify such an hypothesis.

REFERENCES

1. Bushana-Rao, K.S.P., Van Roost, E., Masson, P.L., Heremans, J.F., and Andre, F., *Biochim. Biophys. Acta 317*, 286 (1973).
2. Meyer, F.A., Eliezer, N., Silberberg, A., Vered, J., Sharon, N., and Sade, J., *Bull. Physio-Pathol. Respir. 9*, 259 (1973).
3. Chambers, R.E. and Clamp, J.R., *Biochem. J. 125*, 1009 (1971).
4. Reisfeld, R.A., Lewis, V.J., and Williams, P.E., *Nature (London) 195*, 281 (1962).
5. Gibbons, R.A., *Protides Biol. Fluids Proc. Colloq. 16*, 299 (1968).
6. Masson, P.L. *in* "Cervical Mucus in Human Reproduction" (Elstein, M., Moghissi, K.S., Borth, R., eds.) p. 82 Scriptor, Copenhagen (1973).

Quantitative Study of the β-Elimination Reaction on Glycoproteins

Frederick Downs, Christine Peterson, Varahabhotla L.N. Murty, and Ward Pigman

The carbohydrate side chains of mucus-type glycoproteins are *O*-glycosycally bound between an *N*-acetylgalactosamine residue and the hydroxyl groups of the serine and threonine residues in the protein core. In order to evaluate accurately the extent of *O*-glycosyl linkages in a glycoprotein, quantitative information on both the changes of amino acids and carbohydrates involved in this bond is required. The only method that satisfies these objectives is the alkaline borohydride procedure. The purpose of this work was to determine the reliability and accuracy of this method on several glycoproteins. Using bovine, ovine, and porcine submandibular mucus glycoproteins, we compared the loss of seryl and threonyl residues with the recovery of their reduced derivatives, alanine and 2-aminobutanoic acid, respectively. In addition, the loss of *N*-acetylglucosamine and *N*-acetylgalactosamine was compared with the formation of the corresponding amino sugar alcohols. The experimental procedures concerning these parameters were given by Downs *et al.* (1).

The effect of the β-elimination reaction on bovine submandibular glycoprotein is shown in Table I. The loss of serine and threonine, and their conversions into alanine and 2-aminobutanoic acid, respectively, are in excellent agreement, and the β-elimination is essentially complete in 10 h. Results for the loss of hexosamines and their recovery, as amino alcohols, are also given in this Table. The recovered hexosaminitols are in excellent agreement with the reduced amino acid derivatives, as are the results observed for the ovine and porcine submandibular glycoproteins.

ISBN 0-12-301301-1

Table I. Results of the β-Elimination and Reduction Reactions on Bovine Submandibular Glycoprotein[a]

(mol/100 mol amino acid)

Time (h)	Loss			Increase			Loss			Increase		
	GalN	GlcN	Total	GalNol	GlcNol	Total	Ser	Thr	Total	Ala	Aba[b]	Total
0	0	0	0	0	0	0	0	0	0	0	0	0
0.25	5.7	0.8	6.5	5.4	0.7	6.1	4.7	1.5	6.2	4.7	1.5	6.2
0.50	10.5	1.4	11.9	9.8	1.0	10.8	8.0	3.1	11.1	8.0	3.1	11.1
0.75	13.4	1.3	14.7	13.4	1.2	14.6	10.0	4.5	14.5	9.9	4.4	14.3
1.0	16.2	1.4	17.6	15.6	1.4	17.0	11.0	5.4	16.4	10.9	5.3	16.2
2.0	19.6	1.6	21.2	19.1	1.5	20.6	13.2	7.9	21.1	13.1	7.9	21.0
5.0	23.5	1.8	25.3	23.0	1.6	24.6	14.2	10.0	24.2	14.3	10.0	24.3
10.0	23.6	1.9	25.5	23.2	1.6	24.8	14.6	11.1	25.7	14.6	11.1	25.7
20.0	24.3	2.1	26.4	24.0	1.8	25.8	15.0	11.5	26.5	14.6	10.8	25.4

[a] Solutions of bovine submandibular glycoprotein (5mg/ml) in aqueous 0.1 M NaOH and 0.6 M $NaBH_4$ were incubated at 45°C for the times indicated. This was followed by the use of an improved reduction system, which involved the simultaneous additions of 0.66 M $NaBH_4$ and 0.016 M $PdCl_2$, in 0.8 M HCl

[b] Aba, 2-aminobutanoic acid

The data obtained for bovine submandibular glycoprotein has additional interest, because an alkaline treatment for 15 min converts ~10% of the *N*-acetylglucosamine into *N*-acetylglucosaminitol residues. *N*-Acetylglucosamine has not been reported to be involved in an *O*-glycosyl linkage to serine and threonine; the formation of glucosaminitol in the reaction products has been attributed as arising from the secondary peeling reaction of the oligosaccharide side-chain, although the recovery of galactosaminitol and glucosaminitol equals the destructions of hydroxy amino acids (Table I). Consequently, the occurrence of some *N*-acetylglucosamine residues involved in *O*-glycosyl bond to serine and threonine cannot be ruled out.

The rate of release of the oligosaccharide side-chains from serine is faster than that from threonine during the β-elimination for the three submandibular glycoproteins investigated (Fig. 1).

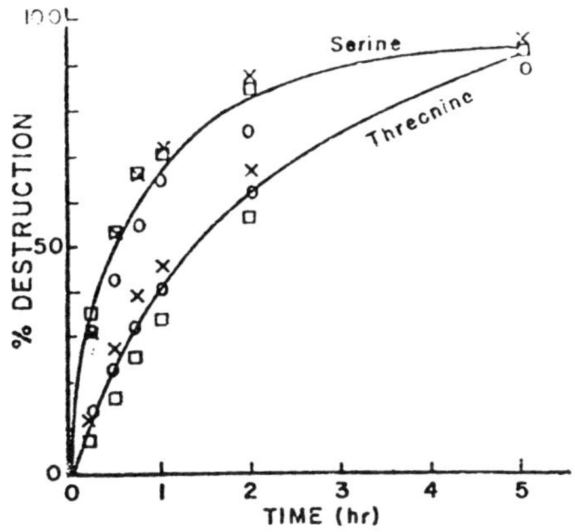

Fig. 1. Relative rates of destruction of serine and threonine during the alkali-catalyzed β-elimination reaction in the presence of sodium borohydride for porcine (□), ovine (x), and bovine (o) submandibular glycoproteins. % Destruction = mmol amino acid destroyed/mmol amino acid destroyed after 20 h x 100.

The amount of unsaturated amino acids formed during the alkali-catalyzed β-elimination of glycoproteins has also been monitored by measuring the increase in absorbance at 240 nm (2). The results shown in Fig. 2 indicate an increase in absorbance for bovine submandibular glycoprotein. However, a sample containing equal amounts of bovine serum albumin (a protein containing no carbohydrate), *N*-acetylgalactosamine, and sialic acid also showed an increase in absorbance at 240 nm. It is, therefore, obvious that 240-nm absorbance readings are not specific for measuring the extent of the β-elimination reaction.

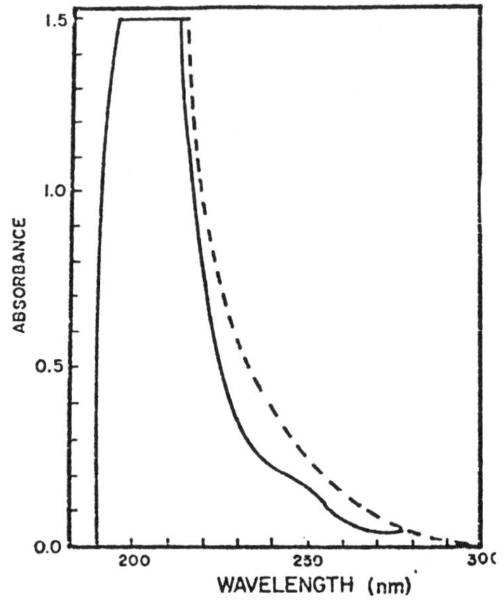

Fig. 2. Alkali-catalyzed β-elimination of bovine serum albumin (———) and bovine submandibular glycoprotein (----). Bovine serum albumin and bovine submandibular glycoprotein were dissolved in 0.5 N NaOH, at a concentration of 0.8 mg/ml and 2.0 mg/ml, respectively, at 0°C. The solutions were incubated at 50°C. After 30 min, the absorbance was measured on a 0.05-ml sample diluted to 1.0 ml with water. The samples were scanned against a water blank, by use of a Perkin-Elmer Model 402 recording spectrophotometer.

REFERENCES

1. Downs, F., Peterson, C., Murty, V.L.N., and Pigman, W., *Int. J. Pept. Prot. Res. 10,* 315 (1977).
2. Planter, J.J., and Carlson, D.M., *Anal. Biochem. 65,* 153 (1975).

Structure of Ten Glycopeptides from α_1-Acid Glycoprotein

Bernard Fournet, Gérard Strecker, Geneviève Spik, Jean Montreuil, Karl Schmid, J. Paul Binette, Lambertus Dorland, Johan Haverkamp, Bernard L. Schut, and Johannes F. G. Vliegenthart

The heterogeneity of the carbohydrate moiety of the human plasma α_1-acid glycoprotein has been studied extensively (1-11). However, some ambiguity still exists with regard to the precise structures of the various carbohydrate chains of this glycoprotein. Recently, the careful isolation of 26 glycopeptides in the asialo-form was described (11) and the present study deals with the structure determination of 10 typical representatives of these by means of permethylation analysis, partial acetolysis, and 360-MHz ^{1}H-n.m.r. spectroscopy.

MATERIALS AND METHODS

α_1-Acid glycoprotein was isolated from a pool of normal human plasma (12). Asialoglycopeptides were fractionated as described in a previous paper (11). The molar ratios of hexoses and *N*-acetylhexosamines were determined by g.l.c. (13). Methylation was performed according to Hakomori (14), and the partial methylated monosaccharides were identified as their acetylated methyl glycoside derivatives (15,16). Acetolysis (17) was carried out with a reaction time of 3 h. After reduction and permethylation, the oligosaccharides were identified by g.l.c.-m.s. (mass fragmentography at *m/e* 236, 260, and 277, Riber model apparatus) (18). ^{1}H-N.m.r. spectroscopy was performed at 360 MHz on a Bruker HX-360 spectrometer operating in the Fourier-transform mode at probe temperatures of 25°C and 60°C. Chemical shifts are given relative to sodium 2,2-dimethyl-2-silapentane-5-sulfonate (indirectly to acetone in D_2O: 2.225 p.p.m.).

ISBN 0-12-301301-1

Table I. Carbohydrate Composition of 10 Pronase-glycopeptides Derived from Desialylated Reduced, and Carboxymethylated α_1-acid Glycoprotein

Glycopeptide	Carbohydrate compound[a]						
	Fuc	Gal	Man	Gal/Man[b]	GlcN	Hex/HexN[b]	Total
Class A							
GP II-6	0	1.9(2)	3.0(3)	0.7	3.97(4)	1.25	9
Class B							
GP II-5	0	2.7(3)	3.02(3)	1	5.1(5)	1.2	11
GP III-7	0	3.0(3)	3.0(3)	1	4.94(5)	1.2	11
GP V-5	0	3.0(3)	3.0(3)	1	4.99(5)	1.2	11
Class C							
GP III-6	0	3.82(4)	3.0(3)	1.3	5.99(6)	1.17	13
GP V-4	0	3.54(4)	3.0(3)	1.3	5.73(6)	1.17	13
Class D							
GP III-5	0.85(1)	4.09(4)	3.0(3)	1.3	5.64(6)	1.17	14
GP V-2	0.91(1)	3.5(4)	3.0(3)	1.3	6.2(6)	1.17	14
GP V-3	1.12(1)	4.06(4)	3.0(3)	1.3	5.8(6)	1.17	14
GP V-1	0.70(1)	3.86(4)	3.0(3)	1.3	4.76(5)	1.4	13

a Expressed in mol of monosaccharides per mol of glycopeptides
b Molar ratio of hexose to hexosamine

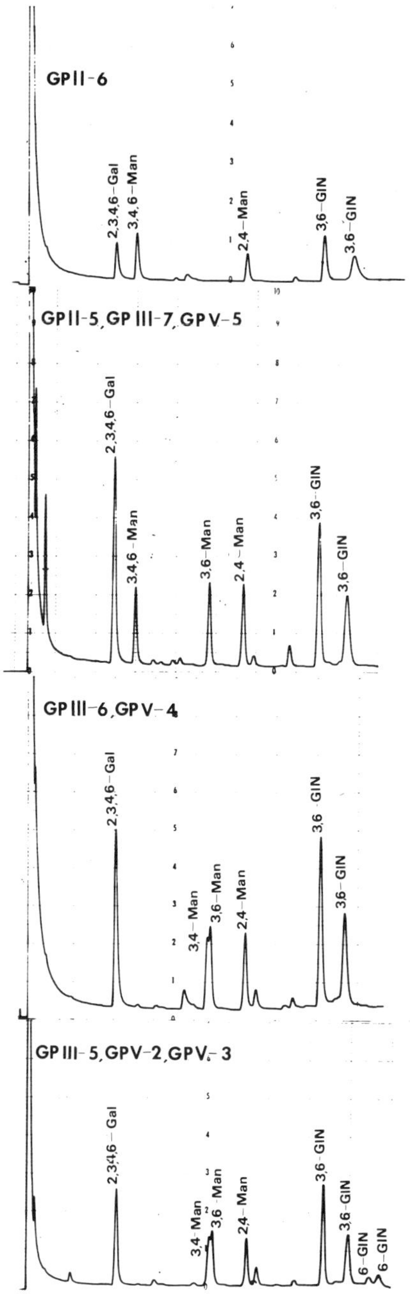

Fig. 1. Gas chromatography of acetylated monosaccharide methyl ethers obtained by methanolysis of permethylated glycopeptides and peracetylation. Aerograph 1200 apparatus, glass column (0.2 x 300 cm), 3% Carbowax 6000 on Chromosorb W-HMDS (60-80 mesh), 110°-200°C (2°C/min), nitrogen: 30 ml/min.

Table II. Identification and Molar Ratios of Monosaccharide Methyl Ethers Present in the Methanolyzates of Ten Permethylated Asialoglycopeptides from α_1-Acid Glycoprotein

Glycopeptide	Methylated monosaccharides			
	2,3,4-tri-O-methyl-fucose	2,3,4,6-tetra-O-methyl-galactose	3,4,6-tri-O-methyl-mannose	3,4-di-O-methyl-mannose
Class A				
GP II-6	0	1.88(2)	1.88(2)	0
Class B				
GP II-5	0	3.06(3)	1.02(1)	0
GP III-7	0	2.85(3)	1.07(1)	0
GP V-5	0	3.12(3)	1.2(1)	0
Class C				
GP III-6	0	3.76(4)	0	0.95(1)
GP V-4	0	3.85(4)	0	0.97(1)
Class D				
GP III-5	0.84(1)	3.83(4)	0	0.88(1)
GP V-2	0.68(1)	3.81(4)	0	1.01(1)
GP V-3	0.82(1)	3.75(4)	0	1.04(1)
GP V-1	0.51(1)	4.3(4)	0	0.87(1)

a Expressed in mol of monosaccharides per mole of glycopeptides

2,4-di-O-methyl-mannose	3,6-di-O-methyl-mannose	3,6-di-O-methyl-glucosamine	6-mono-O-methyl-glucosamine	Total
1.0(1)	0	4.2(4)	0	9
0.97(1)	0.89(1)	4.55(1)	0	11
1.05(1)	1.09(1)	4.5(5)	0	11
1.0(1)	0.88(1)	4.99(5)	0	11
1.0(1)	0.99(1)	5.54(6)	0	13
1.0(1)	0.81(1)	5.5(6)	0	13
1.0(1)	0.96(1)	4.94(5)	0.80(1)	14
1.0(1)	0.91(1)	5.2(5)	0.6(1)	14
0.88(1)	0.77(1)	4.86(5)	0.52(1)	14
1.0(1)	0.84(1)	3.87(4)	0.43(1)	13

RESULTS AND DISCUSSION

The molar carbohydrate compositions of the 10 compounds are reported in Table I. According to the number of the various monosaccharide constituents, the glycopeptides can be divided into 4 classes A, B, C, and D (A=GP II-6; B=GP II-5, GP III-7, and GP V-5; C=GP III-6 and GP V-4; D=GP III-5, GP V-2, and GP V-3). GP V-1 is a reducing oligosaccharide. The data of methylation analysis are represented in Table II and Fig. 1. The same methylated neutral sugars in identical molar ratios were found in all glycopeptides belonging to one class. The oligosaccharides isolated from the acetolyzate of GP V-5 are: β-Gal-(1→4)-GlcNAc, β-GlcNAc-(1→2)-Man and β-GlcNAc-(1→4)-α-Man-(1→3)-Man. The presence of the trisaccharide demonstrates that the third residue of *N*-acetyllactosamine is linked to the mannose residue 4.

The glycopeptide structures of the classes A-D would be characterized by 360-MHz ^{1}H-n.m.r. spectroscopy. Some relevant data are given in Table III.

Table III. Chemical Shifts of H-1 and H-2 Mannose Residues of the Bi-, Tri-, and Tetraantennary Structures

Structure	*H-1 of D-mannose*			*H-2 of D-mannose*		
	3	*4*	*4'*	*3*	*4*	*4'*
Biantennary	*4.77*	*5.12*	*4.93*	*4.24*	*4.18*	*4.11*
Triantennary	*4.76*	*5.12*	*4.93*	*4.21*	*4.21*	*4.11*
Tetraantennary	*4.77*	*5.12*	*4.86*	*4.22*	*4.22*	*4.09*

Class A glycopeptide has a biantennary structure that is identical to the asialoglycan of human serotransferrin (19,20). The characteristic n.m.r. features of the mannotrioside branching-core substituted by GlcNAc residues 2, 5, and 5' are the chemical shifts of the mannose residue H-1 and H-2, as given in Table III. Glycopeptides of classes B and C have tri- and tetraantennary structures, respectively, which means the attachment of one or two additional *N*-acetyllactosamine residues to the mannotrioside core of the biantennary structure. The occurrence of these extra branches is expressed in the shift increments of the mannose residue H-1 and H-2 (Table III).

Class D glycopeptides consist of a tetraantennary glycan chain with an extra fucose residue. The type of linkage of this fucose residue to a GlcNAc residue 7 follows from the chemical shift of the fucose residue H-5 (4.84 p.p.m.) and from the shift increment of the *N*-acetyl signal of GlcNAc residue 7 (2.080→2.068 p.p.m.).

The combination of the results of methylation analysis, partial acetolysis, and 360-MHz ^{1}H-n.m.r. spectroscopy leads to the conclusions that in the triantennary structures the additional *N*-acetyllactosamine is coupled *via* a β-(1→4) linkage to a mannose residue 4 whereas in the tetraantennary structures the additional *N*-acetyllactosamine residues are coupled *via* β-(1→4) and β-(1→6) linkages to mannose residues 4 and 4', respectively (see J. Montreuil and J. F. G. Vliegenthart in this volume).

REFERENCES

1. Jeanloz, R. W., *in* "Glycoproteins" (A. Gottschalk, ed.), p. 565, American Elsevier Publish. Co., New York (1972).
2. Spiro, R. G., *Adv. Prot. Chem. 27*, 350 (1973).
3. Kornfeld, R., and Kornfeld, S., *Annu. Rev. Biochem. 45*, 217 (1976).
4. Montreuil, J., *Pure Appl. Chem. 42*, 431 (1975).
5. Wagh, P. V., Bornstein, I., and Winzler, R. J., *J. Biol. Chem. 244*, 658 (1969).
6. Hatcher, V. B., and Jeanloz, R. W., *Colloq. Int. C.N.R.S. 221*, 329 (1973).
7. Schwarzmann, G., Reinhold, V., and Jeanloz, R. W., *Colloq. Int. C.N.R.S. 221*, 85 (1973).
8. Fournet, B., Takerkart, G., Brohon, J., and Montreuil, J., *Bull. Soc. Chim. Biol. 50*, 1352 (1968).
9. Bayard, B., and Montreuil, J., *Arch. Int. Physiol. Biochim. 77*, 564 (1969).
10. Bayard, B., and Fournet, B., *Carbohydr. Res. 46*, 75 (1976).
11. Schmid, K., Nimberg, R. B., Kimura, A., Yamaguchi, H., and Binette, J. P., *Biochim. Biophys. Acta 492*, 291 (1977).
12. Schmid, K., *in* "The Plasma Proteins" (F. W. Putman, ed.), vol. 1, p. 184, Academic Press, New York (1975).
13. Zanetta, J. P., Brechenridge, W. C., and Vincendon, G., *J. Chromatogr. 69*, 291 (1972).
14. Hakomori, S. I., *J. Biochem. (Tokyo) 55*, 205 (1964).
15. Fournet, B., Leroy, Y., Montreuil, J., and Mayer, H., *Colloq. Int. C.N.R.S. 221*, 111 (1973).
16. Fournet, B., Thesis, Lille (1973).
17. Bayard, B., and Montreuil, J., *Carbohydr. Res. 24*, 427 (1972).

18. Fournet, B., Dhalluin, J. M., Strecker, G., and Montreuil, J., in preparation.
19. Spik, G., Bayard, B., Fournet, B., Strecker, G., Bouquelet, S., and Montreuil, J., *FEBS Lett.* *50*, 296 (1975).
20. Dorland, L., Haverkamp, J, Schut, B. L., Vliegenthart, J. F. G., Spik, G., Strecker, G., Fournet, B., and Montreuil, J., *FEBS Lett.* *77*, 15 (1977).

GLYCOCONJUGATE RESEARCH / Volume I
Proceedings of the Fourth International Symposium on Glycoconjugates

Studies on the Proteoglycans from Bovine Cornea

Leonardo Galligani, Pietro Speziale, Maria Cristina Sosso, and Cesare Balduini

Most of the studies on the structure of proteoglycans have been carried out on cartilage tissue. It has been shown that these macromolecules display a large polydispersity and heterogeneity, and are able to form assemblies of high-molecular weight (aggregates), which have probably a great significance in the overall organization of the extracellular matrices (1).

We have concentrated our attention on the structure of proteoglycans extracted from corneal tissue, which, although constituting only a small percentage of the dry weight of the tissue, play an important role in its structural organization (2).

METHODS

Proteoglycans were extracted from fresh bovine corneas for 30 h at room temp. with 10 vol. of 0.05 M Tris-HCl buffer (pH 7.5), containing 4 M guanidine hydrochloride (GuaHCl) and 10 mM EDTA.

For dissociative density gradient centrifugation (4 M GuaHCl), solid CsCl was added directly to the extract to obtain a final density of 1.5 g/ml (3). For associative density gradient centrifugation (0.4 M GuaHCl), the 4 M GuaHCl extract was previously dialyzed overnight against 9 vol. of 0.05 M Tris-HCl buffer (pH 7.5), containing 10 mM EDTA, then

ISBN 0-12-301301-1

solid CsCl was added to a final density of 1.5 g/ml. After density gradient centrifugation for 44 h at 20°C in 5-ml tubes, five 1-ml fractions were collected from each tube with a Pasteur pipette. Densities were determined by weighing in a 50-microliter constriction pipette.

Gel chromatography of the proteoglycans was carried out on a column of Sepharose 2B (0.9 x 50 cm) in 0.05 M acetate buffer containing 0.5 M GuaHCl. Fractions of 1 ml were collected. Fractionation on DEAE Sephadex A-25 was performed at 4°C by stepwise elution with H_2O and the following solutions of NaCl in 0.05 M phosphate buffer at pH 6.9: 0.25 M, 0.50 M, 0.75 M, 1.00 M, 1.25 M, 1.50 M, 1.75 M, 2.00 M, and 3.00 M.

Hexuronate was determined by the carbazole method of Bitter and Muir (4), neutral sugars by the anthrone method (5), and proteins by the method of Lowry *et al.* (6) with serum albumin as standard. Aminoacids and hexosamines were analyzed on a Hitachi-Perkin-Elmer chromatography apparatus after hydrolysis at 100°C for 24 h in 6 N HCl and for 6 h in 4 N HCl, respectively, under nitrogen in sealed tubes.

RESULTS AND DISCUSSION

The results of the different analyses performed on the proteoglycans extracted and purified from bovine cornea are summarized in Table I. Three different types of proteoglycan were found, which, on the basis of their physiochemical properties, have been designated as aggregating proteo-chondroitin sulfate (P-CS), aggregating proteokeratan sulfate (P-KS A) and non-aggregating proteokeratan sulfate (P-KS B). In fact, following CsCl-density gradient centrifugation under associative conditions two peaks were found, one banding at the density of 1.44 g/ml, enriched in glucosamine, the other banding at the density of 1.59 g/ml, enriched in galactosamine and hexuronate. When this last peak was further fractionated by chromatography on DEAE-Sephadex, two types of proteo-glycans were separated, one, eluted at a NaCl concentration of 0.75 M, bearing almost exclusively chondroitin sulfate chains, the other, eluted with 1.25 M NaCl, bearing mostly keratan sulfate chains, as it results from the different ratios reported in Table I. CsCl-density gradient centri-fugation under dissociative conditions yielded only one peak banding at the density of 1.44 g/ml with coincidence of galactosamine, glucosamine, and hexuronate. When the proteoglycans banding at higher density under associative conditions were recentrifuged under dissociative conditions, a shift to the lower density (1.44 g/ml) in their sedimentation pattern was found.

Table I. Proteoglycans of Bovine Cornea

Property	*Non-Aggregating P-KS B*	*Aggregating P-KS A*	*Aggregating P-CS*
Density of banding under associative conditions	*1.44*	*1.59*	*1.59*
Density of banding under dissociative conditions (g/ml)	*1.44*	*1.44*	*1.44*
DEAE-Sephadex (Molarity of NaCl for elution)	*1.25*	*1.25*	*0.75*
Sepharose 2 B (Kav)	*0.955*	*0.842*	*0.674*
Molar ratio:			
GlcUA/GalN	*1.70*	*a*	*1.00*
GlcUA/GlcN	*0.41*	*0.37*	*a*
GalN/GlcN	*0.23*	*0*	*a*
Galactose/GlcN	*1.00*	*1.18*	*0*

[a] *Denominator is 0*

Gel filtration experiments conducted with the fractions purified from the DEAE-Sephadex column confirmed that the three types of proteoglycan have different hydrodynamic volumes as indicated by the different *Kav* values reported in Table I. When the aggregating P-CS and P-KS A were combined so as to obtain the same glucosamine/galactosamine ratio present in the original aggregate (1.15), a complete reaggregation occurred as it was proved by the behavior on Sepharose 2B (*K*av 0.449). The non-aggregating P-KS B showed, under the same conditions, a very low degree of interaction with both proteoglycans.

Amino acid analysis showed that small differences exist between the acid and neutral amino acid content of the three proteoglycans, while significant variations occur in the basic aminoacid content (proteokeratan sulfates P-KS A and B have a much higher content of histidine, lysine, and arginine than proteochondroitin sulfate P-CS). Proteoglycans extracted from cornea seem to be quite different in their physiochemical properties from those extracted from cartilage. The aggregates formed by P-CS and P-KS A under associative conditions have a markedly lower buoyant density and hydrodynamic volume when compared to those found in cartilage tissues. The significance of these differences, which may in part be due to the different chemical composition of the proteoglycan subunits, is not yet understood, but it is probably related to the different organization of the extracellular matrix of cornea *vs.* cartilage.

REFERENCES

1. Muir, H., and Hardingham, T. E., *in* "Biochemistry of Carbohydrates" (W. J. Whelan, ed.), vol. 5, p. 153, Butterworths, London, (1976).
2. Bettelheim, F. A., and Plessy, B., *Biochim. Biophys. Acta 381*, 203 (1975).
3. Hascall, V. C., and Sajdera, S. W., *J. Biol. Chem. 244*, 2384 (1969).
4. Bitter, R., and Muir, H., *Anal. Biochem. 4*, 330 (1962).
5. Trevelyand, W. E., and Harrison, J. S., *Biochem. J. 50*, 298 (1952).
6. Lowry, O. H., Rosebrough, N. J., Farr, A. L., and Randall, R. J., *J. Biol. Chem. 193*, 265 (1951).

Synthesis of Glycopeptides Containing the 2-Acetamido-*N*-(L-aspart-4-oyl)-2-deoxy-β-D-glucopyranosylamine Linkage

Hari G. Garg and Roger W. Jeanloz

Synthetic model compounds containing 2-acetamido-*N*-(L-aspart-4-oyl)-2-deoxy-β-D-glucopyranosylamine linkage which is characteristic of a major class of glycoproteins (1), are of great interest for the study of the mechanism of the biosynthesis of the carbohydrate chain and its attachment to the protein backbone. In a previous report (2), we have described the synthesis of glycopeptides derived from beef ribonuclease B containing amino acid sequence 34-38.

This report includes the synthesis of glycopeptides containing a 2-acetamido-2-deoxy-β-D-glucopyranosyl residue and peptides containing the sequence 33-38 of ribonuclease B. 2-Acetamido-3,4,6-tri-*O*-acetyl-2-deoxy-β-D-glucopyranosylamine (3,4) (I) was condensed with *N*-(benzyloxycarbonyl)-L-nitroarginyl-L-aspartic anhydride (5), obtained by treating *N*-(benzyloxycarbonyl)-L-nitroarginyl-L-aspartic acid (5) with acetic anhydride, to give 2-acetamido-3,4,6-tri-*O*-acetyl-*N*-[*N*-(benzyloxycarbonyl)-L-nitroarginyl-L-aspart-4-oyl]-2-deoxy-β-D-glucopyranosylamine (II). Compound III was coupled in the presence of *N*-ethyl-5-phenylisoxazolium-3'-sulfonate (6) or 2-ethoxy-*N*-ethoxycarbonyl-1,2-dihydroquinoline (7) with *N*-(benzyloxycarbonyl)-L-leucyl-L-threonine methyl ester (VII) (8), *N*-(benzyloxycarbonyl)-L-leucyl-L-threonyl-N^{ε}-tosyl-L-lysine *p*-nitrobenzyl ester (7) (VIII), and *N*-(benzyloxycarbonyl)-L-leucyl-L-threonyl-N^{ε}-tosyl-L-lysyl-L-aspartic *p*-nitrobenzyl 2,4-diester (IX) to give 2-acetamido-3,4,6-tri-*O*-acetyl-*N*-[*N*-(benzyloxycarbonyl)-L-nitroarginyl-L-aspart-1-oyl-(L-leucyl-L-threonine methyl ester)-4-oyl]-2-deoxy-β-D-glucopyranosylamine (III), 2-acetamido-3,4,6-tri-*O*-acetyl-*N*-[*N*-(benzyloxycarbonyl)-L-nitroarginyl-L-aspart-1-oyl-(L-leucyl-L-threonyl-N^{ε}-tosyl-L-lysine *p*-nitrobenzyl ester)-4-oyl]-2-deoxy-β-D-glucopyranosyl-

ISBN 0-12-301301-1

amine (IV), and 2-acetamido-3,4,6-tri-*O*-acetyl-*N*-[*N*-(benzyloxycarbonyl)-L-aspart-nitroarginyl-L-aspart-1-oyl-(L-leucyl-L-threonyl-N^{ε}-tosyl-L-lysyl-L-aspartic *p*-nitrobenzyl 2,4-diester)-4-oyl]-2-deoxy-β-D-glucopyranosylamine (V), respectively. Physical characteristics of these glycopeptides are summarized in Table I.

Table I. Physical Characteristics of the Synthetic Glycopeptide[a]

Compound	*Melting points*[b]	*T.l.c.* R_F *(solvent)*[c]	$[\alpha]_D^{21}$ (°)[d]
II	*175-176° (dec.)*	*0.2 (A)*	*+15.5 (c 1.0 MeOH)*
III	*224-225° (dec.)*	*0.4 (B)*	*-10.4 (c 0.4, MeOH)*
IV	*217-219° (dec.)*	*0.31 (C)*	*-13.2 (c 0.6, MeOH)*
V	*214-215° (dec.)*	*0.3 (C)*	*-14.0 (c 0.22, MeOH)*

[a] *The microanalyses were performed by Dr. W. Manser, Zurich, Switzerland. All compounds reported gave satisfactory elemental analyses*

[b] *Melting points were determined with a Mettler FP-2 apparatus and correspond to "corrected melting points"*

[c] *Solvents (v/v): A (7:3; chloroform-methanol); B (9:1; chloroform-methanol); C (9:1; chloroform--ethanol); the spots were detected by spraying the plates with 20% sulfuric acid and heating at 200°C for a few min*

[d] *Rotations were determined for solutions in 1-dm semimicro-tubes with a Perkin-Elmer No. 141 polarimeter*

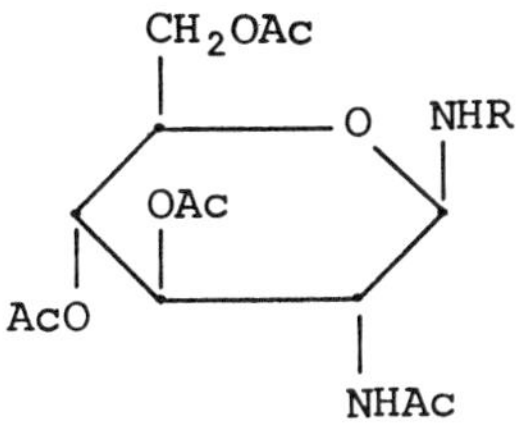

I, R = H

```
                  NO2
                  |     |
II,   R = Cbz-L-Arg-L-Asp-OH
                  NO2
                  |     |
III,  R = Cbz-L-Arg-L-Asp-L-Leu-L-Thr-OMe
                  NO2                     Tos
                  |     |                 |
IV,   R = Cbz-L-Arg-L-Asp-L-Leu-L-Thr-L-Lys-ONBzl
                  NO2                     Tos   ONBzl
                  |     |                 |     |
V,    R = Cbz-L-Arg-L-Asp-L-Leu-L-Thr-L-Lys-L-Asp-ONBzl
              NO2
              |     ┌───┐
VI,   Cbz-L-Arg-L-Asp-O
VII,  Cbz-L-Leu-L-Thr-OMe
                        Tos
                        |
VIII, Cbz-L-Leu-L-Thr-L-Lys-ONBzl
                        Tos   ONBzl
                        |     |
IX,   Cbz-L-Leu-L-Thr-L-Lys-L-Asp-ONBzl
```

NBzl = *p*-$NO_2C_6H_4CH_2$-

ACKNOWLEDGMENT

This work was supported by a grant (AM-03564) from the National Institute of Arthritis, Metabolism and Digestive Diseases, National Institutes of Health.

REFERENCES

1. Marshall, R. D., and Neuberger, A., *in* "Glycoproteins", (A. Gottschalk, ed.), p. 453, 2nd ed., Elsevier, Amsterdam, (1972).
2. Garg, H. G., and Jeanloz, R. W., *Carbohydr. Res. 32,* 37 (1974).
3. Marks, G. S., Marshall, R. D., and Neuberger, A., *Biochem. J. 87,* 274 (1963).
4. Bolton, C. H., and Jeanloz, R. W., *J. Org. Chem. 28,* 3228 (1963).
5. Woodward, R. B., Olofson, R. A., and Mayer, H., *J. Am. Chem. Soc. 83,* 1010 (1961).
6. Izumiya, N., and Makisumi, S., *Nippon Kagaku Zasshi 78,* 1768 (1957); *CA. 54,* 1341 (1960).
7. Belleau, E., and Malek, G., *J. Am. Chem. Soc. 90,* 1651 (1968).
8. Ali, A., and Weinstein, B., *J. Org. Chem. 36,* 3022 (1971).

GLYCOCONJUGATE RESEARCH/Volume I
Proceedings of the Fourth International Symposium on Glycoconjugates

Electron Microscopy of the Extracellular Protein-Polysaccahide from the Red Alga, *Porphyridium cruentum*

Joy Heaney-Kieras and Hewson Swift

The unicellular red alga, *Porphyridium cruentum,* releases a protein-polysaccharide into the medium in liquid, aerated cultures. This high mol. wt. heteropolymer consists of 42% of hexose, 30% of pentose, 8.5% of uronic acid, 9% of sulfate, and 1.5% of amino acids (1). The protein is covalently attached to the polysaccharide by *O*-glycosyl linkages between serine and threonine in the protein, and xylose in the polysaccharide; these linkages can be cleaved in alkali by a β-elimination reaction (2). The protein has a mol. wt. of less than 10 000. In this report, we present an electron-microscopic study of the protein-polysaccharide and correlate the electron-microscopic results with other physical data.

METHODS

The culture of *Porphyridium cruentum* and the isolation of the protein-polysaccharide have been described elsewhere (1). The protein-polysaccharide was separated from the cells by centrifugation. After precipitation as the cetylpyridinium complex, the protein-polysaccharide was purified by a series of steps, which alternated solubilization in salt with precipitation from ethanol. The protein was removed from the polysaccharide by alkali treatment (0.5 N NaOH, 22°C, 24 h), and the polysaccharide was re-isolated by the procedure just described (2).

Electron microscopy was carried out according to Rosenberg *et al.* (3), who applied the technique of Kleinschmidt (4) to bovine nasal proteoglycan. Grids containing the protein-polysaccharide (or the polysaccharide) were rotary shadowed at an angle of 12° with a 4:1 platinum-palladium mixture. A map measurer was used to trace particles on photographic enlargements.

ISBN 0-12-301301-1

Viscometric measurements in an Ubbelohde, low-shear Cannon Semi-micro Dilution viscometer, and calculation of intrinsic viscosities were carried out as described before (2).

RESULTS AND DISCUSSION

Individual molecules of the extracellular protein-polysaccharide were visualized by the Kleinschmidt technique after rotary shadowing. The best results were obtained when the polymer concentration was 4 μg per ml. Most molecules were linear (>90%). Molecules of two contour lengths were observed. The major group had a contour length of *ca.* 50 nm and the minor group (~20%) of *ca.* 1000 nm. The mol. wt. of the protein-polysaccharide may be estimated, if the mol. wt. of a representative disaccharide is assumed to be 500 and the average length of a disaccharide is taken as 0.4-1.0 nm. The more frequent group of molecules would have a mol. wt. of $2.5\text{-}6.3 \times 10^5$, and the less frequent group of $0.5\text{-}1.0 \times 10^6$.

The polysaccharide was examined after removal of the protein by alkali treatment, and the same contour lengths and distribution were found as in the protein-polysaccharide. The intrinsic viscosity of the protein-polysaccharide was 21.0 dl/g in 0.1 M sodium phosphate, pH 7.0, and 20.3 dl/g in 0.5 M guanidinium-Cl, pH 7.0. The values for the polysaccharide in the same solvents were 20.5 dl/g and 20.8 dl/g, respectively. These observations lead us to propose that the structure of the protein-polysaccharide is a long, linear polysaccharide chain whose reducing end is joined to a protein.

REFERENCES

1. Kieras, J.H., Kieras, F.J., and Bowen, D.V., *Biochem. J. 155,* 181 (1976).
2. Heaney-Kieras, J., Rodén, L., and Chapman, D.J., *Biochem. J. 165,* 1 (1977).
3. Rosenberg, L., Hellmann, W., and Kleinschmidt, A.K., *J. Biol. Chem. 245,* 4123 (1970).
4. Kleinschmidt, A.K., *Methods Enzymol. 12B,* 361 (1968).

N-Acetylglucosamine-Containing Oligosaccharides. Synthesis and Methylation Analysis

Elizabeth F. Hounsell, Michael B. Jones, and John A. Wright

Chemical syntheses of disaccharides containing *N*-acetylglucosamine (2-acetamido-2-deoxy-D-glucose) as the reducing sugar gave products useful as models for structural determination of oligosaccharide chains. For the synthesis of the disaccharide 2-acetamido-2-deoxy-4-*O*-β-D-mannopyranosyl-D-glucopyranose [β-D-Man-(1→4)-GlcNAc], condensation under classical Koenigs-Knorr conditions were studied. Reactions were between 2,3,4,6-tetra-*O*-benzyl-D-manno- and D-glucopyranose, having non-participating groups at C-2 (1), with a derivative of 2-acetamido-2-deoxy-D-glucopyranose, having only a free OH-4 group. The preparation of the latter compound gave several intermediates, which were also useful for preparing standard partially methylated alditol acetates for structural analysis by g.l.c.-m.s.

The preparation of 2,3,4,6-tetra-*O*-benzyl-D-glucopyranose was achieved as described by Gigg and Gigg (2). The preparation of 2,3,4,6-tetra-*O*-benzyl-D-mannopyranose followed the same route. This compound has recently been described by Koto *et al.* (3). The reducing sugars were converted *via* their 1-*p*-nitrobenzoates (4) to the halides 2,3,4,6-tetra-*O*-benzyl-D-glucopyranosyl bromide and 2,3,4,6-tetra-*O*-benzyl-D-mannopyranosyl chloride, which were used directly for condensation with benzyl 2-acetamido-3,6-di-*O*-benzyl-2-deoxy-α-D-glucopyranoside. This compound was synthesized as described by Jacquinet *et al.* (5) and as shown in the reaction sequence illustrated in Scheme 1.

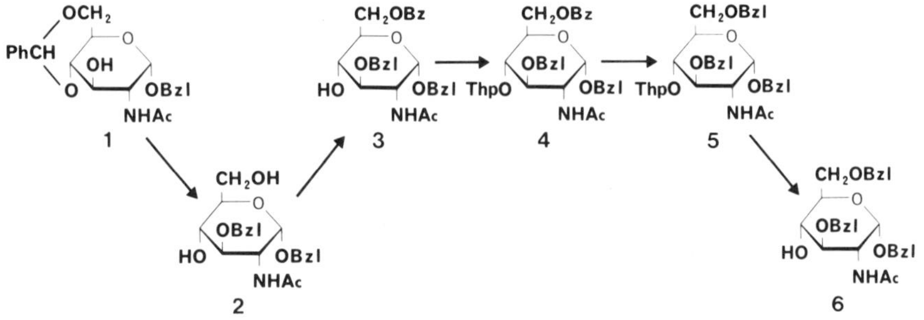

Scheme 1.

Details of the condensation reactions, purification of products, and their identification will be published elsewhere. Two benzylated glucopyranosyl disaccharides [β-Glc-(1→4)-GlcNAc and α-Glc-(1→4)-GlcNAc] and one mannopyranosyl disaccharide were obtained. The data suggest that the latter product was the α anomer, and thus it appears that this facile method does not lead to formation of the disaccharide β-Man-(1→4)-GlcNAc, which has recently been reported by two groups (6,7).

The mannosyl disaccharide obtained was used as a standard for methylation analysis of carbohydrate chains containing (1→4) linkages to 2-acetamido-2-deoxy-D-glucopyranose. Hydrogenation (atmospheric pressure, catalyst platinum on charcoal, solvent acetic acid) gave the free nonbenzylated disaccharide, which was converted to partially *O*-methylated alditol acetates of D-mannopyranose and 2-deoxy-2-(*N*-methylacetamido)-D-glucopyranose by the method of Stellner *et al.* (8). These derivatives were analyzed by combined g.l.c.-m.s. (3% OV-225 on Gas Chrom Q; temperature programme 170°C, 40 min, 2°C/min to 220°C, held at this temperature; carrier gas, helium, 50 ml/min; AEI MS 30 at 70 eV, 500 microamp, separator temperature, 150°C). The alditol acetate obtained from 2-deoxy-2-(*N*-methylacetamido)-3,6-di-*O*-methyl-D-glucopyranose ran with a g.l.c. retention time of 6 relative to the alditol acetate from 2,3,4,6-tetra-*O*-methyl-D-mannopyranose.

The series of all possible partially *O*-methylated derivatives of 2-deoxy-2-(*N*-methylacetamido)-D-glucopyranose were prepared by submitting the intermediates shown in the reaction sequence to methylation, followed by hydrogenation. In this way, compound *1* gave the 3-*O*-methyl and 3,4-di-*O*-methyl derivatives, the latter by treatment with acid to remove the 4,6-di-*O*-benzylidene group and protection of the OH-6 by selective tritylation (5). Compound 2 gave the 4,6-di-*O*-methyl derivative; compound 4 gave the 6-*O*-methyl derivative followed by treatment with acid to remove the Thp group; and compound 6 gave the 4-*O*-methyl derivative.

Compound *2* was also employed in condensation reactions with 2,3,4,6-tetra-*O*-benzyl-D-glucopyranosyl bromide and -mannopyranosyl chloride to give the benzylated trisaccharides Glc-(1→4)-[Glc-(1→6)]-GlcNAc and Man-(1→4)-[Man-(1→6)]-GlcNAc and the partially benzylated disaccharides Glc-(1→6)-GlcNAc and Man-(1→6)-GlcNAc. Due to the far greater reactivity of OH-6 over OH-4 of 2-acetamido-2-deoxy-D-glucopyranose derivatives under the conditions used, these disaccharides had a free OH-4 (detected by acetylation and methylation analysis).

Analysis of the alditol acetates of the compounds described showed that the g.l.c. conditions employed were capable of giving good separation of the mono- and di-*O*-methylated alditol acetates of 2-deoxy-2-(*N*-methylacetamido)-2-deoxy- D-glucose at 220°C. Phthalate compounds (*m/e* 149), which are contaminants from all plastic-ware, were mostly eluted between 170°C and 220°C, and partially methylated neutral sugar alditol acetates mainly at 170°C. The m.s. analysis showed that the amino sugar derivatives partially decomposed under the conditions used, making their definitive identification by m.s. fragmentation patterns difficult. The availability of a series of standards aided the identification of these amino sugar derivatives by the application of g.l.c.

REFERENCES

1. Wulff, G., and Röhle, G., *Ang. Chem. Int. Ed. Eng.* *13*, 157 (1974).
2. Gigg, J., and Gigg, R., *J. Chem. Soc., C*, 82 (1966).
3. Koto, S., Morishima, N., Miyata, Y., and Zen, S., *Bull. Chem. Soc. Jpn. 49*, 2639 (1976).
4. Austin, P.W., Hardy, F.E., Buchanan, J.G., and Baddiley, J., *J. Chem. Soc.*, 2128 (1964).
5. Jacquinet, J.-C., Petit, J.-M., and Sinaÿ, P., *Carbohydr. Res. 38*, 305 (1974).
6. Shaban, M.A.E., and Jeanloz, R.W., *Carbohydr. Res. 52*, *115* (1976).
7. Johnson, G., Lee, R.T., and Lee, Y.C., *Carbohydr. Res.* *39*, 271 (1975).
8. Stellner, K., Saito, H., and Hakomori, S., *Arch. Biochem. Biophys. 155*, 464 (1973).

GLYCOCONJUGATE RESEARCH / Volume I
Proceedings of the Fourth International Symposium on Glycoconjugates

Characteristics of Goblet Cell Mucin of Human Small Intestine

Inderjit Jabbal, David I.C. Kells, Gordon G. Forstner, and Janet F. Forstner

Goblet cell mucin was prepared from human small intestine by homogenization of mucosal scrapings in 5 mM EDTA-NaOH pH 7.0, differential centrifugation, and application of the post-microsomal supernatant solution to Sepharose 4B columns. Over 85% of the hexose applied was recovered in the void-volume peak after elution with 0.1 M K_2HPO_4 - KH_2PO_4, pH 7.0. The void-volume peak was pooled, dialysed, and centrifuged at 30 000*g* for 30 min, and a portion of the supernatant saved for analyses. The rest of the supernatant was further purified by application in 6 M urea to DEAE Bio-Gel (A) and elution with 6 M urea in 0.05 M phosphate buffer, pH 6.0. The first peak contained over 70% of the applied hexose, and was pooled, dialysed, and concentrated for analyses. Subsequent results are given for "supernatant", which is the supernatant solution of the Sepharose 4B void volume peak, and "Peak I", which is the first peak eluted from DEAE Bio-Gel (A) columns.

Analytical ultracentrifugation on cesium chloride density gradient (1) was performed in a Beckman model E ultracentrifuge run for 48 h at 44 000 rpm with 42% CsCl in phosphate buffer (pH 6.8) as solvent. Fig. 1 shows the typical Schlieren optical patterns and the buoyant densities of "supernatant" and "peak I" mucin fractions at equilibrium. A major high density glycoprotein was present in both fractions, whereas in the supernatant fraction there was, in addition, a small amount of a second glycoprotein species of lower density. No protein contaminants were detected in either sample.

ISBN 0-12-301301-1

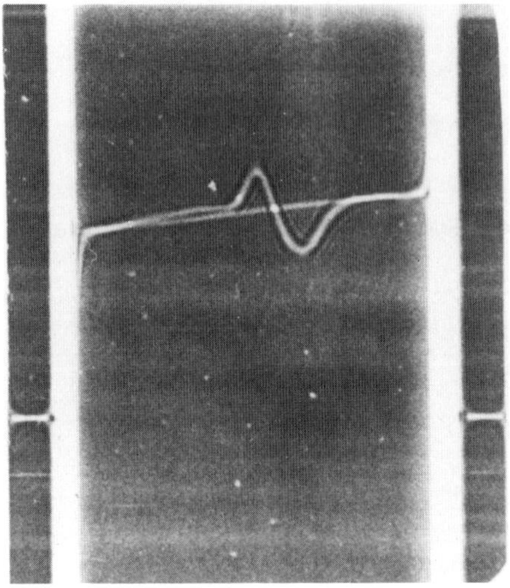

(a)

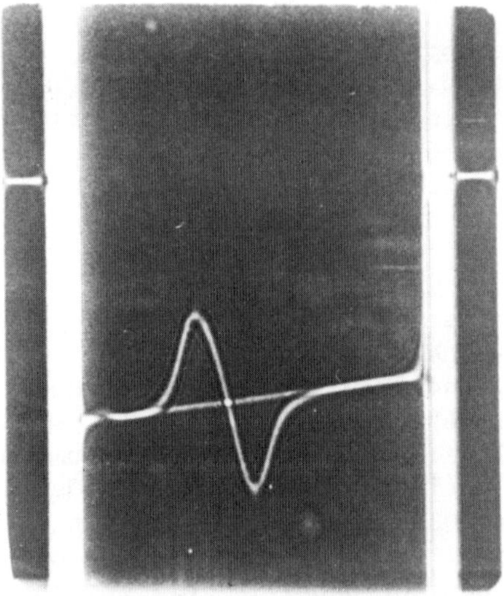

(b)

Fig. 1. Ultracentrifugation on cesium chloride density gradient of human intestinal mucin.(a) Supernatant solution of void volume peak eluted from Sepharose 4B columns. Density of the heaviest species is 1.4760. (b) Peak I of DEAE-Bio-Gel (A) chromatography of (a). Density is 1.4653.

Antibodies were prepared in rabbits against supernatant fractions of one human intestine. Despite measured ABO and Le blood-group differences between mucin samples of three individuals, a reaction of identity was observed for all human supernatant and peak I fractions. Whole human serum and human serum albumin gave no reaction. Thus, the antibody made in rabbits in response to human mucin was not specific for blood-group determinants in the mucin, nor for serum proteins that, theoretically, might be present as minor contaminants in mucin preparations.

Amino acid analyses, performed on supernatant and peak I fractions, were found to have very similar profiles, with serine, threonine, and proline making up about 50% of residues. Less than 1 mol % of cysteine was found in either fraction. Sugar composition was analyzed by g.l.c. and revealed the presence of galactose, *N*-acetylgalactosamine, *N*-acetylglucosamine, fucose, and sialic acid. In supernatant fractions from various patients, differences in Gal or GalNAc content could be attributed to ABO blood-group specificity. Differences in sialic acid content (Table I) are not well explained.

Table I. Composition and Properties of "Supernatant" and "Peak I" Mucin Fractions

Components and properties	Supernatant fractions			Peak I fractions	
	Patient			Patient	
	1	2	3	1	2
Blood-group specificity	A,Le^b	B,Le^b	O,Le^a	A,Le^b	B,Le^b
Galactose/ protein (w/w)	1.69	2.67	0.67	5.54	8.6
Sulfate (% of dry wt.)	1.3	0.8		4.1	3.4
NeuAc (mol%)	5.3	4.4	18.5	0.8	0

In contrast with supernatant fractions, peak I fractions contained more total sugar, more sulfate, and virtually no *N*-acetylneuraminic acid (Table I).

Dissociation of the two glycoprotein species in supernatant fractions was examined, by band ultracentrifugation of PAS-stained samples, with a model E Beckman ultracentrifuge and monitoring light absorption at 555 nm (2). Intact mucin fractions usually gave a major peak (S^o 38-40) and a trailing shoulder. Samples that had been aged at 4°C for 6 months to 2 years, or had been incubated in 6 M guanidine HCl, 42% cesium chloride, or 10 mM dithiothreitol, underwent partial dissociation. The S^o value of the major species did not change, but the trailing shoulder became more prominent. In contrast, peak I samples showed no evidence of dissociation after treatment with 10 mM dithiothreitol (with or without 6 M guanidine HCl) (Fig. 2).

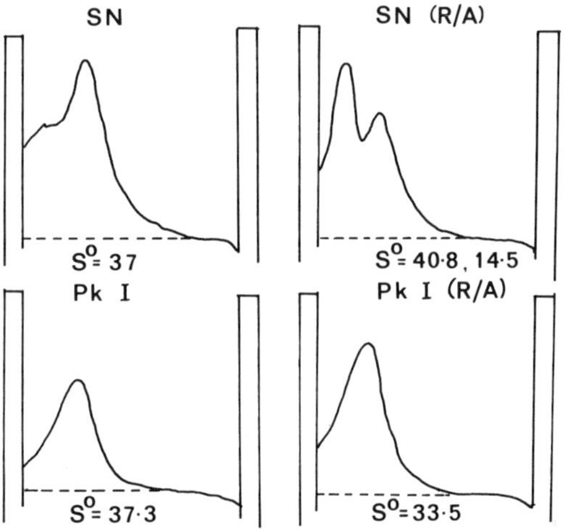

Fig. 2. Effect of reduction and alkylation on band ultracentrifugation of mucin "supernatant" (SN) and "Peak I" (PkI) fractions. Absorption is plotted at 555 nm for 32-min scans. S^o values are sedimentation coefficients. R/A refers to samples that were reduced with 10 mM dithiothreitol in 6 M guanidine HCl, and alkylated with iodoacetamide.

In summary, isolated human goblet cell mucins were immunologically identical, and appeared to contain one major high-density species plus a minor, lighter-density species of glycoprotein. The two species were partially separated on cesium chloride density gradients, column chromatography on DEAE-Bio-Gel, and by band ultracentrifugation. Dissociation of the two species was favored by aging, guanidine, HCl, CsCl, and S-S reducing agents. The major mucin species (peak I) was sulfated, highly glycosylated, poor in sialic acid, and was not dissociated by S-S reducing agents. These findings suggest that the major mucin species is not a polymer comprised of discrete S-S bonded subunits. It is possible, however, that the lighter-density glycoprotein may serve as a noncovalent cross-linking agent within mucin aggregates.

REFERENCES

1. Creeth, J.M., and Denborough, M.A., *Biochem. J. 117*, 879 (1970).
2. Jabbal, I., Forstner, G., and Kells, D.I.C., *Anal. Biochem. 69*, 558 (1975).

GLYCOCONJUGATE RESEARCH / Volume I
Proceedings of the Fourth International Symposium on Glycoconjugates

Fractionation of Sponge Structural-Glycoproteins by Affinity Chromatography on Lectins

Simone Junqua and Ladislas Robert

Structural glycoproteins, one of the four families of macromolecules of the intercellular matrix, were isolated from various vertebrate and invertebrate tissues (for a review, see 1 and 2). As they are associated with collagen or elastin fibers, their separation could be carried out only by relatively harsh procedures, such as by partial hydrolysis with hot trichloroacetic acid or digestion with collagenase to eliminate polymeric collagen.

We wish to report here a procedure enabling the separation of structural glycoproteins from collagen by avoiding such rough treatments. The isolation of structural glycoproteins of *Spongia officinalis* was achieved by use of a dissociating buffer containing 1% SDS in 0.01 M EDTA, pH 8.0 (3). Collagen was eliminated by ultracentrifugation, and the supernatant was applied to a Sepharose 4B column and eluted with a 0.01 M EDTA buffer (pH 8.0) containing 0.1% of SDS. The major peak material, eluted almost at the level of BSA, contained glycoproteins. After reduction and alkylation, it was passed through a Dowex AG 1-X2 column in order to remove SDS.

Affinity chromatography of SDS-free glycoproteins on a ConA-Sepharose column followed by a WGA-Sepharose column resulted in five fractions (Table I). Fraction A, which was not bound to ConA, was separated on a WGA-Sepharose column in two sub-fractions (AI and AII). Fraction B showed affinity for both lectins. Fraction C bound to ConA was further fractionated on a WGA-column into two sub-fractions (CI and CII).

ISBN 0-12-301301-1

Table I. Binding to Lectins and Composition of the Five Glycoprotein Fractions

Fractions	ConA binding	WGA binding	Proteins (%)[a]	Hex (%)[b]	HexN (%)[c]	Man/ GlcNAc ratio[d]
AI	-	-	37	20		
AII	-	+	28	2.4	2.6	0.34
B	+	+	29	3.1	2.6	0.65
CI	+	-	4.6	6.0	1.4	0.86
CII	+	+	1.4	5.5	2.0	0.67

[a] As % of total proteins. Proteins were determined by the Lowry method

[b] As % of total proteins. Hexoses were determined by the orcinol method

[c] As % of total proteins. Hexosamines were determined by the modified Elson-Morgan procedure

[d] The molar ratio of mannose to N-acetylglucosamine was determined by g.l.c

The amino acid compositions of the five fractions thus obtained were quite similar and might even be identical for Fractions AII and B on the one hand, and for Fractions CI and CII on the other hand (Table II).

Table II. Comparison of some Amino Acid Contents of the Five Glycoprotein Fractions Isolated on Lectins

	Res. per 1000 res. Fractions				
	AI	AII	B	CI	CII
OH-Pro	0	0	0	0	0
Dicarboxylic amino acids	235	211	209	182	181
Gly	88	87	83	79	78
Ala	80	80	78	89	84
Ileu + Leu	130	143	147	157	155
Basic amino acids	119	119	118	128	130

The absence of hydroxyproline and hydroxylysine, and the high content of dicarboxylic amino acids and of hydrophobic amino acids (Ala, Val, Ileu, and Leu) are in good agreement with values previously reported for the composition of invertebrate and vertebrate structural glycoproteins (1).

On the contrary, the heterogeneity of the carbohydrate composition of sponge glycoproteins is clearly indicated by the different affinities of the glycoproteins for ConA and WGA, respectively (Table I). It is of interest that Fraction AI, which possesses the highest carbohydrate content, has no affinity for lectins. However, this result is in good agreement with some recent reports (4).

Fraction AII, unbound to ConA, has a mannose content of about half of that of Fraction B, which was bound to ConA. These data could find an explanation in the observations of Ogata *et al.* (5), who suggested that at least two mannosyl residues were required for binding to ConA.

In the same way, Fraction CI appears not to have the minimal number of *N*-acetylglucosamine residues required for binding to WGA. Both Fractions CI and CII contain glucose and were strongly bound to ConA (eluted with borate buffer). The other components of the carbohydrate moiety are common to all 4 fractions bound to lectins: *N*-acetylglucosamine, galactose, mannose, fucose, arabinose, and hexuronic and sialic acids (6).

All the glycoprotein fractions gave a precipitation line with a rabbit antiserum obtained by immunization with Fraction B, and exhibited by tandem immunoelectrophoresis an immunochemical identity (6). Sponge structural-glycoproteins were shown previously to possess immunochemical species-specificity (3).

ACKNOWLEDGMENTS

Supported by CNRS (GR N°40), DGRST (76.7.0980.01) and INSERM (20.75.43). The competent help of Dr. Lemonnier (Laboratoire de Biochimie du Professeur Bourillon) in the g.l.c.-determination is thankfully acknowledged.

REFERENCES

1. Robert, L., Junqua, S., and Moczar, M., *Front. Matrix Biol. 3*, 113 (1976).
2. Anderson, J.C., *Int. Rev. Connect. Tissue Res. 7*, 251 (1977).
3. Jungua, S., Fayolle, J., and Robert, L., *Protides Biol. Fluids 22*, 337 (1975).
4. Krusius, T., *FEBS Lett. 66*, 86 (1976).
5. Ogata, S.L., Maramatsu, T., and Kobata, A., *J. Biochem. 78*, 687 (1975).

Characterization of Glycophosphosphingolipids from Tobacco Leaves

Karan Kaul, Thomas C.-Y. Hsieh, Roger A. Laine, and Robert L. Lester

Inositol-containing glycophosphingolipids have not been reported to exist in animals but seem to be widespread in fungi and higher plants (1,2). A concentrate composed of such lipids was prepared from leaves of *Nicotiana tabacum* (3). Partial chemical structures of the two most abundant components of this concentrate (PSL-I and PSL-II) were reported (3): PSL-I, GlcNAc-(GlcUA, *myo*-inositol)-1-P-ceramide; and PSL-II, GlcN-(GlcUA, *myo*-inositol)-1-P-ceramide. We now present the complete chemical characterization of the trisaccharide moiety of PSL-I.

Sequence and linkages of carbohydrate residues were determined by g.l.c.-m.s. analysis of partially methylated alditol and inositol acetates derived from the carboxyl-reduced trisaccharide. Carboxyl-reduced PSL-I was prepared as reported earlier (3). The trisaccharide was obtained by alkaline hydrolysis of the carboxyl-reduced lipid and alkaline phosphatase treatment of the water-soluble products. The trisaccharide was *N*-acetylated (4) and purified by chromatography on a mixed-bed resin, followed by gel filtration on a Biogel P-2 column. Permethylation of the trisaccharide and subsequent processing was performed as described by Björndal *et al.* (5) and Stellner *et al.* (6). The only carbohydrate-derived products identified by electron impact and chemical ionization (CH_4) g.l.c.-m.s. analysis were:

ISBN 0-12-301301-1

1,5-di-*O*-acetyl-2-deoxy-3,4,6-tri-*O*-methyl-2-(*N*-methylacetamido)glucitol, 1,4,5-tri-*O*-acetyl-2,3,6-tri-*O*-methylglucitol, and mono-*O*-acetyl-penta-*O*-methyl-*myo*-inositol. The presence of these products is consistent with the presence of terminal GlcNAc*p*, 4-linked Glc*p*, and monosubstituted *myo*-inositol residues in the carboxyl-reduced PSL-I trisaccharide. (2,3,6-Tri-*O*-methylglucitol could conceivably arise from a 5-linked furanose form of glucose.)

Earlier experiments (3) had shown that the phosphate ester linkage in PSL-I is at C-1 of *myo*-inositol. In order to determine the position of linkage of the glucuronic acid residue on *myo*-inositol, PSL-I was oxidized with $NaIO_4$. The products were reduced with $NaBD_4$ and were subjected to acid hydrolysis. The dideuterated alcohol phosphate derived from *myo*-inositol was isolated from the hydrolyzate by anion-exchange chromatography. After alkaline phosphatase treatment, the product was acetylated. By g.l.c.-m.s.-chemical ionization (CH_4) analysis, this product was identified as erythritol-d_2, thus indicating that *myo*-inositol in PSL-I is substituted at C-2.

ANALYSIS OF THE ANOMERIC CONFIGURATION

The acetylated PSL-I trisaccharide obtained from the carboxyl-reduced lipid was purified on a Porasil column, dissolved in chloroform-*d*, and analyzed by 80-MHz p.m.r. spectroscopy. Two equal peaks were obtained at τ 3.46 and 3.58 with $J_{1,2}$ < 2 Hz, indicating the presence of two α anomeric protons. Hoffman *et al.* (7) described a procedure for determination of the anomeric configuration of sugar residues in acetylated oligo- and poly-saccharides based on the difference in resistance of α- and β-glycosides to oxidation by CrO_3 in acetic acid. The acetylated trisaccharide was subjected to CrO_3 oxidation under the conditions described by Laine and Renkonen (8), and the remaining carbohydrates were analyzed as their alditol acetates by g.l.c. GlcNAc was also quantitatively determined by colorimetry (9). Both GlcNAc and Glc were found to be resistant to CrO_3 oxidation, indicating that GlcNAc as well as GlcUA residues are in α-anomeric configuration in PSL-I. Based on these results we propose the following structure for PSL-I:

α-D-GlcNAc*p*-(1→4)-α-D-GlcUA*p*-(1→2)-
myo-inosit-1-yl-(1-ceramide) phosphate

Our studies on the more complex components of the glycophosphosphingolipid concentrate of tobacco leaves have shown that these lipids can be fractionated into two groups by anion-exchange chromatography. Lipids of the PSL-I group contain 1 mol of GlcNAc per mol of P while lipids of the PSL-II group contain 1 mol of GlcN per mol of P. PSL-I and PSL-II represent the simplest components of the two respective groups. The more complex lipids of both groups contain arabinose and galactose residues, in addition to the constituents of PSL-I or PSL-II.

REFERENCES

1. Carter, H. E., Johnson, P., and Weber, E. J., *Annu. Rev. Biochem. 34,* 109 (1965).
2. Lester, R. L., Smith, S. W., Wells, G. B., Rees, D. C., and Angus, W. W., *J. Biol. Chem. 249,* 3388 (1974).
3. Kaul, K., and Lester, R. L., *Plant Physiol. 55,* 120 (1975).
4. Roseman, S., and Daffner, I., *Anal. Chem. 28,* 1743 (1956).
5. Björndal, H., Hellerquist, C. G., Lindberg, B., and Svensson, S., *Angew. Chem. Int. Ed. Engl. 9,* 610 (1970).
6. Stellner, K., Saito, H., and Hakomori, S., *Arch. Biochem. Biophys. 155,* 464 (1973).
7. Hoffman, J., Lindberg, B., and Svensson, S., *Acta Chem. Scand. 26,* 661 (1972).
8. Laine, R. A., and Renkonen, O., *J. Lipid Res. 16,* 102 (1975).
9. Gatt, R., and Berman, E. R., *Anal. Biochem. 15,* 167 (1966).

Isolation of Reduced Carbohydrate Fragments from the Linkage-Region of Cartilage Keratan Sulfate

Fred J. Kieras

The glycosaminoglycan keratan sulfate (KS) is covalently attached to the protein component of cartilage proteoglycans by *O*-glycosyl linkages between *N*-acetylgalactosamine (GalNAc) in the polysaccharide and threonine and serine residues in the protein (1). A simple and rapid method for isolation in high yield of linkage-region fragments of reduced keratan sulfate is described. It is based on the observation that reducing sugars bind tightly to Dowex 1 (OH^-), whereas reduced analogs (sugar alcohols) do not (2).

MATERIALS AND METHODS

The methods used for the preparation of keratan sulfate from bovine nasal septum and for analysis of sugar components have been described (3). The material eluted from Dowex 1 (Cl^-) with 2 M NaCl was used and had the following composition (μmol/mg): Hexose 1.38, glucosamine 0.997, galactosamine 0.112, sialic acid 0.20, and amino acids 0.698. β-Elimination and reduction of this material was carried out as previously described (4) in 0.05 N NaOH, 1 M $NaBH_4$, for 23 h at 50°C. After this treatment, 80% of the hexose and 90% of the glucosamine (GlcN) was recovered, 80% of the original galactosamine destroyed, and 79% of the destroyed galactosamine converted into galactosaminitol. This reduced keratan sulfate was used in the experiments described below for the isolation of linkage-region fragments.

Complete hydrolysis of reduced keratan sulfate was carried out in 4 N HCl, 8 h at 100°C, and mild hydrolysis in 0.2 N trifluoroacetic acid, 3 h, at 100° C. After removal of acid by evaporation at 40°C, the hydrolyzates were applied to Dowex 1 (X-8, 200-400 mesh, OH^-) and eluted with H_2O or HCl. A volume

ISBN 0-12-301301-1

of 10 ml of resin was used for 100 mg of starting keratan sulfate. The reduced fragments were analyzed by paper chromatography and by g.l.c.

RESULTS AND DISCUSSION

Reduced keratan sulfate was completely hydrolyzed to monosaccharides, and after removal of acid the hydrolyzate was applied to a Dowex 1 (OH^-) column. The only sugar found in the water eluate was galactosaminitol, which was recovered in 85% yield based on the amount in the reduced keratan sulfate. No hexose, hexosamine, or reducing sugar was found in this fraction, demonstrating that all of the reducing sugar has been retained by the resin.

Larger, reduced fragments were produced by mild acid hydrolysis as just described. When this hydrolyzate was applied to the Dowex 1 column and the column eluted with water, 72% of the original galactosaminitol was eluted, in addition, 6.2% of the hexose, 0.9% of the glucosamine, and 9.5% of the galactosamine were found in this fraction. These sugars were bound as oligosaccharides terminated by a galactosaminitol residue, and they comprised 50% of the total sugars in this fraction. Further elution of the column with 1 N HCl gave an additional 18% of the total galactosaminitol, and also nonreducing hexose and glucosamine. The overall yield of galactosaminitol in both fractions was 90%. These fragments are probably bound to the Dowex 1 resin by ester sulfate groups that are not completely removed under the conditions of mild acid hydrolysis.

The water fraction was analyzed by paper chromatography and g.l.c., and was shown to contain free *N*-acetylgalactosaminitol, the reduced disaccharide galactosyl-(1→3)-*N*-acetylgalactosaminitol, and larger fragments terminated by a galactosaminitol residue, in agreement with earlier work (3).

These experiments clearly demonstrate the utility of this method for the isolation of keratan sulfate linkage-region oligosaccharides. The method has also been applied to proteoglycan subunit of bovine nasal septum, and approximately 70% of the galactosamine attributable to the keratan sulfate linkage-region was isolated as reduced fragments.

ACKNOWLEDMENT

Supported by Grant AM 18165-03 from the National Institutes of Health, U.S.A.

REFERENCES

1. Bray, B.A., Lieberman, R., and Meyer, K., *J. Biol. Chem.* *242,* 3373 (1967).
2. Yamaguchi, H., Inamura, S., and Makino, K., *J. Biochem.* *(Tokyo) 79,* 299 (1976).
3. Kieras, F.J., *J. Biol. Chem.* *249,* 7506 (1974).
4. Kieras, F.J., *Carbohydr. Res.* *41,* 339 (1975).

GLYCOCONJUGATE RESEARCH / Volume I
Proceedings of the Fourth International Symposium on Glycoconjugates

Electron Microscopic Examination of Isolated Proteoglycan Aggregates

James H. Kimura, Philip Osdoby, Arnold I. Caplan, and Vincent C. Hascall

Mesenchymal cells, isolated from stage 23-24 limb buds of the chick embryo, were cultured under conditions that allowed chondrogenesis to occur *in vitro* (1,2). After 8 days, the culture dishes were completely covered by closely spaced cartilaginous nodules (1). At this time, proteoglycan synthesis was occurring at a maximum rate, and the chemical and physical properties of the proteoglycans were characteristic of those isolated from hyaline cartilage (1,2).

Proteoglycans were extracted from day 8 and from day 16 cultures with 4 *M* guanidinium chloride, an effective extracting solvent which dissociates proteoglycan aggregates (3). The extracts were dialyzed against 0.5 *M* guanidinium chloride to reaggregate the proteoglycans, and associative-equilibrium CsCl density gradients were used to obtain proteoglycan aggregate (A1) preparations (1,2). These procedures were also used to prepare A1 fractions from 13 day old chick embryo epiphyses and from bovine nasal cartilage. Additionally, proteoglycans were extracted from day-8 cultures with 0.5 *M* guanidinium chloride, a solvent concentration that does not dissociate aggregates. The dialysis step was omitted, and a direct CsCl associative gradient was used to prepare the proteoglycan aggregate (a-A1) fraction (4). Protease inhibitors described elsewhere (5) were included in all solvents.

Aliquots of each proteoglycan sample were chromatographed on Sepharose 2B; 60-80% of the proteoglycans in each case were excluded and therefore aggregated. Portions of the excluded peaks were spread in cytochrome c monolayers and processed for electron microscopy essentially as described by Rosenberg *et al.* (6) and Kimura *et al.* (7).

ISBN 0-12-301301-1

Figure 1 shows an example of a micrograph of a proteoglycan aggregate, in this case from bovine nasal cartilage. The central filament of hyaluronic acid is visible (arrows), and the monomers extend out from the hyaluronic acid on both sides. In most cases, the polysaccharide side-chains of the monomers are partially condensed along the core protein, but occasionally they are more extended (asterik). The core lengths of the monomers were measured from their points of attachment to hyaluronic acid, as indicated by examples in the figure. The number of monomers per measured lengths of hyaluronic acid were also measured.

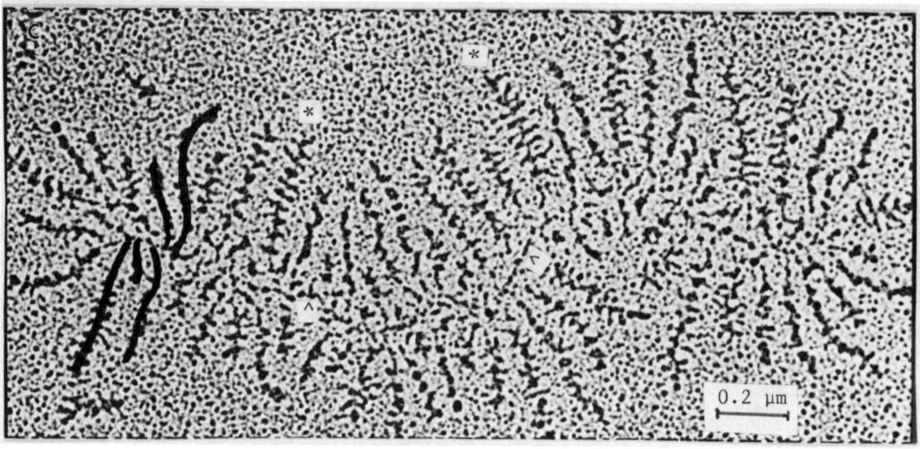

Fig. 1. Micrograph of proteoglycan aggregate from bovine nasal cartilage.

Table I shows the average lengths of monomers in well spread aggregates for each of the proteoglycan samples. In all but the day-16 sample, the length distributions were Gaussian with the indicated standard deviations. Both samples derived from day-8 cultures, and that from the embryonic chick epiphyses contained monomers with statistically identical distributions (*ca.* 310 ± 60 nm). The distribution for the day-16 preparation was similar to the other samples derived from chick tissue, but was skewed slightly toward shorter lengths. The distribution from the bovine nasal cartilage sample had much larger values for the mean and standard deviation (343 ± 95 nm). The values for the average length of hyaluronic acid per monomer are also given in Table I. The value for the A1 sample from day 8 was about 1.5 times that for the a-A1 sample derived from similar cultures.

Table I. Electron Microscopic Measurements of Proteoglycan Aggregates

Source of proteoglycan aggregate	Number of monomers measured	Mean Core-length (nm ± S.D.)	Number of aggregates examined	Mean length of HA/monomer in aggregates (nm ± S.D.)
Bovine nasal cartilage	590	343 ± 95	15	29.3 ± 5.8
Chick embryo epiphyses	549	312 ± 63	18	46.6 ± 7.5
Day-8 cultures extracted with 4 **M** guanidinium chloride	535	313 ± 62	20	47.7 ± 10
Day-8 cultures extracted with 0.5 **M** guanidinium chloride	562	308 ± 58	13	31.7 ± 3.8
Day-16 cultures	530	297 ± 58	9	35.6 ± 3.0

These results suggest the following conclusions: (a) the size distributions of lengths for monomers in aggregates derived from the cultures are very similar to those of monomers derived from chick epiphyses, suggesting that the cultures are synthesizing proteoglycans with core structures similar to those of the proteoglycans present *in vivo;* (b) the size distributions of monomers in the aggregates from bovine nasal cartilage differ significantly from those of the chick preparations, indicating that different cartilages can have proteoglycans with widely different degrees of core-length polydispersity; (c) the size distribution of monomers in aggregates isolated from the day-16 culture was skewed toward sizes smaller than that for the day-8 culture, suggesting that the population of sizes of proteoglycans in the same cartilage matrix may be different at different stages of development or maturation; and (d) the average length of hyaluronic acid per monomer may be greater for aggregates recovered from dissociative extracts than for those isolated without prior disaggregation, suggesting either that reaggregation is not totally efficient or that the dissociative extracts contain additional hyaluronic acid that was not involved in aggregate formation *in situ.*

REFERENCES

1. Hascall, V.C., Oegema, T.R., Brown, M., and Caplan, A.I., *J. Biol. Chem. 251,* 3511 (1976).
2. De Luca, S., Heinegård, D., Hascall, V.C., Kimura, J.H., and Caplan, A.I., *J. Biol. Chem. 252,* 6600 (1977).
3. Hascall, V.C., and Sajdera, S.W., *J. Biol. Chem. 244,* 2384 (1969).
4. Reddi, A.H., Hascall, V.C., and Hascall, G.K., *J. Biol. Chem. 253,* 2429 (1978).
5. Oegema, T.R., Hascall, V.C., and Dziewiatkowski, D.D., *J. Biol. Chem. 250,* 6151 (1975).
6. Rosenberg, L., Hellmann, W., and Kleinschmidt, A.K., *J. Biol. Chem. 250,* 1877 (1975).
7. Kimura, J.H., Osdoby, P., Caplan, A.I., and Hascall, V.C., *J. Biol. Chem. 253,* 4721 (1978).

Differences in Substrate Specificities of *endo*-β-*N*-Acetylglucosaminidases C_{II} and H

Akira Kobata, Katsuko Yamashita, and Tadashi Tai

Endo-β-*N*-acetylglucosaminidases are widely distributed in microorganisms, animals, and plants. Since these enzymes hydrolytically cleave the *N,N'*-diacetylchitobiose moiety of asparagine-linked oligosaccharides of various glycoproteins, releasing the sugar moieties almost intact, they are becoming indispensable tools in the studies of the glycoprotein structure and function. So far, four enzymes as listed in Table I are available in suitably pure form for this purpose.

Table I. Endo-β-N-*acetylglucosaminidases from Microbial Origin*

Name	*Source*
Endo-β-N-*acetylglucosaminidase D*	Diplococcus pneumoniae
Endo-β-N-*acetylglucosaminidase H*	Streptomyces plicatus *and* griseus
Endo-β-N-*acetylglucosaminidase* C_I	Clostridium perfringens
Endo-β-N-*acetylglucosaminidase* C_{II}	Clostridium perfringens

However, exact substrate specificities of these enzymes must be clarified, before their unlimited application for glycoprotein study. Surveys of the action spectra of these enzymes on available glycopeptides indicated that the specificities of *endo*-β-*N*-acetylglucosaminidase D and C_I are exactly the same,

ISBN 0-12-301301-1

and their specificities were elucidated in previous reports (1,2). Glycopeptides having sugar chains with the following structure are all cleaved by these enzymes:

```
                    R                              R'
                    ↓                              ↓
                    6                              6
α-Man-(1→3)-β-Man-(1→4)-β-GlcNAc-(1→4)-GlcNAc-Asn
```

R = α-Man-(1→), or oligosaccharide-α-Man-(1→)
R' = H or α-Fuc-(1→)

C_{II} and H enzymes have quite different action spectra from D and C_I enzymes, acting mainly on high-mannose glycopeptides. During the structural studies of ovalbumin glycopeptides (GP-I, II, III), we found that only a part of the GP-III was cleaved by the C_{II} enzyme, while the H enzyme hydrolyzed completely. In the hope of elucidating the basis of the difference in substrate specificities of these enzymes, the structures of oligosaccharides released from GP-III by C_{II} and H enzymes were studied. The oligosaccharide fraction released from GP-III by C_{II} enzyme was composed of two oligosaccharide components (A and B). Another oligosaccharide (C) was released from the C_{II} enzyme-resistant portion of GP-III by H enzyme digestion. The structures of the three oligosaccharides were elucidated as shown in Scheme 1. From the structural difference of A and C, we can estimate that the α-mannosyl residue located at the C-3 position of the branching β-mannosyl residue, as indicated in Scheme 1 (underlined) is one of the recognition points of the C_{II} enzyme. Substitution of this residue at C-2 by other sugars (as in A and B) does not alter the susceptibility of the glycopeptide to C_{II} enzyme. However, further substitution of this residue at C-4 by another sugar (as in C) completely deprived the glycopeptide of its susceptibility to C_{II} enzyme. H enzyme may not have a strict requirement for the α-mannosyl residue, because it acts on the glycopeptide with C as well as that with A structure. Since C_{II} and H enzymes do not act on the IgG core glycopeptide (Scheme 2), another part of the oligomannosyl moiety of susceptible glycopeptides must also be recognized by these enzymes. GP-VI, which was recently isolated in our laboratory from an exhaustive Pronase digest of ovalbumin, cleared this point. The glycopeptide (Scheme 2) was cleaved by C_{II} enzyme and H enzyme as well as GP-V. Therefore another α-Man-(1→3) residue (Scheme 2, underlined) is essential for a glycopeptide to be susceptible to both C_{II} and H enzymes. That thyroglobulin unit A was readily cleaved by both enzymes indicates that this essential α-mannose residue can be substituted at the C-2 position by

Oligosaccharide A

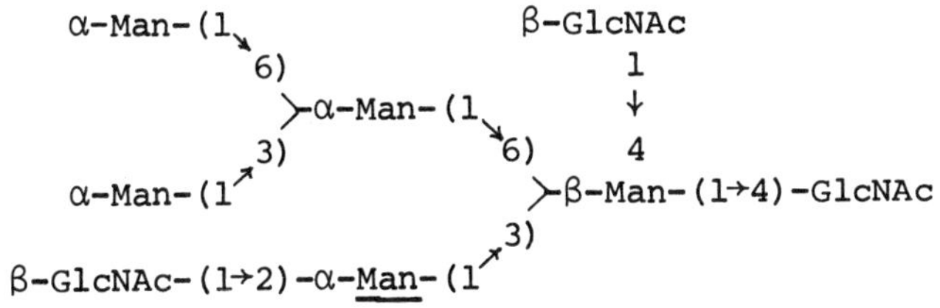

Oligosaccharide B

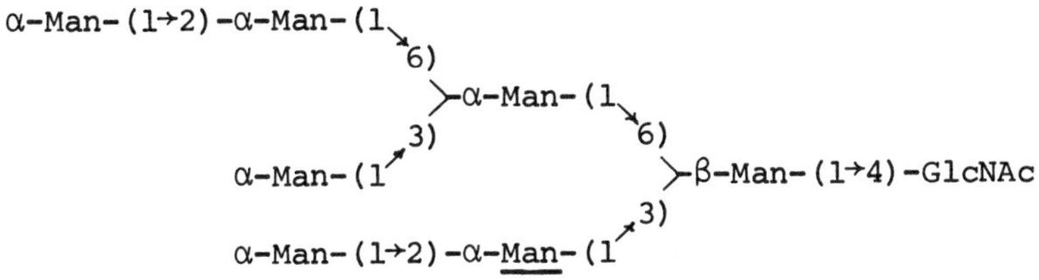

Oligosaccharide C

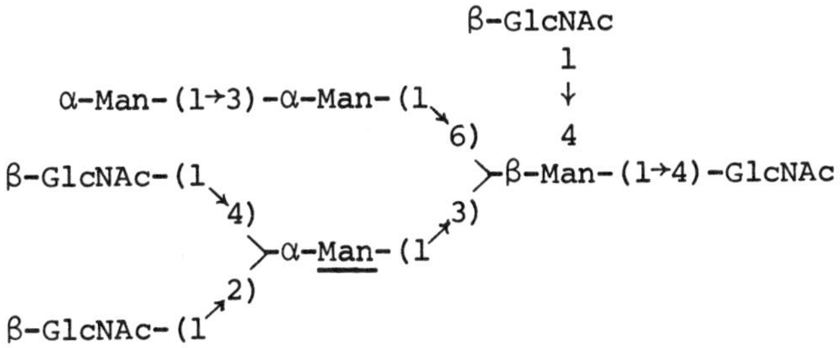

Scheme 1. Structures of three oligosaccharides liberated from ovalbumin GP-III by endo-β-N-*acetylglucosaminidase digestion.*

Glycopeptides		*Endo*- -*N*-acetylglucosaminidase	
Name	Structure	C_{II}	H
Bovine IgG core	α-Man-(1→6) ⟩Man-β-(1→R) α-Man-(1→3)	-	-
Ovalbumin GP-V	α-Man-(1→6) ⟩Man-α-(1→6) α-<u>Man</u>-(1→3) ⟩Man-β-(1→R) α-Man-(1→3)	+	+
Ovalbumin GP-VI	α-<u>Man</u>-(1→3)-α-Man-(1→6) ⟩Man-β-(1→R) α-Man-(1→3)	+	+
Thyroglobulin unit A	α-Man-(1→2)-α-Man-(1→6) ⟩α-Man-(1→6) α-Man-(1→2)-α-<u>Man</u>-(1→3) ⟩Man-β-(1→R) α-Man-(1→2)-α-Man-(1→2)-α-Man-(1→3)	+	+

R = 4GlcNAc-β-(1→4)-GlcNAc→Asn

Scheme 2. Action of endo-β-N-*acetylglucosaminidase on several glycopeptides* other sugars without losing its role. Summarizing all the data presented, it is suggested that the structure requirement of the C_{II} enzyme is (R→2)-α-Man-(1→3)-α-Man-(1→6)-[(R→2)-α-Man-(1→3)]-β-Man-(1→4)-β-GlcNAc-(1→4)-GlcNAc-Asn, while that of the H enzyme is (R→2)-α-Man-(1→3)-α-Man-(1→6)-(R→3)-(R→4)-β-Man-(1→4)-GlcNAc-Asn, in which R represents either hydrogen or sugars.

REFERENCES

1. Tai, T., Yamashita, K., Ogata-Arakawa, M., Koide, N., Muramatsu, T., Iwashita, S., Inoue, Y., and Kobata, A., *J. Biol. Chem. 250,* 8569 (1975).
2. Ito, S., Muramatsu, T., and Kobata, A., *Arch. Biochem. Biophys. 171,* 78 (1975).

GLYCOCONJUGATE RESEARCH / Volume I
Proceedings of the Fourth International Symposium on Glycoconjugates

Studies on the Structure, Distribution, and I Blood-Group Activity of Polyglycosylceramides

Jerzy Kościelak, Ewa Zdebska, and Halina Miller-Podraza

It has been firmly established during the past decade that A, B, and H blood-group specific substances of human erythrocytes are glycosphingolipids (1-4). Other evidence pointed to the presence in erythrocytes also of A, B, H active glycoproteins: for instance Whittemore *et al.* (5) extracted erythrocyte lipids and glycoplipids with aqueous 1-butanol and found that a large portion of total A, B, H blood-group activity of erythrocytes remained insoluble in the reagent. Gardas and Kościelak (6,7), as well as Kościelak *et al.* (8) and Gardas (9), reported that this 1-butanol-insoluble, water-soluble blood-group activity resulted largely from the presence of glycosphingolipids of unusual complexity comprising 30-40 sugar residues per mole. This result has been recently confirmed (M. Dejter-Juszyński, N. Harpaz, H. M. Flowers, and N. Sharon, this volume). These complex glycolipids were designated polyglycosylceramides by us (8). Apart from A, B, H activities, the glycolipids exhibited also a strong I blood-group activity (8,10,11). Careful fractionation of acetylated polyglycosylceramides revealed that the substances constitute a family of compounds having from 22 to 59 sugar residues per mole (8). The predominant molecular species were those with 22 and 30 sugar residues per mole. Fatty acid composition and long-chain base composition were similar to those of "usual" glycosphingolipids. The fractions had type-II chain structure (see 12) with all *N*-acetylglucosamine residues substituted exclusively at C-4 and most galactopyranose residues at C-3, and C-6 (8). The proportion of the latter residues indicates that polyglycosylceramides are highly branched molecules with the number of side-chains increasing in the more complex fractions. The terminal structures were A- and H-active

ISBN 0-12-301301-1

saccharide sequences and, in addition, some β-galactopyranosyl residues. We found that the latter increased proportionally to the molecular complexity of the individual fractions. A, and H *vs.* I activity were inversely proportional, with the former predominating in fractions of low complexity and the latter in fractions of high complexity. The high blood-group I activity of polyglycosylceramides makes them a suitable object for studies aimed at the elucidation of the chemical basis of blood-group I activity. Since our previous work indicated that terminal β-galactopyranosyl groups are not essential for I activity of the glycolipids, we subjected these substances to step-wise Smith degradation, coupled with methylation-m.s. and immunological studies. All the present work was done on two fractions of polyglycosylceramides isolated from blood-group O erythrocytes (Table I).

Table I. Composition of Two Fractions of Polyglycosylceramides from Blood-group O Erythrocytes

Component (%)[a]	*Fraction 1*	*Fraction 2*
Fucose	*10.0 (2.7)*	*9.8 (3.4)*
Galactose	*34.7 (8.7)*	*36.6 (12.6)*
Glucose	*4.0 (1.0)*	*2.9 (1.0)*
N-*Acetylglucosamine*	*48.4 (9.9)*	*47.9 (13.4)*
Sphingosine	*4.0 (0.6)*	*3.2 (0.7)*
Total	*101.1*	*100.4*

[a] *Mol in parentheses. Methods as in ref. 8*

Excellent recoveries of sugars, which approached the theoretical values, were due to the improved method of depolymerization, which was essentially the acetolysis procedure of Yang and Hakomori (13) but performed at 90°C instead of 80°C (see Table II).

Table II. Recovery of Carbohydrates from Polyglycosylceramides as Obtained by Different Methods of Depolymerization

Methods	*Recovery of carbohydrate (%)*[a]
Acetolysis at 80°C[b]	*50–80*
Methanolysis with 1.5 M HCl at 80°C for 20 h	*40–60*
Formolysis and acid hydrolysis	ca. *50*
Acid hydrolysis with 4 M HCl at 100°C for 3 h	ca. *50*
Acetolysis at 90°C	*95–97*

[a] *Weight of all carbohydrates divided by total weight of sample x 100*
[b] *According to Yang and Hakomori (13)*

With this modified technique, we found that polyglycosylceramides contained some excess of *N*-acetylglucosamine over galactose residues. This was in contrast to our previous study in which we found an excess of galactose over *N*-acetylglucosamine residues, but these results had been obtained with the unmodified technique of acetolysis. Presence of large amounts of *N*-acetylglucosamine in the glycolipids as well as their resistance toward hydrolysis suggest the occurrence of GlcNAc→GlcNAc sequences. This was confirmed by Smith degradation of the glycolipids (see Table III). The intact material did not contain any terminal *N*-acetylglucosamine group, but this residue appeared after the first cycle of Smith degradation (3,4,6-*O*-Me_3-GlcNAcMe-ol, Table III). The second cycle of Smith degradation degraded the terminal *N*-acetylglucosamine residue with a proportional decrease of the branching unit of galactose (2,4-*O*-Me_2-Galol in Table III). This finding suggests that some *N*-acetylglucosamine residues are linked directly to galactopyranosyl residues which are branching points. Some new terminal *N*-acetylglucosamine residues were, however, exposed after two cycles of Smith degradation, indicating that GlcNAc-(1→4)-GlcNAc sequences might be present in the glycolipids. The third cycle of Smith degradation did not result in the total destruction of terminal *N*-acetylglucosamine residues, and some 2,4-*O*-Me_2-Galol structures were still present after methylation. We cannot exclude, however, some overoxidation because a slight excess of glucose (2,3,6-*O*-Me_2-Glcol) was found at this stage.

Table III. Methylation Study of Smith Degradation Products of Fraction 2[a]

O-Methyl derivatives[b]	*Intact fraction 2*	*Oxidation cycle*		
		1st	*2nd*	*3rd*
2,3,4-O-Me_3-Fucol	*2.9*	*0*	*0*	*0*
2,3,4,6-O-Me_4-Galol	*3.1*	*0.7*	*0.6*	*+*
3,4,6-O-Me_3-Galol	*1.5*	*0*	*0*	*0*
2,4,6-O-Me_3-Galol	*3.9*	*3.0*	*3.2*	*++*
2,3,6-O-Me_3-Glcol	*1.0*	*1.0*	*1.0*	*+*
2,3,4-O-Me_3-Galol	*0*	*0*	*0*	*0*
2,4-O-Me_2-Galol	*4.4*	*4.2*	*1.4*	*+*
3,4,6-Me_3-GlcNAcMe-ol	*0*	*4.7*	*1.9*	*+*
3,6-O-Me_2-GlcNAcMe-ol	*13.4*	*7.9*	*6.0*	*++*
Total	*30.2*	*21.5*	*14.1*	

[a] *Each cycle of Smith degradation consisted of the following steps: oxidation with 0.1 M sodium periodate for 24 h at room temperature, reduction, second oxidation with sodium periodate for 24 h, reduction, and mild acid hydrolysis. The first two cycles were repeated and essentially identical results obtained with oxidation steps being carried out for 5 days at 4°C, instead of 24 h at room temperature. Methylation procedure and identification of partially methylated alditol and hexosaminitol acetates were carried out as previously reported (8)*

[b] *Analyzed as acetates*

At no stage of the Smith degradation was found any 2,3,4-*O*-Me_3-Galol in the hydrolyzates of permethylated glycolipids. Therefore, all side-chains in the glycolipids are linked to C-6 of the branching galactopyranosyl units. Chromium trioxide degradation of polyglucosylceramides (15) indicated that, with the exception of fucose, all sugars were β linked. Thus, our results indicate the presence in polyglycosylceramides of two types of side chains: β-Gal-(1→4)-β-GlcNAc-(1→6)-(R→3)-Gal, and β-Gal-(1→4)-β-GlcNAc-(1→4)-β-GlcNAc-(1→R).

The main chain is most probably formed of alternating residues of 3-substituted galactose and 4-substituted *N*-acetylglucosamine residues. The presence of di-*N*-acetylchitobiose sequences in the glycolipids was also suggested by the isolation, from partial acid hydrolysates of de-*N*-acetylated glycolipids, of an oligosaccharide with the chromatographic mobility on paper identical to that of an authentic sample of di-*N*-acetylchitobiose. This has been demonstrated in three different solvent systems. Immunodiffusion studies were carried out with 4 different anti-I sera, which had been described by Dzierzkowa-Borodej *et al.* (15) (see Fig. 1). The reactivity with three of them persisted throughout two cycles of Smith degradation, whereas one of them (serum Woj) reacted only with the intact material. Since the latter antiserum reacts better with O_h than O_H erythrocytes (15), it is directed against β-Gal-(1→4)-GlcNAc terminal groups of the glycolipids, and thus resemble in specificity anti-Type XIV pneumococcus serum (see 12). Other anti-I antibodies reacted most probably with *N*-acetylglucosamine-rich branching-regions of the glycolipids.

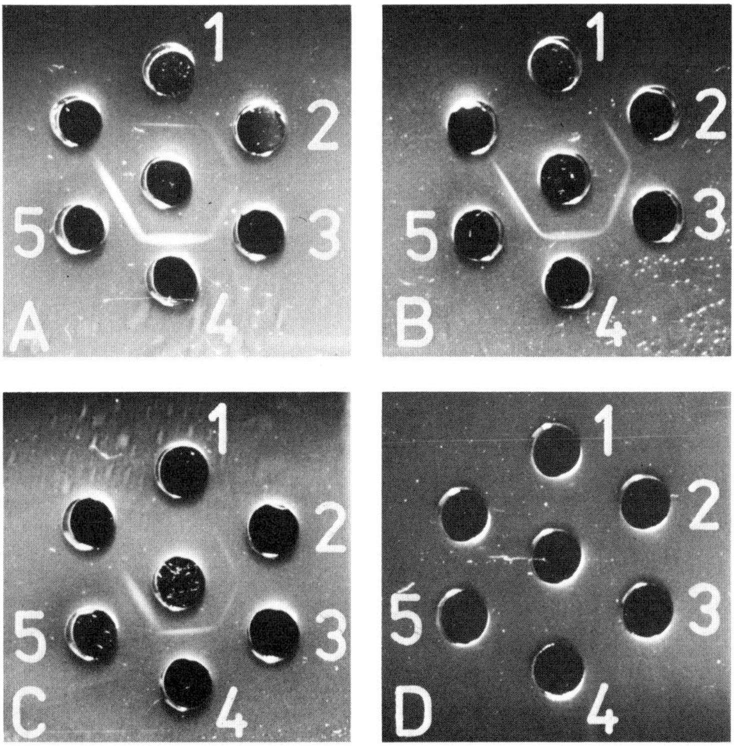

Fig. 1. Immunoprecipitation patterns of intact and degraded polyglycosylceramides with anti I sera: "Woj" (1); Zg (2,3); Pyt (5); Gaj (6,15). Center wells filled with intact glycolipids (A) and with those subjected to one, two, and three cycles of Smith degradation, respectively, (B, C, D).

Our preliminary studies on the occurrence of polyglycosylceramides in other cells and tissues indicate that they may be present in hog intestinal mucosa. Indeed, similar complex glycosphingolipids were found in hog gastric mucosa by Slomiany and Slomiany (17). We have not found polyglycosylceramides in bovine brain.

ACKNOWLEDGMENT

This work was supported by NIH PLO Grant No. 05-043-1 and Polish Government Grant No. 239.

REFERENCES

1. Hakomori, S., Stellner, K., and Watanabe, K., *Biochem. Biophys. Res. Commun.* *49*, 1061 (1972).
2. Kościelak, J., Piasek, A., Górniak, H., Gardas, A., and Gregor, A., *Eur. J. Biochem.* *37*, 214 (1973).
3. Stellner, K., Watanabe, K., and Hakomori, S., *Biochemistry* *12*, 656 (1973).
4. Ando, S., and Yamakawa, T., *J. Biochemistry (Tokyo)* *73*, 387 (1973).
5. Whittemore, M. B., Trabold, M. C., Reed, C. F., and Weed, R. J., *Vox Sang.* *17*, 289 (1969).
6. Gardas, A., and Kościelak, J., *Eur. J. Biochem.* *32*, 178 (1973).
7. Gardas, A., and Kościelak, J., *FEBS Lett.* *42*, 101 (1974).
8. Kościelak, J., Miller-Podraza, H., Krauze, R., and Piasek, A., *Eur. J. Biochem.* *71*, 9 (1976).
9. Gardas, A., *Eur. J. Biochem.* *68*, 177 (1976).
10. Gardas, A., and Kościelak, J., *Vox Sang.* *26*, 227 (1974).
11. Gardas, A., *Eur. J. Biochem.* *68*, 185 (1976).
12. Watkins, W. M., *in* "Glycoproteins, their Composition Structure and Function" (A. Gottschalk, ed.), p. 830, Elsevier, Amsterdam, New York, (1972).
13. Yang, H-J., and Hakomori, S., *J. Biol. Chem.* *246*, 1192 (1971).
14. Laine, R. A., and Renkonen, O., *J. Lipid. Res.* *16*, 102 (1975).
15. Dzierzkowa-Borodej, W., Seyfried, H., and Lisowska, E., *Vox Sang.* *28*, 110 (1975).
16. Slomiany, B. L., and Slomiany, A., *FEBS Lett.* *73*, 175 (1977).

GLYCOCONJUGATE RESEARCH / Volume I
Proceedings of the Fourth International Symposium on Glycoconjugates

A Simple Method for Preparation of Polyacrylamide Gels or Polymers Containing Thioglycoside Ligands

Yuan Chuan Lee, Stephanie Cascio, and Reiko T. Lee

The role of carbohydrate in glycoconjugates can be studied with synthetic glycoside ligands attached to proteins (1) or insoluble matrices (2,3). Of the many available insoluble matrices suitable for attaching carbohydrate ligands, the polyacrylamide gel system proved to be most versatile. In the past, we have developed a system of polyacrylamide gel that contains an active ester group for incorporation of ω-aminoalkyl glycosides (4,5). An alternative method is to acryloylate ω-aminoalkyl glycosides, and to use the resulting acrylamido derivative as a comonomer in the acrylamide polymerization mixture (6). Although both glycosides and thioglycosides have been used in either system, thioglycosides are generally preferable, because of their higher resistance to enzymatic hydrolysis and their ease of cleavage with mercuric ion under mild conditions (7).

We have now developed a simple method to prepare thioglycosides containing acrylamido groups at the terminal position of aglycon, which can be readily polymerized with acrylamide to form a gel. The following scheme outlines the reactions involved:

$$(Ac)_n\text{-glycosyl-SH} + CH_2{=}CHCONH(CH_2)_nNHCOCH{=}CH_2$$
$$\rightarrow (Ac)_n\text{-glycosyl-S-}(CH_2)_2CONH(CH_2)_nNHCOCH{=}CH_2$$
$$\rightarrow (H)_n\text{-glycosyl-S-}(CH_2)_2CONH(CH_2)_nNHCOCH{=}CH_2$$

ISBN 0-12-301301-1

An illustrative procedure is described below: 2,3,4,6-tetra-*O*-acetyl-1-thio-D-galactose was prepared from thiopseudourea derivative by a slight modification of the procedure described by Černý *et al.* (8). A syrupy thiogalactose preparation (obtained from 10 mmol of thiopseudourea derivative) was dissolved in 10 ml of 50% ethanol and mixed with *N,N'*-methylenebis-(acrylamide) (2.08 g, 20 mmol) dissolved in 50 ml of 50% ethanol. To this mixture was added 1 ml of pyridine and 6 mg of hydroquinone and the mixture was kept overnight at room temperature. The reaction mixture was concentrated by evaporation to about 1/2 of the original volume and then shaken with a mixture of 75 ml each of water and chloroform. The water phase was extracted once with 50 ml of chloroform, and the combined chloroform layers were extracted with water (4 x 50 ml) to remove excess *N,N'*-methylenebis(acrylamide). The chloroform layer was dried (Na_2SO_4), filtered, and evaporated. The resulting syrup was deacetylated in 20 ml of 0.01 M sodium methoxide in methanol for 2 h at room temperature. The desired product, 2-(acrylamidomethylaminocarbonyl)ethyl 1-thio-β-D-galactopyranoside crystallized upon storage of the reaction mixture in the cold overnight (yield 2.3 g, 6.54 mmol.). Addition of benzene and petroleum ether (b.p. 35-60) to the filtrate produced additional crystalline product (0.14 g, 0.4 mmol).

The usefulness of these compounds is demonstrated by the following experiment. Soluble, linear copolymers of acrylamide and 2-(acrylamidomethylaminocarbonyl)ethyl 1-thio-β-D-galactopyranoside (1:1 molar ratio) were trapped in 12% acrylamide gel with 10% cross linking, essentially as described by Sutoh *et al.* (9). The gel was ground, defined, packed in a column, and washed with 0.15 M NaCl. Precolumn lectin (9) was applied to the column, which was eluted with 0.15 M NaCl, and then with 0.05 M galactose in 0.15 M NaCl. Elution with 0.15 M NaCl produced a large protein peak with no hemagglutination activity, and 0.05 M galactose a small, sharp protein peak, which after dialysis showed a single band in gel electrophoresis and the same specific activity as the peanut lectin purified by another method (9).

REFERENCES

1. Krantz, M.J., Holtzman, N.A., Stowell, C.P., and Lee, Y.C., *Biochemistry 15,* 3963 (1976).
2. Chipowsky, S., Lee, Y.C., and Roseman, S., *Proc. Nat. Acad. Sci. (USA) 70,* 2309 (1973).
3. Weigel, P., Schnaar, R., and Kuhlenschmidt, M., *Fed. Proc. 28,* 2014 (1977).
4. Schnaar, R., and Lee, Y.C., *Biochemistry 14,* 1535 (1975).
5. Schnaar, R., Weigel, P., Lee, Y.C., and Roseman, S., *Methods Carbohydr. Chem.*, in press.

6. Weigel, P., Schnaar, R., Lee, Y.C., and Roseman, S., *Methods Carbohydr. Chem.*, in press.
7. Krantz, M.J., and Lee, Y.C., *Anal. Biochem. 71,* 318 (1976).
8. Černý, M., Staněk, J., and Pacák, J., *Monatsh. 94,* 290 (1963).
9. Sutoh, K., Rosenfeld, L., and Lee, Y.C., *Anal. Biochem. 79,* 329 (1977).

GLYCOCONJUGATE RESEARCH/Volume I
Proceedings of the Fourth International Symposium on Glycoconjugates

Glycosphingolipids in Chicken Egg Yolk

Yu-Teh Li, Chin Chin Wan, Jow-Long Chien, and Su-Chen Li

It has been shown that the glycosphingolipid composition in animal cells reflects tissue and species specificity. Avian eggs are specialized cells which differ considerably from other types of cells. We found that fresh, unfertilized chicken egg-yolk contained sialosylgalactosylceramide, monosialosyllactosyl-ceramide, and disialosyllactosylceramide. Besides gangliosides, it also contained galactosylceramide as the major neutral glyco-sphingolipid. These glycosphingolipids were isolated and purified according to Scheme 1.

The yield of these glycosphingolipids from one dozen of egg yolks was 22 mg of sialosylgalactosylceramide, 25 mg of mono-sialosyllactosylceramide, 4 mg of disialosyllactosylceramide, and 15 mg of galactosylceramide. Fig. 1 shows the purified gangliosides obtained from egg yolk.

Sequential enzymic hydrolysis and permethylation analysis established their structures to be α-NeuAc-(2→3)-β-Gal(1→1)-Cer for monosialosylgalactosylceramide, α-NeuAc-(2→3)-β-Gal-(1→4)-β-Glc-(1→1)-Cer for monosialosyllactosylceramide, and NeuAc-α-NeuAc-(2→3)-β-Gal-(1→4)-β-Glc-(1→1)-Cer for disialosyllactosyl-ceramide. Table I summarizes the fatty acid and sphingosine composition of the egg-yolk glycosphingolipids.

It is most intriguing to find in egg yolk those glycolipids associated with myelin (monosialosylgalactosylceramide and galactosylceramide), retina (disialosyllactosylceramide), and extraneural tissues (sialosyllactosylceramide). Under the same condition, no glycosphingolipid was detected in the egg white, vitelline membrane, and egg shell membrane. The unique glyco-sphingolipid pattern in egg yolk may have a special physiolog-ical function. We found that disialosyllactosylceramide rapid-ly disappeared from the egg yolk during the ontogeny of the chick embryo.

ISBN 0-12-301301-1

It should not be overlooked that the easy availability of chicken eggs makes them a convenient source for the above mentioned glycosphingolipids.

Egg-yolk acetone powder
↓
Extraction with tetrahydrofuran - 0.01 M KCl (4:1)
↓
add 0.6 N NaOH in methanol
↓
add one-half volume of acetone
↙ Filtrate (discarded) ↓ Precipitate
↓
wash and extract with C:M (1:1), C:M (1:2)
↓
Dialysis
↓
DEAE-cellulose acetate column

0 NaOAc	0.01 N NaOAc	0.04 N NaOAc
↓ Neutral glycolipid	↓	↓
↓ Bio-Sil A column		
↓ Preparative t.l.c.	Preparative t.l.c.	Preparative t.l.c.
↓ Purified glycosphingolipids (galactosyl ceramide)	↓ Purified G7 and GM3	↓ Purified GD3

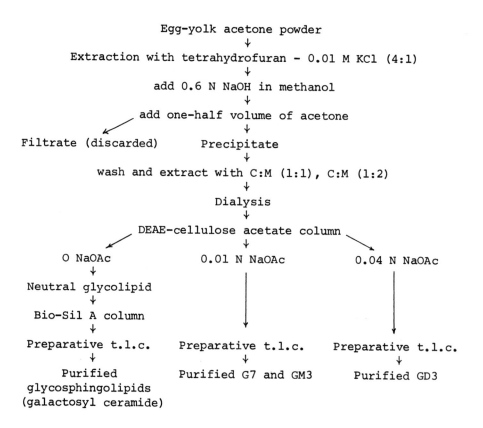

Scheme 1. Procedure for Isolation of Egg-yolk Ganglioside.

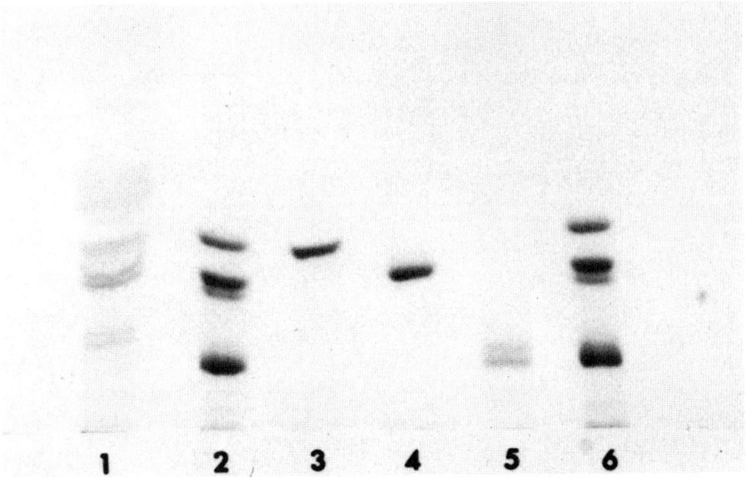

Fig. 1. T.l.c. in chloroform-methanol-0.25% KCl (60:32:7) of gangliosides from egg yolk: (1) total mixture; (2) standards G7, GM3, and GD3; (3) egg G7; (4) egg GM3; (5) egg GD3; and (6) standards G7, GM3, and GD3.

Table I. Fatty Acid and Sphingosine Composition of Egg Yolk Glycosphingolipids

Component	Gal-Cer	G7	GM3	GD3
Fatty acids				
C 16:0	43.2	96.7	45.8	55.7
C 18:0	14.6	3.3	12.0	35.4
C 18:1	30.6		36.4	8.9
C 18:2	11.6		5.8	
Sphingosine (%)				
d 18:1	82.2	84.8	57.5	73.2
d 18:0	11.5	9.9	14.6	15.6
t 18:0	7.3	5.3	28.9	11.2

ACKNOWLEDGMENT

This work was supported by Grants NS 09626 and RR 00164 from the National Institute of Health and Grant PCM 76-16881 from the National Science Foundation.

Structure of the Carbohydrate Unit of Soybean Agglutinin

Halina Lis and Nathan Sharon

Digestion of soybean agglutinin with purified Pronase, followed by gel filtration on Sephadex G-50 and ion-exchange chromatography on a long column of Dowex 50 X-2, afforded the carbohydrate moiety of the glycoprotein as $Man_9GlcNAc_2Asn$ (I), in an apparently homogenous form and in high yield. Estimation of mol. wt. by ultracentrifugation and gel filtration on Sephadex G-50 gave values very close to the calculated one (1990), but different from the value originally reported by us (1). Treatment of I with *endo*-β-*N*-acetylglucosaminidase H cleaved it into two fragments, GlcNAc-Asn and $Man_9GlcNAc$ (II), the latter having an *N*-acetylglucosamine residue at the reducing end. Sequential sodium borohydride reduction-periodate oxidation-reduction of II, followed by acid hydrolysis gave xylosaminitol as the only amino sugar alcohol. Acetolysis of I yielded mannobiose, mannotriose, and $Man_{\sim 5}GlcNAc_2Asn$ (III) in nearly equimolar proportions, in addition to small amounts of mannose. The mannobiose and -triose were rapidly hydrolyzed to mannose by α-mannosidase. The mannose residues of III were released by incubation with α-mannosidase followed by β-mannosidase. Reduction of II with NaB^3H_4, followed by acetolysis and exhaustive digestion with α-mannosidase, afforded a disaccharide identified as β-MaN-(1→4)-^{3}H-GlcNAcol.

Methylation studies in conjunction with g.l.c.-m.s. revealed the presence of the 2,3,4,6-tetramethyl, 3,4,6-trimethyl, and 2,4-dimethyl ethers of mannose and the 3,6-dimethyl ether of *N*-acetylglucosamine in hydrolyzates of permethylated I. Digestion of I with purified jack bean α-mannosidase resulted in a rapid (5-7 h) release of up to 6 mannose residues; methylation studies of enzymically degraded I revealed the presence of the 2,3,4,6-tetramethyl, 2,4,6-trimethyl, 2,3,4-trimethyl, and 2,4-dimethyl ethers of mannose, and the complete absence of the

ISBN 0-12-301301-1

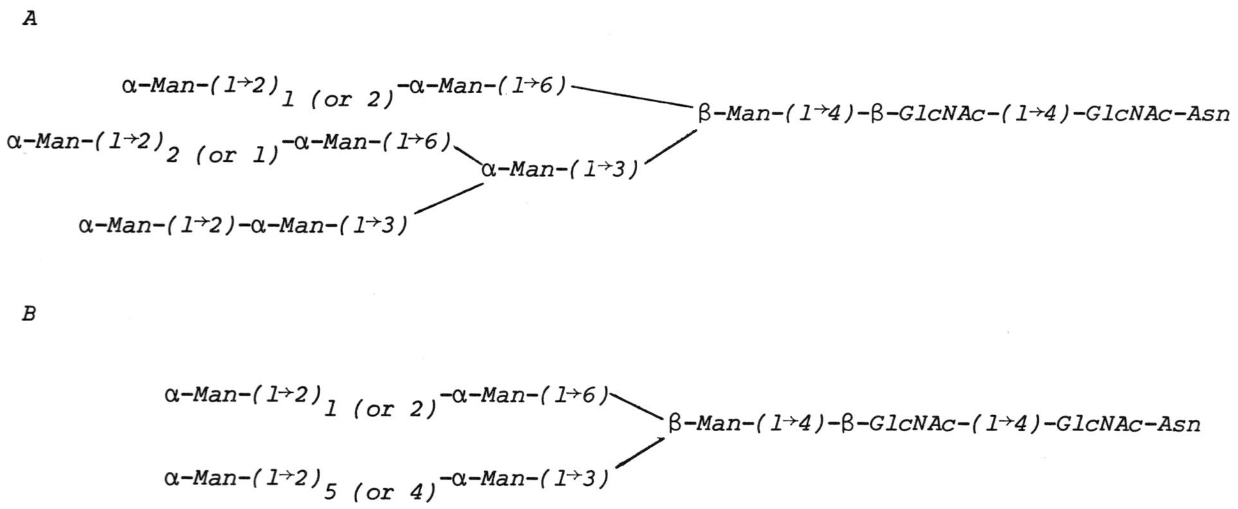
A
α-Man-(1→2)1 (or 2)-α-Man-(1→6)
β-Man-(1→4)-β-GlcNAc-(1→4)-GlcNAc-Asn
α-Man-(1→2)2 (or 1)-α-Man-(1→6)
α-Man-(1→3)
α-Man-(1→2)-α-Man-(1→3)
B
α-Man-(1→2)1 (or 2)-α-Man-(1→6)
β-Man-(1→4)-β-GlcNAc-(1→4)-GlcNAc-Asn
α-Man-(1→2)5 (or 4)-α-Man-(1→3)

Scheme 1.

3,4,6-trimethyl ether. On the basis of these and other results, we propose that I is a mixture of Asn-oligosaccharides having the structures A and B (see Scheme 1).

This is the first demonstration of the presence, in a plant glycoprotein, of the branched core α-Man-(1→3)-[α-Man-(1→6)]-β-Man-(1→4)-β-GlcNAc-(1→4)-GlcNAc, previously found in many animal glycoproteins as well as in those from fungi and yeasts.

REFERENCE

1. Lis, H., Sharon, N., and Katchalski, E., *J. Biol. Chem.* *241*, 684 (1966).

Soluble Proteoglycans and Glycoproteins of Brain

Richard U. Margolis, Renée K. Margolis, Wei-Lai Kiang, and Christine P. Crockett

We have previously reported that the sulfated glycosaminoglycans of brain (chondroitin sulfate and heparan sulfate) are present as proteoglycans, in which the polysaccharide chains are glycosidically linked to the hydroxyl group of serine residues in the protein moiety (1). A recent preliminary report outlined the isolation and properties of a soluble chondroitin sulfate proteoglycan from rat brain (2), and we present here further data concerning the composition of this brain proteoglycan, which appears to be a cytoplasmic constituent of neurons and glia.

Homogenization of rat brains with 5 mM phosphate-buffered saline (pH 7.2) followed by high-speed centrifugation (2 h, 140 000*g*) solubilized 40% of the total glycosaminoglycans, while most of the "particulate" hyaluronic acid and chondroitin sulfate was closely associated with certain low-density microsomal membrane subfractions, from which it can be easily dissociated by various types of mild washing procedures (3,4). A proteoglycan accounting for 97% of the chondroitin sulfate present in the original high-speed supernatant solution (2-3% of the soluble brain protein and 50% of the total brain chondroitin sulfate) was then isolated by ion-exchange chromatography on DEAE-cellulose (2). The resulting crude proteoglycan was reduced and alkylated (10 mM DTT followed by 40 mM iodoacetamide in 4 M guanidine,

ISBN 0-12-301301-1

50 mM Tris buffer, pH 8.5), dialyzed into 0.2 M acetate buffer (pH 5.6), and separated from lower-molecular size proteins and nucleic acid by gel filtration on Sepharose CL-6B. The purified proteoglycan is composed of 75% protein, 13% glycoprotein carbohydrate, and 12% glycosaminoglycans, consisting of 83% of chondroitin 4-sulfate, 5% of chondroitin 6-sulfate, 6% of heparan sulfate, and 6% of hyaluronic acid. The glycoprotein oligosaccharides contain glucosamine, galactosamine, mannose, fucose, galactose, and sialic acid, in the molar ratio of 1.00:0.20:0.68:0.29:0.93:0.60, as well as ester sulfate residues. The major amino acids (in descending order of concentration) are glutamic acid, serine, aspartic acid, glycine, and leucine, which together account for over half of the total amino acids. *S*-Carboxymethylcysteine (2.1 residues/100) was also found.

Because of its high protein content most of the proteoglycan remained at the top of a cesium chloride density-gradient run under dissociative conditions (4 M guanidinium chloride), although as expected the chondroitin sulfate polysaccharide chains sedimented to the bottom fractions after release by alkali from the protein moiety (2). Owing to the presence of mannose-containing glycoprotein oligosaccharides, the proteoglycan was also largely adsorbed on a column of Concanavalin A-Sepharose, from which it could be eluted with methyl α-D-glucopyranoside. Sulfate-labeled proteoglycan did not enter 4% SDS-polyacrylamide gels, and no other proteins staining with Coomassie Blue were observed to migrate in these gels. Proteoglycan labeled in its hexosamine and sialic acid residues emerged as a single peak of protein and radioactivity after gel filtration on Sepharose CL-2B in 4 M guanidinium chloride, and it was not retarded on Sepharose CL-6B after removal of the chondroitin sulfate polysaccharide chains by chondroitinase ABC. In the analytical ultracentrifuge, the proteoglycan demonstrated a single sharp peak with a sedimentation coefficient of 6.1 S. In view of our inability to separate the various components of this macromolecular complex on the basis of either size or density under a wide variety of conditions, we suggest that the soluble proteoglycans of brain may consist of sulfated glycoprotein oligosaccharides and 2-3 different types of glycosaminoglycan polysaccharide chains, all of which are covalently linked to a common protein core.

In a recent report, Branford White and Hudson (5) have described a procedure for the isolation of a proteoglycan (claimed to be composed exclusively of chondroitin sulfate) by dissociative extraction of a lipid-free protein residue of brain. Since yields and other relevant data were not included in the publication, we have tested this procedure and found that the fraction reported to contain only chondroitin sulfate is in fact largely composed of hyaluronic acid, and that the purified product accounts for only 2.7% of the chondroitin sulfate present in brain (6). It therefore does not appear from our experience that dissociative extraction of a brain protein residue under these conditions yields either a significant proportion or a representative sample of proteoglycans.

REFERENCES

1. Margolis, R. U., Margolis, R. K., and Atherton, D. M., *J. Neurochem. 19,* 2317 (1972).
2. Margolis, R. U., Lally, K., Kiang, W.-L., Crockett, C., and Margolis, R. K., *Biochem. Biophys. Res. Commun. 73,* 1018 (1976).
3. Margolis, R. K., Margolis, R. U., Preti, C., and Lai, D., *Biochemistry 14,* 4797 (1975).
4. Kiang, W. -L., Crockett, C. P., Margolis, R. K., and Margolis, R. U., *Biochemistry 17,* 3841 (1978).
5. Branford White, C. J., and Hudson, M., *J. Neurochem. 28,* 581 (1977).
6. Margolis, R. K., Crockett, C. P., and Margolis, R. U., *J. Neurochem. 30,* 1177 (1978).

GLYCOCONJUGATE RESEARCH / Volume I
Proceedings of the Fourth International Symposium on Glycoconjugates

Association of a Major Tumor Glycoprotein, Epiglycanin, with Glycosaminoglycan

Douglas K. Miller and Amiel G. Cooper

The allogeneically transplantable TA3-Ha ascites cell of murine mammary adenocarcinoma has been shown to contain on its surface a high mol. wt. (>500 000) glycoprotein, termed epiglycanin, not found on the syngeneically transplantable TA3-St subline (1). Epiglycanin is thought to play an important role in aiding the allogeneic transplantation of the TA3-Ha subline by masking H-2 histocompatibility antigens (1).

Epiglycanin has been found to be released *in vivo* into ascites fluid and serum (2) and *in vitro* into culture medium (3). It can be detected in submicrogram amounts by a sensitive, highly specific hemagglutination-inhibition assay using a lectin from *Vicia graminea* seeds (4), which binds to Gal→GalNAc→Ser(Thr) moieties (5). Fractionation on Sepharose 4B of culture media from TA3-Ha cells labeled for 24 h with glucosamine indicated the presence of previously characterized, soluble ascites epiglycanin (1,3) (Fig. 1A, Peak B) as well as another highly labeled void-volume component (Peak A) that contains about half as much *Vicia graminea*-inhibitory glycoprotein as Peak B but has 3.5 times as much Lowry protein. The TA3-St cell, in contrast, released into the medium neither *Vicia*-inhibitory glycoproteins nor glucosamine-labeled Peaks A or B (Fig. 1B).

ISBN 0-12-301301-1

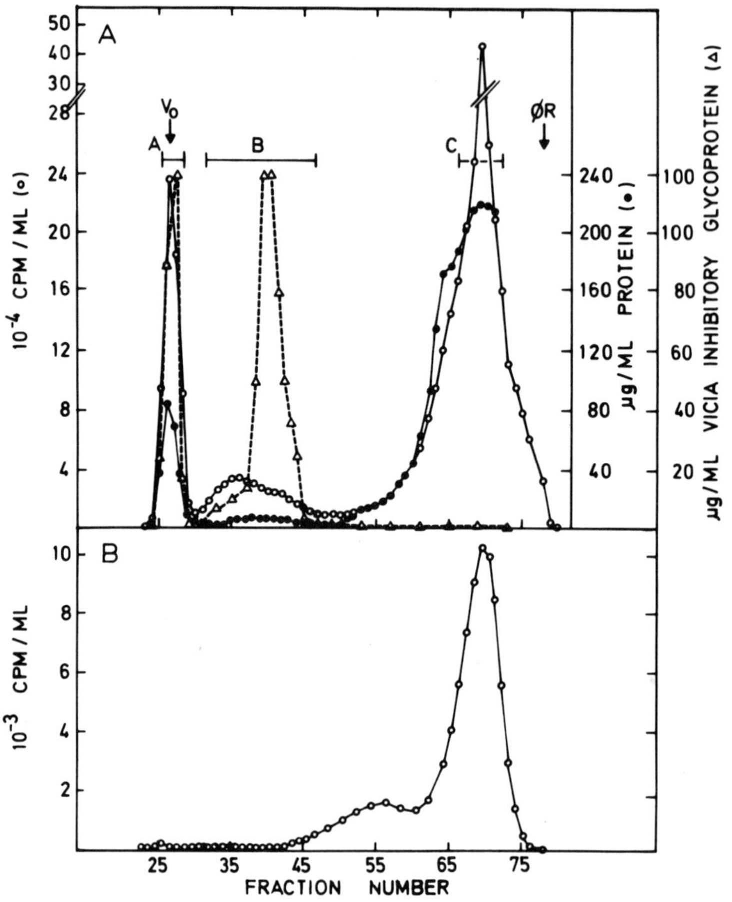

Fig. 1. Sepharose 4B fractionation of glucosamine-labeled glycoproteins released into the medium from TA3-Ha and Ta3-St cells. (A) TA3-Ha cells (4 x 10^9) were grown for 16 h in 1 liter of MEM medium for suspension culture with Gentamycin (50 µg/ml) and [^{3}H]glucosamine (100 µCi) in the absence of serum. The medium was dialyzed, lyophilized, and fractionated over a 2.5 x 95 cm Sepharose 4B column which was eluted against water, with phosphate-buffered saline (PBS). Aliquots of fractions were analyzed for radioactivity (o), Vicia-*inhibitory glycoprotein (Δ), and protein (•) measured by an automated adaptation of the method of Lowry* et al. *(9). (B) TA3-St cells (3 x 10^8) were grown for 21 h in 365 ml of MEM-spinner medium containing Gentamycin (50 µg/ml), 10% heat-inactivated fetal calf serum, and [^{14}C]glucosamine (50 µCi). The medium was treated with perchloric acid (0.6 M final conc.), dialyzed against water, lyophilized, fractionated over the same column as in A, and counted for radioactivity (o).*

In this study, we have investigated the possible association in Peak A of epiglycanin-like material with glycosaminoglycans. To this end, we initially treated purified glucosamine-labeled Peak A with a crude testicular hyaluronidase preparation (300 U/mg) which completely digested the void volume Peak A to give a new labeled peak containing all the *Vicia* activity, which emerged in a position identical to that of Peak B. In addition, most of the counts emerged in a low mol. wt. peak near the solvent volume. These results suggest that the *Vicia* activity of Peak A might be an association of Peak B epiglycanin with hyaluronic acid.

Treatment of Peak A with purified *Streptomyces* (Fig. 2B) or testicular (15 000-20 000 U/mg) hyaluronidase (Fig. 2C) in the presence of unlabeled hyaluronic acid, however, indicated that the *Vicia*-inhibitory material in Peak A was not affected. The radioactive material was mostly eluted after digestion in a position identical to that of the glucuronic acid from the carrier hyaluronic acid. In addition, a third broad, labeled peak was formed which contained a small amount of *Vicia* activity and emerged in the position of Peak B *Vicia* activity (see Fig. 2B and C). Thus, the hyaluronic acid found in Peak A was incidental to the location of the primary *Vicia*-active component in Peak A. Further treatment of the hyaluronidase-resistant Peak A with chondroitinase ABC (Fig. 2D) released no additional radioactivity or *Vicia* activity, despite the fact that marker chondroitin 4-sulfate was digested.

In contrast, treatment of labeled Peak B with *Streptomyces* hyaluronidase caused essentially no release of radioactivity or change in the *Vicia* inhibitory profile. Descending paper chromatography (Whatman No. 1; 1-butanol-acetic acid-1 M NH_4OH, 2:3:1, v/v) of Peaks A and B after treatment with chondroitinase ABC further indicated the presence of only hyaluronic acid, since the radioactive label was found in spots only in the position of unsaturated hyalobiuronic acid and not in spots corresponding to digested chondroitin sulfate.

To determine whether the *Vicia*-inhibitory activity in Peak B were exchangeable to Peak A, we incubated [^{3}H]glucosamine-labeled Peak B with [^{14}C]glucosamine-labeled Peak A in 4 M guanidinium chloride, dialyzed overnight against 0.5 M guanidinium chloride, and fractionated the resultant mixture on a column of Sepharose 4B eluted with phosphate-buffered saline (pH 7.2). No radioactivity or *Vicia*-inhibition activity was shifted or exchanged. Similarly, fractionation of Peak A on Sepharose CL-2B after prior treatment with Leo testicular hyaluronidase indicated that both the radioactivity and *Vicia*-inhibitory activity remained in the void volume.

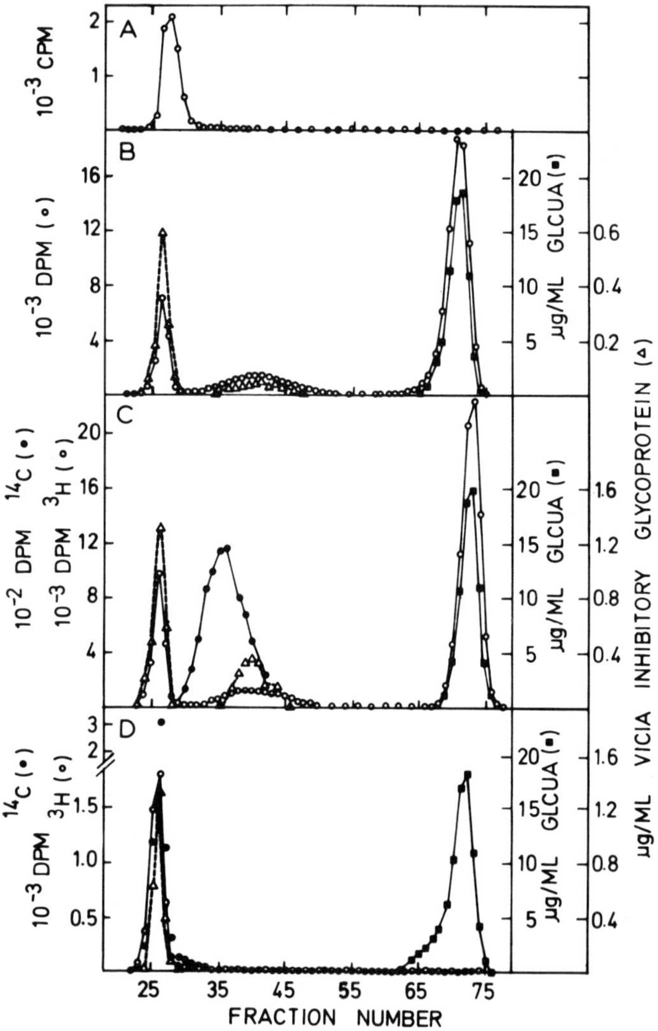

A
B
C
D
10^{-3} CPM
10^{-3} DPM (○)
10^{-2} DPM ^{14}C (●)
10^{-3} DPM ^{3}H (○)
μg/ML GLCUA (■)
μg/ML VICIA INHIBITORY GLYCOPROTEIN (△)
FRACTION NUMBER

As has been shown (6), trypsin cleavage of intact, glucosamine-labeled TA3-Ha cells gave a broad, labeled peak containing essentially all of the *Vicia*-inhibitory activity, which was eluted from the Sepharose 4B column after that of intact, spontaneously released Peak B (Fig. 3A *vs*. Fig. 1A). Treatment of purified Peak B with trypsin brought about almost complete cleavage of the molecule to a form that was eluted on Sepharose 4B in a position similar to that of the *Vicia*-inhibitory glycoprotein released by trypsin from the cell surface (Fig. 3B). Similar treatment of Peak A resistant to hyaluronidase and chondroitinase ABC (Fig. 3C) produced a complex elution profile including a broad, labeled region containing the activity. The latter peak was eluted slightly before the *Vicia*-containing peak of the cell surface and Peak B glycopeptides.

Fig. 2. Sepharose 4B fractionation of untreated and enzyme-treated Peak A. (A) [^{3}H]Glucosamine- or [^{14}C]glucosamine-labeled Peak A with no enzyme treatment. (B) [^{3}H]Glucosamine-labeled Peak A (1.5 ml) incubated for 3 h at 37°C with 2 mg of hyaluronic acid, 0.6 U of Streptomyces *hyaluronidase (Calbiochem), and NaCl-Na acetate buffer (pH 5.0) in a total volume of 2.5 ml. (C) [^{3}H]Glucosamine-labeled Peak A was incubated for 1.5 h at 37° with 1500 U of testicular hyaluronidase (15 000-20 000 U/mg; Leo) as in Part B, to which 5 µl of [^{14}C]glucosamine-labeled Peak B was added in the cold, immediately prior to loading on the Sepharose 4B column. The majority of the* Vicia *activity seen in the region of Peak B comes from the added ^{14}C-Peak B. (D) [^{14}C]Glucosamine-labeled Peak A after testicular hyaluronidase incubation and Sepharose 4B chromatography (Part B), and [^{3}H]glucosamine-labeled Peak A after testicular hyaluronidase incubation and Sepharose 4B chromatography (Part C) were combined and incubated for 2.3 h at 37°C, with 0.5 U of chondroitinase ABC (Miles Labs) and 2.6 mg of chondroitin 4-sulfate in NaCl-Tris HCl (pH 8.0). Fractions were analyzed for ^{3}H(o) and ^{14}C(●) radioactivity,* Vicia*-inhibitory glycoprotein (Δ), and glucuronic acid (■) according to a variation of the Heinegård modification of the carbazole procedure (10).*

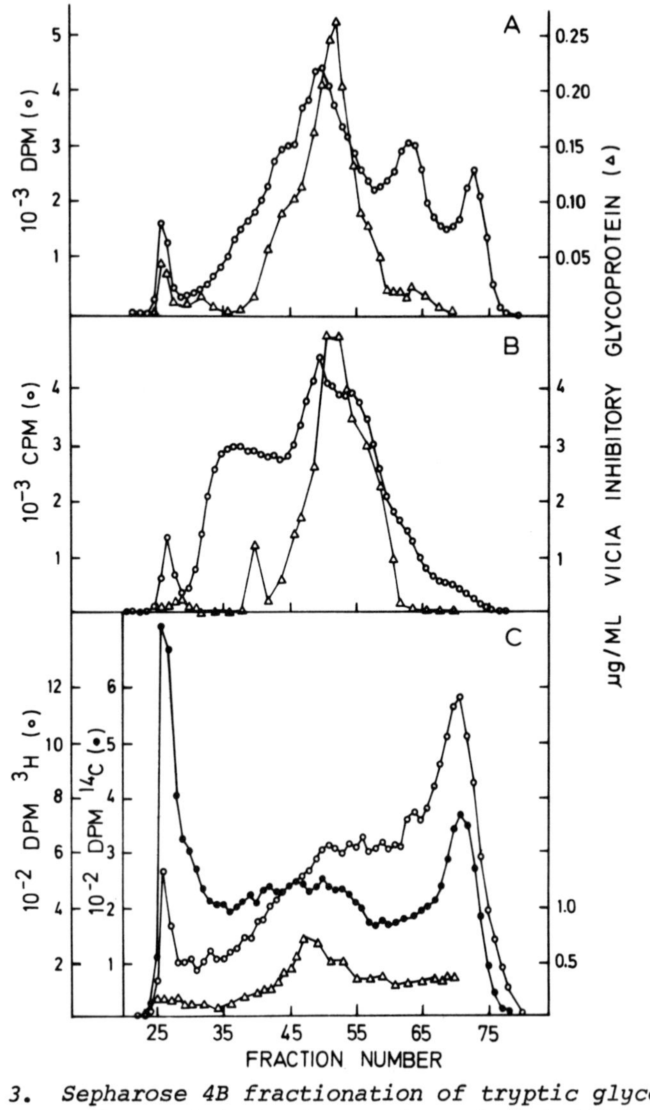

Fig. 3. Sepharose 4B fractionation of tryptic glycopeptides removed from intact glucosamine-labeled TA3-Ha cells and from released, glucosamine-labeled Peaks A and B. (A) Tryptic glycopeptides removed from 16-h [^{3}H]glucosamine-labeled TA3-Ha cells (5 x 10^5 cells/0.5 µCi/ml PBS) by 2-20 min incubations with TPCK-trypsin (18 µg/10^7 cell/ml) at 5°C. (B) [^{3}H]Glucosamine-labeled Peak B treated with 100 µg/ml of trypsin for 1 h at 37°C. (C) [^{3}H]- and [^{14}C]Glucosamine-labeled Peak A treated with hyaluronidase and chondroitinase ABC (from Fig. 3D) and incubated with 100 µg/ml trypsin for 1 h at 37°C. Fractions were analyzed for ^{3}H (o) and ^{14}C (●) radioactivity and Vicia-inhibitory glycoprotein (Δ).

We have shown that Peak A, but not Peak B epiglycanin, is associated with hyaluronic acid. The *Vicia* activity in Peak A appears to reside in a molecule that has some similarity to Peak B epiglycanin. Impure hyaluronidase appears to convert the Peak A, *Vicia*-active molecule into a molecule being eluted from Sepharose 4B identically with Peak B epiglycanin, while trypsin treatment degrades much of the lectin activity into glycopeptides with a peak of lectin activity only slightly larger than similar glycopeptides from Peak B material. On the other hand, Peak A *Vicia* activity is found in a much larger molecule which, when treated with trypsin and chromatographed on Sepharose 4B, has a much more diffuse elution profile of radioactivity and *Vicia* activity, perhaps due to a variable proteolytic cleavage of a large, proteoglycan-like molecule to which long carbohydrate chains are attached (7,8).

ACKNOWLEDGMENTS

We thank Dr. Jeremiah E. Silbert for his many helpful suggestions, Dr. Brian Toole for his gift of testicular hyaluronidase, and Professor H. C. D. De Wit for his gift of *Vicia graminea* seeds. This research was supported by USPHS (CA19987) and American Cancer Society (BC 201) research grants and Career Development Award (CA70780) to A. G. C.

REFERENCES

1. Codington, J. F., *in* "Cellular Membranes and Tumor Cell Behavior (M. D. Anderson Hospital and Tumor Institute, Houston)", p. 399, Williams and Williams, Baltimore, (1975).
2. Cooper, A. G., Codington, J. F., and Brown, M. C., *Proc. Natl. Acad. Sci. U.S.A. 71*, 1224 (1974).
3. Miller, D. K., and Cooper A. G., *J. Biol. Chem. 253*, 8798 (1978).
4. Cooper, A. G., and Brown, M. C., *in* "Automated Immunoanalysis" (R. T. Ritchie, ed.), Marcel Dekker, New York, p. 471, (1978).
5. Uhlenbruck, G., and Dahr, W., *Vox Sang. 21*, 338 (1971).
6. Miller, D. K., Cooper, A. G., and Brown, M. C., *J. Biol. Chem. 253, 8804* (1978).
7. Mathews, M. D., *Biochem. J. 125*, 37 (1971).
8. Kleinman, H. K., Silbert, J. E., and Silbert, C. K., *Connect. Tissue Res. 4*, 17 (1975).
9. Lowry, D. H., Rosebrough, N. J., Farr, A. L., and Randall, R. J., *J. Biol. Chem. 193*, 265 (1951).
10. Heinegård, E., *Chem. Scr. 4*, 199 (1973).

Protein-Sugar Interaction. Binding Properties of Wheat Germ Agglutinin

Michel Monsigny, Jean-Philippe Grivet, Annie-Claude Roche, Francis Delmotte, and Régine Maget-Dana

Wheat germ agglutinin (WGA, *Triticum vulgare* lectin) with a specificity in the order tri-*N*-Ac-chitotriose > di-*N*-Ac-chitobiose > GlcNAc, has been studied intensively during the past years by several groups. This protein, which is very easily prepared by affinity chromatography (1), contains 16 disulfide bridges and 3 tryptophan residues per monomer. At physiological pH, WGA is a dimer (mol. wt. 36 000) (2) with four binding sites (3). The binding of WGA to di-*N*-acetyl-chitobiose (but not to GlcNAc) enhances the tryptophan fluorescence, and the intensity maximum is shifted towards low wavelength (4,5). Using fluorescence quenching measurements with heavy ions as I^- (6), pulse fluorimetry (7), and chemical modifications (8), we were able to show that two over three tryptophan residues per monomer are fluorescent and are close to the binding sites. However, we have not been able to establish whether the fluorescent tryptophan residues are in or near the binding site or to localize the tryptophan residues in the binding site.

WGA was also suggested to bind selectively *N*-acetylneuraminic acid (9) and sialoglycoproteins. In the present communication, we present data (a) localizing the tryptophan residue in the subsite C of the three binding-site subsites, (b) showing that the indole group and ligand are in close contact, (c) showing that the subsite B binds specifically the *N*-acetyl group, and (d) showing that the sialic acid group is not involved in the binding of glycoconjugate.

ISBN 0-12-301301-1

INHIBITION OF RED BLOOD CELL AGGLUTINATION

In a comparative study, it was found that *N*-acetylneuraminic acid does inhibit agglutination by WGA, but that *N*-glycolylneuraminic acid does not. However the *N*-acetylneuraminic acid concentration required to inhibit one unit was found 10 000 times higher than that of di-*N*-acetylchitobiose, 200 times higher than that of GlcNAc, and even 6 times higher than that of *N*-acetylamino acids such as *N*-acetylphenylalanine. Furthermore, it was shown that various preparations of *N*-acetylneuraminic acid-containing glycoconjugates, sialyllactose (Sigma), and bovine submaxillary mucin glycopeptides were as active as the corresponding sialic acid-free glycoconjugates (hydrolysis of sialic acid either by 0.1 M H_2SO_4, 80°C, 1 h; or by *Vibrio cholerae* neuraminidase). When limulin was used instead of WGA, such compounds as bovine submaxillary mucin lost entirely their inhibiting power after desialylation. By use of gangliosides extracted from horse red blood cells (after Pronase treatment), WGA was shown not to aggregate vesicles containing these gangliosides while limulin did aggregate (10).

N.M.R. STUDIES

By use of di-*N*-acetyl β-chitobioside, it was shown that the *N*-acetamido group of the nonreducing terminal group was involved in the binding process. The n.m.r. peak of the proton of the *N*-acetyl group of the reducing residue was not broadened and was slightly shifted upfield, but that proton of the *N*-acetyl group of the nonreducing residue was broadened and almost disappeared upon binding to wheat germ agglutinin. Free GlcNAc or methyl 2-acetamido-2-deoxy-β-D-glucopyranoside binds WGA but does not affect the tryptophan fluorescence of WGA, the disaccharide affects the tryptophan fluorescence, and the tryptophan residue is localized in the subsite C (see further) according to the n.m.r. data. For these reasons, we propose that free GlcNAc mainly binds subsite B, and that di-*N*-acetylchitobiose binds subsites B and C.

FLUORESCENCE AND PHOSPHORESCENCE STUDIES

It is well known that heavy atoms quench the fluorescence and enhance the intersystem crossing. In order to localize the tryptophan residue in the binding site, we synthesized the thiomercuribenzoate derivatives of β-D-glucopyranose (Glc-Hg), 2-acetamido-2-deoxy-β-D-glucopyranose (GlcNAc-Hg), di-*N*-acetyl-β-chitobiose (CB-Hg), and tri-*N*-acetyl-β-chitotriose (CT-Hg) by treating the isothiouronium derivatives (11,12) with hydroxymercuribenzoate. GlcNAc-Hg, CB-Hg, and CT-Hg quenched the tryptophan fluorescence of WGA, this quenching being completely reversed by addition of di-*N*-acetylchitobiose. Glc-Hg had no effect at all. The changes of tryptophan phosphorescence at 70 K were found to be very important and depending upon the size of the ligand. The phosphorescence of WGA was increased twice upon binding CT-Hg, five times upon binding GlcNAc-Hg, and six times upon binding CB-Hg and was not affected by Glc-Hg. The lifetimes of WGA, and of CT-Hg-WGA, CB-Hg-WGA, and GlcNAc-Hg-WGA complexes were found to be 5.6, 4.4, 0.3, and 0.3 (90%) sec, respectively. Thus, in the CB-Hg-WGA and GlcNAc-Hg-WGA complexes, the Hg atom is in close contact with the indole group of the tryptophan residue, which is in the subsite C; the GlcNAc residue of GlcNAc-Hg is bound in the binding site B; and the two GlcNAc residues of subsites A, B, and C and the Hg atom, which is far away from the indole group according to the phosphorescence data, is out of the binding site. These results are also consistent with previous data on the binding of methylumbelliferyl derivatives of the same sugars (3), and with the location of Trp-21 by X-ray diffraction (13). These data support the hypothetical scheme of the WGA binding site (Fig. 1).

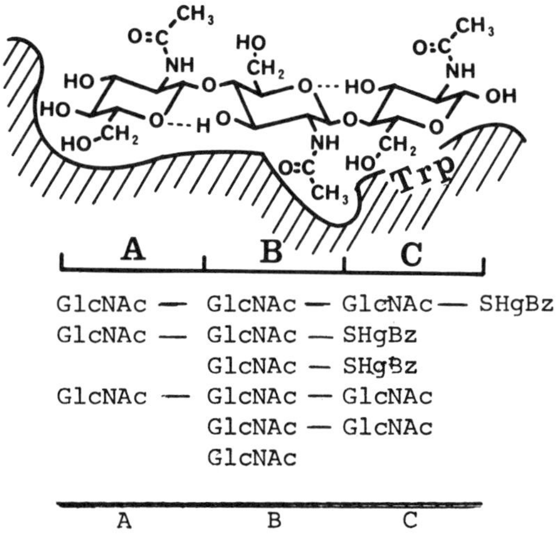

Fig. 1. Hypothetical scheme of WGA binding site.

REFERENCES

1. Bouchard, P., Moroux, Y., Tixier, R., Privat, J. P., and Monsigny, M., *Biochimie 58*, 1247 (1976).
2. Rice, R. H., and Etzler, M. E., *Biochemistry 59*, 4093 (1975).
3. Privat, J. P., Delmotte, F., and Monsigny, M., *FEBS Lett. 46*, 224 (1974).
4. Lotan, R., and Sharon, N., *Biochem. Biophys. Res. Commun. 55*, 1340 (1973).
5. Privat, J. P., Delmotte, F., Mialonier, G., Bouchard, P., and Monsigny, M., *Eur. J. Biochem. 47*, 5 (1974).
6. Privat, J. P., and Monsigny, M., *Eur. J. Biochem. 60*, 555 (1975).
7. Privat, J. P., Wahl, Ph., Monsigny, M., and Auchet, J. C., *Eur. J. Biochem. 68*, 573 (1976).
8. Privat, J. P., Lotan, R., Bouchard, P., Sharon, N., and Monsigny, M., *Eur. J. Biochem. 68*, 563 (1976).
9. Burger, M. M., and Goldberg, A. R., *Proc. Natl. Acad. Sci. U.S.A. 57*, 359 (1967); Greenaway, P. J., and LeVine, D., *Nature New Biol. 241*, 191 (1973).
10. Maget-Dana, R., Roche, A. C., and Monsigny, M., *FEBS Lett. 79*, 305 (1977).
11. Rafestin, M. E., Obrenovitch, A., Oblin, A., and Monsigny, M., *FEBS. Lett. 40*, 62 (1974).
12. Delmotte, F., Privat, J. P., and Monsigny, M., *Carbohydr. Res. 40*, 353 (1975).
13. Wright, C. S., *J. Mol. Biol. 111*, 439 (1977).

GLYCOCONJUGATE RESEARCH/Volume I
Proceedings of the Fourth International Symposium on Glycoconjugates

Rat Colonic Mucus Glycoprotein

Varahabhotla L.N. Murty, Fred Downs, and Ward Pigman

Recently, considerable interest has been shown in the biosynthesis and structural aspects of secretory glycoproteins and their possible alterations in pathological conditions. Changes in the synthesis of carbohydrate chains in mucus glycoprotein of colonic cancer tissue were reported (1). Studies are undertaken to see if any structural differences exist between normal and tumor-bearing rat colonic glycoproteins. It is, therefore, imperative to isolate these glycoproteins in a pure state without degradation. We describe in this paper the first part of our study, the isolation, purification, and characterization of a glycoprotein from normal rat colonic mucosa.

Pooled rat colonic mucosal scrapings, obtained as described by Perret *et al.* (2), were suspended in cold distilled water (10 ml/g) and homogenized. The homogenized material was kept in boiling water for 5 min, cooled in ice, briefly sonicated, and dialyzed at 4°C against distilled water. The nondialyzable material was centrifuged at 46 000*g* for 1 h and the clear supernatant solution lyophilized. The recovery of sialic acid in the supernatant solution was 90% of that of the whole mucosa. The glycoprotein was isolated by a first treatment with lithium diiodosalicylate (3), followed by three successive treatments with hydroxyapatite (4). The latter treatment resulted in a 40-fold increase in the sialic acid-to-protein ratio.

The purified rat colonic glycoprotein (40%, based on sialic acid) was homogeneous by analytical ultracentrifugal criteria; it had a mol. wt. of 9.0×10^5 (sed. equil.); no subunits were detected in the presence of dodecyl sodium sulfate. The glycoprotein contained 85% of carbohydrate and 15% of protein. The principal sugars were *N*-acetylglucosamine, *N*-acetylgalactosamine, sialic acid, galactose, fucose, and mannose. The glycoprotein contained sulfate but no uronic acid residues (Table I).

ISBN 0-12-301301-1

Table I. Chemical Composition of Rat Colonic Glycoprotein

Components	*g/100 g*	*Mole ratio*[a]
Protein	*13.7*	
Sialic acid	*15.0*	*2.2*
NeuAc	*11.7*	*1.8*
NeuGl	*3.3*	*0.4*
GalNAc[a] *(total)*	*10.1*	*2.2*
GalNAc[a] *(terminal)*	*4.6*	*1.0*
GalNAc[a] *(internal)*	*5.5*	*1.2*
GlcNAc[a]	*8.4*	*1.8*
Gal	*37.4*	*10.2*
Fuc	*15.3*	*4.7*
Man	*1.3*	*0.4*
Sulfate	*5.37*	*2.7*

[a] *Based on the number of GalNAc residues linked to Ser and Thr residues in the core protein*

The occurrence of sulfate, in the absence of uronic acid, suggests that some of the amino sugar residues are sulfated, as in the sheep colonic glycoprotein (5). The sialic acids were found to be a mixture of *N*-acetyl- and *N*-glycolyl-neuraminic acids in a molar ratio of 4:1, and had an *O*-acetyl content of 1 mol/mol of sialic acid. The amino acid composition with a high amount of threonine and serine, and a low content of aromatic and traces of sulfur-containing amino acids (Table II), is similar to that found in mucous glycoproteins, but it differs by its low proportion of Asx and Glx. The carbohydrate-to-protein linkage was shown by the alkaline β-elimination reaction (6) to consist of an *O*-glycosyl linkage between *N*-acetylhexosamine and peptidyl seryl and threonyl residues. The molar ratio of the sum of fucose and sialic acid to *N*-acetylgalactosamine is 3.1:1.0. The first two sugars are known to occur in terminal position. Since *N*-acetylgalactosamine is linked to peptidyl serine or threonine residues, the side chains (which average 22 units) must be branched. High mol. wt. oligosaccharide side-chains were also reported in other glycoproteins (7).

Table II. Amino Acid Composition of Rat Colonic Glycoprotein[a]

Amino acid	*mol/100 mol*	*Amino acid*	*mol/100 mol*
Lys	*2.9*	*Pro*	*9.6*
Arg	*1.6*	*Gly*	*8.5*
Asx[b]	*10.2*	*Ala*	*6.9*
Thr	*18.1*	*Val*	*5.1*
Ser	*11.7*	*Ile*	*3.0*
Glx[c]	*15.7*	*Leu*	*4.6*
		Phe	*1.7*

[a] *The following amino acids constitute <1% of the glycoprotein: His, Tyr, Cys 1/2, Met*
[b] *Aspartic acid or asparagine*
[c] *Glutamic acid or glutamine*

ACKNOWLEDGMENT

This study was supported by NIH Grant No. 5R26CA17168-03.

REFERENCES

1. Kim, Y.S., Isaacs, R., *Cancer Res. 35,* 2092 (1975).
2. Perret, V., Lev, R., and Pigman, W., *Gut 18,* 382 (1977).
3. Marchesi, V.T., and Andrews, E.P., *Science 74,* 12 (1971).
4. Tettamanti, G., and Pigman, W., *Arch. Biochem. Biophys. 124,* 41 (1968).
5. Kent, P.W., and Marsden, J.C., *Biochem. J. 87,* 38P (1963).
6. Downs, F., Peterson, C., Murty, V.L.N., and Pigman, W., *Int. J. Pept. Prot. Res. 10,* 315 (1977).
7. Rovis, L., Anderson, B., Kabat, E.A., Gruezo, F., and Liao, J., *Biochemistry 12,* 1955 (1973); Rovis, L., Kabat, E.A., Pereira, M.E.A., and Feizi, T., *Biochemistry 12,* 5355 (1973).

^{13}C-NMR Analysis of the Effect of Calcium on the Structure of a Hyaluronic Acid Matrix

Mary A. Napier and Nortin M. Hadler

It is likely (1) that orderedness, reminiscent of that in ordered films, is present between the primarily flexible chains of the polymer of hyaluronic acid (HA) as it exists in solution. In the studies reported here we have utilized natural abundance ^{13}C-n.m.r. to demonstrate orderedness in an HA matrix. Such orderedness can be ablated either by enzymatic depolymerization of the matrix or by the addition of calcium salts. Perturbing the matrix in this way reduces the translational diffusivity of lysine and glucose (Hadler and Napier, in this volume).

^{13}C-N.m.r. spectra for 2.5% HA matrices were recorded at ambient temperature, 27°C, on a Varian XL-100-12 spectrometer operating at 25.16 MHz in the pulsed, Fourier transform-mode (pulse angle, 23.3°). Spectra were collected with wide-band proton decoupling, pulse acquisition time of 0.4 sec, and exponential weighing time constant of -0.04 sec.

For matrices dissolved in D_2O, the chemical shifts of the individual carbon atoms of the glucuronate and *N*-acetylglucosamine residues correlate with the observed chemical shifts of the analogous hexose residue. The resonance position of the carbon atoms in the matrix appears unshifted relative to the hyaluronidase-digested material. But the matrix spectrum exhibits broadened and poorly resolved resonance peaks, most notably for the side-chain C-6' methylene group, and the ring carbon atoms (C-1-C-5 and C-1'-C-5'). The anomeric carbon C-1' is less resolved than C-1. Depolymerization of HA results in a dramatic increase in resolution and narrowing of resonance peaks.

ISBN 0-12-301301-1

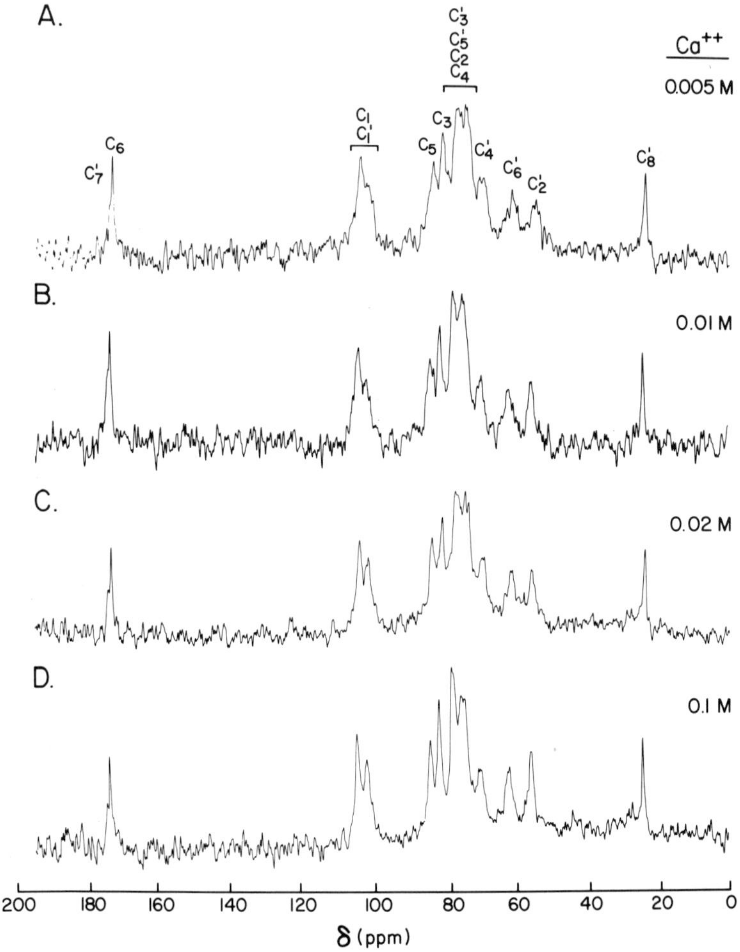

Fig. 1. ^{13}*C-N.m.r. spectra (25.16 Hz) at ambient temperature of 2.5% HA at the calcium concentrations indicated in* D_2O*. Spectra were obtained after 118,357 (A), 119,236 (B), 117,310 (C), and 119,400 (D) transients of 0.4 sec sweep-time.*

From Fig. 1 it is seen that at high Ca^{2+} concentration (0.1 M), the resonance peaks narrow considerably, but no shifts in resonance positions are observed. Peak width at half-height for C-8' decreases from approximately 30 Hz at 5 mM Ca^{2+} to approximately 17 Hz at 0.1 M Ca^{2+}. For carbon atoms C-1', C-2', C-6', and C-7' dramatic narrowing is observed in the presence of 0.1 M Ca^{2+}. The changes in spectral line shape appear to occur gradually with increasing Ca^{2+} concentration. No change in peak morphology is observed in buffered *vs.* non-buffered solutions (data not shown). Spectra equivalent to A or B are observed for 0.1 M KCl and for 0.5 M NaCl, 0.01 M phosphate buffer, pH 7.0 (data not shown).

The effect of Ca^{2+} on the natural abundance ^{13}C-n.m.r. spectrum of HA suggests that interchain stabilization in a 2.5% matrix at 5 mM Ca^{2+} or less occurs in a fashion similar to that seen in the putty studied crystallographically. These observations are consistent with interchain stabilization in regions of the 2.5% HA matrix presumably mediated by H-bonding at C-6', and subsequent loss of flexibility at *N*-acetylglucosamine residues. The dramatic sharpening of all carbon peaks that follows hyaluronidase depolymerization suggests as well that sizable chain lengths are required for interchain interactions of this nature.

With the ^{13}C-n.m.r. analysis we have demonstrated the presence of interacting segments of adjacent chains in regions of the HA matrix at physiologic concentrations of calcium. At higher calcium concentrations such regions become disordered and the ability of the matrix to enhance the translational diffusivity of glucose and lysine is lost. It is possible that these regions of interacting segments retain the hexad motif documented in hydrated HA putty studied by x-ray diffraction (3). In that case, these regions could have the configuration of corrugated sheets with highly structured solvation shells, and might well provide a high conductance surface.

REFERENCES

1. Hadler, N.M., and Napier, M.A., *Semin. Arthritis Rheum.* *7,* 141 (1977).
2. Sheehan, J.K., Atkins, E.D.T., and Nieduszynski, I.A., *J. Mol. Biol. 91,* 153 (1975).

GLYCOCONJUGATE RESEARCH /Volume I
Proceedings of the Fourth International Symposium on Glycoconjugates

The Chemical Structure of a Glycoprotein from the Cervical Mucus (Premenstrual Phase) of *Macaca radiata*

Nasir-ud-Din, Roger W. Jeanloz, Vernon N. Reinhold, James D. Moore, and Janet W. McArthur

Cervical mucus is a hydrophilic, gel-like, dynamic epithelial secretion, which performs a key role in mammalian reproductive process. The mucus displays distinct differences in biophysical and physiological properties during the ovulatory cycle, and these changes are accompanied by alterations in carbohydrate composition (1,2). The bonnet monkey, whose menstrual cycle is very similar to that of the human cycle and which produces large amounts of mucus was used for this study.

The mucus was purified (3), fractionated (4), and the oligosaccharides from the major glycoprotein were prepared by reductive β-elimination as described by Iyer and Carlson (5). The oligosaccharides were separated on a column of Bio-Gel P-4 and the main oligosaccharide component (Table I) was investigated by methylation studies.

The results of the methylation studies showed the presence of terminal fucose; terminal and 2-linked, and 6- and 3-linked galactose; 4-linked *N*-acetylglucosamine; and 6-linked, and 3- and 6-linked *N*-acetylgalactosamine residues.

Acid cleavage of sialic acid residues from a glycoprotein is generally accomplished by treatment with 50 mM sulfuric acid at 80° for 60 min. Mild acid treatment of the methylated oligosaccharide under these conditions resulted in the removal of sialic acid residues, as demonstrated by subsequent methylation and introduction of a trideuteromethoxyl group at C-3 of D-galactose chain residues. The introduction of a trideuteromethoxyl group (∼10%) at C-3 of terminal D-galactose residues was also observed. This incorporation of a trideuteromethoxyl group at terminal residues arising from partial removal of L-fucose from D-galactose residues is unlikely, as the ratio of L-fucose to terminal D-galactose, before and after acid treatment, was very similar. Furthermore, the incorporation of the trideutero-

ISBN 0-12-301301-1

methoxyl group at C-3 of the D-galactose residue only shows that the cleavage of the sialic acid residues was selective.

Table I. Composition of the Major Glycoprotein Separated on Sepharose 2B(I) and of Alkali-borohydride-treated Oligosaccharides (II)

Components[a]	*I*		*II*	
	%	*Molar ratio*[b]	%	*Molar ratio*[b]
L-Fucose	*6.0*	*0.90*	*8.5*	*0.76*
D-Galactose	*16.0*	*2.10*	*25.0*	*2.04*
N-*Acetyl-galactosamine*	*14.0*	*1.50*	*11.0*	*0.73*
N-*Acetyl-glucosamine*	*9.2*	*1.00*	*15.0*	*1.00*
N-*Acetyl-neuraminic acid*	*10.0*	*0.77*	*11.2*	*0.53*
N-*Acetyl-galactosaminitol*			*14.5*	*0.96*

[a] *Determined by gas-liquid chromatography*
[b] *Molar ratio relative to* N-*acetylglucosamine*

Structural studies of the oligosaccharides obtained from the peri-ovulatory glycoprotein have shown that D-galactose residues are linked at C-3, C-2, and C-6 (5,6). Similarly, *N*-acetyl-galactosamine residues were shown to be linked at C-3 and C-6, and the sialic acid residues linked to the C-6 of *N*-acetyl-galactosamine residues (6).

From the structural studies of the peri-ovulatory phase and methylation studies of the pre-menstrual-phase mucus glycoprotein, it is concluded that the linkages of D-galactose and *N*-acetylgalactosamine, and the point of attachment of sialic acid residues are different in these two glycoproteins (see Table II).

Table II. Linkages of Carbohydrate Components of the Main Glycoproteins of Peri-ovulatory and Premenstrual Phases of Monkey Cervical Mucus

Components	Linkages in glycoproteins	
	Peri-ovulatory phase	Premenstrual phase
D-Galactose	L-Fucp-(1→2)-D-Galp D-Galp-(1→3)-D-Galp D-GlcNAcp-(1→4)-D-Galp	L-Fucp-(1→2)-D-Galp D-Galp-(1→6)-D-Galp NeuAcp-(2→3)-D-Galp
N-Acetylglucosamine	D-Galp-(1→4)-D-GlcNAcp	D-Galp-(1→4)-D-GlcNAcp
N-Acetylgalactosamine	NeuAcp-(2→6)-D-GalNAcp D-Galp-(1→3)-D-GalNAcp D-GlcNAcp-(1→3)-D-GalNAcp	D-Galp- or D-GlcNAcp-(1→3,6)-D-GalNAcp

Variations in the points of attachment of sugar residues in glycoproteins of secretions, particularly in the blood-group active glycoproteins (7), are well known. The role played by these changes in the glycoprotein structures during the ovulatory cycle is not known, nor whether changes are restricted to the carbohydrate component of the glycoprotein.

The pre-menstrual glycoprotein was reduced with dithiothreitol and alkylated with iodoacetic acid. This procedure was repeated four times, and the alkylated glycoprotein was chromatographed on DEAE-cellulose; two components containing different proportions of cysteine and *S*-carboxymethylcysteine were obtained, indicating incomplete reduction and *S*-carboxymethylation, and clearly suggesting that cysteine is a component of the glycoprotein.

REFERENCES

1. Iacobelli, S., Garcea, N., and Angeloni, C., *Fertil. Steril. 22,* 727 (1971).
2. Hatcher, V.B., Schwarzman, G.O.H., Jeanloz, R.W., and McArthur, J.W., *Fertil. Steril. 28,* 682 (1977).
3. Hatcher, V.B., Schwarzmann, G.O.H., Jeanloz, R.W., and McArthur, J.W., *Biochemistry 16,* 1518 (1977).
4. Nasir-ud-Din, Jeanloz, R.W., and Reinhold, V.N., *Fed. Proc. 25,* 1443 (1976).
5. Iyer, R.N., and Carlson, D.M., *Arch. Biochem. Biophys. 142,* 101 (1971).
6. Jeanloz, R.W., Nasir-ud-Din, and Reinhold, V.N., *Adv. Exp. Med. Biol. 89,* 524 (1977).
7. Spiro, R.G., *Adv. Protein Chem. 27,* 399 (1973).

GLYCOCONJUGATE RESEARCH / Volume I
Proceedings of the Fourth International Symposium on Glycoconjugates

Distribution and Conformation of a Highly Crystalline α-Glucan in *Aspergillus* Hyphal Walls

John H. Nordin, Thomas F. Bobbitt, Michel Roux, Jean-François Revol, and Robert H. Marchessault

Nigeran, a hot-water soluble, unbranched D-glucan (DP 175-200) composed of alternating α-(1→3) and α-(1→4) linkages (1) is a hyphal wall constituent (2) of several *Aspergillus* and *Penicillium* species. Gold *et al.* (3) reported that nitrogen deprivation stimulates nigeran deposition 5- to 10-fold and that, once in the wall, the polysaccharide does not display detectable turnover.

Single crystals of pure nigeran have been grown from aqueous solution, and X-ray diffraction utilized to ascertain its unit cell dimensions and "corrugated ribbon" chain-conformation (4). Because it seemed probable that nigeran could exist *in situ* as a crystalline wall component and could serve as a general model for studying assembly of crystalline cell-wall polymers, we have employed a combination of X-ray diffraction analysis, enzymolysis, and electron microscopy to gain further insight into nigeran's supramolecular structure and organization in the hyphal wall.

Nigeran normally constitutes 4-6% of the wall weight of *Aspergillus awamori* hyphae grown submerged in a complete medium, but increases to 25-30% when the cells are starved of nitrogen for 96 h. Shadowed preparations of hyphae containing 25% of nigeran were found to be opaque when viewed in the electron microscope. Fragmented hyphal ends suggested that the material responsible for the opacity was associated with the external portion of the wall. Walls containing 5% of nigeran or walls with 25% of nigeran, which had been subjected to boiling, enzymatic hydrolysis, or cold 0.5 N NaOH extraction, were found not to be opaque to the electron beam. This demonstrated that the wall component responsible for the opacity was extractable. However, when thin sections were observed by electron microscopy, increased

ISBN 0-12-301301-1

polymer deposition was not found to be accompanied by significant wall thickening, suggesting that extra nigeran is intercalated between existing wall components or replaces them during nitrogen depletion. In both cases the wall thickness was 250-300 nm.

X-Ray diffraction studies demonstrated that nigeran occurs *in situ* in a hydrated, crystalline conformation identical with that of the polysaccharide crystallized from aqueous solution, and is the predominant crystalline species in the wall. Diffraction patterns for nigeran were found in hyphal walls even after a 36-h extraction with boiling water or cold 0.5 N NaOH, although both procedures resulted in the diminution of the X-ray diffraction pattern intensity. However, hot-alkali extraction removed all traces of the nigeran from the walls along with certain other glucan components.

Digestion of *A. awamori* and *A. niger* cell walls *in situ* at 40°C with the specific α-glucanase mycodextranase (5) revealed that a certain fraction was susceptible to direct enzymic degradation. Another portion was attacked after its release from the wall with boiling water, while a third fraction was hydrolyzed only after treatment with hot dilute alkali.

Enzymic hydrolysis experiments conducted over a range of temperatures from 20°C to 50°C with both pure crystalline nigeran and hyphal walls indicated that the polymer's "susceptibility to digestion" was in fact temperature dependent. The crystalline polymer was almost totally resistant to attack at 20°C, with increased levels of digestion occurring with increased temperature. This suggested that the crystal lattice structure of nigeran impeded attack by the enzyme, and, as the system's kinetic energy increased, larger areas of the lamellar structure were exposed to degradation. The crystallinity of cell-wall polysaccharides may be important in organism defenses against enzymic attack and after environmental stress.

REFERENCES

1. Reese, E. T., and Mandels, M., *Can. J. Microbiol. 10,* 103 (1964).
2. Tung, K. K., and Nordin, J. H., *Biochem. Biophys. Res. Commun. 28,* 519 (1967).
3. Gold, M. H., Mitzel, D. L., and Segel, I. H., *J. Bacteriol. 113,* 856 (1973).
4. Sundararajan, P. R., Marchessault, R. H., Quigley, G. J., and Sarko, A., *J. Am. Chem. Soc. 95,* 2001 (1973).
5. Tung, K. K., Rosenthal, A. H., and Nordin, J. H., *J. Biol. Chem. 246,* 2722 (1971).

Use of Glycosyltransferases and Glycosidases in Structural Analysis of Oligosaccharides

James C. Paulson, Lowrie R. Glasgow, Thomas Beyer, Catherine Lowman, Michael J. Holroyde, and Robert L. Hill

Glycosidases are widely recognized as valuable tools for elucidating the structure of oligosaccharides. In principle, glycosyltransferases should be equally valuable, but these enzymes as well as many glycosidases are difficult to obtain in pure form with defined enzymatic properties. Recently galactosyl-(GT) (1), sialyl-(ST) (2), *N*-acetylgalactosaminyl (GNT) (3), and fucosyl transferase-(FT) (4) have been purified to homogeneity on nucleotide affinity adsorbents, and their enzymatic properties examined. In addition, five of the six glycosidases from the culture filtrate of *Streptococcus (Diplococcus) pneumoniae* were obtained virtually free of contaminating enzymes by ion-exchange and affinity chromatography (5). These include a neuraminidase (α-N), a β-D-galactosidase (β-G), an *endo-N*-acetyl-α-D-galactosaminidase, an *N*-acetyl-β-D-glucosaminidase (β-GLN), and an *endo-N*-acetyl-β-D-glucosaminidase (E-β-GLN). In this report, the pure glycosidases and glycosyltransferases are used together to obtain structural information about glycoprotein oligosaccharides.

α_1-Acid glycoprotein (AGP) has been used as a model protein to examine the suitability of glycosidases for the sequential digestion of glycoprotein carbohydrate groups. By composition, the five oligosaccharides of AGP have, on the average, four terminal chains with the sequence NeuAc→Gal→GlcNAc attached to a branched core containing mannose and *N*-acetylglucosamine. Sequential treatment of AGP with α-N, β-G, β-GLN, and finally E-β-GLN resulted in removal of over 85% of the total carbohydrate. Two glycosyltransferases have been particularly useful for monitoring the extent of the digestion by the first three glycosidases, since removal of the sialic acid exposes galactose, a substrate for ST, and removal of both sialic acid and galactose exposes *N*-acetylglucosamine, a substrate for the GT.

ISBN 0-12-301301-1

Table I. Glycosylation of Glycosidase-Treated Derivatives of α_1-Acid Glycoprotein

α_1-Acid Glycoprotein derivative	Residues removed[b]			Reglycosylation by:[a]	
				Sialyltransferase	Galactosyltransferase
	NeuAc[c]	Gal[c]	GlcNAc[c]	(Mol NeuAc/mol)	(Mol Gal/mol)
Native				3.3	0.0
Asialo	13.5			14.6[b]	0.1
Asialo-agalacto	13.5	18.3		0.0	17.6[d]
Asialo-agalacto-ahexosamino	13.5	18.3	18.7	0.0	1.2

[a] Expressed as mol monosaccharide/mol α_1-acid glycoprotein (44 000 mol. wt.)

[b] Equivalent to 80% of the total galactose residues susceptible to hydrolysis by S. pneumoniae β-galactosidase

[c] Mol/mol

[d] Equivalent to 94% of the total N-acetylglucosamine residues susceptible to hydrolysis by S. pneumoniae N-acetylglucosaminidase

The ST and the GT were used to glycosylate derivatives of AGP before and after treatment with glycosidases, as shown in Table I.

As indicated, the glycosidases were able to selectively expose, and then efficiently remove acceptors for the two transferases. Although the β-G efficiently removed all terminal galactose residues that served as acceptors for the sialyltransferase, direct quantitation of galactose in asialoagalacto-AGP revealed that 12% of the total galactose is not removed. The nature of the galactose residues removed was better defined by determining the substrate specificities of the two enzymes (6). The ST and the β-G were found to have strict substrate specificities for β-D-galactoside residues β-(1→4)-linked especially when *N*-acetylglucosamine was the penultimate sugar. β-D-Galactoside residues (1→3 or →6)-linked were virtually inactive (data not shown). The strict specificities of these two enzymes make it possible to quantitatively determine the β-Gal-(1→4)-GlcNAc sequence found in asialo-glycoproteins. Based on the incorporation of sialic acid with the sialyltransferase, the total galactose found in β-Gal-(1→4)-GlcNAc-linkage in asialo-α_1-acid glycoprotein was 88%, >90% in the asialo-galactoside-binding lectin of rabbit liver, and 63% in the human Factor VIII involved in blood coagulation. In each case, treatment with β-G abolished the ability of ST to incorporate sialic acid. Attempts to remove the remaining 12% of the galactose residues in asialoagalacto-α_1-acid glycoprotein with β-G from jack bean, E. coli, and bull testes were unsuccessful, even though these enzymes hydrolyze β-D-galactoside residues (1→3 or →6)-linked Others have suggested that fucose may be attached to galactose, thus blocking release by β-G. Pure GNT from porcine submaxillary glands has a strict acceptor specificity for α-L-Fuc-(1→2)-galactosides and can be used to detect this sequence in glycoproteins. When pure FT was used to attach α-L-Fuc residues (1→2)-linked to the D-galactose residues of asialo-α_1-acid glycoprotein, as little as 0.02 nmol of the product could be detected with the GalNAcT. But in unmodified asialo-α_1-acid glycoprotein, there was not sufficient α-Fuc-(1→2)-Gal to block even 0.1% of the β-G resistant galactose residues, the nature of which will require further study (6).

REFERENCES

1. Barker, R., Olsen, K.W., Shaper, J.H., and Hill, R.L., *J. Biol. Chem. 247*, 7135 (1972).
2. Paulson, J.C., Beranek, W., Hill, R.L., *J. Biol. Chem. 252*, 2356 (1977).
3. Schwyzer, M., and Hill, R.L., *J. Biol. Chem. 252*, 2338 (1977).
4. Beyer, T.A., Prieels, J-P, Hill, R.L., *Biochem. Soc. Trans. 5*, 838 (1977).
5. Glasgow, L.R., Paulson, J.C. and Hill, R.L., *J. Biol. Chem. 252*, 8615 (1977).
6. Paulson, J.C., Prieels, J.-P., Glasgow, L.R., and Hill, R.L., *J. Biol. Chem. 253*, 5617 (1978).

The Molecular Structure of Some Novel Antigenic Glycans from Group D *Streptococci*

John H. Pazur and L. Scott Forsberg

Type specific carbohydrates from many strains of pneumococci and Groups A and C streptococci have been studied extensively, and their purification, structure, and immunological properties have been described (1,2). Further, a scheme for the serological classification of these organisms based on the type of carbohydrate in the cell walls has been developed and is employed for the diagnosis of infectious diseases (3). The study of the type-specific carbohydrates of the Group D streptococci is a more recent development. The isolation of a glycan from a group D organism, *Streptococcus faecalis,* strain N, was first described in 1965 (4), and more recently a number of other interesting glycans from different strains of this group have been prepared (5-7).

In this report, structural and immunological studies on two diheteroglycans and two tetraheteroglycans from two group D organisms, *S. faecalis* strain N and *S. boris* strain C3, are recorded. From the *S. faecalis* cell-walls, a diheteroglycan of D-glucose and D-galactose, and a tetraheteroglycan of L-rhamnose, D-glucose, D-galactose, and *N*-acetyl-D-galactosamine were obtained. From the *S. bovis* cell-walls, a diheteroglycan of D-glucose and L-rhamnose, and a tetraheteroglycan of L-rhamnose, D-galactose, D-glucuronic acid, and the unusual sugar 6-deoxy-L-talose, were obtained. The tetraheteroglycan from *S. faecalis* also contains phosphate as a structural component.

Antisera with antibodies directed against these glycans were obtained from rabbits immunized intravenously with vaccines of nonviable cells. In agar diffusion tests, the antisera yielded single precipitin bands with the individual glycans and double precipitin bands with a mixture of the glycans. The antibodies in such antisera have been useful in the elucidation of the structure of the glycans by the hapten-inhibition method. In

ISBN 0-12-301301-1

addition, antibodies with specificity for single carbohydrate residues were prepared by affinity chromatography on various types of glycosyl Sepharoses (8) and by electrofocusing techniques (9).

In the isolation procedures, the glycans were extracted from the cell walls of the organisms under mild conditions, either with 10% trichloroacetic acid at 4°C for 24 h or with 0.05 M KCl and 0.01 M HCl solution of pH 2 at 100°C for periods of 5 to 60 min. The antigenic glycans in such extracts were precipitated from solution by addition of acetone, and then separated into individual components by fractional precipitation with ethyl alcohol, Bio-Gel filtration, or chromatography on DEAE-cellulose.

The mol. wts. of the glycans were determined by density gradient centrifugation of the compounds and reference glycans, and by Bio-Gel filtration. The values were 15 000 and 5000 for the di- and tetra-heteroglycans from *S. faecalis*, and 12 000 and 6000 for the di- and tetra-heteroglycans from *S. bovis*. Quantitative values for the monosaccharide constituents of the glycans were obtained by suitable colorimetric methods. These values and the mol. wt. values were used to calculate the number of residues of each monosaccharide in a typical molecule of the glycans. The values for the four glycans are recorded in Table I.

Table I. Number of Residues per Mol of Heteroglycan of Type-Specific Carbohydrates from Two Streptococcal Strains

Component	S. faecalis		S. bovis	
	Di-	*Tetra-*	*Di-*	*Tetra-*
6-deoxy-L-talose	*0*	*0*	*0*	*8*
L-Rhamnose	*0*	*8*	*48*	*16*
D-Glucose	*57*	*16*	*32*	*0*
D-Galactose	*34*	*4*	*0*	*8*
N-*Acetyl-D-galactosamine*	*0*	*4*	*0*	*0*
D-Glucuronic acid	*0*	*0*	*0*	*8*

```
→4)-β-D-Glcp-(1→4)-β-D-Glcp-(1→4)-β-D-Galp-(1→
                          6
                          ↑
                          1
                       β-D-Glcp
                          4
                          ↑
                          1
                       β-D-Galp
```

Scheme 1.

```
→2)-L-Rhap-(1→3)-L-Rhap-(1→2)-L-Rhap-(1→
                       2
                       ↑
                       1
                     D-Glcp
                       6
                       ↑
                       1
                     D-Glcp
```

Scheme 2.

```
→3)-6-deoxy-L-Talp-(1→3)-L-Rhap-(1→3)-D-Galp-(1→2)-L-Rhap-(1→
                           4
                           ↑
                           1
                       D-GlcUAp
```

Scheme 3.

The sugar sequences of the glycans have been determined by an integrated analytical scheme based on methylation, g.l.c., m.s., periodate oxidation, enzymic hydrolysis, and chemical degradation (10,11). Paper chromatography was employed extensively for the identification of the constituents of the glycans and for isolation of hydrolytic products from these glycans. Of special value was the periodate oxidation reaction in the determination of the structure of the diheteroglycan of glucose and galactose. This glycan was oxidized exclusively at the galactose residues by periodate because of hemi-acetal bond formation between the aldehydic group of the oxidized galactose residues and the 3 position of the glucose residues (12). As a result, a glucosyl trisaccharide could be isolated from a mild acid hydrolysate of the oxidized glycan. On methylation analysis, this trisaccharide yielded 2 mol of 2,3,4,6-tetra-*O*-methylglucose and 1 mol of 2,3-di-*O*-methylglucose, which after reduction and acetylation were identified by g.l.c.-m.s.

Additional structural data were obtained by acetolysis, acid fragmentation, enzymic hydrolysis, and elimination reactions. On the basis of these results, the total structure for the diheteroglycan of *S. faecalis* has been determined, and the structure for the repeating unit is shown in Scheme 1. The configuration of the linkages is β, as revealed by enzymic tests. The molecular structures of the di- and tetra-heteroglycans from *S. bovis* have also been deduced, and the proposed structures of repeating units for these glycans are shown in Schemes 2 and 3. In the diheteroglycan from *S. bovis,* the repeating units are joined by alternating (1→3)- and (1→2)-linkages. The complete structure for the tetraheteroglycan from *S. faecalis* has not yet been determined. However, it appears that this glycan consists of a main chain of rhamnose, galactose, and *N*-acetylgalactosamine with many side-chains of glucosyl phosphate units linked to the rhamnose units of the main chain.

REFERENCES

1. Heidelberger, M., *in* "Lectures in Immunology", Academic Press, New York (1956).
2. McCarty, M., *Harvey Lect. 65,* 73 (1971).
3. Wannamaker, L.W., and Matsen, J.M., *in* "Streptococci and Streptococcal Diseases", Academic Press, New York (1972).
4. Pazur, J.H., *Abstr. Papers Am. Chem. Soc. Meet. 31C* (1965).
5. Kane, J.A., Karakawa, W.W., and Pazur, J.H., *J. Immunol. 106,* 103 (1972).
6. Pazur, J.H., Cepure, A., Kane, J.A., and Hellerquist, C.G., *J. Biol. Chem. 248,* 279 (1973).
7. Pazur, J.H., Dropkin, D.J., Dreher, K.L., Forsberg, L.S., and Lowman, C.S., *Arch. Biochem. Biophys. 176,* 257 (1976).

8. Pazur, J.H., Miller, K.B., Dreher, K.L., and Forsberg, L.S., *Biochem. Biophys. Res. Commun. 70,* 545 (1976).
9. Pazur, J.H., and Dreher, K.L., *Biochem. Biophys. Res. Commun. 74,* 818 (1977).
10. Bjorndal, H., Hellerquist, C.G., Lindberg, B., and Svensson, S., *Angew. Chem. Int. Ed. Engl. 9,* 610 (1970).
11. Pazur, J.H., and Forsberg, L.S., *Carbohydr. Res. 60,* 167 (1978).
12. Pazur, J.H., and Forsberg, L.S., *Carbohydr. Res. 58,* 222 (1977).

GLYCOCONJUGATE RESEARCH / Volume I
Proceedings of the Fourth International Symposium on Glycoconjugates

Rat α-Lactalbumin: A Glycoprotein

Rajani Prasad, Billy G. Hudson, Ralph Butkowski, and Kurt E. Ebner

Rat α-lactalbumin is a glycoprotein (1), whereas, the α-lactalbumin from other species contains no or only a small amount of carbohydrate. The apparent mol. wt. from molecular-sieve columns and electrophoresis in sodium dodecyl sulfate was 26-28 000. The mol. wt. in guanidine hydrochloride was 16 000 and by sedimentation equilibrium was 15 400 ± 5%. The percent carbohydrate was 13.4. Three charge forms are present in gel electrophoresis and these charge forms may be separated on DEAE cellulose by using a linear gradient of KCl in 0.01 M Tris, pH 7.1. Form II represents about 80% of the α-lactalbumin. Treatment of the various forms with neuraminidase (*Vibro cholera*) results in the conversion of all three forms into a single, slower migrating species indicating that the charge differences are due to sialic acid.

Rat α-lactalbumin (70 mg) was reduced, alkylated with ethylenimine, and digested with trypsin (2). A major glycopeptide was isolated by chromatography on Bio-Gel P-10 and DEAE cellulose. The mol. wt. was 3976 by amino acid analysis and 4000 by thin layer chromatography. The amino acid and carbohydrate composition of this major glycopeptide is presented in Table I. The glycopeptide was sequenced in a Beckman sequencer using a DMAA program. The PTH amino acids were identified by high-pressure liquid chromatography

ISBN 0-12-301301-1

Table I. Amino Acid and Carbohydrate Composition of Tryptic Glycopeptide of Rat α-Lactalbumin

Component	Composition (res./mol) A^a	B^b
Arginine	1.00	1.00
Aspartic acid	3.45	3.0
Threonine	1.11	1.0
Glutamic acid	2.2	2.0
Glycine	2.4	2.0
Isoleucine	1.22	1.0
Leucine	1.10	1.0
Tyrosine	1.16	1.0
Phenylalanine	1.27	1.0
Mannose	3.25	
Galactose	1.85	
Fucose	0.95	
Sialic acid	1.37	
N-Acetylglucosamine	3.8	

[a] *Values based on 1 residue of arginine*
[b] *Values obtained from sequence data*

and back hydrolysis. Native rat α-lactalbumin was sequenced through the first 50 residues, and on the basis of these studies the glycopeptide was positioned between residue 44 and 57. The carbohydrate unit was located at Asn-45.

The amino acid sequence of the rat glycopeptide was compared to that of the bovine (Scheme 1), and it was apparent that there is a high degree of homology. The secondary structure of the rat and comparable peptide from the bovine was calculated according to the Chou and Fasman method (3). The rat glycopeptide has a β turn beginning at residue 44, whereas the β turn in the bovine begins at residue 45. Asn-45 in bovine lactalbumin is in

	44			47			50		
Bovine:	Asn-	Asn-	Gln-	Ser-	Thr-	Asp-	Tyr-	Gly-	Leu
Rat:	Asn-	Asn-	Gly-	Ser-	Thr-	Glu-	Tyr-	Gly-	Leu
		CHO							

	53			56		58
Bovine:	Phe-	Gln-	Ile-	Asn-	Asn-	Lys
Rat:	Phe-	Gln-	Ile-	Ser-	Asn-	Arg

Scheme 1.

the i position, which is rarely glycosylated (4), whereas Asn-45 in the rat is in the i + 1 position, which is more readily glycosylated. The β turn in the rat is more extended since it contains 5 residues instead of the normal 4 residues found in most glycoproteins (4).

REFERENCES

1. Brown, R. C., Fish, W. W., Hudson, B. G., and Ebner, K. E., *Biochim. Biophys. Acta, 441*, 82 (1977).
2. Brew, K., Castellino, F. J., Vangman, T. C., and Hill, R. L., *J. Biol. Chem., 245*, 4570 (1970).
3. Chou, P. Y., Adler, A. J., and Fasman, J., *J. Mol. Biol., 96*, 22 (1975).
4. Beeley, J. G., *Biochem. Biophys. Res. Commun., 76*, 1051 (1977).

Glycoprotein T: A Soluble Glycoprotein from Calf Thymus

Peter R. Rabin, G.S. Mason, and Edwin H. Eylar

A soluble glycoprotein, referred to as Glycoprotein T, has been isolated from calf thymus glands.

Calf thymus (2.5 kg) was homogenized in 10 liters of 0.15 M sodium acetate at 4°C. This was filtered through cheesecloth and centrifuged at 10 000*g* in a Sorvall GSA rotor for 30 min. The pink supernatant solution (Fraction I) was heated in glass flasks, in a 90°C water-bath, for about 20 min or until a yellow flocculent precipitate appeared. The flasks were cooled rapidly in an ice bath and centrifuged at 100 000*g* for 20 min. The yellow supernatant solution (Fraction II) was then filtered under vacuum through a layer of Celite 545 over Whatman 114 paper. This clear, yellow filtrate was ultrafiltered, in an Amicon TCIR, through a PM 10 filter, which had a 10 000-mol. wt. cutoff point. The ultrafiltrate (Fraction IIIA) contained many smaller mol. wt. factors including ubiquitin, thymosin, and Bach factor. The retentate (Fraction IIIB) was precipitated with solid ammonium sulphate to a final 50% saturation. This was stirred for 3 h at 4°C, and then centrifuged at 40 000*g* for 30 min.

The precipitate (Fraction IV) was washed with 50% saturated ammonium sulphate (w/v), centrifuged at 40 000*g* for 30 min, and then dissolved in 150 ml of water with stirring. It was dialyzed against distilled water for 2 days through 6000-8000 mol. wt. cutoff tubing. The undissolved material was centrifuged out, and the supernatant was applied to a CM-cellulose column (4.5x20.0 cm) equilibrated with 0.1 mM ammonium acetate, pH 6.0, at 4°C. After being washed with starting buffer to remove unadsorbed protein, Glycoprotein T was eluted with a linear gradient of ammonium acetate (pH 6.0, 1-500 mM). The fraction eluted between 0.21 M and 0.28 M was lyophilized and rechromatographed on a column of DEAE-cellulose (4.5cmx20.0cm), which had been equilibrated with 0.01 M ammonium hydrogencarbonate, pH 8.3.

ISBN 0-12-301301-1

The column was washed with 500 ml of 0.01 M ammonium hydrogencarbonate, pH 8.3. Glycoprotein T was eluted with a linear gradient of ammonium hydrogencarbonate, pH 8.3 (0.01 M-0.40 M). The fraction eluted between 0.15 M and 0.20 M, containing Glycoprotein T, was purified further on two columns of Sephadex G-75 fine. After lyophilization, the pure glycoprotein was stored at -20°C.

Glycoprotein T was shown to be homogeneous by poly(acrylamide) gel electrophoresis, at pH 8.6 and 2.7, after staining with Coomassie Blue or PAS stain. In SDS-PAGE at pH 8.6, dissociation occurs giving mol. wts. of 48 500, 32 000, and 16 000, indicating that the glycoprotein is composed of 3 identical subunits. Table I shows the amino acid composition of native and performic acid-oxidized Glycoprotein T, expressed as mol of amino acid per 100 mol of glycoprotein (mol. wt. 16 000).

Table I. Amino Acid Composition of Glycoprotein[a]

	Glycoprotein T			
Residues	*Unmodified, hydrolyzed for 24h*	*Oxidized, hydrolyzed for 24h*	*48h*	*72h*
Lysine	7.3	8.7	8.3	9.2
Histidine	2.3	2.5	2.2	2.1
Arginine	4.0	5.5	5.5	5.1
Aspartic acid	10.8	8.7	12.0	10.4
Threonine	6.2	4.2	5.6	3.2
Serine	6.6	3.9	6.6	3.5
Glutamic acid	11.3	13.8	15.6	14.1
Proline	7.4	7.0	7.4	8.0
Glycine	6.4	8.6	8.2	10.4
Alanine	10.4	9.5	9.1	7.8
Valine	8.0	8.0	9.0	7.7
Isoleucine	3.9	3.8	5.0	4.8
Leucine	10.1	8.2	10.6	9.5
Tyrosine	2.3	2.1	3.1	2.1
Phenylalanine	4.0	3.5	4.4	3.4
Cysteic acid		2.3	2.7	2.1
Methionine sulphone-sulphoxide		1.8	1.8	1.8

[a] *In mol of amino acid/100 mol of protein*

Table II. Carbohydrate Composition of Glycoprotein T

Residues	*Weight percent*
Galactose	*17.4*
Mannose	*21.9*
Glucose	*4.7*
N-*Acetylglucosamine*	*34.1*
Sialic acid	*21.8*

The carbohydrate composition given in Table II is expressed as weight percent. The total carbohydrate content was 7.8%. Glycoprotein T comprises as much as 10-15% of the soluble proteins from calf thymus glands.

No free NH_2-terminal amino acid was found after dansylation or Edman degradation, indicating a blocked NH_2-terminus. Carboxypeptidase A and B yielded valine as the CO_2H-terminal amino acid.

Antibody to Glycoprotein T was produced in rabbits that had been sensitized with three 1-mg doses of glycoprotein. Double-diffusion experiments showed a single precipitin band, indicating the purity of Glycoprotein T.

GLYCOCONJUGATE RESEARCH / Volume I
Proceedings of the Fourth International Symposium on Glycoconjugates

Structural Determination of Complex Carbohydrate Components by Field Desorption Mass Spectrometry

Vernon N. Reinhold

Molecular ionization in mass spectrometers by chemical, electron-impact, and field-ionization techniques requires that the sample be in the vapor state. Because of large mol. wt., polarity, or a combination of these properties, many sample materials cannot be volatilized without pyrolysis, which complicates subsequent mass spectral interpretation of structure. Low-mol. wt. polar compounds can in many cases be derivatized and analyzed with few complications, and many components of glycoconjugate structure are analyzed in this way.

An additional limitation to the mass spectral analysis of polar materials by use of electron-impact ion sources is the great propensity for bond rupture yielding high-intensity, low-mass fragments, and the absence of important parent ions. To obviate this complication, mass spectrometers with field and chemical ionization sources, sometimes called "soft-ionization" methods, have been developed with considerable success.

Although derivatization and soft-ionization techniques have extended the use of mass spectrometry for structural analysis, many compounds of considerable biological interest cannot be analyzed by these modifications (*i.e.*, salts of onium compounds and sulfate esters). Moreover, derivatization of carbohydrate oligomers adds considerable weight to the sample molecules, which imposes definite limits for instrumental reasons.

A new technique, which desorbs samples from the surface of a specially prepared emitter with the aid of an intense electrical field, is called field-desorption mass spectrometry (FDMS). Field desorption apparently induces ionization from the solid state and does so with a minimum of energy resident on the ionized species (soft ionization), thus providing ample parent ions for the determination of mol. wt. (Fig. 1).

ISBN 0-12-301301-1

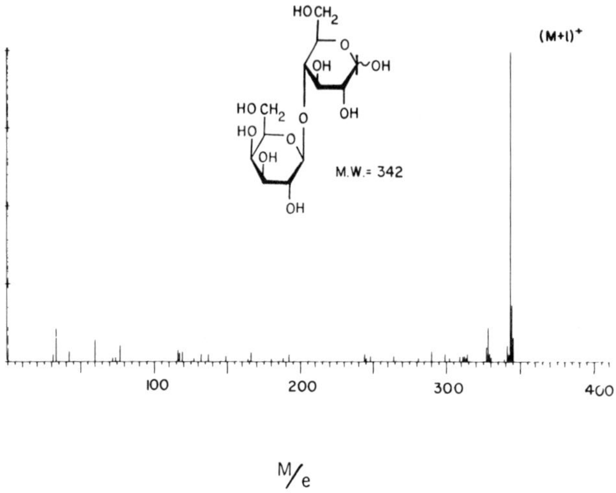

Fig. 1. M.s. of 4-O-β-D-galactopyranosyl-D-glucopyranose at 16 mA,

This technique makes possible the analysis of underivatized carbohydrate oligomers and other fragile compounds.

It is clear that the determination of mol. wt., as made by this technique, by itself has limited value for structural analysis. Desorption at higher emitter currents, however, induces sufficient energy into the desorbing sample that considerable fragmentation can be expected (Fig. 2). Fragmentation appears to proceed along pathways similar to those observed in electron-impact spectra. Ion intensity is very dependent on the heating current of the emitter.

For the mass spectra of carbohydrate oligomers, the labile glycosidic linkage offers an advantage not found in many other polymers, as fragmentation of this bond results in a charge carried by the cyclic oxygen atom of the reducing residue. As a consequence, for the trisaccharide raffinose (Fig. 3), the molecular ion is observed at *m/e* 504, the disaccharide- and monosaccharide-ion fragments at *m/e* 325 and *m/e* 163, respectively. This trisaccharide is a unique case where the (1→2)-glycosidic linkage between the glucose and fructose residues can rupture on either side of the bridging oxygen atom to yield a stable oxonium cyclic ion (*m/e* 343).

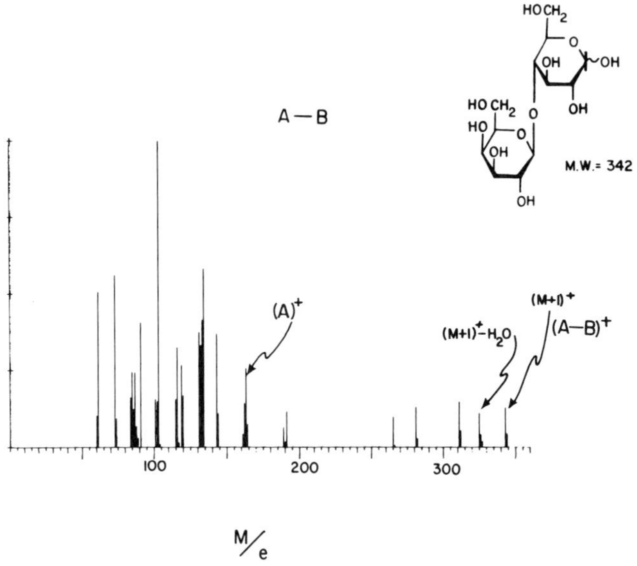

Fig. 2. M.s. of 4-O-β-D-galactopyranosyl-D-glucopyranose at 22 mA.

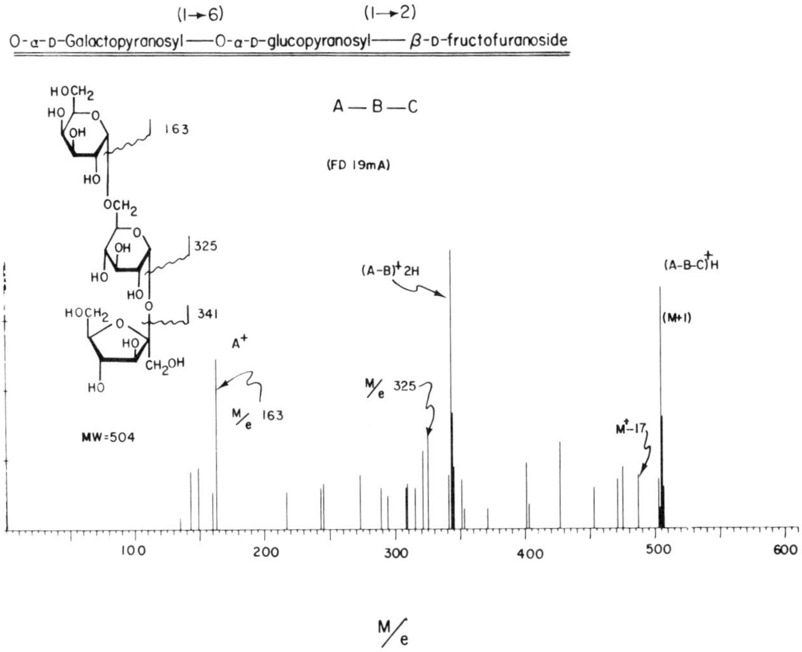

Fig. 3. M.s. of O-α-D-galactopyranosyl-(1→6)-O-α-D-glucopyranosyl-(1→2)-β-D-fructofuranoside.

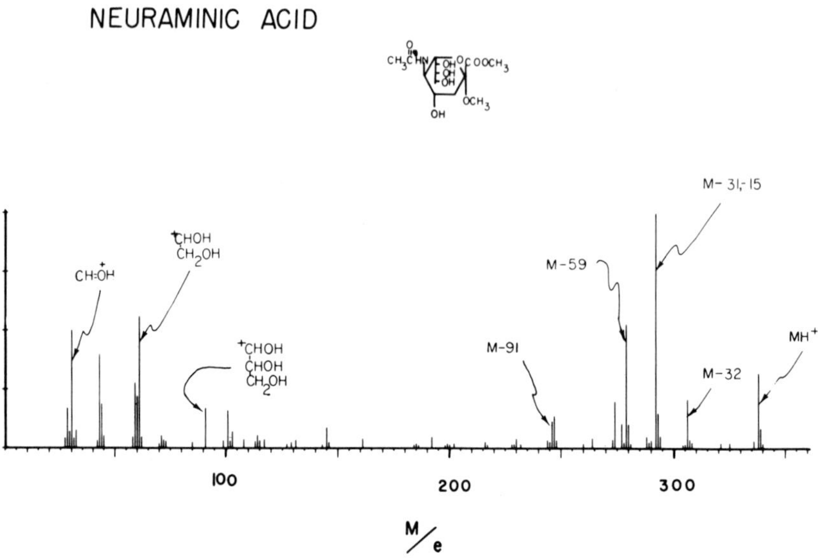

Fig. 4. M.s. of neuraminic acid.

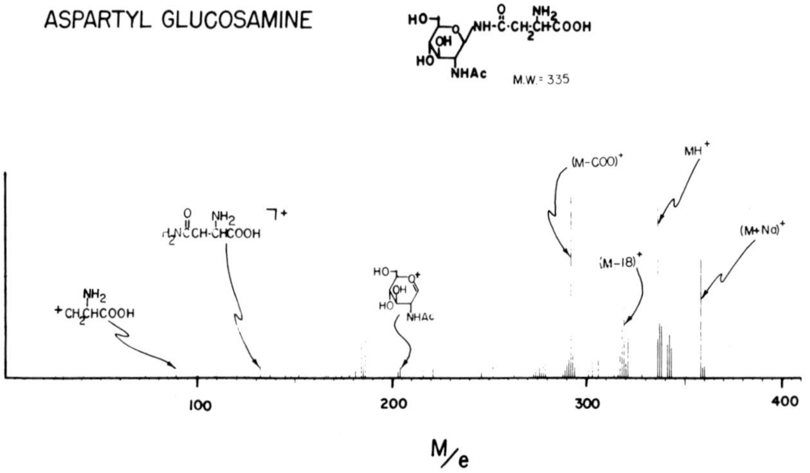

Fig. 5. M.s. of 2-acetamido-1-N-(L-aspart-4-oyl)-2-deoxy-D-glucopyranosylamine.

Desorption at higher emitter currents provides considerable fragmentation for detailed analysis of structure, as shown for neuraminic acid (Fig. 4). Many of the cleavages exocyclic to the pyranose ring observed here are identical with those of the derivatized sample ionized by electron impact.

The carbohydrate-protein linkage residue, *N*-aspartyl-*N*-acetylglucosamine, desorbs to provide considerable structural detail (Fig. 5). In addition to the molecular ion, an ion is observed at 23 atomic mass units higher, (23 amu = Na). The property of co-desorption with alkali metals has been designated cationization and is detected in many field-desorption spectra.

Fig. 6 presents the field-desorbed spectra of a synthetic phosphate diester analyzed in bacterial cell wall studies. The underivatized material desorbed to indicate the mol. wt. and considerable structure detail. Although these studies are preliminary and must be supported by much additional work, field-desorption mass spectrometry offers much interest in the study of glyconjugate structures.

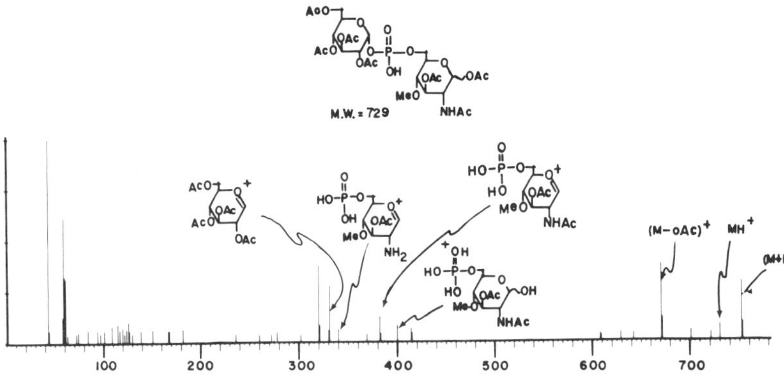

Fig. 6. M.s. of 2-acetamido-1,3-di-O-acetyl-2-deoxy-4-O-methyl-D-glucopyranose 6-(2,3,4,6-tetra-O-acetyl-α-D-glucopyranosyl phosphate).

ACKNOWLEDGMENT

This work was carried out in the Division of Research Resource Mass Spectrometry facility at Massachusetts Institute of Technology (Professor K. Biemann, Principal Investigator).

GLYCOCONJUGATE RESEARCH / Volume I
Proceedings of the Fourth International Symposium on Glycoconjugates

Glycoprotein of Peripheral Nerve (PNS) Myelin

M. Waheed Roomi, Armana Ishaque, Naseem Khan, and Edwin H. Eylar

PNS myelin contrasts markedly with CNS myelin in its protein composition and shows a wide variety of protein bands on SDS-PAGE. Wood and Dawson (1) found only one or two glycoproteins in PNS myelin, Singh and Spritz (2) found two glycoproteins in SDS gel electrophoresis of rat PNS myelin, one at mol. wt. 28 000 and the other at 22 000, and Kitamura *et al.* (3) reported two glycoproteins in bovine root myelin at mol. wt. 28 000 and 13 000. In the present study, we observed 9 protein bands in rabbit PNS myelin, four of which are glycoproteins. Fig. 1 shows the protein profile of defatted rabbit sciatic nerve myelin along with P2 and myelin basic proteins. The major protein bands are P0, P3, P4, BP, P5, and P2, with mol. wts. 28 000, 23 000, 18 000, 15 000, and 13 000, respectively. The gel on the right was stained with PAS reagent. Three major glycoproteins were found at the P0, P4, and BP bands, with a minor glycoprotein at the P6 band. The mol. wts. are 28 000 (P0), 23 000 (P4), 19 000 (BP), and 60 000 (P6). The dark, rapidly migrating band is a glycolipid, because no protein band was seen at this position in the Coomassie Blue-stained gel. A similar pattern was seen in human and chicken PNS myelin; thus is is likely that at least four glycoproteins exist in PNS myelin.

ISBN 0-12-301301-1

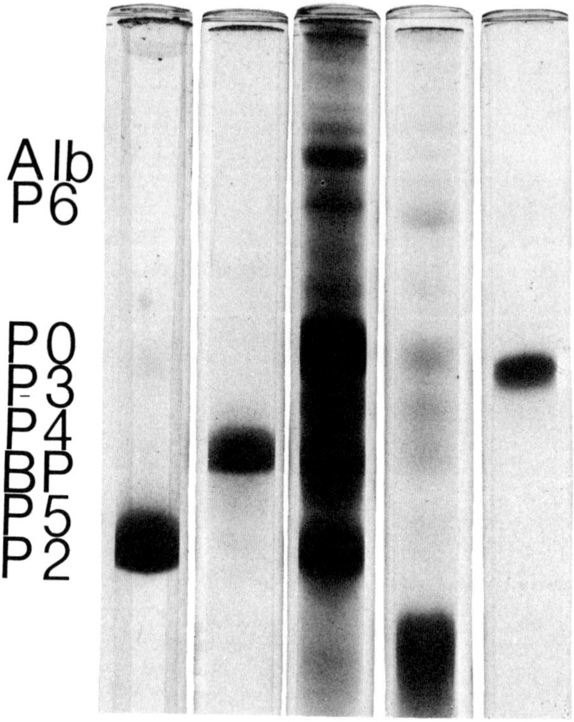

Fig. 1. Protein profile of various myelin proteins on SDS-PAGE. From left to right: P2 protein (20 μg), myelin BP (20 μg), defatted rabbit PNS myelin (100 μg), defatted rabbit PNS myelin (PAS stained, 300 μg), and P0 protein (15 μg).

The isolation and characterization of the major glycoprotein (P0) comprising 50% or more of the total glycoproteins was carried out by gel filtration on 0.5% Agarose in 2% SDS (0.1 M phosphate buffer, pH 7.2) of the acid insoluble residue of the defatted myelin. A solution of the insoluble residue (about 100 mg) of defatted myelin in 2 ml of 0.1 M sodium phosphate buffer (pH 7.2) containing 10% SDS was incubated at 55°C for 30 min and centrifuged. The supernatant was applied to the Agarose column and eluted with 0.1 M sodium phosphate buffer (pH 7.2) containing 2% SDS, 2.5-ml fractions being collected. The protein profile was monitored at 280 nm. Appropriate tubes were evaluated by SDS polyacrylamide gel electrophoresis and suitably pooled. The protein was precipitated by the addition of saturated NaCl, processed according to the method of Wood and Dawson (1), and rechromatographed twice to give pure PO protein. The isolated PO protein, devoid of detergent, is highly insoluble in aqueous solvent but appears homogeneous on PAGE-SDS, giving a mol. wt. of 28 000 (Fig. 1). PO is a glycoprotein as it stained with Coomassie Blue and PAS reagent on gels. It contains 6.3% of carbohydrate, which was determined as trifluoroacetate derivatives of methyl glycosides by g.l.c.: fucose 10.6%, galactose 8.1%, mannose 28.9%, *N*-acetylglucosamine 35.7%, and sialic acid 16.6%. The NH_2 terminal residue is isoleucine, determined as the dansyl derivative. The protein composition was calculated as mol %: Lys (9.0), His (2.5), Arg (7.0), Asp (8.9), Thr (4.3), Ser (6.9), Glu (9.0), Pro (4.6), Gly (10.0), Ala (6.5), Cys (1.6), Val (8.9), Met (1.2), Ile (4.4), Leu (7.4), Tyr (4.4), and Phe (3.9). These data indicate that the amino acid composition of PO protein is different from that of other myelin proteins, such as the P2, myelin BP, and proteolipid. Thus the suggestion of Wood and Dawson (1) that the PO and P2 proteins are related does not seem tenable. The NH_2 terminal sequence has been determined for 14 residues and is as follows: NH_2-Ile-Val-Val-Tyr-Thr-Asp-Pro-Glu-Val-X-Gly-Ala-Val-Gly-. This peptide region is quite hydrophobic and is in keeping with the main character of the PO protein.

REFERENCES

1. Wood, T. G., and Dawson, R. M. C., *J. Neurochem. 22,* 627 (1974).
2. Singh, H., and Spritz, M., *Biochim., Biophys. Acta 351,* 379 (1974).
3. Kitamura, K., Suzuki, M., and Uyemura, K., *Biochim. Biophys. Acta 455,* 806 (1976).

GLYCOCONJUGATE RESEARCH/Volume I
Proceedings of the Fourth International Symposium on Glycoconjugates

Glycoprotein Constituents of Lung Mucus Gel and Their Polypeptide and Carbohydrate Interactions

Mary Callaghan Rose, William S. Lynn, and Bernard Kaufman

The mucus layer that coats the surface epithelium of the airways contains macromolecules secreted by goblet and by submucosal glandular cells. These materials interact to form a gel that is insoluble in water and physiological saline. Copious amounts of lung mucus gel (LM-gel) are produced in diseases such as asthma, bronchitis, and cystic fibrosis. Our studies are concerned with the resolution and chemical characterization of the components of LM-gel, and with the conditions affecting their aggregation and gel-forming properties.

PURIFICATION

LM-gel, after extraction of soluble components by saline, contained: (a) high-mol. wt. mucin glycoproteins that do not enter 2% polyacrylamide-0.5% agarose gels, and (b) a number of lower mol. wt. proteins (150 000-13 000). LM-gel was solubilized by reduction-carboxymethylation in 1% sodium dodecyl sulfate (SDS), and the components were separated by chromatography on Sepharose 4B. Whereas LM-gel could be partially solubilized in 9 M urea or 6 M guanidinium chloride, the soluble components were not separated by subsequent chromatography.

The mucins were eluted in the void volume (V_O) of the Sepharose 4B chromatogram, comprised ≃25% by weight of the LM-gel, and contained 85% carbohydrate (glucosamine, galactosamine, galactose, fucose, and sialic acid, but no mannose). The mucins were included on a Sepharose 2B column and were further fractionated *via* hydroxylapatite (HA) chromatography. By use of 0.1% SDS and a stepwise phosphate gradient, 90-100% of the hexose content was recovered, whereas only 60-80% was recovered by DEAE chromatography or $BaSO_4$ precipitation.

ISBN 0-12-301301-1

The principal fraction (~80%) was eluted with 0.3 M phosphate and yielded a single peak on rechromatography. A minor component was obtained with 0.5 M phosphate.

POLYPEPTIDE INTERACTIONS INVOLVED IN GEL FORMATION

The effect of thiol agents on interactions of the components of LM-gel was investigated. LM-gel, reduced but not carboxymethylated, was chromatographed on Sepharose 4B. The pooled fractions were dialyzed extensively *vs.* 1 M urea-1 M NaCl and then water, lyophilized, and redissolved in the original volume of saline. The 13 000 mol. wt. proteins did not form a gel, but remained as an insoluble granular precipitate. The mucin fraction (V_0) formed a gel after removal of SDS and reducing agent, and could only be dissolved in a solvent containing both agents. The reduced, carboxymethylated mucin fraction, however, was soluble in saline alone at a concentration of 2 mg/ml, but formed a voluminous gel at concentrations of 8-10 mg/ml. The gel properties of the mucins were lost on proteolytic digestion with Pronase. Trypsin digestion of the reduced, carboxymethylated mucin(s) yielded a major glycopeptide and two minor hexose-containing fractions that were included in a Sepharose 4B column.

In summary: (a) The mucin components are clearly capable of forming a gel in the absence of the other protein components of LM-gel. (b) While thiol interactions are clearly implicated in aggregation and gel formation, additional polypeptide interactions are also involved. (c) After proteolytic digestion, the gel properties of mucins are lost and a series of glycopeptides are produced.

CARBOHYDRATE INTERACTIONS INVOLVED IN GEL FORMATION

One would anticipate that the carbohydrate chains of lung mucins are also involved in the formation of mucus gel, since carbohydrate comprises 85% of the mucin molecule and since the physical structure of at least one mucin, sheep submaxillary mucin, has been shown to be altered by incubation with purified neuraminidase. In order to determine whether the carbohydrate chains of human lung mucins (which contain very little sialic acid) are indeed involved in the formation or stabilization of the mucus gel, one must have a mechanism for specifically perturbing the carbohydrate structure. This, in turn, requires knowledge of the structure(s) of the oligosaccharide chains. Thus, we have examined some of the features of the oligosaccharide chains and the linkage to the polypeptide chain.

All the glycoproteins in the major lung mucin possess an oligosaccharide unit containing the blood-group hapten, as indicated by the constant blood-group activity (units/nmol of hexose) throughout sequential adsorption on *Phaseolus lunatus*-Sepharose 4B. Further evidence for the purity of the 0.3 M HA fraction was provided by the constant ratio of amino acids/glycine in the proteins being eluted in different segments of the peak. While there was no significant variation in amino acid composition across the peak, there was a variation in the total oligosaccharides present between the individual glycoprotein fractions, as shown by a variation in the content of glucosamine and galactosamine composition.

The behavior of lung mucin(s) under nonreductive β-elimination conditions was monitored by loss of blood-group activity and hexose. After 4 h, about 65% of the hexose remained in the nondialyzable fraction, although the blood-group activity had decreased by more than 99%. Under these conditions, the erythrocyte blood-group lipids are stable, and both the blood-group activity and hexose of pig submaxillary mucin (PSM) were essentially completely eliminated at the same rate. Under reductive β-elimination conditions, ~90% of the oligosaccharide chains of PSM were eliminated in 16 h, whereas ~50% of the oligosaccharide chains of lung mucins were eliminated. A second reductive β-elimination resulted in the concomitant production of galactosaminitol and 2-aminobutyric acid, indicating that the oligosaccharides in the nondialyzable fraction were also linked to serine (or threonine) residues. While the dialyzable material contained a trace proportion of amino acids, the content of serine and threonine was so low that only an insignificant amount of carbohydrate could be present in the dialyzate in the form of glycopeptides.

In summary: (a) There are apparently two classes of oligosaccharides associated with the lung mucins, one that contains the blood-group activity and is eliminated rapidly, and one that is eliminated much more slowly. (b) The production of 2-amino-butyric acid concomitant with the release of oligosaccharides containing galactosaminitol, in both the first and second elimination, suggests that some structural features, other than the presence of a second type of linkage (*e.g.*, asparaginyl), are affecting the rate of elimination of mucins.

GLYCOCONJUGATE RESEARCH / Volume I
Proceedings of the Fourth International Symposium on Glycoconjugates

Isolation and Chemical Characterization of Glycoproteins from Canine Tracheal Pouch Mucus

Goverdhan P. Sachdev, Owen F. Fox, Gary Wen, Terry Schroeder, Ronald C. Elkins, and Raoul Carubelli

Canine tracheal mucus was collected under aseptic conditions by weekly aspiration from surgically formed subcutaneous tracheal pouches of Beagle dogs (1). A mucus preparation suitable for column chromatography was obtained by disruption of the disulfide bonds with dithiothreitol in 6 M guanidine hydrochloride, followed by alkylation of the free sulfhydryl groups with iodoacetic acid.

The reduced, *S*-carboxymethylated mucus was first fractionated by gel filtration on Sephadex G-200. Chromatography of the excluded Sephadex G-200 fraction on Bio-Gel A-15m (see Flow Diagram) yielded three high-molecular-weight glycoprotein fractions. Following rechromatography on the same column, the main fraction (IIb) showed a single PAS-positive band on SDS-electrophoresis using composite gels containing 2% polyacrylamide and 0.5% agarose. This fraction behaved as a high-molecular-weight (*ca.* 500 000) glycoprotein with a high carbohydrate content (80%); the homogeneity of this glycoprotein was further established by the identification of arginine as its only amino-terminal amino acid (2). Chromatography of the included Sephadex G-200 fraction on DEAE-cellulose column, eluted with a NaCl gradient (0.1 to 0.8 M) in 5 mM sodium phosphate buffer, pH 7.0, yielded two homogeneous glycoproteins (SDS-electrophoresis on 5% polyacrylamide gels) of lower molecular weight, IIIa 30 000, and IIIb 26 500 (3). A single amino-terminal amino acid was detected for each glycoprotein: alanine in IIIa and glycine in IIIb.

ISBN 0-12-301301-1

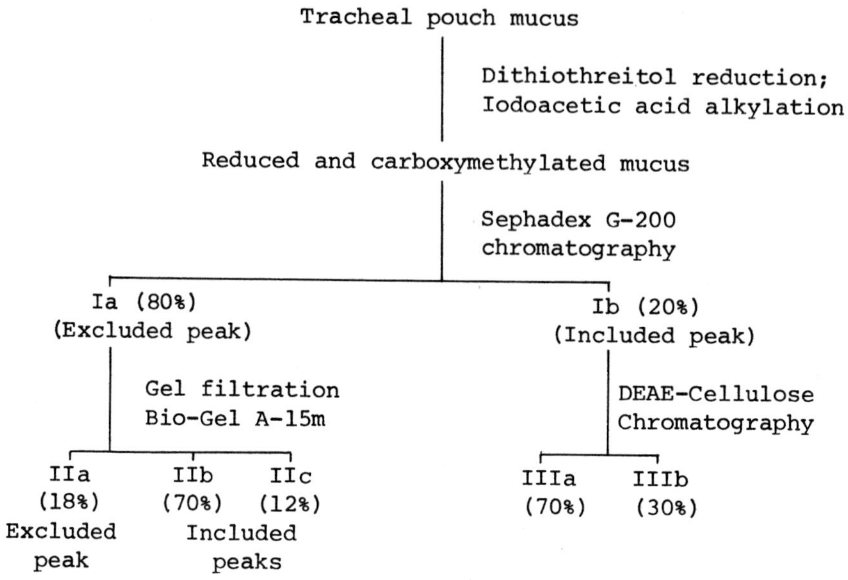

Flow Diagram

Table I. Composition of Canine Tracheal Mucus and of Three Homogeneous Glycoprotein Fractions[a]

Composition	*SCM-Mucus*[b]	*IIb*	*IIIa*	*IIIb*
Protein (4)	*50.00*	*14.50*	*61.00*	*70.50*
Fucose (5)	*11.52*	*14.31*	*5.75*	*4.77*
Galactose (5)	*15.51*	*15.71*	*7.73*	*5.99*
N-Acetylgalactosamine (5)	*6.90*	*30.10*	*8.73*	*5.80*
N-Acetylglucosamine (5)	*12.30*	*23.49*	*8.56*	*11.46*
Sialic Acid (6)	*6.00*	*2.10*	*7.00*	*5.40*
Sulfate (7)	*2.70*	*3.70*	*4.05*	*3.77*

[a] *Expressed in %*
[b] *SCM-Mucus: Reduced and carboxymethylated mucus*

These three glycoproteins contained fucose, galactose, *N*-acetylgalactosamine, *N*-acetylglucosamine, sialic acid, and sulfate monoester (see Table I). The high-molecular-weight glycoprotein (IIb) had a higher content of neutral hexoses, hexosamine, sialic acid, and sulfate, per mg protein, than the low-molecular-weight glycoproteins (IIIa and IIIb). Linkage of the sugar side-chains to hydroxy amino acids through *O*-glycosyl bonds was established by reductive β-elimination with 0.05 M NaOH containing 1 M $NaBH_4$ (8). A 60% decrease in serine plus threonine with concomitant equivalent gains in alanine plus α-aminobutyric acid, and a 60% loss of galactosamine accompanied by the appearance of an equivalent amount of galactosaminitol were observed.

ACKNOWLEDGMENT

This research was supported in part by Contract NO1-HR-52956 from the National Heart, Lung and Blood Institute.

REFERENCES

1. Wardell, J. R., Jr., Chakrin, L. W., and Payne, P. J., *Am. Rev. Resp. Dis. 101*, 741 (1970).
2. Woods, K. R., and Wang, K. T., *Biochim. Biophys. Acta 133*, 369 (1967).
3. Weber, K., and Osborn, M., *J. Biol. Chem. 244*, 4406 (1969).
4. Lowry, O. H., Rosebrough, N. J., Farr, A. L., and Randall, R. J., *J. Biol. Chem. 193*, 265 (1951).
5. Wang, C. S., Burns, R. K., and Alaupovic, P., *J. Bacteriol. 120*, 990 (1974).
6. Svennerholm, L., *Biochim. Biophys. Acta 24*, 604 (1957).
7. Terho, T. T., and Hartiala, K., *Anal. Biochem. 41*, 471 (1971).
8. Iyer, R. N., and Carlson, D. M., *Arch. Biochem. Biophys. 142*, 101 (1971).

Glycopeptides of Influenza Virus

Ralph T. Schwarz, Michael F. G. Schmidt, and Hans-Dieter Klenk

The carbohydrate moiety of the glycoprotein subunits HA_1 (mol. wt. 50 000) and HA_2 (mol. wt. 30 000) of the hemagglutinin of an influenza A virus (fowl plague virus, strain Dutch) has been analyzed. Analysis of glycopeptides obtained after digestion with Pronase indicates that the hemagglutinin has at least 2 different types of carbohydrate side-chains (Fig. 1). The side chain of type I, found on HA_1 and HA_2, is composed of glucosamine, mannose, galactose, and fucose. The corresponding glycopeptide has a mol. wt. of 2 600, suggesting the presence of about 12 monosaccharide units in the chain. The side chain of type II contains a high amount of mannose and is found exclusively on HA_2. The corresponding glycopeptide has a mol. wt. of about 2 000, suggesting the presence of about 9 monosaccharide units. The number of side chains per HA molecule appears to be in the range 5-6. Host specific variations in the carbohydrate content of the hemagglutinin are due to differences in size, and not in number of side chains.

The glycopeptides obtained from the other viral glycoprotein, the neuraminidase (mol. wt. 76 000), have also been analyzed. Both types of carbohydrate side-chains present in the hemagglutinin have also been detected in this glycoprotein (1).

ISBN 0-12-301301-1

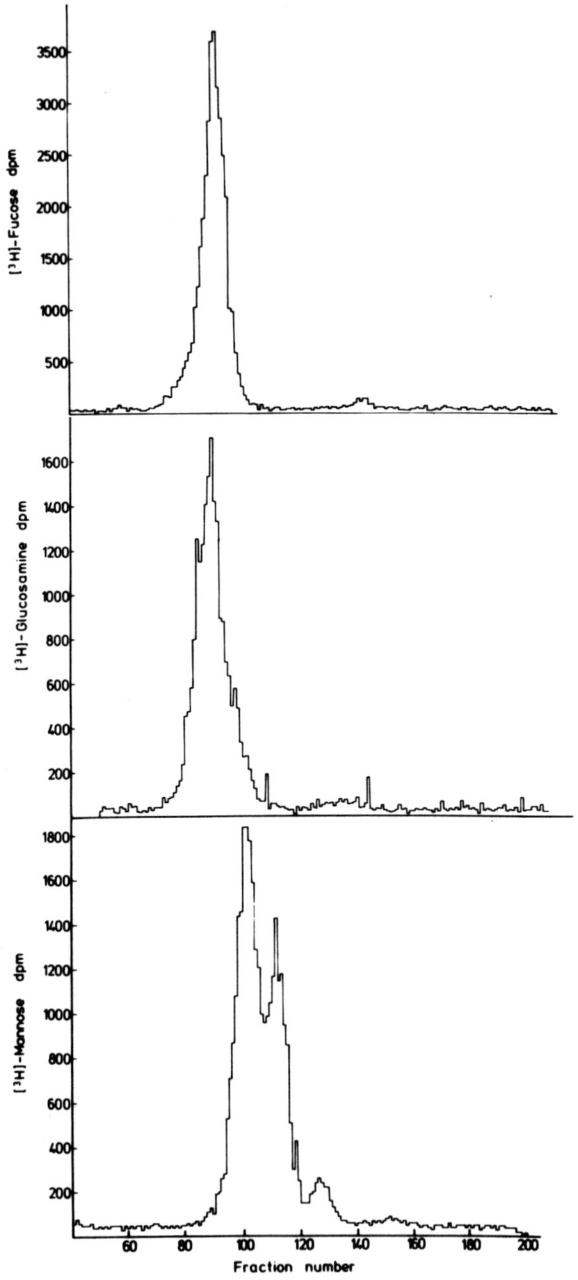

Fig. 1. Glycopeptides obtained from virions labeled with [1-^{3}H]fucose, [6-^{3}H]glucosamine, and [2-^{3}H]mannose. The fowl plague virus was grown in the presence of the radioactive

Figure 1 continued:

sugars in chick embryo cells. The purified virus was subjected to Pronase digestion and chromatography on Biogel P-6. After labeling with fucose and glucosamine, only one major peak representing type I glycopeptide could be detected. After labeling with mannose, an additional peak appeared at Fraction 114, corresponding to type II glycopeptide.

Under conditions of partial inhibition of glycosylation, virus particles have been isolated that contained glycoprotein with reduced carbohydrate content. Glycopeptide analysis indicated that this reduction was due to a lack of whole carbohydrate side-chains and not the incorporation of incomplete ones. This observation suggests that glycosylation of the viral glycoproteins involves block transfer of the core sugars (glucosamine and mannose) to the polypeptide chains (2, 3, M. F. G. Schmidt, L. Lehle, and R. T. Schwarz in this volume). Cell-fractionation studies indicate that this step of the glycosylation takes place in the rough endoplasmic reticulum. After migration of the hemagglutinin to the smooth endoplasmic reticulum, glycosylation is completed by the attachment of the peripheral sugars [galactose and fucose (4)].

REFERENCES

1. Schwarz, R. T., Schmidt, M. G. F., Anwer, U., and Klenk, H.-D., *J. Virol. 23*, 217 (1977).
2. Krag, S. S., and Robbins, P. W., *J. Biol. Chem. 252*, 2621 (1977).
3. Schwarz, R. T., Schmidt, M. F. G., and Lehle, L., *Eur. J. Biochem. 85*, 163 (1978).
4. Klenk, H.-D., Schwarz, R. T., Schmidt, M. F. G., and Wöllert, H., *in* "Topics in infectious diseases", vol. 3, Springer-Verlag, New York (1978).

The Exposure of the Carbohydrate of Ovalbumin

Virginia Shepherd and Rex Montgomery

It is well known that the asparaginyl carbohydrates from chicken ovalbumin (AC) can be fractionated into five fractions, AC-A through -E, by chromatography on Dowex 50W-X2 (1,2). Two of these fractions, AC-C and -D, have been further separated by chromatography of the borate complexes on Durrum DA-4 anion-exchange resin, each into two subfractions, AC-C_1 and -C_2, and AC-D_1 and -D_2 (3). The order of elution from a borate column showed that those fractions of AC with no terminal α-D-mannopyranosyl residues (AC-B and AC-C_1) formed very weak complexes with borate. Those fractions with one terminal mannosyl group (AC-A, -C_2, and -D_1) as determined by hydrolysis with α-D-mannosidase formed a complex with borate but to a lesser extent than AC-D_2 and -E that had several mannosyl residues external to the linkage core.

Compositional analysis of the new fractions showed that the principal components have the following mole ratios of residues: AC-C_1, $Asn_1GlcNAc_5Man_3$; AC-C_2, $Asn_1GlcNAc_4Man_6$; AC-D_1, $Asn_1GlcNAc_4Man_4$; and AC-D_2, $Asn_1GlcNAc_2Man_6$.

A comparison of the binding of the seven fractions of AC to concanavalin A was made using the procedure of Bessler *et al.* (4) in which the competition for the binding site with *p*-nitrophenyl α-D-mannopyranoside was measured spectrophotometrically. The binding affinities, relative to methyl α-D-mannopyranoside, are summarized in Table I, where Oval I and Oval II are fractions of native ovalbumin that do not bind (Oval I) or do bind (Oval II) to an affinity column of concanavalin A on Sepharose. Oval I contains the carbohydrate groups represented in AC-B, -C_1, and -D_1 and is eluted with the void volume. Oval II contains the carbohydrate components found in AC-A, -C_2, -D_2, and -E and was eluted with methyl α-D-glucopyranoside.

ISBN 0-12-301301-1

Table I. Relative Binding Capacities of Ovalbumin and AC Fractions to Concanavalin A

Ligand	Conc. to give 50% inhibition ($M \times 10^{-5}$)	Relative binding capacity[a]
Ovalbumin	3.0	5.3
Oval I		2.0[b]
Oval II	0.1	160.0
$Oval^{Man}$		1.2[b]
Oval II^{Man}		c
Ac-A	3.0	5.3
-B	4.5	3.6
$-C_1$		c
$-C_2$	2.0	8.0
-C	3.0	5.3
$-D_1$		2.5[b]
$-D_2$	1.8	8.9
$-D^{Man}$	2.0	8.0
-D		1.0[b]
-E	2.2	7.3
Methyl α-D-mannopyranoside	16.0	1.0
Methyl 2-acetamido-2-deoxy-β-D-glucopyranoside	70.0	0.2

[a] Relative to the concentration of methyl α-D-mannopyranoside to give 50% inhibition = 1.0

[b] Calculated from a ratio of K_L/K_M, where K_M equals the association constant for methyl α-D-mannopyranoside

[c] For the concentrations studied no binding was observed

It is noted that the smallest binding affinities are found in those ligands with no terminal mannosyl residues (Oval I, and AC-B, $-D_1$, and $-C_1$). The binding increases with the number of terminal mannosyl residues, somewhat parallel to the complex formation with borate. The binding of Oval II to the carbohydrate site of concanavalin A is about twenty times that of its component carbohydrate groups. It is proposed that the competition for the binding site is statistically enhanced in Oval II by protein-protein interactions.

Exhaustive enzymic hydrolysis of ovalbumin with α-D-mannosidase removed approximately 50% of the mannose that was hydrolyzed from the AC mixture by similar treatment. Such hydrolysis in either case eliminated the binding with concanavalin A at the carbohydrate site, in spite of the fact that there were terminal mannose residues in the degraded ovalbumin. It appears, therefore, that the most external mannose residues contribute most to the binding of ovalbumin to concanavalin A and that a portion of the mannose oligosaccharide is "buried" in the protein structure where it is not accessible for interaction with these proteins.

ACKNOWLEDGMENT

This research was supported by Grant GM 14013 from the U.S. Public Health Service.

REFERENCES

1. Cunningham, L.W., Ford, J.D., and Rainey, J.M., *Biochim. Biophys. Acta 101,* 233 (1966).
2. Huang, C.-C., Mayer, H., and Montgomery, R., *Carbohydr. Res. 13,* 127 (1970).
3. Shepherd, V., and Montgomery, R., *Abst. Am. Chem. Soc. Meet. 174 Biol.,* 74 (1977).
4. Bessler, W., Shafer, J.A., and Goldstein, I.J., *J. Biol. Chem. 249,* 2819 (1974).

GLYCOCONJUGATE RESEARCH/Volume I
Proceedings of the Fourth International Symposium on Glycoconjugates

Branched Ceramide Hepta- and Octasaccharides as Forssman Hapten Variants of Dog Gastric Mucosa

Amalia Slomiany and Bronislaw L. Slomiany

Forssman hapten, a heterophilic antigen of glycosphingolipid nature, was isolated from various sources (1,2) and its structure was established as a pentaglycosylceramide (2). A structural variant of Forssman hapten with regard to the internal carbohydrate chain composition has been also reported (3). Our studies on water-soluble glycolipids of dog gastric mucosa revealed the presence of several *N*-acetylgalactosamine-containing glycosphingolipids exhibiting Forssman antigenic activity (4). Three of these glycosphingolipids were immunologically characterized and their structures were determined (4-6).

METHODS

Extraction of lipids from dog mucosa scrapings was performed by the tetrahydrofuran procedure (5). After diethyl ether partition, dialysis, and silicic acid column chromatography, the partially separated, water-soluble glycolipids were applied to DEAE-Sephadex A-25 columns (5). The neutral glycolipid fractions from these columns were subjected to alkaline methanolysis and further purification on thin-layer plates in several solvent systems (5,6).

Methyl glycosides, methyl esters of fatty acids, long-chain bases, and alditol acetates of monosaccharides and of partially methylated sugar derivatives were obtained according to the procedures detailed previously (5,6). The products were analyzed by g.l.c. Partial degradation of the glycolipids was performed with formic acid, methanolic HCl, and trichloroacetic acid (5,6).

ISBN 0-12-301301-1

Sequential degradation of the native glycolipids and products of partial acid hydrolyses were carried out with glycosylhydrolases. The immunological activity of the analyzed compounds was tested by double diffusion micromethod and by hemagglutination-inhibition assay (4,6).

RESULTS AND DISCUSSION

The water-soluble fraction of dog gastric mucosa glycolipids consisted of three major and several minor neutral glycosphingolipids. The major glycosphingolipid components were purified to homogeneity by column and thin-layer chromatography (4-6) and were obtained in a yield of 3.9 mg of Glycosphingolipid I, 2.4 mg of Glycosphingolipid II, and 1.1 mg of Glycosphingolipid III per 100 g of wet mucosa scrapings. The composition, molar ratio of carbohydrates, and relative migration rates of the purified glycosphingolipids are given in Table I.

Table I. Composition, Molar Ratios of Carbohydrates, and Relative Migration Rates of the Native Glycolipids I, II, and III

	Molar ratios				Relative migration in solvent systems		
Glycolipid	*Fuc*	*Gal*	*Glc*	*GalNAc*	*a*	*b*	*c*
I		*1.98*	*1.0*	*1.95*	*0.25*	*0.47*	*0.40*
II		*2.91*	*1.0*	*2.97*	*0.22*	*0.42*	*0.41*
III	*0.98*	*2.95*	*1.0*	*2.91*	*0.18*	*0.15*	*0.22*

[a] *Chloroform-methanol-water 65:35:8 (V/V)*
[b] *Chloroform-methanol-NH_4OH 40:80:25 (V/V)*
[c] *Chloroform-methanol-acetic acid-water 55:45:5:5 (V/V)*

Glycolipid I. Partial acid hydrolysis (5) of this compound gave fragments identified as Gal→Gal→Glc→Cer, Gal→Glc→Cer, and Glc→Cer. The oligosaccharide portion of the glycolipid was resistant to sequential treatment with α-galactosidase, β-galactosidase, and *N*-acetyl-β-hexosaminidase, but was hydrolyzed to tetraglycosylceramide (GalNAc→Gal→Gal→Glc→Cer) by *N*-acetyl-α-galactosaminidase. The resulting glycolipid was cleaved to Glc→Cer by glycosylhydrolases in the sequence *N*-acetyl-β-hexosaminidase, α-galactosidase, and β-galactosidase.

These results together with permethylation studies (5) indicate that Glycolipid I has the structure: α-GalNAc-(1→3)-β-GalNAc-(1→3)-α-Gal-(1→4)-β-Gal-(1→4)-Glc→Ceramide.

Glycolipid II. Partial acid methanolysis resulted in the formation of Gal→Gal→Glc→Cer, Gal→Glc→Cer, and Glc→Cer. Hydrolysis of the native glycosphingolipid with *N*-acetyl-α-galactosaminidase and β-galactosidase resulted in the loss of one out of three *N*-acetylgalactosamine and galactose residues, respectively. Treatment of the native glycolipid with *N*-acetyl-α-galactosaminidase followed by *N*-acetyl-β-hexosaminidase led to the loss of two *N*-acetylgalactosamine residues, whereas cleavage with β-galactosidase followed by *N*-acetyl-β-hexosaminidase released one galactose and one *N*-acetylgalactosamine residue. A total loss of *N*-acetylgalactosamine and one galactose residue occurred when the enzymes were added in the order: *N*-acetyl-α-galactosaminidase, β-galactosidase, and *N*-acetyl-β-hexosaminidase. Enzymic degradation of the remaining portion of Glycolipid II (Gal→Gal→Glc→Cer) was accomplished as described for Glycolipid I. These data and the results of permethylation analysis (5) suggest the following structure for Glycolipid II:

α-GalNAc-(1→3)-β-GalNAc-(1→3)\
                                                  α-Gal-(1→4)-
    β-Gal-(1→3)-β-GalNAc-(1→4)/

-β-Gal-(1→4)-Glc→ceramide

Glycolipid III. Hydrolysis of Glycolipid III in trichloroacetic acid (6) resulted in the liberation of fucose and of a glycolipid having a chromatographic behavior and sugar composition identical with those of Glycolipid II. Sequential enzymatic degradation of the fucose-free compound revealed the identity of this portion of Glycolipid III with that of native Glycolipid II (5,6). Also, the permethylated defucosylated Glycolipid III and the native glycolipid II gave identical sugar derivatives. The data of enzymatic and chemical analyzes suggest that this compound is a branched octaglycosylceramide of the following structure:

                    α-GalNAc-(1→3)-β-GalNAc-(1→3)\
                                                                  α-Gal-(1→4)-
α-Fuc-(1→2)-β-Gal-(1→3)-β-GalNAc-(1→4)/

-β-Gal-(1→4)-Glc→ceramide

The data clearly indicate that Glycolipids I, II, and III share the terminal structure that determines immunological properties of Forssman antigen [α-GalNAc-(1→3)-β-GalNAc-(1→3)...]. Furthermore, all three glycolipids gave a single precipitation line with anti-Forssman antigen and showed identity, about equal reactivity, and weak blood-group A activity. The native Glycolipid III also exhibited blood-group H activity. Removal of fucose from this glycolipid led to loss of H-activity, but had no effect on its reactivity with Forssman antiserum nor on its ability to inhibit hemagglutination in the A-anti A system.

Our results (4-6) together with previously reported findings (3) clearly indicate the existence of numerous structural variants of the Forssman glycolipid hapten.

ACKNOWLEDGMENT

Supported by Grant No. AM-00068-25 from National Institute of Arthritis, Metabolism, and Digestive Diseases.

REFERENCES

1. Siddiqui, B., and Hakomori, S. I., *J. Biol. Chem. 246*, 5766 (1971).
2. Smith, E. L., McKibbin, J. M., Karlsson, K. A., and Pascher, T., *Biochim. Biophys. Acta 388*, 171 (1975).
3. Gahmberg, C. G., and Hakomori, S. I., *J. Biol. Chem. 250*, 2438 (1975).
4. Slomiany, A., Slomiany, B. L., and Annese, C., *FEBS Lett. 81*, 157 (1977).
5. Slomiany, A., and Slomiany, B. L., *Eur. J. Biochem. 76*, 491 (1977).
6. Slomiany, B. L., and Slomiany, A., *Eur. J. Biochem. 83*, 105 (1978).

GLYCOCONJUGATE RESEARCH / Volume I
Proceedings of the Fourth International Symposium on Glycoconjugates

Glyceroglucolipids: The Major Glycolipids of Human Gastric Secretion

Bronislaw L. Slomiany, Amalia Slomiany, and George B. J. Glass

Our studies on glycolipids of human gastric secretion (1,2) indicate that these compounds differ from the glycolipids of gastric mucosa with respect to their sugar composition and the nature of the lipid core. Whereas the lipid core of glycolipids found in the gastric mucosa contains sphingosine, the lipid core of glycolipids from gastric secretion consists of diglyceride (3,4). In this report, we summarize results of structural studies on the glyceroglucolipids and findings on the nature of ABH blood-group antigens in human gastric secretion.

MATERIALS AND METHODS

The human gastric secretion obtained from healthy individuals by gastric incubation was dialyzed, lyophilized, and extracted with chloroform-methanol (2:1). The extract was concentrated and fractionated on a DEAE-Sephadex column into neutral and acidic fractions (3-6). Further fractionation of the neutral and acidic lipids was accomplished on silicic acid columns (3,4). The columns were developed first with chloroform, followed by acetone, acetone-methanol (9:1), and methanol. T.l.c. in several solvent systems (5,6) was used as the final step in glycolipid purification.

The procedures used in the analysis of glycolipid components and the methods applied in structural studies were given previously (5,6). The conditions of alkaline degradation of the native and delipidated gastric secretion samples, and the assays of ABH blood-group activities were the same as reported (7,8).

ISBN 0-12-301301-1

RESULTS AND DISCUSSION

The lipids constitute 16-28% of dry weight of dialyzed and lyophilized gastric secretion. Of the lipid portion, 30% are glycolipids. Seven individual glycolipids have been purified from the neutral lipid fraction and two from the acidic. All neutral glycolipids are composed of glucose, glyceryl ether, and fatty acid groups. The glycolipids of the acidic fraction contain glucose, glyceryl ether, fatty acid, and sulfate groups. Glycosphingolipids were not detected among the purified compounds nor in the crude glycolipid fractions. Thus far, the structures of one acidic and three neutral glycolipids have been elucidated:

α-Glc-(1→3)-diglyceride
α-Glc-6-SO_4-(1→6)-α-Glc-(1→6)-α-Glc-(1→3)-diglyceride
α-Glc-(1→6)-α-Glc-(1→6)-α-Glc-(1→6)-α-Glc-(1→6)-α-Glc-(1→6)-α-Glc-(1→3)-diglyceride
α-Glc-(1→6)-α-Glc-(1→6)-α-Glc-(1→6)-α-Glc-(1→6)-α-Glc-(1→6)-α-Glc-(1→6)-α-Glc-(1→6)-α-Glc-(1→3)-diglyceride.

The diglyceride portion of these glycolipids consists mainly of 2-*O*-acyl-1-*O*-alkylglycerol. Our recent data indicate that glyceroglucolipids are also present in the human saliva and in the secretions from the dog Heidenhain-fundic pouch and rat stomach (9-11). Therefore, one may speculate that the mucous secretions of the pulmonary and reproductive tracts also contain glyceroglucolipids.

In another series of experiments, we have determined the blood type and its titer in gastric secretion of ten individuals of Se status. The lipids were then extracted and both delipidated residue and lipid fractions were analyzed for the presence of appropriate blood-group antigens. The native activity (0.8-0.4 μg/0.1 ml) persisted in the delipidated residue, but was not detectable in the total lipid extracts nor in the glycolipids purified therefrom. The alkaline degradation (7,8) of the native and delipidated samples led to a total loss of blood-group activity. However, similar alkaline treatment of the A-active glycosphingolipid, isolated from gastric mucosa, had no effect on its antigenic property. These data strongly indicate that blood-group antigens in human gastric secretion are exclusively of glycoprotein nature and that blood-group antigens of gastric mucosa are entirely of glycosphingolipid nature.

ACKNOWLEDGMENT

Supported by Grant No. AA-00312-4 from NIAAA and Grant No. AM-00068-25 from NIAMDD, NIH, PHS.

REFERENCES

1. Slomiany, B. L., Slomiany, A., and Glass, G. B. J., *Fed. Proc. 36*, 978 (1977).
2. Slomiany, A., Slomiany, B. L., and Glass, G. B. J., *J. Am. Oil Chem. Soc. 54*, 218 (1977).
3. Slomiany, A., and Slomiany, B. L., *Biochim. Biophys. Res. Commun. 76*, 115 (1977).
4. Slomiany, B. L., Slomiany, A., and Glass, G. B. J., *FEBS Lett. 77*, 47 (1977).
5. Slomiany, B. L., Slomiany, A., and Glass, G. B. J., *Eur. J. Biochem. 78*, 33 (1977).
6. Slomiany, B. L., Slomiany, A., and Glass, G. B. J., *Biochemistry 16*, 3954 (1977).
7. Slomiany, A., Slomiany, B. L., and Glass, G. B. J., *Blood 50*, 302 (1977).
8. Slomiany, A., Slomiany, B. L., and Glass, G. B. J., *Biochim. Biophys. Acta*, *540*, 278 (1978).
9. Slomiany, B. L., and Slomiany, A., *Biochem. Biophys. Res. Commun. 79*, 61 (1977).
10. Slomiany, B. L., Slomiany, A., and Glass, G. B. J., *Eur. J. Biochem. 84*, 53 (1978).
11. Slomiany, A., and Slomiany, B. L., *J. Am. Oil Chem. Soc. 55*, 239A (1978).

GLYCOCONJUGATE RESEARCH/Volume I
Proceedings of the Fourth International Symposium on Glycoconjugates

The Heterogeneity and Polydispersity of Articular Cartilage Proteoglycans

David A. Swann, Susan Powell, and Stuart Sotman

Proteoglycans (PG) and glycoproteins (GP) were extracted from bovine articular cartilage with 4 M GuaHCl, 0.05 M sodium acetate pH 5.8, and 5 mM benzamidine at 4°C for 48 h. After centrifugation (92 000*g*, 30 min, 40°C), the supernatant extract was fractionated by sedimentation into three density gradients (CsCl, initial density 1.60 g/ml, 170 000*g*, 64 h at 4°C) under dissociative conditions (1). Based upon the distribution of uronic acid, the first gradient (G-I) fractions were pooled to yield fractions with high (B), medium (M), and low (T) buoyant densities. The G-I-B, G-I-M, and G-I-T fractions were each refractionated into a second gradient (G-II) and they were then refractionated in a third gradient (G-III), each time B,M, and T fractions being collected. During this preparative procedure, the fractions were not exposed to conditions of low ionic-strength and lyophilization. The fractions were dialyzed and lyophilized prior to analysis as previously described (2).

The major portion of the extracted constituents was distributed into three fractions: A high-density proteoglycan fraction (PG-I); a medium-density proteoglycan fraction (PG-II) (these fractions did not contain detectable quantities of GP); and a low-density fraction (LDF) that contained PGs and GPs. The yields and abbreviated chemical compositions of these fractions are given in Table I.

The PG-I and PG-II fractions were then fractionated by gel-permeation chromatography on a Bio-Gel A-50 M column (100 x 4.5 cm) eluted with 0.5 M GuaHCl. The results are shown in Fig. 1 and the chemical composition of the fractions in Table II. Mol. wts. were determined by sedimentation equilibrium measurements on solutions in 0.5 M GuaHCl and 0.05 M sodium acetate (pH 5.8). Dithiothreitol (200 mM) was included for analysis of the PG-II A sample.

ISBN 0-12-301301-1

Table I. Yield and Composition of Fractions PG-I, PG-II, and LDF

		% w/w					Partial amino acid composition Residues/1000 residues						
Fraction	Yield[a]	Uronic acid	GlcN	GalN	SO_4	Amino acids	Asp	Thr	Ser	Glu	Pro	Gly	Arg
PG-I	61	19.5	7.7	16.3	16.2	16	64	65	116	151	104	111	34
PG-II	10	9.5	8.7	7.0	13.0	27	81	72	78	151	97	102	41
LDF	15	3.5	4.3	4.6	1.1	72	123	49	67	109	76	73	53

[a] Expressed as a percentage of total extracted constituents

Table II. Chemical Properties of Fraction PG-I and PG-II Fractions

Properties	PG-I					PG-II				
	A	B	C	D	E	A	B	C	D	E
Mol. wt. x 10^{-6}	1.85		1.27		1.02	1.39				
Protein (%)[a]	12.4	14.9	15.5	16.0	16.4	28.0	30.6	24.9	40.1	39.3
GlcN (%)[a]	4.2	5.8	6.5	8.0	9.6	7.7	8.5	8.2	5.1	2.8
GalN (%)[a]	17.5	16.0	12.5	8.3	7.4	8.2	8.9	6.5	6.6	7.4
Amino acid composition[b]										
Asp	69	62	65	70	67	81	83	82	110	118
Thr	63	64	62	71	69	65	64	67	59	53
Ser	128	124	111	103	94	86	87	85	75	78
Glu	148	151	152	164	154	141	142	143	124	123
Pro	90	100	103	108	106	98	102	110	83	75
Gly	139	124	113	110	103	97	92	88	79	81
Cys/2	11	6	10	9	14	17	11	15	18	21
Arg	37	40	37	43	37	42	48	50	46	44

a w/w
b Residue/1000 residue

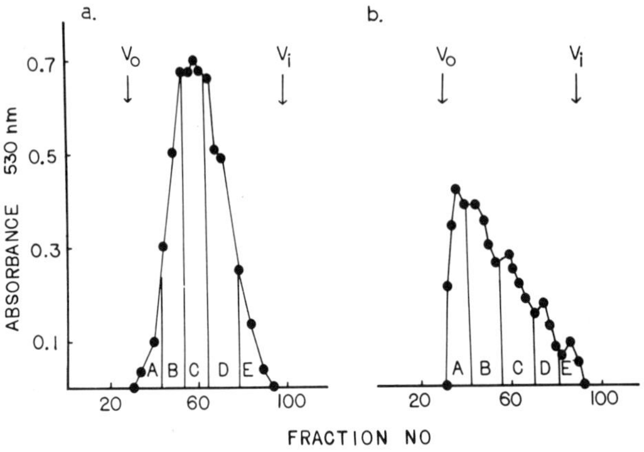

Fig. 1. Fractionation on a Bio-Gel 4-50 M column of fractions (a) PG-I and (b) PG-II.

Further fractionation of the LDF fraction (Table I) by associative density-gradient sedimentation, differential precipitation with CPC, and dissociative gel-chromatography yielded a proteoglycan fraction (PG-III) containing no detectable GP. PG-III had a composition similar to those of the PG-II-D and -E fractions (Table II). The three PG fractions differed in their ability to interact with HA. PG-I formed stable aggregates in the ultracentrifuge (2), whereas only extremely small quantities of a complex were observed with PG-II under identical conditions. No complex formation was observed with PG-III. These results indicate that articular cartilage contains three pools of proteoglycans having distinct physical and chemical properties.

The polydispersity of the high-density PG (PG-I) appears to be due both to differences in the length of the protein core (3-5) and to the degree of substitution with chondroitin sulfate (CS) chains (4). The calculated mol. wts. of the protein moiety in PG-1 fractions A, C, and E are 229 400, 197 000, and 167 000, respectively (Table II). The GlcN content per PG molecule in each fraction was constant, whereas the GalN content and the GalN/Serine ratio decreased with decreasing mol. wt. These data also support the view (3-5) that the CS and KS chains are present in different regions of the PG molecule. The similarity in the amino acid compositions of PG-I-A and PG-I-E fractions (apart from Ser and Gly), despite an almost two-fold change in the mol. wt. of the core protein, however, does not agree with the conclusion (3) that PG contains a 60 000 dalton HA-binding region with a composition significantly different from that of the remainder of the core protein.

ACKNOWLEDGMENT

Supported by funds from the Shriners Burns Institute, Boston Unit and NIH Grant No. AM 19834.

REFERENCES

1. Hascall, V.C., and Sadjera, S., *J. Biol. Chem. 241,* 2384 (1969).
2. Swann, D.A., Powell, S., Broadhurst, J., Sordillo, E., and Sotman, S., *Biochem. J. 157,* 503 (1976).
3. Heinegard, D., and Hascall, V.C., *J. Biol. Chem. 249,* 4250 (1974).
4. Hardingham, T.E., Ewins, R.J.F., and Muir, H., *J. Biol. Chem. 157,* 127 (1976).
5. Rosenberg, L., Wolfenstein-Todel, C., Margolis, R., Pal, S., and Strider, W., *J. Biol. Chem. 251,* 6439 (1976).

Subunit Structure of Rat Glomerular Basement Membrane

Sarah A. Taylor and Robert G. Price

The glomerular basement membrane (G.B.M.), the only continuous barrier between the plasma and the urinary filtrate, seems to act as a mechanical filter preventing the excretion of molecules with mol. wt. in excess of 65 000. Glomerular disease often results in thickening of the G.B.M., the selectivity of the membrane is lost, and serum proteins are excreted in the urine. Knowledge of structure and biosynthesis is a necessary prerequisite for the investigation of the mechanism of changes in G.B.M. in disease.

We have prepared normal rat G.B.M. by the method of Price and Spiro (1). The results we have obtained for the amino acid and sugar analyses of rat G.B.M. agree well with those found in the cow (2) and dog (3). The G.B.M. contains 10% of carbohydrate and is characterised by the presence of hydroxyproline, hydroxylysine, and a large number of glycine residues, indicating that G.B.M. is closely related to the collagen family of molecules. In addition to glucose and galactose, the G.B.M. contains a significant level of mannose, fucose, glucosamine, and galactosamine.

When the G.B.M. was solubilised in 0.1 M sodium phosphate buffer (at pH 7.1) containing 1% of SDS and 1% of 2-mercaptoethanol for 16 h at 37°C with shaking, a residue was obtained which accounted for about 20% of the membrane. The insoluble fraction was separated by low-speed centrifugation,

ISBN 0-12-301301-1

and the soluble fraction in the supernatant fraction was prepared for hydrolysis by extensive dialysis to remove the buffer, SDS, and mercaptoethanol. Table I shows the amino acid and sugar analyses of the soluble and insoluble fractions when compared to normal G.B.M. These data indicate that the insoluble fraction is more collagen-like, having higher levels of hydroxyproline, hydroxylysine, proline, and glycine than either the soluble fraction or whole G.B.M. The soluble fraction contains more polar glycoproteins with higher levels of aspartic acid, tyrosine, and lysine. However, although the soluble fraction has reduced levels of hydroxyproline and hydroxylysine when compared to whole G.B.M. and the insoluble fraction, these amino acids are still present. The disaccharide containing glucose and galactose increases in the insoluble fraction while decreasing in the soluble fraction. Glucosamine and galactosamine, two of the components of the heteropolysaccharide unit, were present in greater amounts in the soluble fraction than in the insoluble fraction.

Table I. Composition of Rat G.B.M., and Insoluble and Soluble Membrane Fractions[a]

	Composition (residues/1000)		
Component	*G.B.M.*	*Insoluble fraction*	*Soluble fraction*
Hyp	*64*[b]	*103*[b]	*28*[b]
Asp	*73*	*55*	*80*
Pro	*109*	*147*	*113*
Gly	*185*	*257*	*163*
½ Cys	*14*	*7*	*11*
Tyr	*19*	*10*	*16*
Lys	*27*	*16*	*30*
Hyl	*16*	*38*	*8*
GlcGalHyl	*12*	*22*	*2*
GlcN	*7*	*3*	*10*
GalN	*1*	*0.6*	*2*

[a] *Only values for key components are given*
[b] *Determined by Woessner method (4)*

We have separated twelve major and ten minor components from the soluble fraction by SDS electrophoresis on 5% polyacrylamide gels. These components were in the mol. wt. range 25 000 to 240 000. The principal bands were isolated and analysed for amino acid composition. The amount of each component amino acid varied considerably between the bands. In three of the bands, glycine values were similar to those found in the insoluble fraction, suggesting the presence of collagen-like components. However, these bands showed values for hydroxyproline lower than expected. This finding was confirmed by the Woessner (4) procedure. We therefore found a greater range of hydroxyproline to glycine ratios in the soluble components of rat G.B.M. than Sato and Spiro (5) reported for the components present in the subunits of the soluble fraction of bovine G.B.M.

It is hoped that these studies of the structure of normal rat G.B.M. will prove a useful basis for future investigations into the changes of this membrane during disease.

ACKNOWLEDGMENT

This work was supported by the Wellcome Foundation and the National Kidney Research Fund, U.K.

REFERENCES

1. Price, R. G., and Spiro, R. G., *J. Biol. Chem. 252,* 8597 (1977).
2. Spiro, R. G., *J. Biol. Chem. 242,* 1915 (1967).
3. Kefalides, N. A., *Biochemistry 5,* 702 (1966).
4. Woessner, J. F., *Arch. Biochem. Biophys. 93,* 440 (1961).
5. Sato, T., and Spiro, R. G., *J. Biol. Chem. 251,* 4062 (1976).

GLYCOCONJUGATE RESEARCH / Volume I
Proceedings of the Fourth International Symposium on Glycoconjugates

Chemical Characterization of Bovine Erythrocyte Glycolipids

Keiichi Uemura and Tamotsu Taketomi

Several glycosphingolipids were isolated from bovine erythrocyte stroma. Two major neutral glycolipids were characterized as: lactosylceramide, and α-D-Gal-(1→3)-β-D-Gal-(1→4)-β-D-GlcNAc-(1→3)-β-D-Gal-(1→4)-β-D-Glc-(1→1)-Cer. Two major gangliosides were NeuGl-(2→3)-β-D-Gal-(1→4)-β-D-Glc-(1→1)-Cer and NeuGl-(2→3)-β-D-Gal-(1→4)-β-D-GlcNAc-(1→3)-β-D-Gal-(1→4)-β-D-Glc-(1→1)-Cer. The ceramide moiety of each glycolipid contained predominantly $d_{18:1}$ sphingosine, and normal fatty acids of $C_{16:0}$, $C_{22:0}$, $C_{24:0}$, and $C_{24:1}$.

Glycosphingolipids of mammalian erythrocytes differ from species to species. Little is known about glycolipids of bovine erythrocytes, except the presence of glucosamine-containing ganglioside (1) and several glycolipids reported recently (2, 3). In this paper we describe the attempted isolation and chemical characterization of major glycolipids of bovine erythrocyte stroma.

MATERIALS AND METHODS

Isolation. Lipids were extracted from bovine erythrocyte stroma with chloroform-methanol as described (4), and crude sphingolipid was fractionated by DEAE-Sephadex column chromatography into neutral and acidic sphingolipids (5). Five neutral glycopipids and three gangliosides were isolated

ISBN 0-12-301301-1

by subsequent chromatography on silicic acid-Hyflo Super Cel columns and purified by rechromatography.

Analyses. Each preparation was examined by t.l.c. and i.r. spectra. Sugar, fatty acid, and long-chain base compositions were determined by g.l.c. (6), and sialic acids by the method of Yu and Ledeen (7). The sugar sequence was determined by partial acid hydrolysis, the linkage by methylation analysis (8) and periodate oxidation (9), and the anomeric configuration by CrO_3 oxidation (10) and specific rotation analysis.

RESULTS AND DISCUSSION

The carbohydrate compositions and partially methylated sugars of neutral glycolipids (N_1-N_5) and gangliosides (A_6-A_8) are summarized in Table I. No galactosamine was detected in any glycolipid, and gangliosides contained both NeuGl and NeuAc (17:3). Preparations obtained from five individual cattle showed similar distribution of glycolipids, though an intraspecies variation was reported (11). The structures of major glycolipids were identified as:

N_2 β-Gal-(1→4)-β-Glc-(1→1)-Cer

N_5 α-Gal-(1→3)-β-Gal-(1→4)-β-GlcNAc-(1→3)-β-Gal-(1→4)-β-Glc-(1→1)-Cer

A_6 NeuGl-(2→3)-β-Gal-(1→4)-β-Glc-(1→1)-Cer
NeuAc-(2→3)-β-Gal-(1→4)-β-Glc-(1→1)-Cer

A_7 NeuGl-(2→3)-β-Gal-(1→4)-β-GlcNAc-(1→3)-β-Gal-(1→4)-β-Glc-(1→1)-Cer
NeuAc-(2→3)-β-Gal-(1→4)-β-GlcNAc-(1→3)-β-Gal-(1→4)-β-Glc-(1→1)-Cer

The sugar sequence was derived from the g.l.c. analysis of products obtained by partial acid hydrolysis. When *N*-deacylated N_5 or *N*-deacylated asialo A_7 was treated with $NaIO_4$, the glucosamine unit was oxidized. A linkage β-D-Gal-(1→4)- rather than β-D-Gal-(1→3)-D-GlcNAc is probable. The results of the CrO_3 oxidation indicated that all sugars have the β-D configuration, except the terminal galactose unit of N_5. The specific rotation of N_5 also showed the presence of the α-D configuration. The optical rotation $[\alpha]_D$ of N_2, N_5, A_6, and A_7, was -3.4° (*c* 0.93, pyridine), +25.5° (*c* 0.94, pyridine), +4.1° (*c* 0.79, pyridine), and -2.8° (*c* 0.81, methanol), respectively. Compound N_5 has the same carbohydrate moiety as the glycolipid isolated from rabbit erythrocytes (12, 13). The major ganglioside (A_7) of bovine erythrocyte

Table I. Carbohydrate Composition and Partially Methylated Sugars of N_{1-5} and A_{6-8}

Composition	N_1	N_2	N_3	N_4	N_5	A_6	A_7	A_8
Proportion of total glycolipids				%				
	2.1	49.3	3.0	2.3	18.4	6.5	17.9	0.6
Sugar component				mol ratio				
Glc	1.00	1.00	1.00	1.00	1.00	1.00	1.00	1.00
Gal	0.29	1.02	1.86	1.85	3.06	1.06	1.96	2.72
GlcNAc	0	0	0.24	0.89	0.99	0	0.88	1.82
GalNAc	0	0	0	0	0	0	0	0
Sialic acid	0	0	0	0	0	1.11	1.25	2.01
Sialic acids				relative %				
NeuAc	0	0	0	0	0	28.2	10.1	34.0
NeuGl	0	0	0	0	0	71.8	89.8	66.0
Methylated sugars (as methyl glycosides)								
2,3,6-TriMe-Glc		+			+	+	+	
2,3,6-TriMe-Gal		-			-	-	-	
2,4,6-TriMe-Gal		-			+	+	+	
2,3,4,6-TetraMe-Gal		+			+	-	-	

was reported by Kuhn and Wiegandt (1). Minor glycolipids were not fully characterized, but the results suggest that N_1 consists of Glc-Cer and Gal-Cer, N_3 of Gal-Gal-Glc-Cer and GlcNAc-Gal-Glc-Cer, and N_4 of Gal-GlcNAc-Gal-Glc-Cer.

REFERENCES

1. Kuhn, R., and Wiegandt, H., *Z. Naturforsch, Ser. B. 19,* 80 (1964).
2. Moskal, J. R., Chien, J.-L., Basu, M., and Basu, S., *Fed. Proc. 34,* 645 (1975).
3. Chien, J.-L., Li, S.-C., Laine, R., and Li, Y.-T., *Fed. Proc. 36,* 731 (1977).
4. Taketomi, T., and Kawamura, N., *J. Biochem. (Tokyo) 72,* 791 (1972).
5. Ledeen, R. W., Yu, R. K., and Eng, L. F., *J. Neurochem. 21,* 829 (1973).
6. Uemura, K., Hara, A., and Taketomi, T., *J. Biochem. (Tokyo) 79,* 1253 (1976).
7. Yu, R. K., and Ledeen, R. W., *J. Lipid Res. 11,* 506 (1970).
8. Hakomori, S., *J. Biochem. (Tokyo) 55,* 205 (1964).
9. Taketomi, T., Hara, A., Kawamura, N., and Hayashi, M., *J. Biochem. (Tokyo) 75,* 197 (1974).
10. Laine, R. A., and Renkonnen, O., *J. Lipid Res. 16,* 102 (1975).
11. Yamakawa, T., Irie, R., and Iwanaga, M., *J. Biochem. (Tokyo) 48,* 490 (1960).
12. Eto, T., Ichikawa, Y., Nishimura, K., Ando, S., and Yamakawa, T., *J. Biochem. (Tokyo) 64,* 205 (1968).
13. Stellner, K., Saito, H., and Hakomori, S., *Arch. Biochem. Biophys. 155,* 464 (1973).

GLYCOCONJUGATE RESEARCH / Volume I
Proceedings of the Fourth International Symposium on Glycoconjugates

An Enzymatic Micromethod for the Determination of Hyaluronic Acid in the Presence of Excess Chondroitin Sulfate

Amina Vocaturo, John Baker, Giuliano Quintarelli, and Lennart Rodén

Several excellent methods have been described for the fractionation and quantitation of connective tissue polysaccharides (1). However, in the course of investigations of cartilage proteoglycans, we were unable to quantitate the small amounts of hyaluronic acid that are present in proteoglycan aggregates by any of the existing methods. In the present communication, we report the use of *Streptomyces hyalurolyticus* hyaluronidase, discovered by Ohya and Kaneko (2), for the enzymatic determination of hyaluronic acid in the presence of as much as a 1000-fold excess of chondroitin sulfate. The method is based upon an earlier procedure by Hatae and Makita (3).

MATERIALS AND METHODS

Streptomyces hyaluronidase [2000 turbidity units (TRU) per mg of protein] was obtained from Miles Laboratories, Elkhart, Indiana. Hyaluronic acid was isolated from human umbilical cords by digestion with papain and fractionation with cetylpyridinium chloride as described previously (1). It contained 46.3% uronic acid and 36.6% hexosamine. Uronic acid was determined by a carbazole method (4) and hexosamine by the procedure of Boas (5), omitting the resin treatment. Chondroitin sulfate was prepared from bovine nasal septa (1), and cartilage proteoglycan aggregate was isolated as described by Heinegård (6). Reducing terminal *N*-acetylglucosamine residues were determined by the Morgan-Elson procedure (7) modified as described below.

For the assay of hyaluronic acid, samples of this polysaccharide (2.5-25 μg) and chondroitin sulfate (0-10 mg) were digested with 3 TRU of *Streptomyces* hyaluronidase at 60°C for 3 h, in a final volume of 0.1 ml of 0.05 M sodium acetate, pH 5.0.

ISBN 0-12-301301-1

After incubation, 20 μl of 0.8 M potassium tetraborate, pH 9.1 (pH adjusted with HCl), were added to each tube, and the mixtures were heated to 100°C for 3 min, followed by immediate cooling. Ehrlich's reagent (0.6 ml) was then added, and the tubes were incubated at 37°C for precisely 20 min. After being cooled in ice-water, tubes containing chondroitin sulfate were centrifuged at 7000 r.p.m. for 10 min to remove the precipitate that had formed upon addition of Ehrlich's reagent, while the others were kept in ice-water. The absorbance of the solutions was measured at 585 nm. Standards contained 0.5 to 5.0 μg of *N*-acetylglucosamine.

RESULTS

Exhaustive digestion of hyaluronic acid in amounts ranging from 2.5 to 100 μg allowed subsequent quantitation of the polysaccharide up to approximately 80 μg, with a linear response to about 30 μg. The content of reducing terminal *N*-acetylglucosamine residues in the limit digest, which contains mainly tetra- and hexa-saccharides, amounted to 19.5% of the dry-weight of the polysaccharide.

Upon digestion of hyaluronic acid (10 μg) with 3 TRU of enzyme at 60°C, maximal liberation of terminal *N*-acetylglucosamine residues occurred after 2 h. Larger amounts of enzyme (5 or 10 TRU), or addition of fresh enzyme after 2 h, did not increase the extent of digestion. On the basis of these results, digestion with 3 TRU of enzyme at 60°C for 3 h was chosen as standard procedure, and the amount of hyaluronic acid was generally kept at 25 μg or less.

The presence of 10 mg of chondroitin sulfate in the reaction mixture did not affect the digestion of hyaluronic acid. However, the formation of a precipitate upon addition of Ehrlich's reagent necessitated centrifugation of the samples prior to the colorimetric determination. Furthermore, the chondroitin sulfate sample used in the present study contributed to the color in the Morgan-Elson reaction (the absorbance of 10 mg chondroitin sulfate was 0.096). When subtracted from the absorbance of the mixed digest, the value was similar to that for hyaluronic acid alone (0.25 for 10 μg). Application of the present method to the determination of the hyaluronic acid content of cartilage proteoglycan aggregate (A1) gave a value of 0.7%. This value is in agreement with values previously reported for the hyaluronic acid content of A1 (8).

While this work was in progress, a micromethod for the estimation of hyaluronic acid was reported by Hardingham and Adams (9). This method takes advantage of the highly specific association of hyaluronic acid with cartilage proteoglycans. The method described in the present paper has potential for greatly increased sensitivity, if the reaction products are reduced with borotritide of known specific activity, separated from undigested material, and quantitated. However, it is not presently known whether the fractionation of the reaction mixture could be readily accomplished.

ACKNOWLEDGMENTS

This work was supported by Grants DE 02670 and HL 11310 from the National Institutes of Health.

REFERENCES

1. Rodén, L., Baker, J.R., Cifonelli, J.A., and Mathews, M.B., *Methods Enzymol. 28,* 73 (1972).
2. Ohya, T., and Kaneko, Y., *Biochim. Biophys. Acta 198,* 607 (1970).
3. Hatae, Y., and Makita, A., *Anal. Biochem. 64,* 30 (1975).
4. Bitter, T., and Muir, H., *Anal. Biochem. 4,* 330 (1962).
5. Boas, N.F., *J. Biol. Chem. 204,* 553 (1953).
6. Heinegård, D., *Biochim. Biophys. Acta 285,* 181 (1972).
7. Reissig, J.L., Strominger, J.L., and Leloir, L.F., *J. Biol. Chem. 217,* 959 (1955).
8. Hardingham, T.E., and Muir, H., *Biochem. J. 139,* 565 (1974).
9. Hardingham, T.E., and Adams, P., *Biochem. J. 159,* 143 (1976).

Crystal Structure of α-D-Mannopyranosyl-(1→3)-β-D-manno-pyranosyl-(1→4)-2-acetamido-2-deoxy-D-glucose

Vincent Warin, F. Baert, R. Fouret, Gérard Strecker, Geneviève Spik, Bernard Fournet, and Jean Montreuil

The trisaccharide α-D-Man*p*-(1→3)-β-D-Man*p*-(1→4)-GlcNAc is a sequence found in the common pentasaccharide core, α-D-Man*p*-(1→3)-[α-D-Man*p*-(1→6)]-β-D-Man*p*-(1→4)-β-D-GlcNAc*p*-(1→4)-D-GlcNAc present in most of the glycans *N*-glycosically linked to asparagine (1). As the spatial conformation of the two "antennae" [neuraminyl (or fucosyl)-N-acetyllactosamine or oligomannoside residues conjugated to the pentasaccharide core] depends on the spatial conformation of the oligosaccharide residue located at the branching position, we have investigated the spatial conformation of the trisaccharide α-D-Man*p*-(1→3)-β-D-Man*p*-(1→4)-D-GlcNAc. This trisaccharide was first characterized in the urine of patients with mannosidosis (2). It was isolated from urine according to Strecker *et al.* (3) and crystallized by slow evaporation of a dioxane-water solution.

EXPERIMENTAL

The dimensions of the crystal were $0.2 \times 0.4 \times 0.7$ mm^3 and the crystal data are reported in Table I. The intensities for 4246 reflections were collected on an automatic Philips diffractometer operated in the Θ-2Θ scan mode at a minimum rate of 1°· min^{-1} with an invariable scan-width; 2099 reflections having intensities greater than one-third of the error σ (calculated according to counting statistics) were considered.

ISBN 0-12-301301-1

Table I. Crystallographic Data

Molecular formula:	$C_{20}H_{35}O_{16}N$
Molecular weight :	*545.49*
Crystal system :	*monoclinic*
Space group :	$P2_1$; $Z = 2$
Density :	$D_M = 1.50\ g.cm^{-3}$
λ *Mok*α :	*0.7107 Å*
Cell dimensions :	a = 9.894 (5) Å b = 10.372 (6) Å c = 11.816 (6) Å β = 95.03 (6) Å

STRUCTURE DETERMINATION AND RESULTS

The structure was solved by the direct method employing the Multan Program (4) with an extension of the magic-integer approach (5). Difference Fourier maps allowed to locate all the H atoms. The remaining H atoms were inserted at their expected positions. A last refinement of all the parameters of the atoms, except the isotropic thermal factor of the H atoms, converged to an R final factor of 0.059, $R = \Sigma(|Fobs-Fcalc|)/\Sigma(Fobs)$. The molecular structure is illustrated in Fig. 1. The bond lengths and bond angles involving carbon and oxygen atoms are drawn in Figs. 2 and 3. The Fourier maps indicate a partial (≃ 30%) substitution of the α-D anomer configuration of the GlcNAc molecule by the β-D configuration. The Man*p* and Glc*p* residues have the normal $^4C_1$ chair conformation. The intramolecular hydrogen-bond O-33 --- H --- O-25 (distance O-33 to O-25 = 2.686 Å and angle O-33 --- H --- O-25 = 170°), which stabilizes the cyclic D-GlcNAc*p* residue with regard to the cyclic β-D-Man*p* residue is observed. On the basis of these results, we propose a "T-conformation" rather than a "Y-conformation" for the glycan moiety of human serotransferrin (see Montreuil and Vliegenthart, this volume).

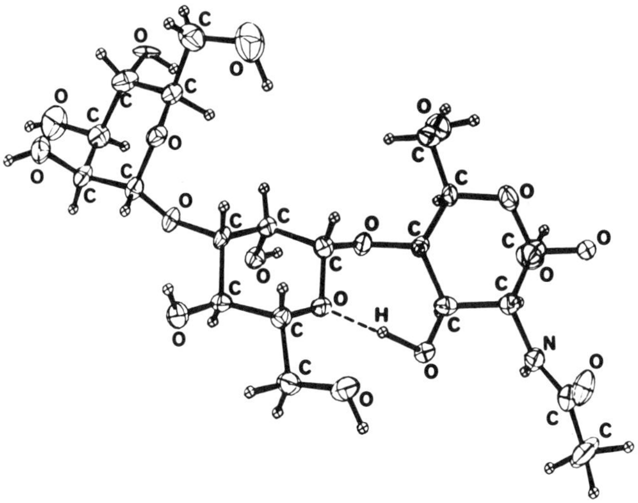

Fig. 1 Molecular conformation of α-D-Manp-(1→3)-β-D-Manp-(1→4)-D-GlcNAc.

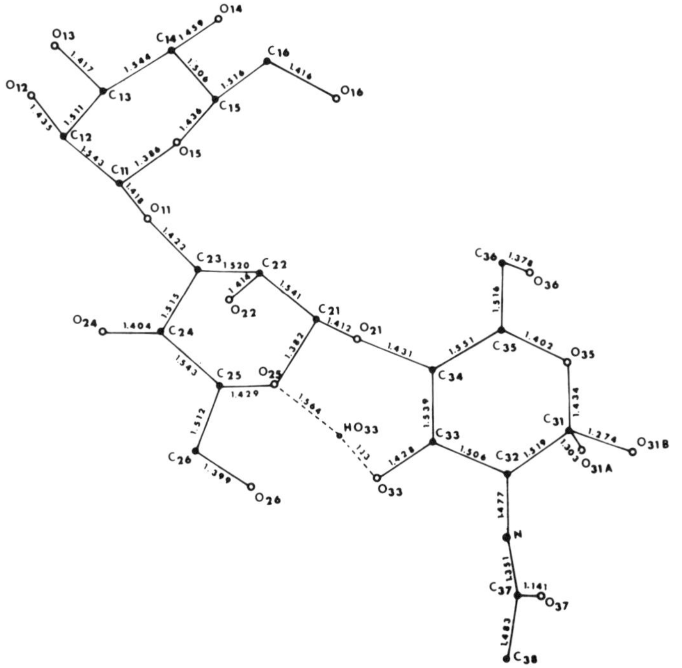

Fig. 2 Intramolecular bond distances (Å) of α-D-Manp-(1→3)-β-D-Manp-(1→4)-D-GlcNAc.

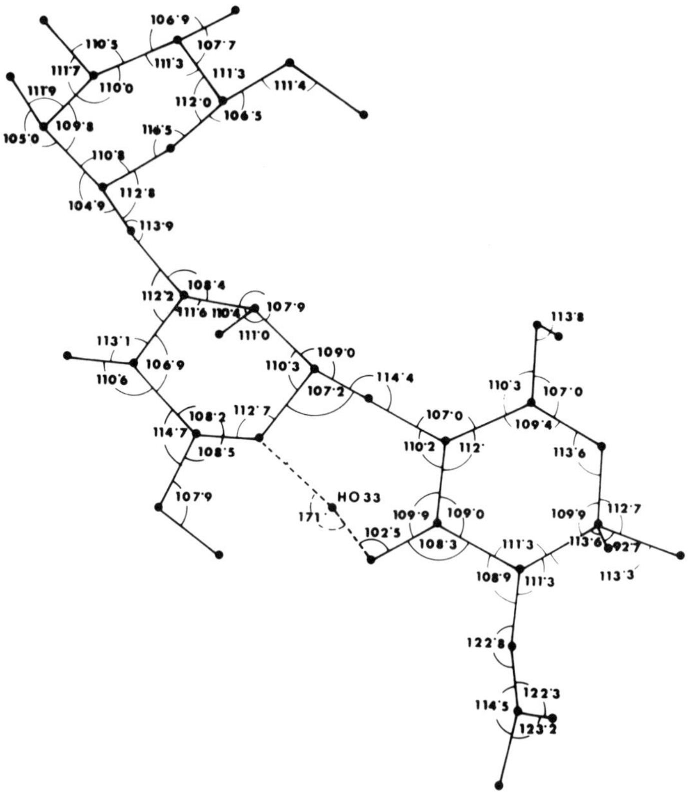

Fig. 3. Intramolecular bond angles (°) of α-D-Manp-(1→3)-β-D-Manp-(1→4)-D-GlcNAc.

REFERENCES

1. Montreuil, J., *Pure Applied Chem. 42*, 431 (1975).
2. Nordén, N.E., Lundblad, A., Svensson, S., Ockerman, P.A., and Autio S., *J. Biol. Chem. 248*, 6210 (1973).
3. Strecker, G., Fournet, B., Bouquelet, S., Montreuil, J., Dhondt, J.L., and Farriaux, J.P., *Biochimie 58*, 579 (1976).
4. Germain, P., Main, P., and Woolfson, M., *Acta Cryst. Ser. B 26*, 274 (1970).
5. Declercq, J.P., Germain, G., and Woolfson, M., *Acta Cryst. Ser. A 31*, 367 (1975).

GLYCOCONJUGATE RESEARCH / Volume I
Proceedings of the Fourth International Symposium on Glycoconjugates

Secondary and Tertiary Structure of Glycosaminoglycans and Proteoglycans

William T. Winter and Struther Arnott

Hyaluronic acid (HA), chondroitin sulfate (CS), and dermatan sulfate (DS) can all be expected to have similar, minimum energy conformations, since they share a common sequence of alternating (1→3) and (1→4) linkages joining successive pyranose rings. If we neglect the linkage trisaccharide at the reducing end of CS and DS, then the chemical differences between the molecules are restricted to the configuration at C-4 of the hexosamine residue, the presence of an ester linked sulfate group at C-4, C-6, or both of the hexosamine residue, and, in the case of DS, the configuration at C-5 of the uronic acid (residue). These variations in geometry and substituent-group location are reflected in altered charge distributions along the polyanion and hydrogen bonds within it, and are likely to manifest themselves in both packing and conformational changes. As a further corollary, the fact that chemical differences among HA, CS, and DS involve the charged groups, primarily suggests that the response of each polyanion to different cations might also be quite different. Equilibrium dialysis (1) and spectroscopic studies (2) have already indicated differential binding of cations by glycosaminoglycans in solution. Recent X-ray fiber diffraction studies on mono- and di-valent metal ion salts of HA (3-5) and CS (6,7) permit more detailed comparisons of the substituent group and cationic influences on conformation and packing.

HA and 4-(CS-4) and 6-sulfated (CS-6) chondroitins all undergo conformational transitions when small quantities of divalent cation, insufficient to balance the negative charge on the polyanion, are added to the polymer solution before film casting. For HA, the transition is from a helix with 4 disaccharide residues per turn (4-fold helix) to a 3-fold helix with a concomitant change from 0.85 to 0.95 nm in *h*, the height of

ISBN 0-12-301301-1

each disaccharide residue projected onto the helix axis. For CS-4, the change is from a 3-fold helix to a 2-fold helix with a slight increase in h from 0.95 to 0.98 nm, and for CS-6 the transition is from a helix with 8 disaccharide residues in 3 turns to a 3-fold helix, with h remaining nearly constant at 0.96 nm. The exchange of calcium for sodium by dialysis does not result in any further conformational or packing change, and the transition is not reversible by extensive dialysis against sodium salts. With HA, a similar conformational change has been effected using strontium salts in place of calcium (5). However, regular packing of the polyanion chains on a hexagonal net is not apparent with the strontium salts, suggesting that the ionic radius of the counterion may be an important parameter of these interactions.

Although the 3-fold helical conformers of both HA and CS-6 are stabilized by calcium ions, the addition of this ion to CS-4 results in the disappearance of the 3-fold helix. A probable explanation for this difference is that the orientation of the chondroitin sulfate molecules is sufficiently perturbed from the orientation typical of HA that sharing of a divalent ion by two anionic sites on the same or different molecules is precluded. These two orientations are shown schematically in Fig. 1. The possible biological relevance of these conformational changes arising from substituent group and cationic effects remains a subject for speculation.

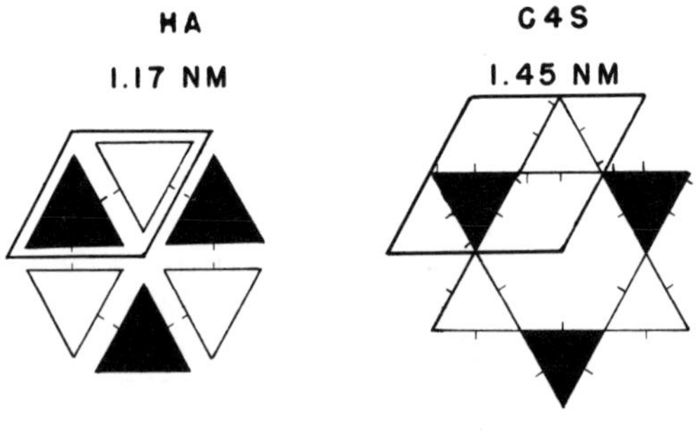

Fig. 1. A schematic representation of the packing of 3-fold helices in hyaluronic acid and chondroitin 4-sulfate. The view is down the helix axis, and alternate chains are shaded or outlined to indicate that they pass through the cell in opposite directions. The lines at the middle of each face denote the carboxylate group and those near the corner the sulfate group.

REFERENCES

1. Dunstone, J.R., *Biochem. J. 85,* 336 (1962).
2. Mathews, M.B., *Arch. Biochem. Biophys. 104,* 394 (1964).
3. Guss, J.M., Hukins, D.W.L., Smith, P.J.C., Winter, W.T., Arnott, S., Moorhouse, R., and Rees, D.A., *J. Mol. Miol. 95,* 359 (1975).
4. Winter, W.T., Smith, P.J.C., and Arnott, S., *J. Mol. Biol. 99,* 219 (1975).
5. Winter, W.T., and Arnott, S., *J. Mol. Biol. 117,* 761 (1977).
6. Cael, J.J., Winter, W.T., and Arnott, S., *J. Mol. Biol. 125,* 21 (1978).
7. Winter, W.T., Arnott, S., Isaac, D.H., and Atkins, E.D.T., *J. Mol. Biol. 125,* 1 (1978).

Cartilage Proteoglycans:

Symposium in Honor of

Martin B. Mathews

GLYCOCONJUGATE RESEARCH/Volume I
Proceedings of the Fourth International Symposium on Glycoconjugates

Introduction to Symposium

It is a special pleasure to chair this symposium for several reasons. Foremost is the fact that we are honoring a colleague with whom I have had the joy of interacting for more than 30 years. Additionally, we have assembled the leaders who have contributed to our understanding of the structure of proteoglycans. I believe that with the exception of Dr. Maxwell Schubert, there are present at this meeting all of the living scientists who have been responsible for significant work on this topic.

I first met Martin Mathews when he came to our laboratory to use equipment for Cohn fractionation of proteins. At that time, he was a graduate student with Dr. Birgit Vennesland. From this encounter, a friendship and collaboration grew which has continued to this day. When Martin joined our group, we were initially concerned with such problems as the purification of testicular hyaluronidase and the demonstration that this enzyme hydrolyzes chondroitin sulfate. Martin soon began to develop his productive studies on the physical chemistry of polyelectrolytes. He became concerned with the structure and function of chondroitin sulfate *in vivo*. From these studies emerged an appreciation of the importance of the chondroitin sulfate proteoglycan and the now classical proposal for the structure of this macromolecule. Although a precise physical chemist, Martin developed a keen interest in biology with particular emphasis on the evolutionary aspects of glycosaminoglycans. His unique studies on comparative biochemistry have been elegantly summarized in his recent book. This brief summary of Martin's studies of connective biochemistry only tell a small part of the role he has played in all of the work that has emerged from our laboratory. Over the years, he has been a major contributor to the overall planning and operations of the laboratories. His ideas, advice, and criticism have been of inestimable value to

ISBN 0-12-301301-1

all of our students and fellows as well as the senior members of the group. His contribution of samples and methods has made the work of others possible. Together with Tony Cifonelli and Lennart Rodén, the preparation and distribution of glycosaminoglycan standards have aided investigators throughout the world.

It is easy to summarize the impact of a scientist on the basis of their publications, but science has meaning only as a human activity. Those of us who have had an opportunity to work with Martin can better appreciate the character of this man. Martin is concerned with all the problems of mankind. He has devoted himself with great energy and self-sacrifice to many causes. In recent years, his concern for the alleviation of suffering has led to the development of a successful health center in the University of Chicago community. He has first and foremost always maintained an understanding for all people, distinguished and unknown.

For your contributions to science and to the well being of man, Martin, we salute you and thank you. With appreciation and affection, we dedicate this symposium to you.

GLYCOCONJUGATE RESEARCH / Volume I
Proceedings of the Fourth International Symposium on Glycoconjugates

The Link Proteins

John Baker and Bruce Caterson

The presence of protein or glycoprotein associated with protein-polysaccharides from cartilage was recognized by Mathews and Lozaityte (1). Later, Hascall and Sajdera (2) using equilibrium density gradient procedures isolated a fraction rich in this component (their 'GPL' fraction) and proposed a role for it in the aggregation of cartilage proteoglycans. Following the publication by Hardingham and Muir (3) of evidence that suggested an important role for hyaluronic acid in the aggregation of cartilage proteoglycans, interest in the protein component, or link proteins, changed to their possible role in stabilizing such aggregate structures. Studies by Gregory (4) and subsequent work by Hardingham and Muir (5) have served to indicate that the link proteins can perform this function.

A model has been proposed (6) for the structure of cartilage proteoglycan aggregates, which is consistent with most of the available data. In this model, three types of components: proteoglycan monomers, hyaluronic acid, and link proteins are seen as interacting to form stable aggregrate structures.

The work on the link proteins mentioned above was carried out with a link-protein fraction prepared by equilibrium density-gradient centrifugation (2). Accordingly, extracted proteoglycans were purified by equilibrium density-gradient

ISBN 0-12-301301-1

centrifugation (starting density 1.69), under associative conditions. Proteoglycans, including proteoglycan aggregates, were recovered from the bottom of the centrifuge tube [fraction A1 by the terminology introduced by Heinegård (7)]. Recentrifugation of this fraction under dissociative conditions gave a link protein-rich fraction at the top of the tube (A1D4 or A1D5, depending on whether the tube is cut into four or five segments). Such preparations still contain proteoglycan. Refinement of the separation by density-gradient centrifugation led to partial but not total removal of contaminating proteoglycans. The work described below deals with the purification of link proteins and the study of some of their properties and interactions.

RESULTS AND DISCUSSION

An A1D5 fraction, derived from bovine nasal cartilage, upon polyacrylamide-gel electrophoresis in SDS was seen to contain two major protein components of mol. wt. 4.7×10^4 and 5.1×10^4, which were termed link proteins 1 and 2, respectively. A third, fainter band of slightly greater mobility than link protein 2 and termed link protein 3 was also discernible. If the sample of A1D5 was labeled with fluorescamine prior to electrophoresis, other bands especially at the surfaces of stacker and running gels were evident. In contrast, much of this material which appears to be primarily protein-rich proteoglycans was not revealed by subsequent Coomassie Blue staining.

A preparative step that has been effective in separating link proteins from protein-rich proteoglycans in an A1D5 fraction is gel permeation chromatography on Ultrogel 34 in 1% SDS (9). Thus, protein-rich proteoglycans emerged in the void volume and a second included peak contained the purified link proteins 1, 2, and 3.

The two major link-proteins 1 and 2 have been separately isolated by preparative gel electrophoresis in SDS. For this purpose an apparatus based upon the design of Koziarz *et al.* (8), which can fractionate approximately 100 mg of protein, was employed. Fractions were obtained that contained only link proteins 1 or 2 as judged by analytical polyacrylamide-gel electrophoresis in SDS (Fig. 1).

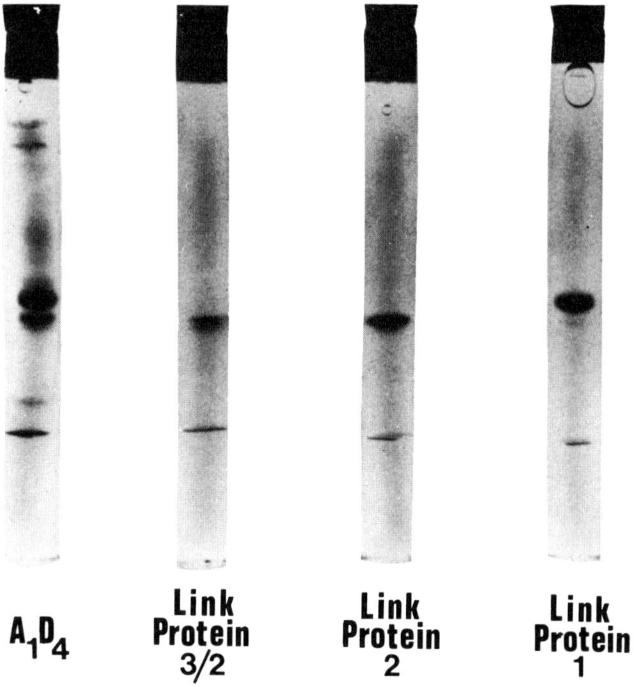

Fig. 1. Preparative polyacrylamide gel electrophoresis of the link proteins (A1D4). Shown are an analytical gel of the starting material, A1D4 (from left), and selected analytical gels through the elution profile from the preparative cell: a fraction enriched in link protein 3, and fractions containing purified link proteins 2 and 1.

Link protein 3 has a mobility too similar to that of link protein 2 to permit its complete separation from the latter. Noticeably, link protein 1 was spread over more fractions than link protein 2, which led us to believe that the former is more heterogeneous. Heterogeneity of glycoproteins commonly resides in their carbohydrate moieties, so the link proteins were stained by the PAS method after separation by polyacrylamide gel electrophoresis in SDS to determine whether link protein 1 was relatively richer in carbohydrate. Indeed, link protein 1 stained strongly and link protein 2 only weakly by this procedure. More convincing were the results of analyzing link proteins 1 and 2 for hexosamine content. From Table I it is evident that link protein 1 contains almost twice as much hexosamine as link protein 2. Otherwise, the amino acid analyses indicate a remarkable similarity in composition between the two link-protein preparations.

Table I. Composition of Link Proteins 1, 2, and 3

Amino acid	*Residues per 1000 residues amino acids* Link proteins 1, 2, and 3	Link proteins 1	Link proteins 2
Aspartic acid	134	135	133
Threonine	51	52	52
Serine	59	62	63
Glutamic acid	82	76	84
Proline	54	48	54
Glycine	98	104	103
Alanine	78	80	77
Valine	62	61	62
Methionine	3	3	2
Isoleucine	30	29	28
Leucine	83	80	82
Tyrosine	64	66	61
Phenylalanine	54	53	52
Histidine	29	29	26
Lysine	58	58	59
Arginine	61	64	62
Glucosamine	23	26	13
Galactosamine	10	10	8
Total hexosamine	33	36	21
GalN to GlcN[a]	0.42	0.37	0.58

[a] *Molar ratio*

It has previously been shown that protein-rich proteoglycans and link proteins after CNBr cleavage give different peptide profiles upon polyacrylamide gel electrophoresis in SDS (3). Particulary characteristic of the CNBr-cleaved link proteins was a sharp, prominent peptide band of mol. wt. approximately 2×10^4. Without reduction with mercaptoethanol after CNBr cleavage, no change in the separation pattern of the link proteins was seen. Further, it has been shown that both link protein 1 and 2 upon CNBr cleavage yield the same major peptide of mol. wt. 2×10^4, indicating that considerable portions of their primary structures are identical (10).

What is known of the distribution of the two major link proteins 1 and 2 in different cartilages is summarized in Table II. Evident from this data is that in tissues that are just beginning to synthesize proteoglycans only link protein 1 is found. In tissues where there is considerable proteolytic activity (*e.g.*, rat chondrosarcoma) only link protein 2 has been identified. The results indicate that possibly only link protein 1 is synthesized and that later, usually partial, proteolytic modification yields link protein 2. The finding that mild trypsin treatment of proteoglycan aggregate from bovine nasal cartilage leads to the recovery of link protein 2 alone supports this view.

Table II. Distribution of Link Proteins 1 and 2

Cartilage	*Link proteins* 1	*Link proteins* 2	*Ref.*
Bovine nasal cartilage	+	+	*15*
Bovine nasal, after mild trypsin treatment	-	+	*12*
Rhesus monkey nasal cartilage	+	+	*16*
Rhesus monkey articular cartilage	+	+	*16*
Sheep articular cartilage	+	±	*16*
Chick epiphyseal cartilage (day 13)	+	±	*16*
Chick limb buds (day 5-21)	+	-	*17*
Rat chondrosarcoma	-	+	*18*

Related to these results is our observation that upon reduction with mercaptoethanol, there is a decrease (approximately 20%) in the amount of link protein 1 and a corresponding increase in link protein 2. An explanation of this finding is illustrated in Fig. 2. A portion of the link proteins that migrate as link protein 1 have a break in their peptide core in a region bridged by a disulfide bond. Then, upon reduction with mercaptoethanol, the main body of the peptide, which is probably identical to link protein 2, is separated from a small, relatively carbohydrate-rich peptide.

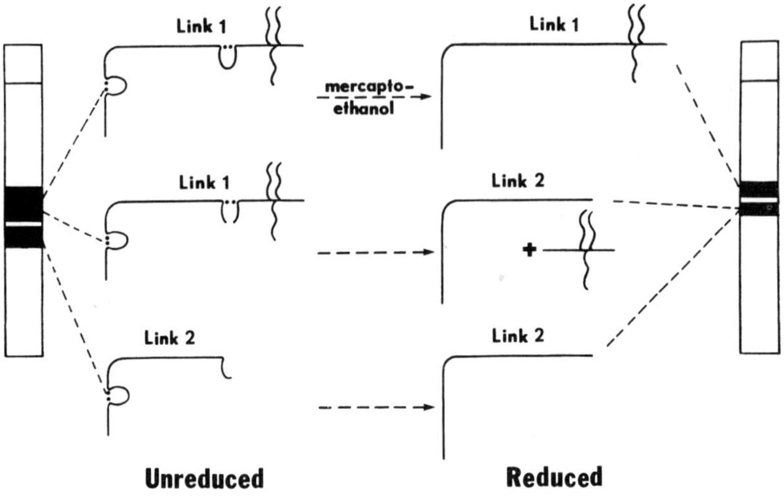

Fig. 2. A possible mechanism of the conversion of some link protein 1 to link protein 2 upon reduction. (Only two disulfide bridges per mol of link protein 1 are shown, although the total number is approximately five).

In Fig. 3 the possible interactions between proteoglycan monomers, hyaluronic acid, and link proteins are represented in the simplest fashion. Link proteins 1 and 2 are shown interacting identically and each in a one to one relationship with proteoglycan monomers. This is in approximate agreement with the relative amounts of link proteins and proteoglycan monomers found in aggregates, but it is not known whether the association of link protein with monomer is strictly one with one.

The interaction between proteoglycan monomers and hyaluronic acid (interaction A, Fig. 3) has been investigated in some detail by Hardingham and Muir (11), and by Hascall and Heinegård (12). Further aspects of this interaction are being investigated currently by Christner *et al.* (13). Heinegård and Hascall (12) have provided the only experimental indication of the interaction between hyaluronic acid and the link proteins (C). At present, there is no direct evidence for the interaction of link proteins with proteoglycan monomers (B), but some experiments as outlined below have been designed to determine whether this type of interaction is likely to occur.

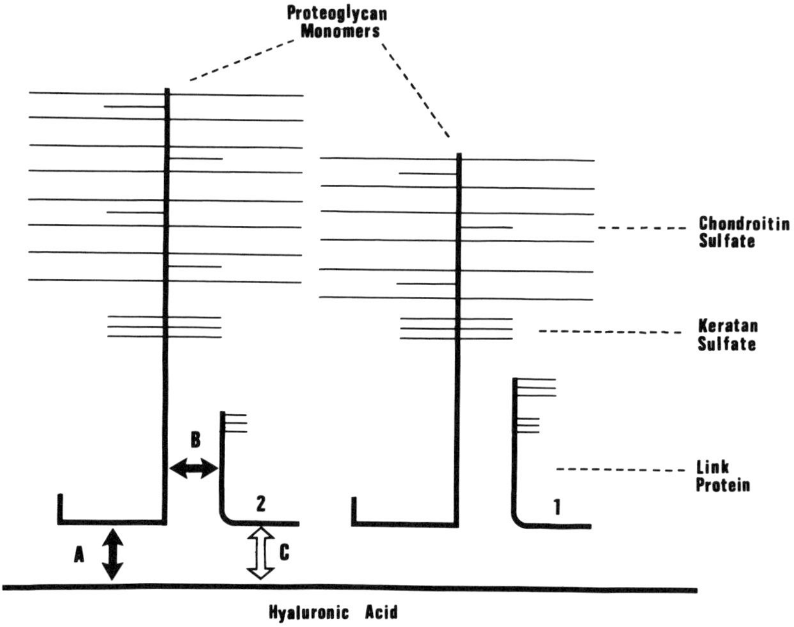

Fig. 3. Diagrammatic representation of proteoglycan monomer, hyaluronic acid, and link protein interactions.

An A1D5 preparation was partially acetylated with [^{14}C]acetic anhydride. The degree of acetylation produced under these conditions is insufficient to affect normal binding of proteoglycan to hyaluronic acid. The product, ^{14}C-A1D5, was chromatographed on a column of Sepharose CL-6B in 4 M guanidine. One major included peak of radioactivity preceded by four minor peaks were located. As judged by analytical polyacrylamide gel electrophoresis in SDS, fractions from the major peak contained only the link proteins and were pooled to give "^{14}C-link proteins". Then, ^{14}C-link proteins and a proteoglycan monomer (A1D1) preparation were mixed in 4 M guanidine (pH 5.8) and brought slowly to associative conditions (0.4 M in guanidine) by overnight dialysis against water. Cesium chloride was added to give a density of 1.66 and the mixture subjected to equilibrium density-gradient centrifugation. The contents of the centrifuge tube were fractionated to give 1.0-ml fractions (numbered 1 to 10 from bottom to top, respectively). Uronic acid analyses showed that the proteoglycan was confined to the bottom fractions (Fig. 4 b,d). A clear indication of proteoglycan-link protein interaction was provided by the close correlation between radioactivity and uronic acid in the eluted fractions (Fig. 4b).

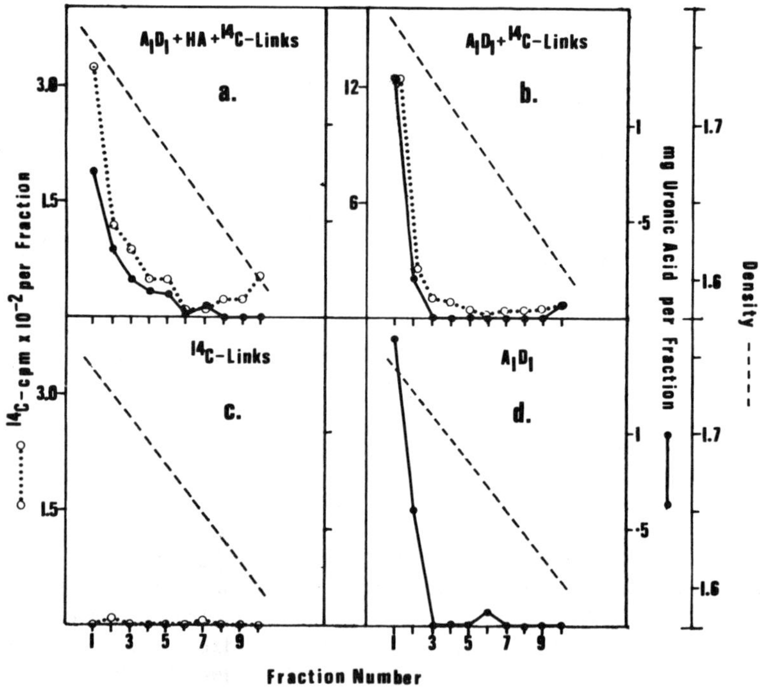

Fig. 4. Density gradient profiles of proteoglycan monomer, hyaluronic acid, and link protein mixtures. (a) A1D1 (400 μg/ml), hyaluronic acid (5 μg/ml) and ^{14}C-link proteins (1.2 μg/ml) in CsCl were centrifuged to equilibrium (starting density 1.66 g/ml). (b) A1D1 and ^{14}C-link proteins were at final concentrations of 400 and 3.2 μg/ml, respectively. (c) and (d) Individual concentrations were as in (b).

In the absence of proteoglycan, radioactivity due to link proteins was virtually undetectable in recovered fractions (Fig. 4c), which is doubtlessly due to the water insolubility of the link proteins. As would be expected, in the presence of proteoglycan and hyaluronic acid, the counts of link protein were closely associated with uronic acid in the bottom fractions (Fig. 4a). This experiment is only a beginning in the study of proteoglycan-link protein interactions. Remaining to be determined are the identities of the interacting sites, and the stoichiometry of the interaction.

Work by others (4,11) has clearly shown that preparations rich in link proteins (A1D4) can give stable aggregates with hyaluronic acid and proteoglycan monomers. It appeared desirable to establish whether purified link proteins could perform this same role, and we have employed Hardingham and Muir's procedure (3) for determining aggregate stability. Accordingly, a mixture of hyaluronic acid and proteoglycan monomers in approximately physiological proportions has a viscosity higher than that of proteoglycan monomers alone, but upon addition of hyaluronic acid decasaccharide, a drop in viscosity with time was observed. The oligosaccharide can successfully compete with hyaluronic acid for proteoglycan binding sites and thus dissociates the complex. In contrast, addition of the oligosaccharide to an aggregate (A1) preparation did not cause a drop in viscosity. Presumably, link proteins in this case stabilized the hyaluronic acid-proteoglycan monomer interaction. In the experiment, purified link proteins, hyaluronic acid, and monomers (A1D1) in 4 M guanidine and in near physiological proportions were dialysed to reach 0.4 M guanidine concentration, the oligosaccharide was added, and relative viscosities were determined at certain times thereafter (Fig. 5). Some drop in viscosity occurred with time, but not to the same extent as observed for hyaluronic acid and monomer without link protein. It remains to be established whether protein-rich proteoglycans can also stabilize the hyaluronic acid-proteoglycan monomer interaction or whether this is a specific property of link proteins.

It has been suggested (14) that the link proteins are markers of cartilage-specific proteoglycans. To prove this experimentally is difficult at present, as the only means of identifying link proteins is rather unsatisfactory, *i.e.*, the recognition of protein bands with characteristic mobilities on polyacrylamide gel electrophoresis in SDS.

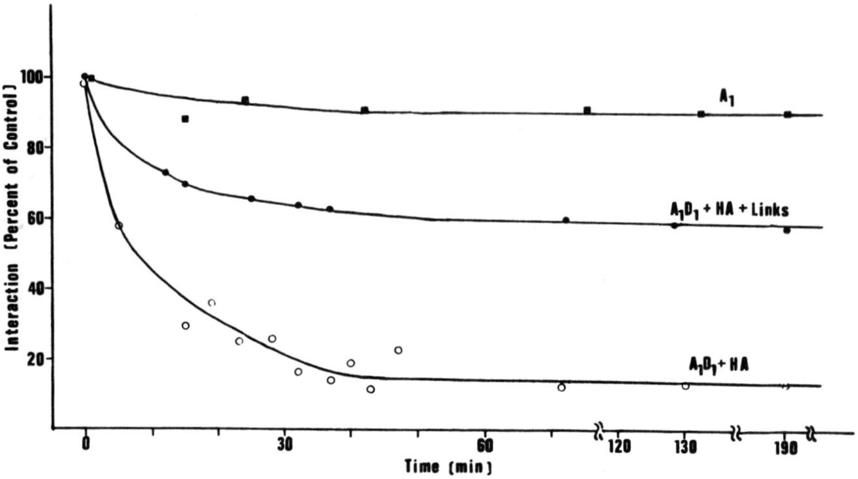

Fig. 5. Stabilization of the proteoglycan-hyaluronic acid interaction by purified link proteins. Each mixture contained proteoglycan monomer (A1D1), link protein, or hyaluronic acid (HA), or both at 1.76 mg/ml, 23.6, and 30 μg/ml concentration, respectively, and the proteoglycan aggregate (A1) was at 2 mg/ml concentration. Hyaluronic acid decasaccharide (20 μg/ml) was added at zero time. All final concentrations are in 0.4 M guanidinium chloride (pH 5.8).

In order to determine with confidence whether the link proteins have a distribution beyond cartilage proteoglycan aggregates, we have initiated immunological studies together with Mr. J. Paslay and Dr. Claude Bennett, in the hope of developing a specific immune assay for the link proteins. Antiserum to link proteins has been obtained from the rabbit. Immunodiffusion of this antiserum against a purified link-protein preparation (Pool 5, Fig. 6a) resulted in the appearance of two precipitin lines (Fig. 6b, well 5). The protein-rich proteoglycan (Pools 1 and 2 from a fractionation of A1D5 on Sepharose CL 6B in 1% SDS, Fig. 6a) gave no precipitin line (Fig. 6b, wells 1 and 2). Pools containing the "hook" region of the proteoglycan (Pools 3 and 4, Fig. 6a) showed one precipitin line (6b, wells 3 and 4), which was coincident with one of the lines from well 5. Both of the lines were evident from link protein 1 (Fig. 6b, well 6), but link protein 2 gave only one obvious line (well 7), which was coincident with one of those from well 6 but not with that from wells 3 and 4. From this data, it appears that the antiserum contains one antibody to both links and another to the "hook" region of the proteoglycan. Efforts are now directed toward purifying the antibody to the link protein.

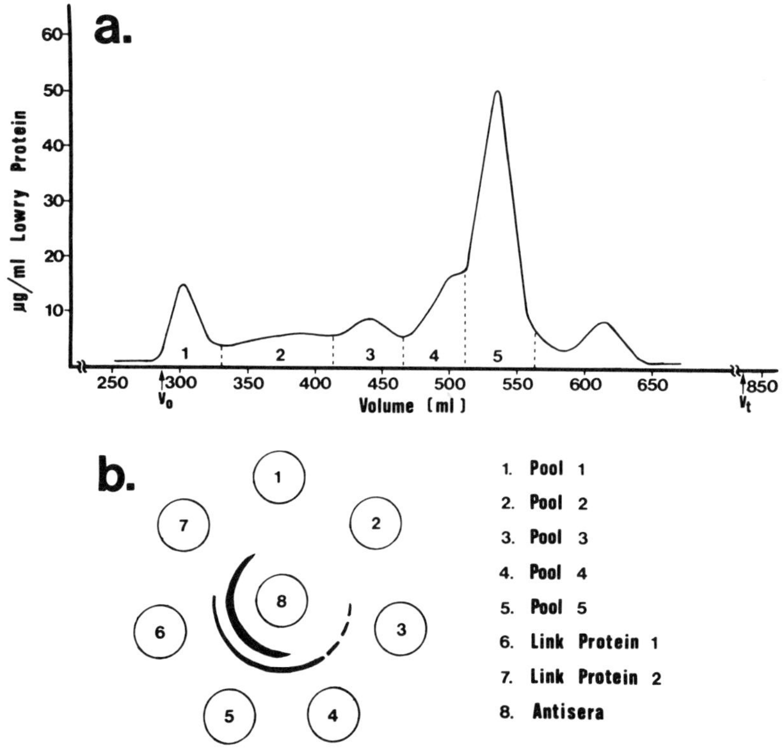

Fig. 6. (a) Fractionation of A1D4 on a column of Sepharose Cl 2B in 4 M guanidine. Fractions were pooled as indicated. (b) Immunodiffusion of Pools 1-5 from (a), link protein 1, and link protein 2 against rabbit antiserum to the link proteins.

ACKNOWLEDGMENTS

This work was supported by NIH Grants DE 02670 and HL 11310. The expert technical assistance of Ms. E. McKenzie, Mr. J. Belcher, and Mr. J. T. Ford is gratefully acknowledged.

REFERENCES

1. Mathews, M. D., and Lozaityte, I., *Arch. Biochem. Biophys. 74*, 158 (1958).
2. Hascall, V. C., and Sajdera, S. W., *J. Biol. Chem. 244*, 2384 (1969).
3. Hardingham, T. E., and Muir, H., *Biochim. Biophys. Acta 279*, 401 (1972).
4. Gregory, J. D., *Biochem. J. 133*, 383 (1973).
5. Hardingham, T. E., and Muir, H., *Biochem. J. 135*, 905 (1973).
6. Heinegård, D., and Hascall, V. C., *J. Biol. Chem. 249*, 4250 (1974).
7. Heinegård, D., *Biochim. Biophys. Acta 385*, 181 (1972).
8. Koziarz, J. J., Köhler, H., and Steck, T. L., *Anal. Biochem. 86*, 78 (1978).
9. Baker, J. R., and Caterson, B., *Biochem. Biophys. Res. Commun. 77*, 1 (1977).
10. Baker, J. R., and Caterson, B., unpublished results.
11. Hardingham, T. E., and Muir, H., *Biochem. J. 139*, 565 (1974).
12. Hascall, V. C., and Heinegård, D., *J. Biol. Chem. 249*, 4242 (1974).
13. Christner, J. E., Brown, M. L., and Dziewiatkowski, D. D., *Biochem. J. 167*, 711 (1977).
14. Vasan, N. S., and Lash, J. W., *Biochem. J. 164*, 179 (1977).

GLYCOCONJUGATE RESEARCH / Volume I
Proceedings of the Fourth International Symposium on Glycoconjugates

Structure of Cartilage Proteoglycans*

Vincent C. Hascall and Dick K. Heinegård

"It is evident, then, from the above considerations that the exact nature of the aggregation is not determinable from the present data. It is reasonable to assume, however, that both lateral and end-to-end aggregation can occur. The variable protein content of the different preparations suggests the possibility that a small amount of extraneous protein may participate in the formation of aggregates."

Mathews and Lozaityte 1958 (1)

I. EARLY DEVELOPMENT OF MODELS FOR PROTEOGLYCAN STRUCTURE

Shatton and Schubert in 1954 (2) were the first to present evidence that chondroitin sulfate in cartilage was firmly attached to noncollagenous protein in large macromolecules referred to subsequently as protein-polysaccharides

* *Dedicated to Dr. Martin B. Mathews, whose critical work has set and continues to set standards and perspectives for the successful development of proteoglycan research.*

ISBN 0-12-301301-1

and currently as proteoglycans. During the next few years, the Schubert group (2-4) devised procedures for isolating and purifying protein-polysaccharide fractions from high speed homogenates of cartilage in water or low concentrations of KCl. Muir in 1958 (5) suggested that serine provided the linkage point for the attachment of the chondroitin sulfate chains to protein on the basis of chemical analyses of chondroitin sulfate fractions isolated from cartilage after extensive proteolysis. The details of this linkage region structure were subsequently determined by the elegant studies of Rodén and coworkers (6-8) and of Anderson, Hoffman, and Meyer (9), which provide a clear understanding of this critical feature of the macromolecular structure of proteoglycans (see Fig. 1).

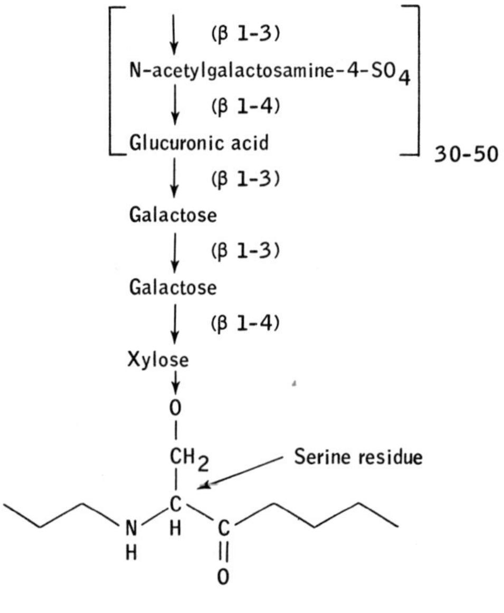

Fig. 1. Structure of the linkage-region oligosaccharide for the covalent attachment of chondroitin sulfate to the core protein of proteoglycans.

In 1958, Mathews and Lozaityte (1) described physical chemical experiments on a proteoglycan preparation isolated by water extraction of ground bovine nasal cartilage and subsequent ethanol precipitation of the proteoglycans, a procedure similar to that of Malawista and Schubert (3). Their material contained about twice as much protein as current proteoglycan preparations. Based on light scattering and viscosity data, they suggested that there was a subunit proteoglycan molecule which had a root-mean-square radius of gyration, R_g, of about 1000 Å and a mol. wt. of about 4 x 10^6. They made an assumption that the macromolecules contained no associated solvent and used the values for weight-average mol. wt. (M_w) and the limiting viscosity number of the sample to calculate an axial ratio of 90 for the macromolecules. They then proposed the first hydrodynamic model for proteoglycans (Fig. 2), in which a number of chondroitin

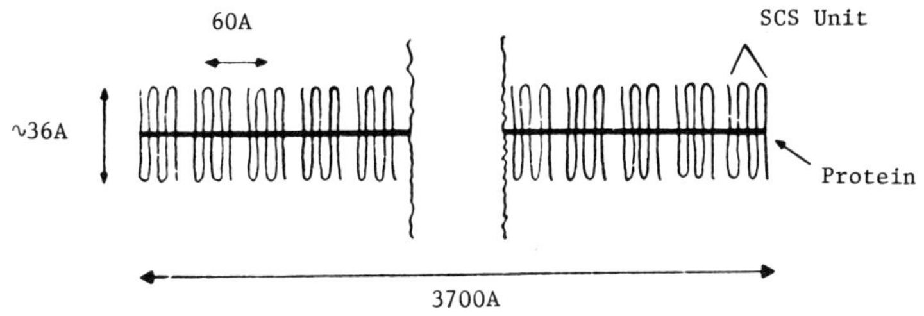

Fig. 2. Model for the proteoglycan unit structure postulated by Mathews and Lozaityte. SCS refers to the chondroitin sulfate chains (From ref. 1).

sulfate chains were attached at one or several points to a core protein with an overall rod-like shape and a length of 3700 Å. Measurements were also done for a proteoglycan preparation isolated from cartilage homogenates by a slightly different procedure. The light scattering results indicated that the macromolecules in this sample had a M_w of 35 x 10^6 and an R_g of 2500-3000 Å. Mathews and Lozaityte (1) suggested that these were aggregated proteoglycan structures because denaturing reagents, 8 M urea and 2 M potassium thiocyanate, reduced the M_w of the sample to about 4 x 10^6. To account for the low value of R_g for the aggregate, they proposed a mechanism for aggregation involving lateral as well as end-to-end interactions for the rod-like proteoglycan subunits. Thus, at this early stage of development of proteoglycan research, work in Mathews' laboratory introduced several insights into proteoglycan structure and aggregation which would become clearer many years later.

At about the same time, critical work in the laboratory of Partridge (10,11) provided additional detail for the emerging proteoglycan model. These investigators proposed that the chondroitin sulfate chains were unbranched with mol. wt. around 28 000 and that they were attached to the protein core at their nonreducing ends.* They also observed that appreciable amounts of glucosamine remained associated with the protein, while the galactosamine, indicative of chondroitin sulfate chains, was removed by treatment with dilute alkaline solutions. For this reason, they suggested that the protein core of cartilage proteoglycans contained keratan sulfate, as well as chondroitin sulfate, and proposed the model shown in Fig. 3. While they had no compelling reason for doing so, they placed the keratan sulfate at one end of the macromolecule, thereby prophetically anticipating the asymmetric distribution of the different polysaccharides along the core protein (see discussion in Section V). At the same time, Gregory and Rodén (12) provided direct evidence that bovine nasal cartilage did contain about 10% keratan sulfate.

* *This latter suggestion was based on the observation that the reducing termini of the chains appeared to be free. Since the procedure used to determine reducing groups involved an alkaline solvent, the reducing ends of the polysaccharide chains were being freed from the protein core by a β-elimination reaction, in which the serine residues with polysaccharide chains release the chains with concurrent formation of unsaturated alanine (9). The reducing groups, then, were being generated by the analytical procedure.*

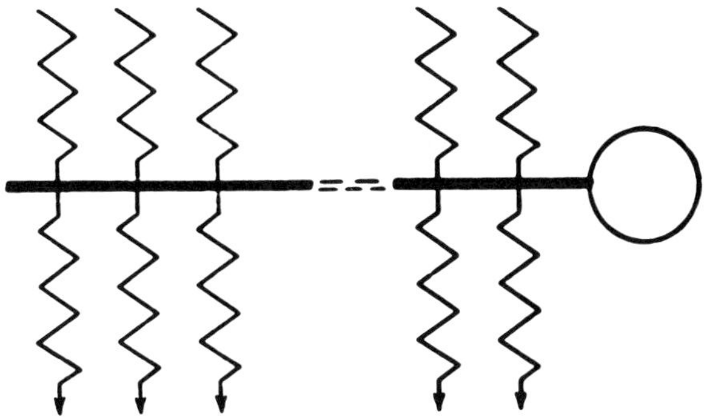

Fig. 3. Model for proteoglycan structure proposed by Partridge and coworkers (10,11). The heavy line represents the protein backbone, the circle the keratan sulfate, and the zig-zag lines the chondroitin sulfate chains with reducing termini indicated by the arrow heads. (From ref. 10).

Based on a suggestion by Anderson *et al.* (9), Mathews (13) subsequently proposed that the chondroitin sulfate chains were not evenly distributed along the core protein. Hydrodynamic measurements of chondroitin sulfate-peptide fragments recovered from trypsin-chymotrypsin digests of proteoglycans suggested that the individual chains were present primarily as doublets in which two chains were separated by a small peptide of less than 10 amino acid residues, and that the doublets were separated by longer peptides. This concept was later extended by the work of Heinegård and Hascall (14) which suggested that the chondroitin sulfate chains were closely spaced in clusters of from 1-10 chains with an average of 4 on peptide regions that were resistant to digestion with trypsin-chymotrypsin. Using electron microscopy, Thyberg *et al.* (15) subsequently identified about 25 side-chain filaments, probably corresponding to the chondroitin sulfate-clusters, in cytochrome c monolayers of monomer proteoglycan molecules prepared from several cartilages.

II. DEVELOPMENT OF DISSOCIATIVE EXTRACTION TECHNIQUES

In the early investigations just discussed, extensive mechanical disruption of cartilage was required to solubilize more than about 20% of the proteoglycans. This suggested that the proteoglycan molecules were tightly bound in cartilage matrices, perhaps as aggregates. Nevertheless, although in retrospect the work of Schubert and coworkers (2-4), and of Franek and Dunstone (16) indicated that proteoglycans isolated by such disruptive procedures contained aggregates, the aggregation phenomenon described by Mathews and Lozaityte (1) was largely overlooked. The work of Sajdera and Hascall (17-20), in 1969, described a novel way to exploit the fact that proteoglycans are present in cartilage primarily as aggregates. Certain concentrations of electrolytes, such as 4 M guanidinium chloride, 3 M $MgCl_2$, or 6 M LiCl effectively extracted proteoglycans in very high yield (up to 85%) from cartilage slices without requiring tissue homogenization. The experiment shown in Fig. 4, for example, shows the difference in the amounts of proteoglycans extracted from cartilage slices by gentle stirring with 4 M guanidinium chloride, a solvent in which aggregates are dissociated, or with 0.15 M KCl, a solvent in which aggregates are stable.

Centrifugal studies (19-21) indicated that proteoglycans sedimented in 4 M guanidinium chloride as a single component whereas a major proportion of the molecules formed large aggregates when the electrolyte concentrations were lowered by dialysis. The low salt concentrations which allow aggregate formation are now referred to as *associative* solvents whereas the higher concentrations, which cause disaggregation, are referred to as *dissociative* solvents. It was proposed that most proteoglycans in hyaline cartilage matrices were aggregated and that effective solvents for extraction were ones which were able to dissociate the aggregates, thereby allowing the monomers to diffuse out of the tissue. Since the fibrillar network of the collagen in hyaline cartilages is highly cross-linked (22), only a few percent of the collagen is solubilized in these solutions. The development of dissociative extraction procedures, most frequently employing 4 M guanidinium chloride but also 2 M $CaCl_2$ or 3 M $MgCl_2$, combined with CsCl isopycnic density gradients under associative conditions (Fig. 5a) (16), or under dissociative conditions (Fig. 5b) (20), allowed investigators to isolate and purify reproducible proteoglycan preparations that contained either a high proportion of aggregates (referred to originally as proteoglycan complex, PGC, and often now as the A1 sample) or as monomers (the A1-D1 or D1 sample).* These preparations have provided the starting point for refining our models for proteoglycan structure and aggregation. The introduction of protease inhibitors in the extraction and purification steps (24,25) has improved the methodology, and A1 samples, from bovine nasal cartilage, that contain up to 85% of the proteoglycans as aggregates can now be recovered.

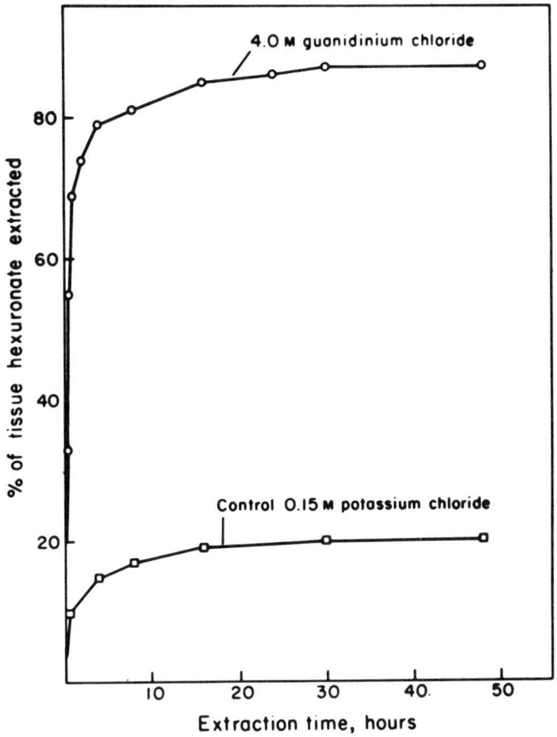

Fig. 4. Cartilage slices from bovine nasal septum were extracted at room temperature with 15 vol. of the indicated solvents. The proportion of total hexuronic acid, indicative of proteoglycans, which was released from the tissue, was monitored at the indicated times (17,19). (From ref. 19).

* *The nomenclature introduced by Heinegård (23) is commonly used presently to define proteoglycan fractions. A1 refers to the bottom fraction isolated from an associative density gradient that contains a high proportion of aggregate. D1 refers to monomer proteoglycan isolated directly from a dissociative extract with a dissociative density gradient, while A1-D1 is monomer proteoglycan isolated from an A1 preparation by a dissociative gradient.*

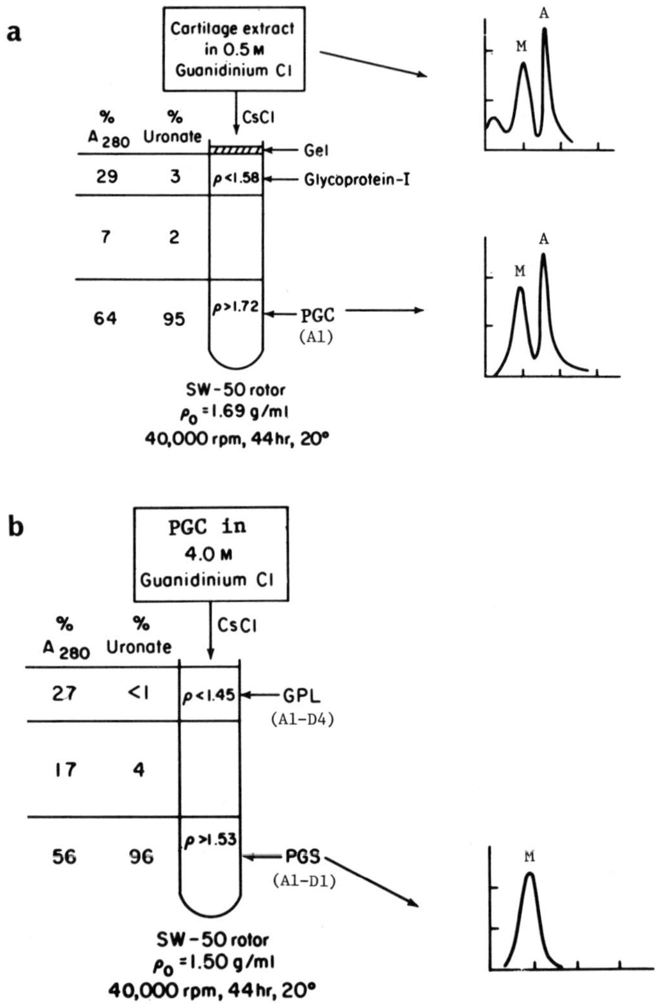

Fig. 5. Schematic diagrams of the associative (a) and dissociative (b) CsCl density gradients originally used to purify proteoglycan fractions (20). The tables to the left of the gradients indicate the approximate proportions of protein (A_{280}) and proteoglycan (uronate) recovered in the indicated fractions after the gradients have been formed. The original as well as the currently used nomenclature for the fractions

III. CURRENT MODEL FOR PROTEOGLYCANS

In early work on the mechanism of proteoglycan aggregation (19-21,26), A1 samples were centrifuged in isopycnic CsCl gradients in the presence of 4 M guanidinium chloride. In these dissociative gradients, the majority of the proteoglycans were recovered as monomers in the bottom, most dense, 2/5 of the centrifuge tubes, because of their high content of high buoyant density sulfated glycosaminoglycans (Fig. 5b). Components in the upper 3/5 of the gradient, which accounted for less than 10% of the weight of the A1 sample, were required in order for the monomers to reaggregate. Because small molecular weight proteins were found at the very top of such gradients, it was suggested that these were responsible for promoting proteoglycan aggregation and they were referred to as the "link" glycoproteins.* Keiser *et al.* (27) demonstrated with SDS-polyacrylamide gel electrophoresis that the top, A1-D4 fraction of the dissociative density gradient contained predominantly two proteins with mol. wt. of *ca.* 40 000 and 45 000, now referred to as "link *a*" and "link *b*", respectively (28). More recent work (29,30) has indicated that link *a* is derived from link *b* by proteolytic cleavage, which releases a small peptide. The suggested role for the link proteins in proteoglycan aggregate formation was put into doubt by the discovery of Hardingham and Muir (31) that cartilage proteoglycans interact specifically with hyaluronic acid (HA). ** Hyaluronic acid was subsequently shown to be present at 0.4-0.8% (w/w) concentration in aggregate (A1) samples (28,32) and to be recovered in the middle fractions of dissociative gradients of A1 samples (32-34) (Fig. 6). It was postulated (31,33) that hyaluronic acid was a critical factor for the formation of aggregates.

Figure 5 continued:

is indicated. The patterns at the right are tracings of ultracentrifugal profiles for the dialyzed 4 M guanidinium chloride extract (upper), the A1 fraction (middle), and the A1-D1 fraction (lower). The monomer components are indicated by M and the aggregate components by A. All centrifugal runs were made in 0.5 M guanidinium chloride.

---

* *Methods for purifying and characterizing the link proteins are discussed in detail by John Baker, this volume.*

** *The characteristics and specificity of this interaction are discussed in detail by Helen Muir, this volume. See also the recent review (30).*

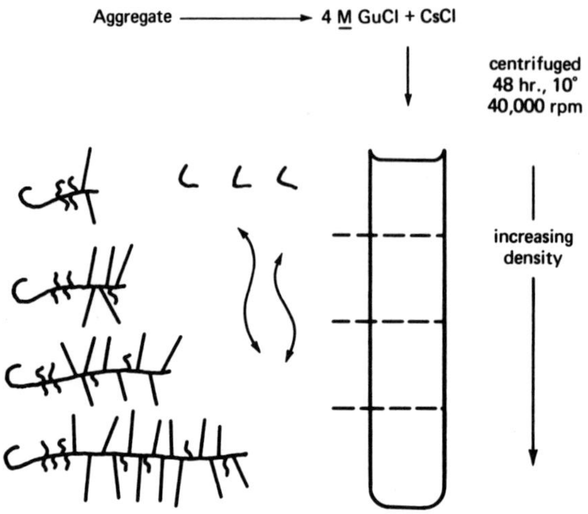

Fig. 6. Schematic diagram indicating the positions of the different components of proteoglycan aggregates in the dissociative density gradients. The link proteins are indicated by L and hyaluronic acid by the double arrow lines.

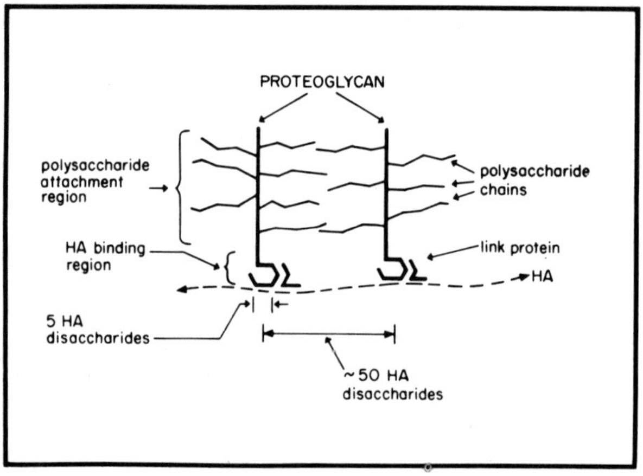

Fig. 7. Schematic model for the noncovalent interactions between proteoglycans, link proteins, and hyaluronic acid responsible for proteoglycan aggregation. (From ref. 35).

Conditions such as dissociative solvents, acid pH, and sulfhydryl modifications, which were able to reverse or prevent aggregate formation, similarly prevented the proteoglycan-HA interaction. A portion of the protein core of proteoglycan molecules (35), now referred to as the HA-binding region, was shown to mediate the interaction with HA (Fig. 7). The interaction was shown to be highly specific for hyaluronic acid since no other biological polymer tested was able to interact (31). HA-oligosaccharides of 5 or more disaccharide repeat units (HA_{10} or greater) were required for effective displacement of proteoglycans from macromolecular HA (36,37). This showed that the interaction site for hyaluronic acid in the HA-binding region of the core protein extends over a length of HA of around 5 disaccharides.

Two experiments done at this time indicated that the proteoglycan-HA interaction alone was not sufficient to explain all the properties of aggregate preparations. While HA_{10} or larger oligomers were effective in displacing proteoglycans from HA, they were unable to displace proteoglycans from intact aggregates, which contain link proteins, under the same solvent conditions (36). Secondly, when intact aggregate samples were digested with chondroitinase and trypsin (35) or more recently with just trypsin (38), a complex was recovered, which contained the HA-binding region and link *a* still bound to HA. Recent experiments indicate that the link *b* molecules are converted to link *a* molecules during the trypsin digest (30). Trypsin digestion of proteoglycan-HA mixtures, in contrast to aggregates, left no protein bound to the HA. On the basis of these experiments, a model was proposed (Fig. 7), in which the link proteins stabilized the proteoglycan-HA complex by interacting with both the HA and a portion of the HA-binding region. The fact that link *a* can interact with HA in the absence of proteoglycan supports this suggestion (35).*

The HA-binding region is located at one end of the core protein and consists of a continuous polypeptide region of about 70 000 mol. wt., which can be isolated (see Section IV). For bovine nasal or tracheal cartilage, the HA-binding region preparation contains about 15% of the total keratan sulfate in the original proteoglycan molecules but none of the chondroitin sulfate. A HA-binding region protein has also been recovered from the Swarm rat chondrosarcoma proteoglycan, which contains little or no keratan sulfate (24). This

* *See also the discussion by John Baker (this volume) which indicates that link molecules can bind to proteoglycan molecules in the absence of HA.*

protein exhibits a single, narrow band on SDS-polyacrylamide gels with a mol. wt. of about 67 000 (unpublished observations).

When protease inhibitos were used in isolation procedures (24,25), up to 85% of the proteoglycans in A1 samples could be recovered as aggregates and therefore contain a functional HA-binding region polypeptide. The rest of the proteoglycans in A1 samples do not appear to be able to aggregate (see Section VII). To date, the capability of proteoglycans to interact with hyaluronic acid has not been demonstrated unequivocally for proteoglycans isolated from any connective tissue other than cartilage, and it is possible that this feature of the core protein in proteoglycans is specific for cartilaginous tissues.

IV. STRUCTURAL METHODS FOR COMPARING PROTEOGLYCANS

A variety of techniques have been developed to define the structure and chemistry of proteoglycans. Combined with reproducible extraction and fractionation procedures, these techniques can now be used to compare the characteristics of proteoglycans isolated from cartilages in different species, from the same cartilage at different stages of tissue development, or from cartilages undergoing pathological changes. The relative contents of glucuronic acid (or galactosamine), glucosamine, and amino acids provide information about the contents of chondroitin sulfate, keratan sulfate, and protein, respectively, in proteoglycan fractions. The enzymatic methods depicted in Fig. 8 have proven useful for structural studies. The excellent studies of Yamagata *et al.* (39) described methods for isolating and characterizing bacterial chondroitinases. These enzymes, particularly chondroitinase ABC, have been used to remove the bulk of the chondroitin sulfate from proteoglycans for subsequent studies of the core structure (40,41). The core preparation of bovine nasal cartilage proteoglycan contains a single population of molecules with variable keratan sulfate to protein ratios. The core molecules are partially retarded on Sepharose 6B (Fig. 8) and have an average M_w of 450 000. Since the preparation contains about 40% protein, the average size of the polypeptide core would be about 180-200 x 10^3. This is consistent with previous physical chemical analyses of intact proteoglycan monomers. They contain about 7% protein and have a M_w of about 2.5 x 10^6 (42,43) with a continuous range between 1 - 4 x 10^6 (42). The chondroitinase enzymes can also be used to determine the relative amounts of the 4-sulfated, 6-sulfated, and unsulfated disaccharides in the chondroitin sulfate chains (39). For example, proteoglycan monomers from

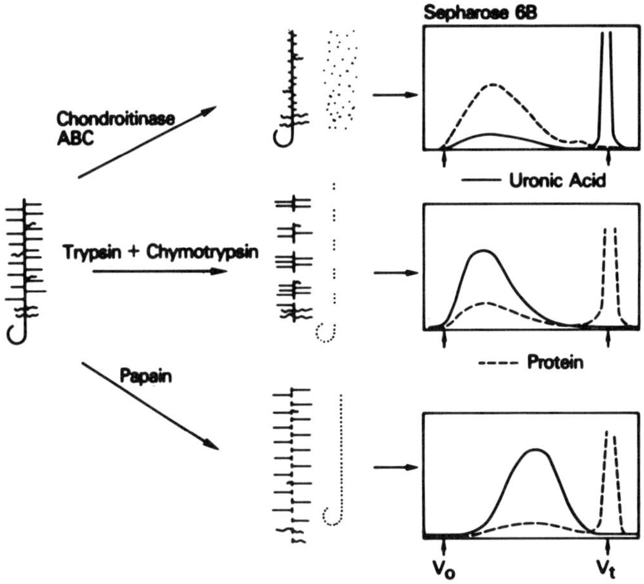

Fig. 8. Schematic diagrams indicating the effects of the different enzyme treatments on the structure of proteoglycan monomer. The elution profiles observed for the digests on Sepharose 6B are given.

bovine nasal and tracheal cartilage, which appear to be identical in almost all other respects, have different chondroitin 4-sulfate and 6-sulfate contents. The preparations from nasal cartilage contain only about 15% of the 6-epimer, whereas those from trachea contain 45% (unpublished data).

Trypsin-chymotrypsin digestion of proteoglycans releases chondroitin sulfate-peptides in which variable numbers (mean 4) of chondroitin sulfate chains remain bound to individual peptides (14). Trypsin digests alone yield chondroitin sulfate-peptide fragments that are somewhat larger and therefore contain an average of greater than four chains per peptide. The hydrodynamic sizes of the chondroitin sulfate-peptide fragments from such digests can be studied by

Sepharose 6B chromatography (Fig. 8). Papain digestion or alkaline β-elimination of proteoglycans, on the other hand, yields single chondroitin sulfate chains with or without a small attached peptide, respectively. Wasteson (44) chromatographed well-characterized chondroitin sulfate samples of different mol. wt. on Sephadex G-200 and Sepharose 6B to establish standard calibration curves for estimating the mol. wt. of chondroitin sulfate samples. This procedure (Fig. 8) has been used to compare chondroitin sulfate chain sizes of proteoglycans from such tissues as articular cartilage at different ages (45,46), or chick limb bud mesenchyme during chondrogenesis and chondrocyte maturation in culture (47) or *in vivo* (48).

Heinegård and Axelsson (38) recently showed that a large proportion of keratan sulfate in proteoglycan molecules is located in a small region of the core protein. Monomer proteoglycan, treated sequentially with chondroitinase ABC, trypsin, and chymotrypsin, was chromatographed on Sepharose 6B (Fig. 9). A retarded peak was observed, which contained about 65% of the keratan sulfate but less than 10% of the residual bound oligosaccharides of the chondroitin sulfate chains. The average mol. wt. of the keratan sulfate-enriched preparation was found to be 125 000, and protein constitutes 20% of the sample. This indicates that most of the keratan sulfate is localized on a portion of the polypeptide (mol. wt. *ca.* 25 000) that contains relatively few chondroitin sulfate chains.

An example of the use of the procedures just described to characterize a proteoglycan is shown in Fig. 10. A monomer sample was isolated from day-8 cultures of chick limb bud mesenchyme cells, which were grown *in vitro* under conditions favorable for chondrogenesis. The cultures were labeled with [^{35}S]sulfate for 6 h before the proteoglycans were isolated. The indicated enzymatic digests were prepared and chromatographed on Sepharose 6B (47). The elution profiles indicate that the M_w of the chondroitin sulfate chains is about 17 000 (Fig. 10c); that these chains are clustered on the core polypeptide (Fig. 10a,b); and that about 8% of the ^{35}S-activity was located in keratan sulfate chains on the keratan sulfate-enriched portion of the core protein (Fig. 10e). These results, then, provide evidence

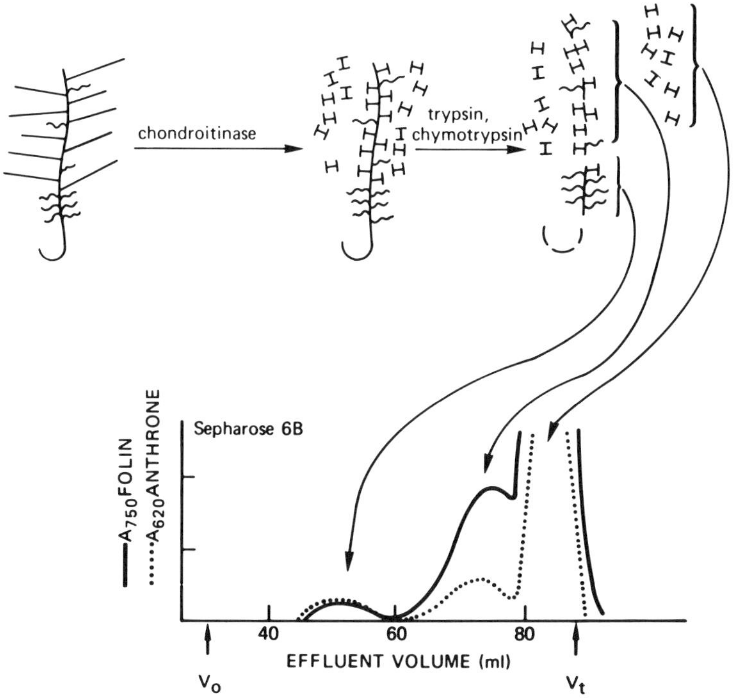

Fig. 9. Schematic diagram of the method developed to isolate the keratan sulfate-enriched peptides. (From ref. 38).

that the proteoglycans synthesized by these cultures are characteristic of cartilage proteoglycans (see also Section VI).

The HA-binding region of proteoglycans can also be studied. A method has been described for estimating the proportion of proteoglycan monomers, in a preparation, that are able to interact with hyaluronic acid (49). Further, the HA-binding region can be purified by a modification (38) of the chondroitinase-trypsin procedure developed originally by Heinegård and Hascall (35). In this procedure, aggregate samples are digested directly with trypsin, and the released

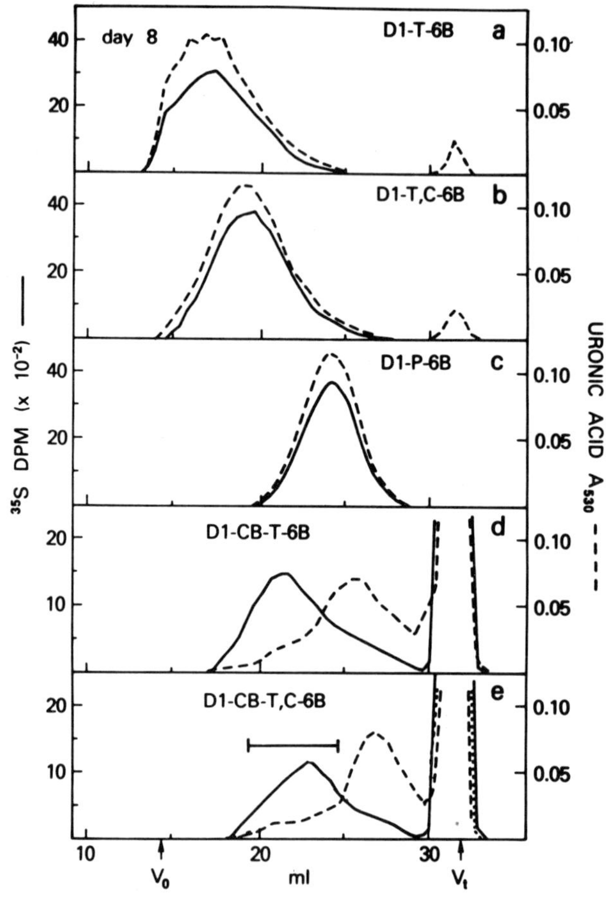

Fig. 10. *Sepharose 6B elution profiles of a sequence of different enzymatic treatments of a [^{35}S]sulfate-labeled monomer (D1) proteoglycan isolated from day 8 cultures of mesenchymal cells of chick limb bud. The D1 sample was excluded from Sepharose 6B. The enzymatic treatments were: (a) trypsin; (b) trypsin plus chymotrypsin; (c) papain; (d) chondroitinase ABC, then trypsin; and (e) chondroitinase ABC, and then trypsin plus chymotrypsin. (From ref. 47).*

chondroitin sulfate-peptide fragments removed by centrifugation in an associative density gradient (Fig. 11). The complex containing HA-binding region, link a protein, and hyaluronic acid is separated from enzyme and small peptides by chromatography on Sepharose 2B or 6B. The components of this complex are dissociated with 4 M guanidinium chloride and separated on Sephadex G-200 in the same solvent. The HA-binding region preparation isolated by this procedure retains its ability to interact with hyaluronic acid (35)

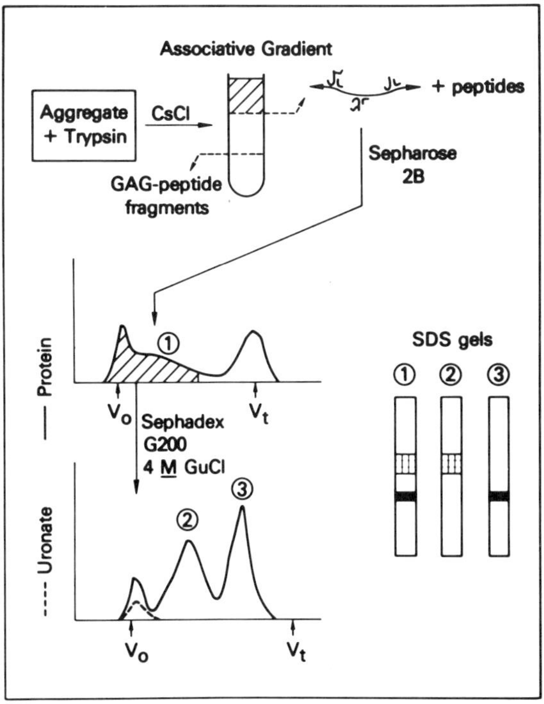

Fig. 11. Schematic diagram of the procedure used to isolate the HA-binding region fraction and the link a fraction from bovine nasal or tracheal cartilage A1 preparations (30,35). (From ref. 30).

and exhibits the same specificity for HA-oligomers as the intact proteoglycan monomers (unpublished observations). Recent work with this preparation has indicated that dansylation but not acetylation of reactive residues, presumably lysine, near the interaction site can prevent the interaction of the HA-binding region polypeptide with hyaluronic acid (50, see also 30).

The use of the procedures just discussed on purified proteoglycans will, in the future, allow us to compile details about the molecular structure of these macromolecules in an analogous manner that the glycosaminoglycan separation and quantification procedures developed over the last 50 years have been used to provide information about the types and relative amounts of glycosaminoglycans present in a wide variety of connective tissues. The monograph on connective tissue biochemistry by Martin Mathews (51) provides an excellent example of the potential usefulness of such information for cartilages.

V. STRUCTURE OF THE POLYSACCHARIDE ATTACHMENT REGION

The experiments depicted in Fig. 12 provided evidence that the keratan sulfate-enriched region was located closer to the HA-binding region than the rest of the polysaccharide attachment region where more than 90% of the chondroitin sulfate chains reside (38,52). Aggregate (A1) samples were treated with hydroxylamine, which cleaves the protein core of the proteoglycan (4). Two fragments were then prepared by Sepharose 2B chromatography (Fig. 12a). The included peak, which contained 80% of the chondroitin sulfate in the original A1 sample, was recovered, digested sequentially with chondroitinase and trypsin, and chromatographed on Sepharose 6B. Only a small amount of the keratan sulfate enriched region was observed (Fig. 12b). The excluded peak from the Sepharose 2B was recovered and then chromatographed on Sephadex G-200 with 4 M guanidinium chloride as the eluent. The proteoglycan fragments, which had remained bound to the hyaluronic acid and therefore contained the HA-binding region, were eluted in the excluded volume, whereas the link protein fraction was well retarded. The proteoglycan fragments excluded from the Sephadex G-200 were

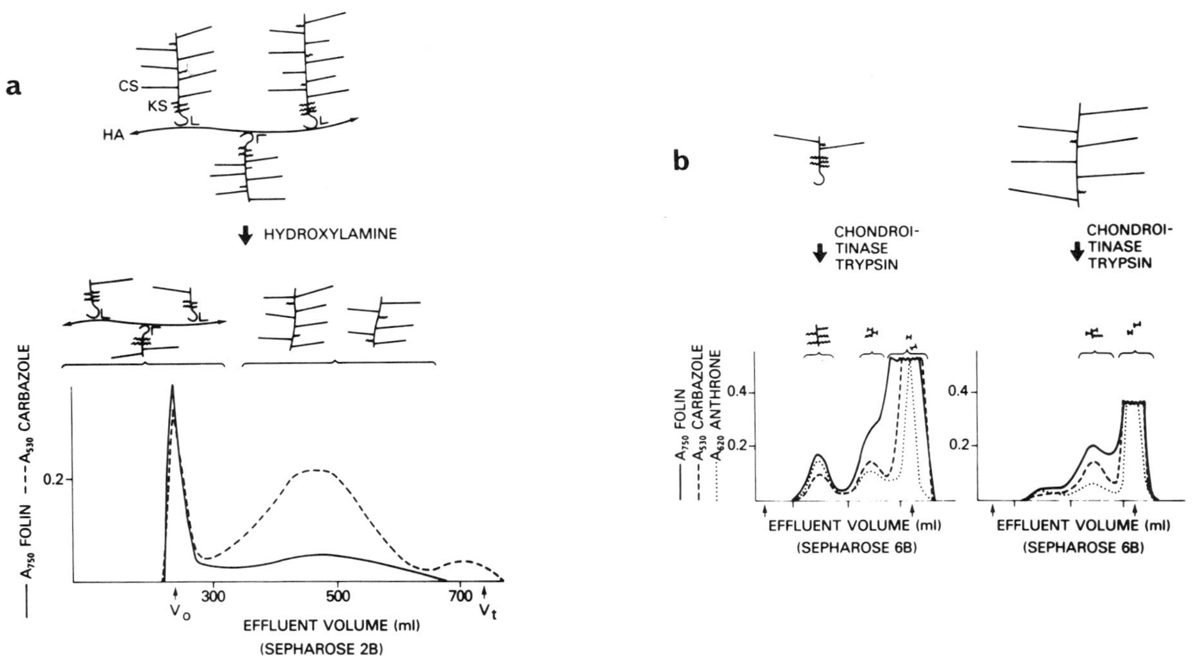

Fig. 12. Summary of experiments in which A1 samples were treated with hydroxylamine to cleave the majority of the chondroitin sulfate peptides from the proteoglycan molecules. See text for details. (From ref. 38).

recovered, digested sequentially with chondroitinase and trypsin, and chromatographed on Sepharose 6B. Most of the keratan sulfate enriched region in the original A1 aggregate was recovered in this fraction (Fig. 12b). Thus, it was concluded that the keratan sulfate-enriched region was located between the HA-binding region and the chondroitin sulfate-enriched region, and a model (Fig. 13), was proposed for the structure of proteoglycan molecules that are able to aggregate.

Several laboratories have subfractionated monomer proteoglycans. Dissociative density gradients have been used to prepare fractions with different average densities (52-55) as indicated in Fig. 6. The following trends have been observed for the composition of the proteoglycans with decreasing buoyant density (Fig. 14): (a) the hydrodynamic size is smaller, as indicated by the increase in K_{av} on Sepharose 2B; (b) the proportion of chondroitin sulfate to protein is smaller; (c) the relative content of glucosamine

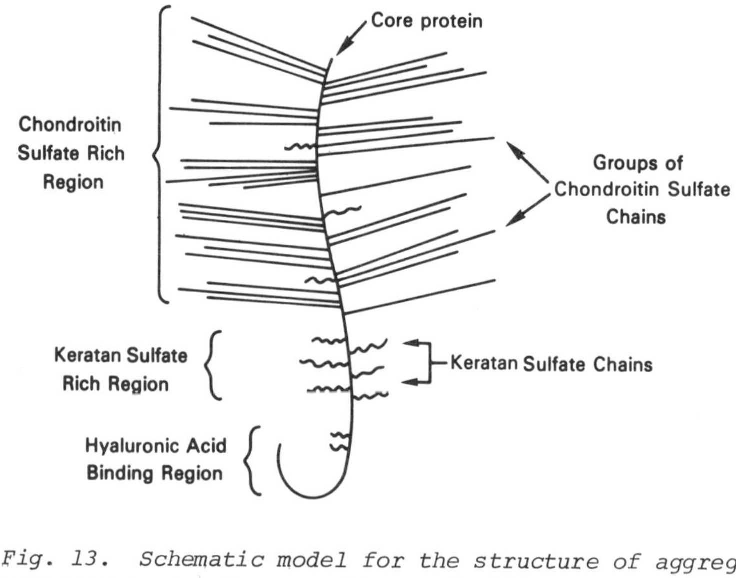

Fig. 13. Schematic model for the structure of aggregating monomer proteoglycans. (From ref. 38).

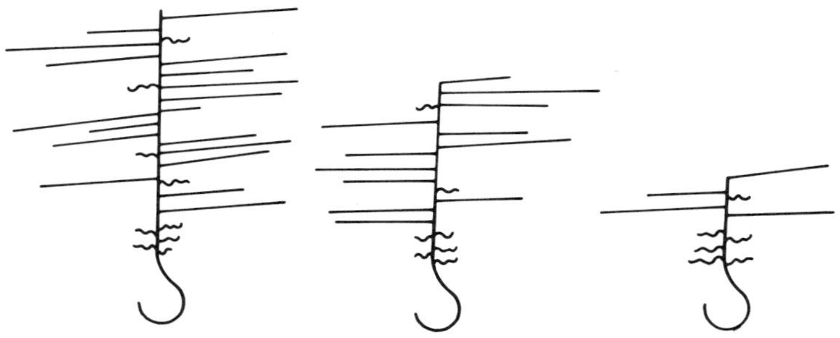

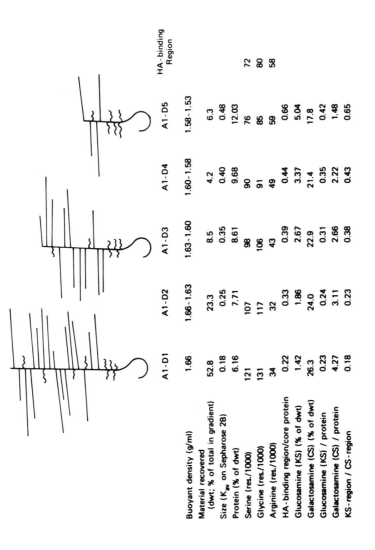

	A1-D1	A1-D2	A1-D3	A1-D4	A1-D5	HA-binding Region
Buoyant density (g/ml)	1.66	1.66-1.63	1.63-1.60	1.60-1.58	1.58-1.53	
Material recovered (dwt; % of total in gradient)	52.8	23.3	8.5	4.2	6.3	
Size (K_{av} on Sepharose 2B)	0.18	0.25	0.35	0.40	0.48	
Protein (% of dwt)	6.16	7.71	8.61	9.68	12.03	
Serine (res./1000)	121	107	98	90	76	72
Glycine (res./1000)	131	117	106	91	85	80
Arginine (res./1000)	34	32	43	49	59	58
HA-binding region/core protein	0.22	0.33	0.39	0.44	0.66	
Glucosamine (KS) (% of dwt)	1.42	1.86	2.67	3.37	5.04	
Galactosamine (CS) (% of dwt)	26.3	24.0	22.9	21.4	17.8	
Glucosamine (KS) / protein	0.23	0.24	0.31	0.35	0.42	
Galactosamine (CS) / protein	4.27	3.11	2.66	2.22	1.48	
KS-region / CS-region	0.18	0.23	0.38	0.43	0.65	

Fig. 14. Chemical characteristics of proteoglycans isolated at different buoyant densities in a dissociative density gradient (52). (from ref. 46).

and of keratan sulfate-enriched region is higher; (d) the proportion of serine and glycine residues, which are enriched at the loci for attachment of chondroitin sulfate chains, is lower while the proportion of cysteine, methionine and arginine residues, which are enriched in the HA-binding region, is higher; and (e) in guinea pig rib cartilage, the ratio of chondroitin 4-sulfate to chondroitin 6-sulfate increased (56). Cyanogen bromide cleavage of proteoglycans releases a large fragment from the HA-binding region (52). This method was used to show that the relative proportion of the HA-binding region increases with decreasing size or buoyancy (Fig. 14). On the basis of such data it was suggested that the monomer proteoglycan from cartilages contains a continuously polydisperse population of macromolecules with the number of chondroitin sulfate chains increasing with the length of the chondroitin sulfate-enriched region of the polypeptide (52,53,57). For bovine nasal proteoglycan, the average molecule with mol. wt. 2.5 x 10^6 would contain almost 100 chondroitin sulfate chains and 50-60 keratan sulfate chains located on the core as suggested by the model in Fig. 13. The polydispersity would range from mol. wt. less than 1 x 10^6 to greater than 4 x 10^6 (42). Such a model would account for most of the known properties of proteoglycans.

VI. RELEVANCE OF THE PROTEOGLYCAN MODEL

Can the structural model for proteoglycans just developed be used to contribute to an understanding of biological processes? Two examples relating to chondrogenesis and to cartilage aging will be described, which suggest that indeed it can.

As discussed in Mathews' book (51), it has been observed that embryonic cartilages contain, in general, little or no keratan sulfate. De Luca *et al.* (47) have studied proteoglycans during the process of chondrogenesis in cultures derived from mesenchymal cells of the chick limb bud at stage 23-24. In this system, the dispersed cells are plated at high initial densities on Petri dishes and cultured under conditions that favor chondrogenic expression (58,59). Within the first 2-3 days in culture, a large proportion of the cells undergo differentiation to the chondrocyte phenotype, and by day 4 the major proteoglycan being synthesized by the cells has characteristics typical of the cartilage-type proteoglycan. Between days 4 and 8 the chondrocytes mature, organize into cartilaginous nodules, and elaborate an extensive extracellular matrix. By day 8-9, when the synthesis of proteoglycans is maximal, the nodules cover the bottom of the Petri dish (59). The proteoglycans synthesized by the newly-differentiated chondrocytes on day 4 have less keratan sulfate than those synthesized by the mature chondrocytes on day 8 or later culture times. This is reflected in the characteristics of the keratan sulfate-enriched region of the proteoglycans (47). In the experiment shown in Fig. 15, monomer proteoglycan fractions recovered from cultures labeled with [^{35}S]sulfate and [^{3}H]serine for 6 h on the indicated days were subjected to the treatment illustrated in Fig. 9. In the day 8 culture, approximately 8% of the total ^{35}S-activity was eluted in the keratan sulfate-enriched peptide region (indicated by the arrow). Chemical analyses verified that this fraction contained keratan sulfate with characteristics similar to preparations isolated from bovine cartilages. The elution position of the keratan sulfate-enriched peptide was progressively later for the proteoglycans isolated from earlier culture times, indicating that the average size of this component was smaller in proteoglycans synthesized by the younger chondrocytes.* Additional experiments indicated that the isolated keratan sulfate chains were significantly smaller for the proteoglycans synthesized by the day 3-5 cultures than for those synthesized in later cultures. This would account for some, and perhaps all, of the size differences observed for the keratan sulfate-enriched peptide fragments. Therefore, changes occur in the keratan sulfate-enriched region of the proteoglycans at the same time that the chondrocytes in the cultures are

* *The larger-sized, ^{35}S-labeled fraction in the day 2 and 3 samples contained heparan sulfate and is possibly derived from proteoglycans being made by cells, in the cultures, that are not yet expressing chondrogenic phenotype.*

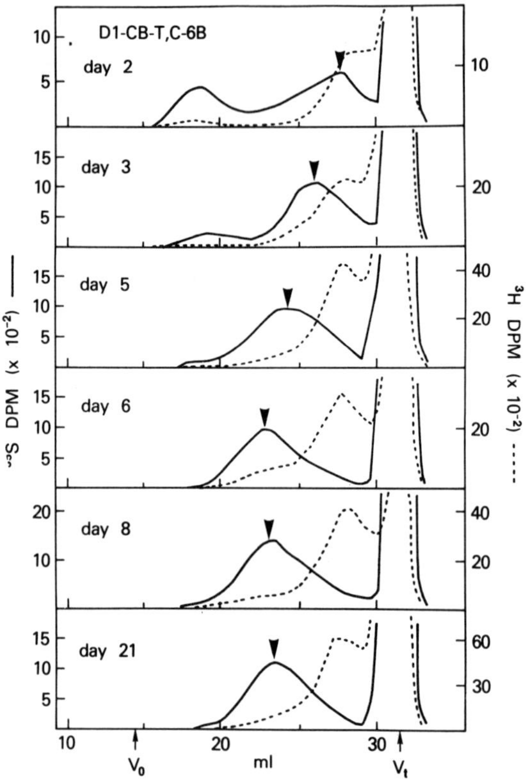

Fig. 15. Characteristics of the keratan sulfate-enriched peptides derived from [^{35}S]sulfate- and [^{3}H]serine-labeled proteoglycans isolated from mesenchymal cells of the chick limb bud, cultured for the indicated number of days. The monomer (D1) fractions were treated with chondroitinase, then trypsin plus chymotrypsin, and chromatographed on Sepharose 6B. See text for details. (From ref. 47).

maturing and organizing an extracellular matrix.* The mechanisms that regulate these post-translational modifications of the proteoglycan structure and their relevance to cartilage development and matrix organization remain to be discovered.

In many hyaline cartilages, the content of keratan sulfate relative to chondroitin sulfate increases as the tissue ages [see discussions in Mathews (51)]. Recently, Inerot *et al.* (46) studied proteoglycans isolated from the articular cartilages of dogs of different ages. Chemical composition of the proteoglycans (Table I) indicated that the relative content

Table I. Compositions of A1 Proteoglycan Samples Isolated from Canine Articular Cartilages of Animals of Different Ages. (From ref. 46).

Age (months)	Chondroitin SO_4 / Keratan SO_4	Chondroitin SO_4 / Protein	Keratan SO_4 / Protein
4.5	15.67	6.72	.43
4.5	10.11	9.45	.93
6	6.69	6.75	1.01
6	10.11	7.29	.72
8	8.09	8.64	1.06
12	4.56	4.83	1.06
17	3.35	5.94	1.77
60	2.85	5.28	1.85
60	2.45	3.90	1.59
96	3.17	5.40	1.70
126	1.50	4.44	2.96
Correlation coefficient	-.89	-.65	

* *The correlation of keratan sulfate content in human chondrosarcoma proteoglycans with the degree of clinical malignancy discussed by Rosenberg in this symposium may well be a reflection of the different degrees of maturation of the chondrocytes in the tumors in an analogous manner to the experiment in the culture system.*

of chondroitin sulfate decreased while the protein and keratan sulfate content increased with increasing age, the changes being more pronounced in the young dogs. It was shown that the changing chondroitin sulfate content was not due to a decrease in the size of the chondroitin sulfate side-chains. As shown in Fig. 16, aggregate (A1) samples from the

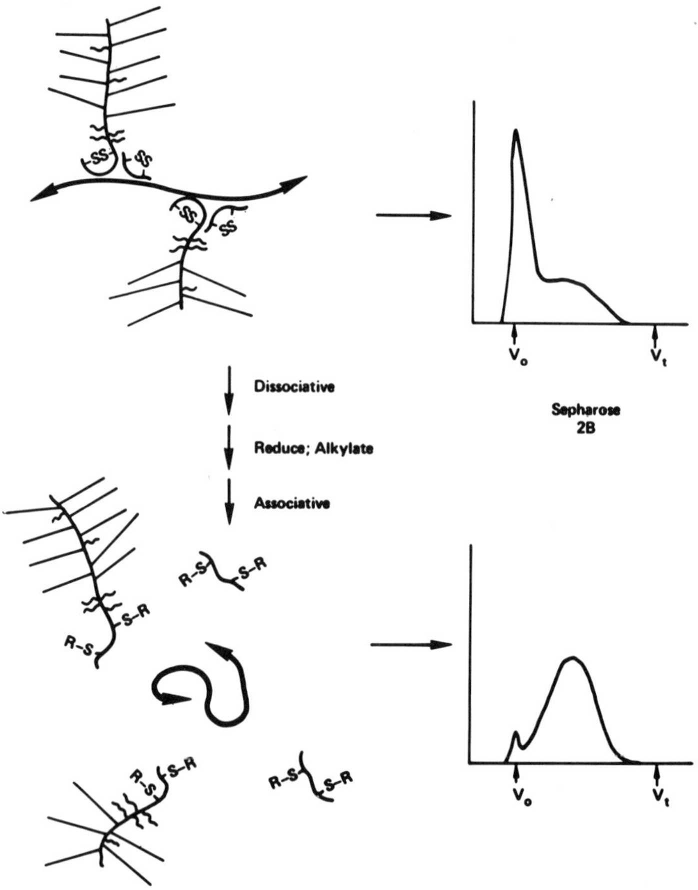

Fig. 16. Schematic model indicating the effect of reduction and alkylation on proteoglycan aggregates.

cartilages were chromatographed on Sepharose 2B before and after reduction and alkylation of disulfide bonds which are necessary for aggregation (19,54). The elution profile of the reduced and alkylated sample, then, provides an analysis of the molecular size distribution of all of the monomer proteoglycans present in the A1 sample, including those that would equilibrate in dissociative density gradients at positions of lower buoyant densities than the A1-D1 fraction. The composite profiles for the A1 samples from the dog cartilages are shown in Fig. 17. While the proportion of aggregate in the A1 samples remained about the same, the average sizes of the monomers decreased appreciably with tissue age. The changes observed were consistent with the suggestion that the proteoglycans from older tissues contained smaller chondroitin sulfate-enriched regions and, hence, smaller numbers of chondroitin sulfate chains per molecule.

Whether the changes in proteoglycan structure with tissue age just described result from biosynthetic or degradative processes, or a combination of both, remains to be determined. It has been shown that chondrocytes in the culture system of the chick limb bud have the capacity to synthesize a wide range of proteoglycan sizes at the same time (60) and that the average size of the proteoglycan population is different at various times in culture (61). Further, Choi and Meyer (62) have shown that the degree of sulfation of keratan sulfate increases in older cartilages, indicating that some of the proteoglycans in the older tissue were synthesized at later times. These results suggest that biosynthetic mechanisms are involved to some extent in the differences observed in the structure of proteoglycans at the different tissue ages.

Whatever the mechanisms for generating the smaller proteoglycans in the older tissue may be, the smaller proteoglycans with their lower content of chondroitin sulfate will be less effective in providing the tissue with the resiliency and compliance required for proper tissue function.

VII. MODEL BUILDING

When attempting to ascribe relevance to models, it is perhaps useful to remember an admonition of Christian deDuve (63). "The biologist ... deals with a reality of such elusive

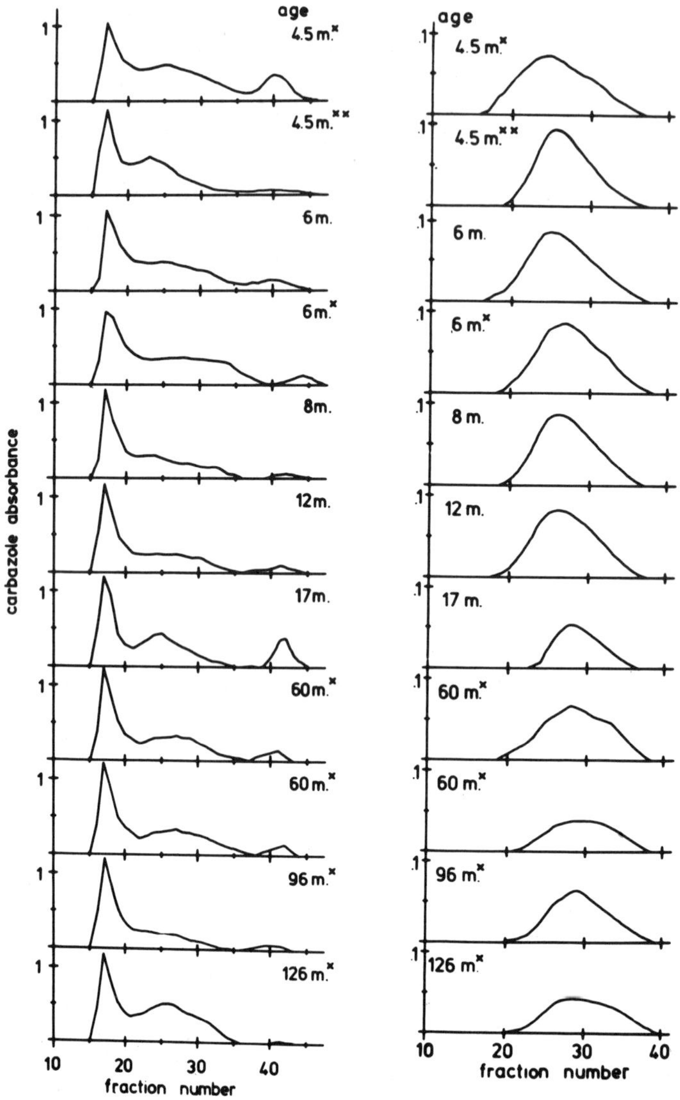

Fig. 17. Sepharose 2B profiles of the A1 samples described in Table I before (at left) and after (at right) reduction and alkylation. (From ref. 46).

complexity that only deliberate simplification can cloak it with the appearance of intelligibility ... But we must accept our concepts for what they are, provisional approximations that are as much fictions of our minds as they are faithful depictions of facts."

The model for proteoglycan structure described here was developed for those molecules that can interact with hyaluronic acid and form aggregates. A1 preparations, however, invariably contain a proportion of proteoglycan molecules that do not aggregate. While some of these may result from endogenous proteolysis (24,25) or from modifications that occur during extraction and purification, it is likely that this population of nonaggregating monomers contains in part a class of proteoglycans with different characteristics.

In a set of experiments (64) nonaggregated proteoglycans were separated from aggregates by three successive sedimentation velocity centrifugation steps, as indicated schematically in Fig. 18. For the second and third steps, the bottom fraction was diluted to the original volume. The nonaggregated molecules from the upper fractions, about 15% of the total, were chromatographed on Sepharose 2B after addition of hyaluronic acid in order to separate those molecules that were unable to interact with hyaluronic acid from those that could. Analytical data for these noninteractive monomer proteoglycans, about 10% of the total A1 sample, show that they were smaller, contained more protein, and had a lower ratio of keratan sulfate to chondroitin sulfate than the aggregating proteoglycan monomers. In spite of the higher protein content, the amino acid composition was typical for the chondroitin sulfate-enriched region, with a high serine and glycine content. Chromatography of trypsin digests indicated that, although the clustering of chondroitin sulfate chains was the same as for the aggregating proteoglycans, a larger proportion of unsubstituted peptides were liberated. The noninteractive proteoglycans, therefore, could not be derived from the aggregating monomers by a simple degradation step. These data indicate that hyaline cartilages contain at least two distinct populations of proteoglycan monomers: aggregating (about 85% of the total for bovine nasal cartilage) and nonaggregating.

Recent experiments in two laboratories suggest that some other features of the model may have to be refined. Pearson

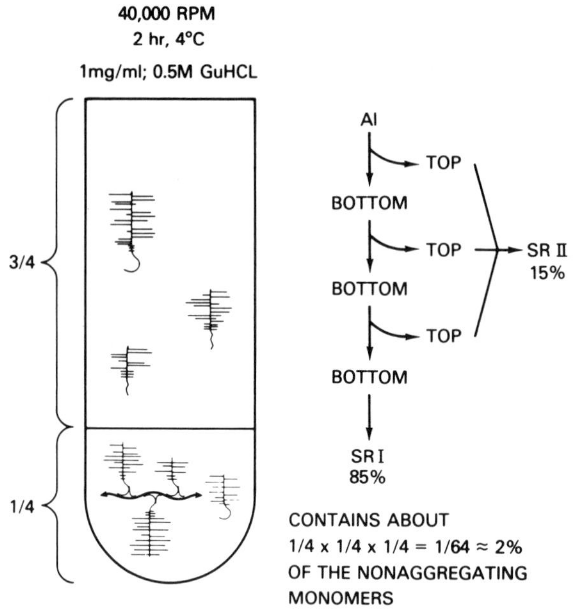

Fig. 18. Schematic diagram indicating the method used to isolate the nonaggregating proteoglycan fraction. See text for details.

and Mason (25) and Stanescu and coworkers (65,66) have used a polyacrylamide-agarose gel electrophoresis system to study proteoglycan monomers. Two separate, closely spaced proteoglycan bands were observed, suggesting that the monomer fraction contains two populations of macromolecules. The relationships between these different proteoglycan species or between nonaggregating and aggregating molecules, if any, remain to be determined.

The exercise of constructing a model for the chemical and physical properties of cartilage proteoglycans received its initial impetus from the pioneering work of Martin Mathews and his associates (1). The present version of the model focuses the attention of investigators on the chondrocytes that synthesize the polydisperse population of proteoglycan molecules and on the extracellular matrix where the molecules provide their function for the tissue. As we continue to modify and refine the model to accomodate new discoveries, we can expect to answer such interesting questions as that posed in 1958 by Mathews and Lozaityte (1):

> "... Recent work indicates that the cartilage cell probably has the major role in synthesis of the components of [proteoglycans]. However, one may only speculate on the level of molecular and structural organization reached within the cell before secretion or extrusion of aggregate precursors into the ground substance. In view of the observation that aggregates of the basic unit are reversibly dissociable under comparatively mild conditions, it is possible that the final stages of aggregation occur external to the cell."

REFERENCES

1. Mathews, M. B., and Lozaityte, I., *Arch. Biochem. Biophys. 74,* 158 (1958).
2. Shatton, J., and Schubert, M., *J. Biol. Chem. 211,* 565 (1954).
3. Malawista, I., and Schubert, M., *J. Biol. Chem. 230,* 535 (1958).
4. Pal, S., Doganges, P. T., and Schubert, M., *J. Biol. Chem. 241,* 4261 (1966).
5. Muir, H., *Biochem. J. 69,* 195 (1958).
6. Gregory, J. D., Laurent, T. C., and Rodén, L., *J. Biol. Chem. 239,* 3312 (1964).
7. Rodén, L., and Smith, R., *J. Biol. Chem. 241,* 5949 (1966).
8. Rodén, L., in "Chemistry and Molecular Biology of the Intercellular Matrix," (Balazs, E. A., ed.) Vol. 2, p. 797, Academic Press, New York, (1970).
9. Anderson, B., Hoffman, P., and Meyer, K., *J. Biol. Chem. 240,* 156 (1965).
10. Partridge, S. M., Davis, H. F., and Adair, G. S., *Biochem. J. 79,* 15 (1961).
11. Partridge, S. M., and Elsden, D. F., *Biochem. J. 79,* 26 (1961).

12. Gregory, J. D., and Rodén, L., *Biochem. Biophys. Res. Commun. 5,* 430 (1961).
13. Mathews, M. B., *Biochem. J. 125,* 37 (1971).
14. Heinegård, D., and Hascall, V. C., *Arch. Biochem. Biophys. 165,* 427 (1974).
15. Thyberg, J., Lohmander, S., and Heinegård, D., *Biochem. J. 151,* 157 (1975).
16. Franek, M. D., and Dunstone, J. R., *J. Biol. Chem. 242,* 3460 (1967).
17. Sajdera, S. W., Doctoral Dissertation, The Rockefeller University, New York, (1969).
18. Hascall, V. C., Doctoral Dissertation, The Rockefeller University, New York, (1969).
19. Sajdera, S. W., and Hascall, V. C., *J. Biol. Chem. 244,* 77 (1969).
20. Hascall, V. C., and Sajdera, S. W., *J. Biol. Chem. 244,* 2384 (1969).
21. Sajdera, S. W., Hascall, V. C., Gregory, J. D., and Dziewiatkowski, D. D., in "Chemistry and Molecular Biology of the Intercellular Matrix," (Balazs, E. A., ed.) Vol. 2, p. 851, Academic Press, New York, (1970).
22. Miller, E. J., *Mol. Cell. Biochem. 13,* 165 (1976).
23. Heinegård, D., *Biochim. Biophys. Acta 285,* 181 (1972).
24. Oegema, T. R., Hascall, V. C., and Dziewiatkowski, D. D., *J. Biol. Chem. 250,* 6151 (1975).
25. Pearson. J. P., and Mason, R. M., *Biochim. Biophys. Acta 498,* 176 (1977).
26. Gregory, J. D., Sajdera, S. W., Hascall, V. C., and Dziewiatkowski, D. D., in "Chemistry and Molecular Biology of the Intercellular Matrix," (Balazs, E. A., ed.) Vol. 2, p. 843, Academic Press, New York, (1970).
27. Keiser, H., Shulman, H. J., and Sandson, J. I., *Biochem. J. 126,* 163 (1972).
28. Hascall, V. C., and Heinegård, D., *J. Biol. Chem. 249,* 4232 (1974).
29. Baker, J. R., and Caterson, B. C., *Biochem. Biophys. Res. Commun. 77,* 1 (1977).
30. Hascall, V. C., *J. Supramol. Struct. 7,* 101 (1977).
31. Hardingham, T. E., and Muir, H., *Biochim. Biophys. Acta 279,* 401 (1972).
32. Hardingham, T. E., and Muir, H., *Biochem. Soc. Trans. 1,* 282 (1973).

33. Gregory, J. D., *Biochem. J. 133,* 383 (1973).
34. Hardingham, T. E., and Muir, H., *Biochem. J. 139,* 565 (1974).
35. Heinegård, D., and Hascall, V. C., *J. Biol. Chem. 249,* 4250 (1974).
36. Hascall, V. C., and Heinegård, D., *J. Biol. Chem. 249,* 4242 (1974).
37. Hardingham, T. E., and Muir, H., *Biochem. J. 135,* 905 (1973).
38. Heinegård, D., and Axelsson, I., *J. Biol. Chem. 252,* 1979 (1977).
39. Yamagata, T., Saito, H., Habuchi, O., and Suzuki, S., *J. Biol. Chem. 243,* 1523 (1968).
40. Hascall, V. C., Riolo, R. L., Hayward, J. H., and Reynolds, C. R., *J. Biol. Chem. 247,* 4521 (1972).
41. Hascall, V. C., and Riolo, R. L., *J. Biol. Chem. 247,* 4529 (1972).
42. Hascall, V. C., and Sajdera, S. W., *J. Biol. Chem. 245,* 4920 (1970).
43. Pasternack, S. G., Veis, A., and Breen, M., *J. Biol. Chem. 247,* 4529 (1974).
44. Wasteson, Å., *J. Chromatogr. 59,* 87 (1971).
45. Hjertquist, S.-O., and Wasteson, Å., *Calc. Tis. Res. 10,* 31 (1972).
46. Inerot, S., Heinegård, D., Audell, L., and Olsson, S.-E., *Biochem. J. 169,* 143 (1978).
47. De Luca, S., Heinegård, D., Hascall, V. C., Kimura, J. H., and Caplan, A. I., *J. Biol. Chem. 252,* 6600 (1977).
48. Kitamura, K., and Yamagata, T., *FEBS Lett. 71,* 337 (1976).
49. Hardingham, T. E., and Adams, P., *Biochem. J. 159,* 143 (1976).
50. Heinegård, D., and Hascall, V. C., Submitted.
51. Mathews, M. B., in "Connective Tissue: Macromolecular Structure and Evolution," (Molecular Biology, Biochemistry and Biophysics, Vol. 19) Springer-Verlag, New York, (1975).
52. Heinegård, D., *J. Biol. Chem. 252,* 1980 (1977).
53. Rosenberg, L., Wolfenstein-Todel, C., Margolis, R., Pal, S., and Strider, W., *J. Biol. Chem. 251,* 6439 (1976).

54. Hardingham, T. E., Ewins, R. J. F., and Muir, H., *Biochem. J. 157,* 127 (1976).
55. Hoffman, P., Mashburn, T. A., Hsu, D.-S., Trivedi, D., and Diep, J., *J. Biol. Chem. 250,* 7251 (1975).
56. Lohmander, S., *Eur. J. Biochem. 57,* 549 (1975).
57. Hascall, V. C., and Heinegård, D., in "Extracellular Matrix Influences on Gene Expression" (Slavkin, H., and Greulich, R., eds.) p. 423, Academic Press, New York (1975).
58. Caplan, A. I., *Exp. Cell Res. 62,* 341 (1970).
59. Hascall, V. C., Oegema, T. R., and Caplan, A. I., *J. Biol. Chem. 251,* 3511 (1976).
60. De Luca, S., Caplan, A. I., and Hascall, V. C., submitted.
61. Kimura, J. H., Osdoby, P., Caplan, A. I., and Hascall, V. C., submitted.
62. Choi, H. U., and Meyer, K., *Biochem. J. 151,* 543 (1975).
63. de Duve, C., in "Lysosomes in Biology and Pathology" (Dingle, J. T., and Fell, H. B., eds.) Vol. 1, p. 3, North Holland Publishing Co., Amsterdam, (1969).
64. Heinegård, D., and Hascall, V. C., In preparation.
65. Stanescu, V., and Maroteaux, P., in "Protides of the Biological Fluids, 22nd Colloquium" (Peeters, H., ed.) p. 201, Pergamon Press, Oxford and New York, (1975).
66. Stanescu, V., Montieux, P., and Sobczak, E., *Biochem. J. 163,* 103 (1977).

GLYCOCONJUGATE RESEARCH / Volume I
Proceedings of the Fourth International Symposium on Glycoconjugates

The Role of Hyaluronic Acid in Proteoglycan Aggregation

Helen Muir and Timothy E. Hardingham

The term proteoglycan (known previously as protein-polysaccharide or chondromucoprotein) refers to macromolecules of connective tissue constructed of a protein-core to which a large number of glycosaminoglycan chains are attached. Generally protein comprises only 10-20% of the weight of the molecule in cartilage proteoglycans, 80-90% being due to chondroitin sulphate with lesser but variable amounts of keratan sulphate. This general structure was first proposed by Martin B. Mathews (1), and our understanding of the structure and properties of cartilage proteoglycans owes a great deal to his work. At about the same time my own studies (2), carried out independently, showed that when cartilage was extracted at neutral pH without the use of proteolytic enzymes, a viscous product was obtained that retained about 10% of protein even when extensively purified. On treatment with proteolytic enzymes, however, the viscosity was rapidly lost, but much of the serine remained attached to the chondroitin sulphate, even after exhaustive proteolysis, indicating that the chondroitin sulphate was attached to the protein core *via* serine residues.

The model proposed by Mathews for the chondroitin-sulphate protein was of a novel bottle-brush shape, with the chondroitin sulphate chains attached lateraly along the protein core. Since the molecular weight of each chondroitin

ISBN 0-12-301301-1

sulphate chain is about 20 000, if 50 or more such chains are attached to a protein core, the composite molecule will have a mol. wt. of the order of a few million. It is a tribute to Mathews' insight that such a molecular model is essentially the same as that accepted today. Recent light scattering measurements (3) indicate the weight average mol. wt. of nasal cartilage proteoglycan to be 2.3×10^6.

In examining the physico-chemical behavior of cartilage proteoglycans, Mathews suggested that they may reversibly form very large aggregates of about 50×10^6 mol. wt. (1). It was not, however, until 1969 that this possibility was investigated further by Hascall and Sajdera (4), who obtained evidence of reversible aggregation of highly purified proteoglycans. They obtained unequivocal results only after they had introduced efficient non-disruptive methods of extraction and density gradient centrifugation in caesium chloride to purify the extracted proteoglycans (5). Since the essential function of proteoglycans in cartilage results from their entrapment in the collagen network, they do not readily diffuse out of the tissue. Sajdera and Hascall (5) found that certain salt solutions at critical concentrations were very effective in extracting proteoglycans. The most extensively used now is 4 M guanidinium chloride. The proteoglycans thus extracted are dialysed to low ionic strength and then purified by equilibrium density-gradient centrifugation in caesium chloride (Fig. 1). Proteoglycans that are rich in carbohydrate are of high buoyant density in caesium chloride and separate at the bottom of the density gradient. Proteoglycans obtained in this way, when examined by analytical ultracentrifugation, showed a bimodal distribution comprised of a faster and more slowly sedimenting species, representing proteoglycan aggregates and monomers, respectively. When subjected to a second density gradient centrifugation in caesium chloride under dissociating conditions in the presence of 4 M guanidinium chloride, proteoglycan aggregates are dissociated, and the fast sedimenting species is no longer observed in the analytical ultracentrifuge. The constituents of the aggregate separate at different densities with dissociated proteoglycans at the bottom. That aggregation is reversible is shown by the fact that, when the separated constituents of the aggregate are

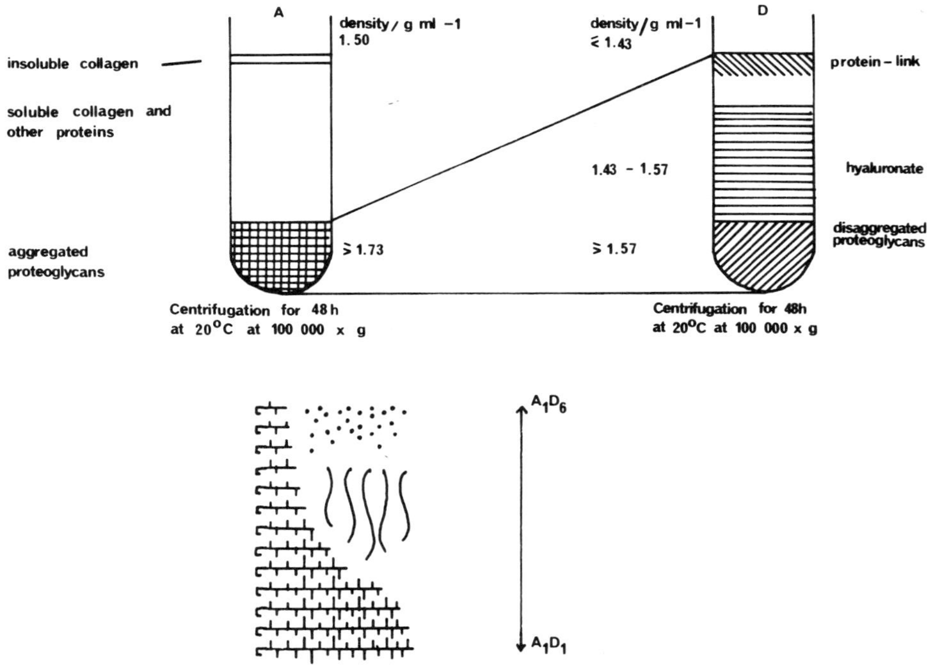

Fig. 1. *Two-stage density-gradient purification of cartilage proteoglycans. (upper) A. Associative gradient in 0.5 M guanidinium chloride, 0.05 M sodium acetate (pH 5.8) containing 0.1 M 6-aminohexanoic acid, 0.005 M benzamidine HCl, and 0.01 M EDTA. D. Dissociative gradient in 4.0 M guanidinium chloride, 0.05 M sodium acetate (pH 5.8) of aggregated proteoglycans isolated from the A gradient. (lower) Distribution of proteoglycans, hyaluronic acid, and protein-link in the dissociative gradient.*

mixed together again, the faster sedimenting species is reformed (4).

It was at first thought that aggregation depended on interaction of proteoglycan with a protein that separated at the top of the second dissociative gradient known as "protein-link" (4, 6, 7). It has since been recognised, however, that aggregation depends upon a highly specific interaction of proteoglycans with hyaluronic acid, first discovered by Hardingham and Muir (8). It was not known at that time that there were significant amounts of hyaluronic acid in cartilage, but it has since been shown to account for 0.5-1.0% of the total uronic acid in laryngeal (9) or nasal (10) cartilage.

HYALURONIC ACID-PROTEOGLYCAN INTERACTION

The interaction of proteoglycan and hyaluronic acid leads to an increase in hydrodynamic size and to a large increase in viscosity (8). Hardingham and Muir (11) found that material in the middle of the dissociative gradient, rather than that at the top, produced the maximum effects on viscosity and gel chromatographic behavior when mixed with dissociated proteoglycan, as would hyaluronic acid from a variety of sources. The buoyant density of hyaluronic acid in caesium chloride was shown by Gregory (12) to be about 1.43 g/ml, close to that of the material in the middle of the dissociative gradient. This material was isolated and shown by analysis to be hyaluronic acid (9).

The stoichiometry of the interaction of hyaluronic acid and proteoglycan was studied by viscometry and gel chromatography (8). As little as 0.01% of hyaluronic acid in a solution of proteoglycan produced detectable effects, and maximum effects were produced when the proportion of hyaluronic acid to proteoglycan (based on weight) was rather less than 1%, *i.e.*, close to the proportion present in cartilage. With such an excess of proteoglycan over hyaluronic acid, it appeared that a large number of proteoglycan molecules were interacting with a single chain of hyaluronic acid. The increase in hydrodynamic size seen by gel chromatography was maximal with the proportion of hyaluronic acid relative to proteoglycan at 1:150 (w/w) respectively. With larger proportions of hyaluronic acid, the effect was diminished. This indicated that proteoglycans possess only a single binding site and do not cross-link hyaluronic acid chains, so that a gel was not produced at higher proportions of hyaluronic acid. Under conditions when proteoglycans are in excess, hyaluronic acid can bind a maximum of about 250 times its weight of proteoglycan.

The interaction is entirely specific to hyaluronic acid. No comparable effects are produced by other polyanions, even close isomers of hyaluronic acid such as chondroitin sulphate, *i.e.*, chemically desulphated chondroitin which differs from hyaluronic acid only in the configuration of the hydroxyl group at C-4 of the hexosamine residues (13), nor does it depend upon the presence of divalent metal cations as it is unaffected by the presence of EDTA (8). Proteoglycans from a variety of anatomical sites including articular cartilage (7, 14, 15) interact with hyaluronic acid, as do those from embryonic cartilage, and this property appears to be an expression of phenotype during chondrogenesis (see Hascall and Heinegård, this volume). There is little evidence that proteoglycans from other types of connective tissue are able to interact with hyaluronic acid.

Oligosaccharides, derived from hyaluronic acid by digestion with testicular hyaluronidase, above a critical size are strong competitive inhibitors of the interaction. Decasaccharides are the smallest oligosaccharides to compete strongly with hyaluronic acid, whereas octasaccharides and smaller oligosaccharides are virtually ineffective (13, 16). The binding of each proteoglycan therefore involves only 5 nm of the hyaluronic acid chain.

Using an average mol. wt. for proteoglycan of 2.5×10^6 and for hyaluronic acid of 5×10^5, a model for the proteoglycan-hyaluronic acid complex was deduced (Fig. 2) based on the stoichiometry of binding. The dimensions agreed reasonably well with those calculated from electron micrographs of proteoglycan aggregates (6, 17). It was calculated that each proteoglycan molecule would occupy a region on the hyaluronic acid chain of about 20 nm in length (18). However, calculations based on the amount of hyaluronic acid in proteoglycan aggregates in laryngeal (19) and nasal (10) cartilage suggest that the average distance between proteo-

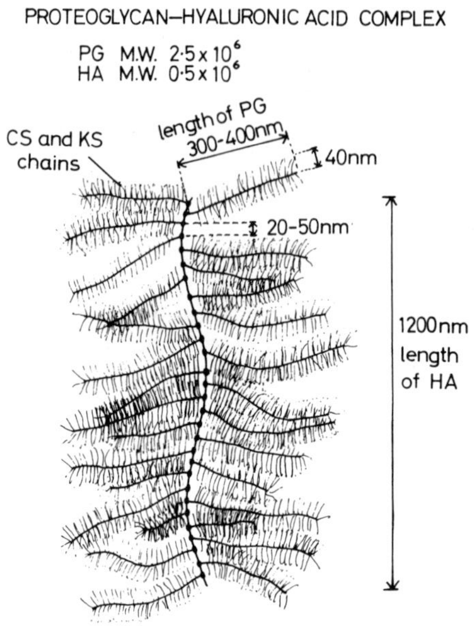

Fig. 2. The proteoglycan-hyaluronic acid complex. The dimensions of the model were deduced from the stoichiometry of the interaction. PG, proteoglycan; HA, hyaluronic acid; CS, chondroitin sulphate; and KS, keratan sulphate (from ref. 18).

glycan molecules in aggregates is much greater than 20 nm, and since only 5 nm of the hyaluronic acid is occupied by each proteoglycan, the hyaluronic acid in cartilage is not fully saturated with proteoglycan. Presumably steric hindrance due to the numerous chondroitin sulphate side-chains along the proteoglycan core-protein limits their closeness of packing on the hyaluronic acid chain. Thus, when much of the chondroitin sulphate on the proteoglycan is removed by digestion with chondroitinase, it can be packed five times more closely along the hyaluronic acid chain (13). Since chondroitinase-treated proteoglycans are able to interact with hyaluronic acid, chondroitin sulphate chains do not participate in the interaction; moreover, aggregates are not dissociated by chondroitinase digestion (10), the hyaluronic acid within the aggregate being largely protected from attack by chondroitinase.

Reduction of disulphide bridges interferes with aggregation (4). Reduction and alkylation of proteoglycan abolishes the interaction with hyaluronic acid, but there is no change in molecular size or protein content. The conformation of the hyaluronic acid binding-region of the protein core of the proteoglycan is maintained by about 5-7 intramolecular disulphide bridges (20). If the reduced proteoglycan is not alkylated, the loss of interaction is largely reversible on re-oxidation with molecular oxygen, indicating that the structure is thermodynamically rather stable. It is also remarkably resistant to denaturation by heating, as the ability to bind to hyaluronic acid was virtually unchanged by heating to 60°C, and the half life was 148 min at 80°C and 14.2 min at 100°C (20).

Although the length of the hyaluronic acid chain occupied by proteoglycan is small, the region of the proteoglycan involved in binding to hyaluronic acid is quite large and has a mol. wt. of about 60 000. It can be obtained by cyanogen bromide cleavage of dissociated proteoglycans, and as it lacks glycosaminoglycan chains, it is of low buoyant density and separates at the top of the caesium chloride density gradient (21). In the intact aggregate, the hyaluronic acid binding-region is relatively protected from proteolysis and can be isolated by mild tryptic digestion from chondroitinase-treated aggregates (22).

The amino acid composition of the hyaluronic acid binding-region differs significantly from that of the whole core protein, particularly in containing more aspartic acid, arginine, methionine, and cystine and less serine, glycine, and proline (21, 22). About 60% of the amino acids of the remainder of the core protein that bears chondroitin sulphate chains consists of approximately equimolar amounts of serine, glycine, proline, and glutamic acid (22).

Although the hyaluronic acid binding-region is resistant to heating, it is very sensitive to specific chemical modification of basic and aromatic amino acids (20). Certain lysine, arginine, and tryptophan residues are essential for interaction with hyaluronic acid. Under carefully selected conditions of reaction with butane-2,3-dione (pH 6.6 for 6 h at 4°C in the dark), about a quarter of the arginine residues reacted with the loss of 87% of binding. Loss of binding resulted also when about 40% of the free amino groups were blocked by acetylation with acetic anhydride. Substitution of primary amino groups, principally ε-amino groups of lysine, with 2-methylmaleic anhydride completely abolished binding, but this was largely regained when the substituent was removed, pointing to the direct participation of the ε-amino groups of lysine in the interaction with hyaluronic acid (20). The lysine residues involved in the interaction are partially protected from chemical substitution (acetylation and dansylation) when hyaluronic acid is present (23). The conformation of the binding site is also dependent on tryptophan residues in the molecule. Reaction with *N*-bromosuccinimide showed a direct correlation with a spectral change and loss of binding. About 82% of the ability to bind to hyaluronic acid was also lost when about one third of the tryptophan residues had reacted with 2-nitrophenylsulphenyl chloride. That tryptophan is not directly involved in the interaction is suggested by the absence of any major change in the fluorescence spectrum of the proteoglycan when interaction with hyaluronic acid took place. It is suggested that tryptophan residues occupy hydrophobic regions in the interior of the binding region. When the ability to bind to hyaluronic acid was abolished, either on reduction and alkylation of disulphide bonds or in the presence of 1% sodium dodecyl sulphate, the tryptophan fluorescence maximum was shifted to longer wavelengths, which would correspond to the tryptophan residues being in a more polar phase (20). The exactness of the conformation of the binding site implies an equally exact conformation of the polar groups on hyaluronic acid. Binding was abolished when 40% of the carboxyl groups were modified, or when they were displaced from the pyranose rings by three carbon atoms (24). *N*-Acetylglucosamine residues are also essential since chondroitin, which differs only in containing *N*-acetylgalactosamine residues, does not compete with hyaluronic acid (13). Thus, the effective binding-site appears to be of a precise shape allowing the maximum number of sub-site interactions to take place.

That oligosaccharides of hyaluronic acid compete with hyaluronic acid and displace it from proteoglycan (13, 16), shows that the interaction is an equilibrium, which nevertheless lies well in favour of complex formation under physiological conditions (25). A preliminary study of labeled oligosaccharides in equilibrium dialysis suggests that the dissociation constant is about 10^{-6} to 10^{-7} M (26). Hardingham (26) has examined the behaviour of the proteoglycan-hyaluronic acid complex in the analytical ultracentrifuge where, under conditions of temperature and concentration normally employed, the complex will be partly dissociated. By raising the relative proportion of hyaluronic acid to proteoglycan from 1% to 2%, however, the amount of associated species increased and did so further when the temperature was reduced to 4°C. The Johnston-Ogston effect (27), which is very pronounced when mixtures of large polyionic molecules are sedimented, also reduces the apparent proportion of the associated species at concentrations of 2-3 mg/ml that are usually used, so that the apparent proportion increases as the concentration is lowered.

THE FUNCTION OF PROTEIN-LINK

Proteoglycan aggregates contain a third constituent, the protein-link, which separates at the top of the dissociative gradient. It is able to bind to hyaluronic acid on its own and appears to be present in the aggregate in a 1:1 ratio with the hyaluronic acid binding-region of the proteoglycan, as deduced from partial degradation of aggregates (22). Protein-link also binds to proteoglycan, but it does not promote aggregation on its own (28). Proteoglycan aggregation as currently envisaged is depicted diagrammatically in Fig. 3.

The protein-link appears to function in stabilising the proteoglycan-hyaluronic acid complex, so that it is no longer in equilibrium with its dissociation products. The behaviour of the complex and aggregate was compared by viscometry under a variety of conditions (25). Although both are similarly dissociated by increasing concentrations of guanidinium chloride (Fig. 4), the aggregate is more resistant to dissociation by urea (< 4 M). With decreasing pH,

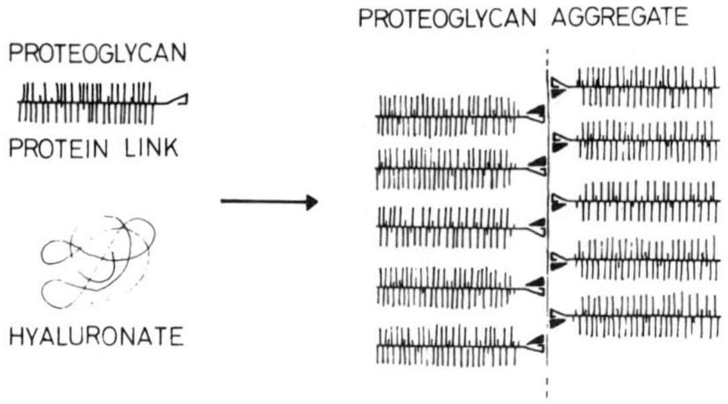

Fig. 3. Schematic representation of cartilage proteoglycan aggregation.

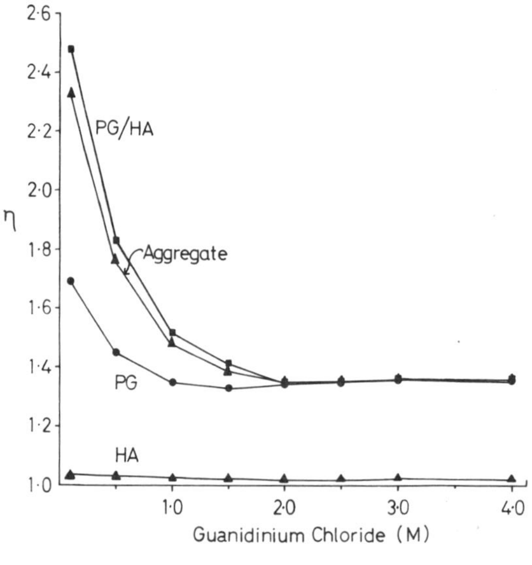

Fig. 4. Variation of relative viscosity with guanidinium chloride concentration of proteoglycan (2.5 mg/ml), proteoglycan aggregate (2.5 mg/ml), proteoglycan-hyaluronic acid complex (PG/HA 100:1, w/w, 2.5 mg/ml) and hyaluronic acid (22 μg/ml). Solutions contained 0.05 M sodium acetate (pH 5.8) at 30°C.

the aggregate remains undissociated until the pH is reduced to 5.5, whereas the hyaluronic acid-proteoglycan complex is increasingly dissociated below pH 8 (Fig. 5). Similarly, the complex dissociates increasingly with increasing temperature, but the aggregate does not do so until the temperature reaches about 55°C (Fig. 6). Above this temperature, an irreversible change takes place so that, on cooling and re-heating, the material behaves like the hyaluronic acid-proteoglycan complex, and it would therefore appear that the protein-link has been denatured. The stabilising role of the protein-link is forcefully illustrated by the fact that aggregates are unaffected by oligosaccharides of hyaluronic acid which dissociate the proteoglycan-hyaluronic acid complex (13) (Fig. 7). Since the aggregate is not in equilibrium with its dissociation products, it does not

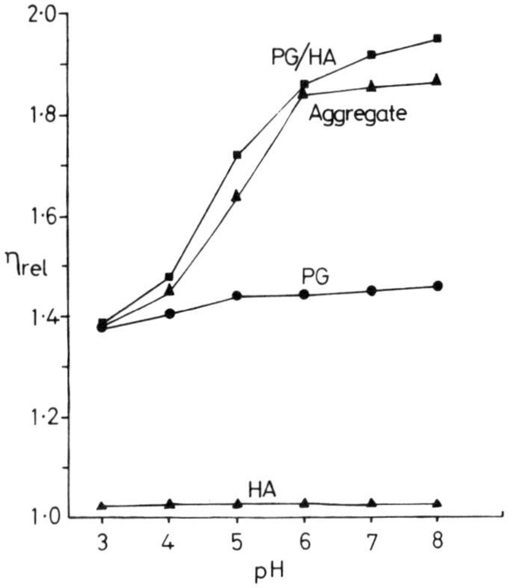

Fig. 5. Variation of relative viscosity with pH of proteoglycan (2.5 mg/ml), proteoglycan aggregate (2.5 mg/ml), proteoglycan-hyaluronic acid (PG/HA 100:1, w/w, conc. 2.5 mg/ml) and hyaluronic acid (25 µg/ml). All solutes in 0.5 M guanidinium chloride buffered at pH 3, 4, 5, and 6 with 0.05 M sodium acetate, and at pH 7 and 8 with 0.05 M Tris HCl at 30°C.

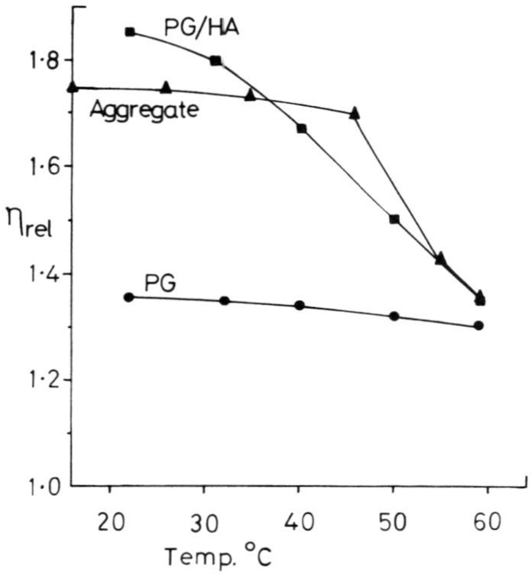

Fig. 6. Variation of relative viscosity with temperature of proteoglycan (2.1 mg/ml), proteoglycan aggregate (2.1 mg/ml) and proteoglycan-hyaluronic acid (PG/HA 100:1, w/w, conc. 2.1 mg/ml) in 0.5 M guanidinium chloride, 0.05 M sodium acetate (pH 5.8) at 30°C.

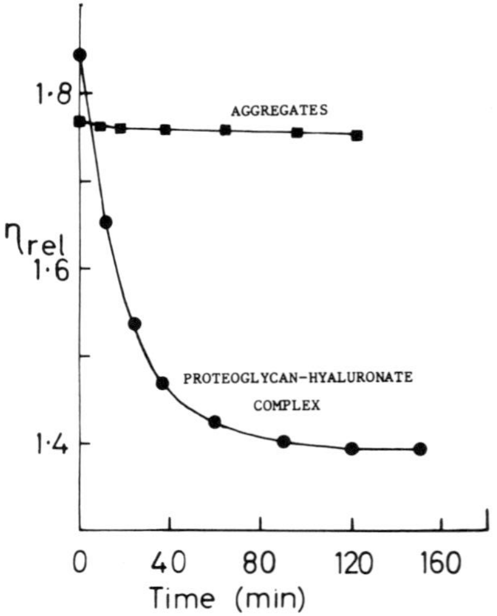

Fig. 7. Variation in relative viscosity of proteoglycan aggregate (2.1 mg/ml) and proteoglycan-hyaluronic acid complex (PG/HA 100:1, w/w, conc. 2.1 mg/ml) with time after the addition of hyaluronic acid oligosaccharides (HA_{10-20}, conc. 30 μg/ml). All solutes in 0.5 M guanidinium chloride, 0.05 M sodium acetate (pH 5.8) at 30°C.

dissociate on analytical ultracentrifugation and is seen irrespective of temperature or concentration, although the apparent proportion of aggregate in a mixture with monomer will be reduced at higher concentrations because of the Johnston-Ogston (27) effect (26).

In the aggregate, hyaluronic acid is partially protected from attack by leech hyaluronidase (an enzyme that specifically degrades hyaluronic acid) (25) and from chondroitinase digestion (22). The hyaluronic acid binding-region is also relatively protected from proteolytic enzymes. That aggregates are present in cartilage is shown by the fact that mild papain treatment gave a product that could be isolated without the use of dissociating solutions, in which protein was still attached to hyaluronic acid, and which resembled the product formed by mild papain treatment of aggregates themselves (10).

Proteoglycans are extremely polydisperse in molecular size and heterogeneous in chemical composition, and hence of wide ranging buoyant density in caesium chloride. A structural model to explain these characteristics has been proposed (10) in which proteoglycans contain an invariant region containing the hyaluronic acid binding-site largely devoid of carbohydrate and a variable region bearing the glycosaminoglycan chains, and considerable evidence now supports this proposal (20-22), (Fig. 8). A minority of proteoglycans are unable to interact with hyaluronic acid and form aggregates. They appear to represent a different population of proteoglycans (9) and their amino acid composition suggests that the hyaluronic acid binding-region is incomplete or absent (20).

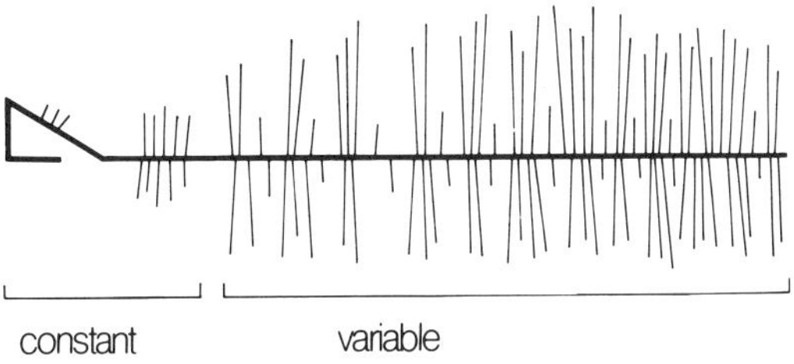

Fig. 8. Proposed structure of cartilage proteoglycans containing constant and variable regions of core-protein (10).

The size of a given aggregate will depend partially on the size of the monomeric proteoglycans, but principally on the length of the hyaluronic acid chain and on the number of proteoglycan molecules that are attached to it. Thus, aggregates prepared with a fixed proportion (0.5% w/w) of hyaluronic acid of decreasing chain-length had progressively lower sedimentation constants (Fig. 9). Increasing the relative proportion of hyaluronic acid to proteoglycan in the aggregating mixture above 0.5% (w/w) produced aggregates of diminishing sedimentation constant, although the amount was largely unaffected (Fig. 10). Sedimentation coefficients

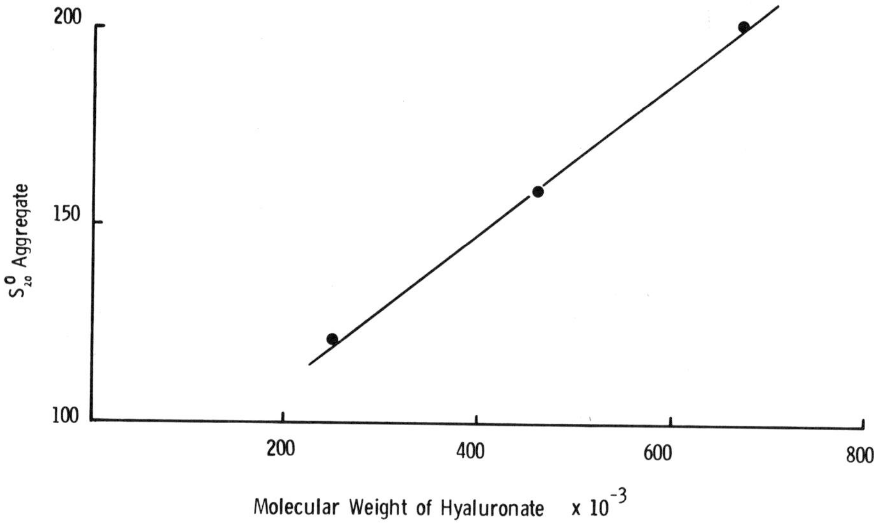

Fig. 9. Variation in sedimentation coefficient (S^{o}) of proteoglycan aggregates formed with hyaluronic acid fractions of different size, (proteoglycan-hyaluronic acid 200:1, w/w). Aggregates were formed by mixing proteoglycan, protein-link fraction, and hyaluronic acid in 4 M guanidinium chloride, 0.05 M sodium acetate (pH 5.8) followed by dialysis to 0.5 M guanidinium chloride, 0.05 M sodium acetate (pH 5.8). Sedimentation velocity determinations were carried out at 20°C in an MSE centriscan; extrapolation to zero concentration was performed by linear regression of a plot of log S vs. concentration.

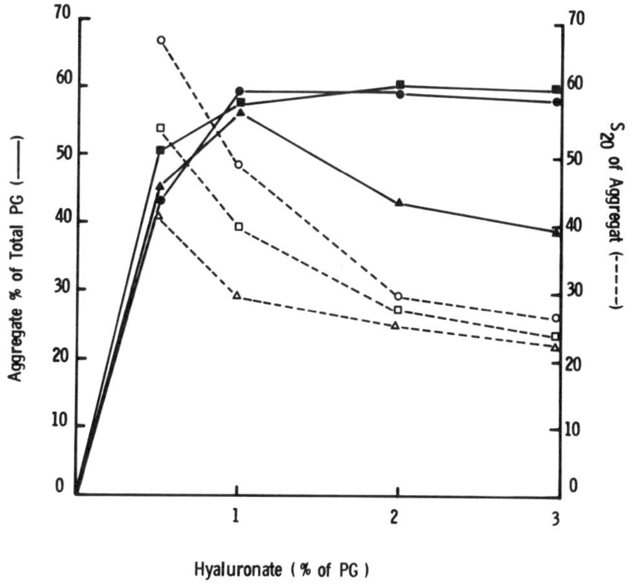

Fig. 10. Variation in the size and proportion of aggregates formed when proteoglycan was mixed with protein-link and varying amounts of hyaluronic acid of three different sizes: mol. wt. 6.7 x 10^5 (o); 4.55 x 10^5 (□), and 2.45 x 10^5 (Δ). Sedimentation-velocity determinations were made with proteoglycan (1.6 mg/ml) in 0.5 M guanidinium chloride, 0.05 M sodium acetate (pH 5.8) at 20°C in an MSE centriscan.

of aggregates and monomers isolated from different anatomical sites vary. The smallest are found in tissues of intervertebral discs, the largest in nasal and laryngeal cartilage, while those of articular cartilage are of intermediate size (26). It remains to be shown which of the three variables is primarily responsible for the differences in size.

SELF-ASSOCIATION OF PROTEOGLYCANS

It has recently been found by light scattering (29) that proteoglycans self-associate in the absence of hyaluronic acid or link protein in 0.15 M sodium chloride but not in solutions significantly above or below physiological ionic-strength. The associated species appeared to consist principally of dimers. These did not dissociate at higher temperatures, in contrast with the behaviour of the hyaluronic acid-proteoglycan complex, and it remains to be established how far dimerisation affects the interaction with hyaluronic acid. It is perhaps appropriate to conclude this article, which is a tribute to Martin Mathews, by pointing out that in his original molecular model (1), for cartilage proteoglycans, he indicated that they might form dimers through some kind of self-association. The wheel has indeed gone full circle for having established the details of a specific aggregation involving heterologous interactions between three different components, we now find renewed evidence of self-association of proteoglycans, which was proposed also by Martin Mathews in his original model for cartilage proteoglycans.

REFERENCES

1. Mathews, M. B., and Lozaityte, I., *Arch. Biochem. Biophys. 74,* 158 (1958).
2. Muir, H., *Biochem. J. 69,* 195 (1958).
3. Pasternack, S. V., Veis, A., and Breen, M., *J. Biol. Chem. 249,* 2206 (1974).
4. Hascall, V. C., and Sajdera, S. W., *J. Biol. Chem. 244,* 2384 (1969).
5. Sajdera, S. W., and Hascall, V. C., *J. Biol. Chem. 244,* 77 (1969).
6. Rosenberg, L., Hellman, W., and Kleinschmidt, A. K., *J. Biol. Chem. 245,* 4123 (1970).
7. Rosenberg, L., Pal, S., and Beale, R. J., *J. Biol. Chem. 248,* 3681 (1970).
8. Hardingham, T. E., and Muir, H., *Biochim. Biophys. Acta 279,* 401 (1972).
9. Hardingham, T. E., and Muir, H., *Biochem. J. 139,* 565 (1974).
10. Hascall, V. C., and Heinegård, D., *J. Biol. Chem. 249,* 4232 (1974).
11. Hardingham, T. E., and Muir, H., *Biochem. Soc. Trans. 1,* 282 (1973).

12. Gregory, J. D., *Biochem. J. 133,* 383 (1973).
13. Hascall, V. C., and Heinegård, D., *J. Biol. Chem. 247,* 4529 (1974).
14. Tsiganos, C. P., and Muir, H., in "Connective Tissue and Ageing", (Vogel, H. G., ed.) Vol. 1, p. 132, Excerpta Medica, Amsterdam (1973).
15. Rosenberg, L., Wolfenstein-Todel, C., Margolis, R., Pal, S., and Strider, W., *J. Biol. Chem. 251,* 6439 (1976).
16. Hardingham, T. E., and Muir, H., *Biochem. J. 135,* 905 (1973).
17. Rosenberg, L., Hellman, W., and Kleinschmidt, A. K., *J. Biol. Chem. 250,* 1877 (1975).
18. Hardingham, T. E., and Muir, H., in "Normal and Osteoarthrotic Articular Cartilage", (Ali, S. Y., Elves, M. W., and Leaback, D. H., eds.) p. 51. Institute of Orthopaedics, London, (1974).
19. Muir, H., and Hardingham, T. E., *M.T.P. Int. Rev. Sci.: Org. Chem. Ser. One, Carbohydr.,* Vol. 5, p. 153, Butterworths, London, (1975).
20. Hardingham, T. E., Ewins, R. J. F., and Muir, H., *Biochem. J. 157,* 127 (1976).
21. Heinegård, D., *J. Biol. Chem. 252,* 1980 (1977).
22. Heinegård, D., and Hascall, V. C., *J. Biol. Chem. 249,* 4250 (1974).
23. Heinegård, D., and Hascall, V. C., Personal communication (1977).
24. Christner, J. E., Brown, M. L., and Dziewiatkowski, D. D., *Biochem. J. 167,* 711 (1977).
25. Hardingham, T. E., and Muir, H., *Ann. Rheum. Dis. 34,* suppl. 2, 26 (1975).
26. Hardingham, T. E., unpublished results.
27. Johnston, J. P., and Ogston, A. G., *Trans. Farad. Soc. 42,* 789 (1946).
28. Tsiganos, C. P., Hardingham, T. E., and Muir, H., *Biochem. J. 128,* 121P (1972).
29. Sheehan, J. K., Nieduszynski, I. A., Phelps, C. F., Muir, H., and Hardingham, T. E., *Biochem. J. 171,* 109 (1978).

Biochemical Assessment of Malignancy in Human Chondrosarcomas

Lawrence Rosenberg, Lin-Heng Tang, and Subhash Pal

Human chondrosarcomas are malignant tumors of cartilage which vary greatly in their degree of malignancy. Some chondrosarcomas are highly malignant tumors. These grow rapidly, destroy bone, invade surrounding soft tissues, and may metastasize within a year after the onset of symptoms. Other chondrosarcomas are of low-grade malignancy. These increase in size slowly, are non-invasive, and may exist almost in harmony with their host for decades without metastasizing.

In a given case, it would be of obvious value to know the degree of malignancy, and the propensity for metastasis, of a particular chondrosarcoma. Such information is essential for prognosis, and could be considered, along with other factors, in deciding between such alternative methods of surgical management as *en bloc* resection of the tumor or amputation. However, the assessment of the degree of malignancy of human chondrosarcomas from histopathologic findings is fraught with difficulties (1-15). This is because only some highly malignant chondrosarcomas show the conventional histologic hallmarks of malignancy (marked cellularity; pronounced cellular and nuclear pleomorphism; numerous mitoses; atypical, bizarre nuclear chromatin patterns). Chondrosarcomas generally show their own special histological signs of malignancy. These signs are subtle, and have been appreciated only relatively recently. Chondrosarcomas of intermediate- or high-grade malignancy may show mild to moderate cellularity and pleomorphism. However, mitoses and overtly atypical nuclei may be absent. Instead, only slight increases in nuclear size ("plump nuclei"), subtle changes in nuclear chromatin patterns (discernible fine or coarse nuclear chromatin pattern), or chondrocytes with two nuclei (binucleate cells) may be found. It is difficult to

ISBN 0-12-301301-1

grade human chondrosarcomas in terms of degree of malignancy precisely and reproducibly on a basis of these histologic changes (2,3,5,6,10-12).

Low-grade chondrosarcomas present an additional problem. The microscopic appearance of low-grade chondrosarcomas and benign cartilage tumors may be similar or identical. Low-grade chondrosarcomas may show no increased cellularity, no pleomorphism, and no binucleate cells. The chondrocytes may contain small, round, and dark-staining nuclei with no discernible chromatin pattern, indistinguishable from those of benign cartilage tumors. Based on histologic findings, it may be difficult or impossible to determine whether a particular cartilage lesion is a benign cartilage tumor or a low-grade chondrosarcoma. The problem of distinguishing between low-grade chondrosarcomas and benign cartilage tumors has been eloquently stated by O'Neal and Ackerman (2):

> "More than most neoplasms, cartilaginous tumors of bone as a group exhibit an unbroken gradation from the completely benign to the highly malignant neoplasm. The zone between the innocent tumors and the overt chondrosarcomas, although the subject of much study and speculation, is still incompletely defined and poorly understood. Underdiagnosis continues to be common because of two features peculiar to many chondrosarcomas: first, a slowly evolving, prolonged clinical course and, second, a histological appearance superficially identical with benign cartilage. In the former instance, short-term follow-up of both treated and untreated cases may give little cause to suspect cancer. The microscopic picture of normal-appearing hyaline stroma containing lacunated cells with no mitoses and little pleomorphism is a special trap for the unwary or inexperienced pathologist. Frequently, it appears that the diagnosis of chondrosarcoma is forced by the clinical course rather than made by adequate pathological assessment."

These problems in the assessment of malignancy of chondrosarcomas from histologic findings lead, in turn, to problems in the management of patients. A problem arises in the young adult who presents with a small, tender, and enlarging peripheral cartilage tumor, which may be a low-grade chondrosarcoma or a benign osteochondroma; or in the adult with a small and mildly symptomatic central cartilage lesion in the proximal femur, which may be a chondrosarcoma or a benign enchondroma. The decision on the part of a surgeon to carry out an *en bloc* radical resection of a symptomatic cartilage tumor should be supported by an unequivocal, objective diagnosis of chondrosarcoma. Yet, in some cases, it may be difficult for an experienced pathologist with a special interest in chondrosarcomas to distinguish between a benign cartilage tumor and a low-grade chondrosarcoma based on cellularity, nuclear size or atypia, or double nuclei. If clinical and radiographic findings do not definitely point to a chondrosarcoma, it may be impossible to arrive at this diagnosis from histologic findings.

At present, the histologic grading of chondrosarcomas in terms of degree of malignancy is a subjective process. Estimates of cellularity, nuclear size, nuclear atypia, and numbers of binucleate cells depend on the subjective evaluation of the individual pathologist, his experience with cartilage tumors, the area of tissue sections examined, and the time spent in studying a cartilage tumor. Differences of opinion are frequently encountered in the grading of chondrosarcomas. A precise, reliable, and objective method is needed for distinguishing between benign and malignant cartilage tumors, and for assessing the degree of malignancy of human chondrosarcomas.

These problems led us to consider the possibility that a biochemical examination of human chondrosarcomas might yield information of value in making clinical decisions. Chondrosarcomas, like cartilages, are highly specialized connective tissues composed of chondrocytes distributed throughout an abundant extracellular substance. The extracellular substance consists of collagen fibers embedded in a gel-like matrix or ground substance. Cartilage ground substance is composed mainly of proteoglycans monomers and proteoglycan aggregates. Cartilage proteoglycan monomer consists of chondroitin sulfate and keratan sulfate chains covalently attached to a protein core. Proteoglycan aggregates are formed by the noncovalent association of proteoglycan monomers with hyaluronic acid and link proteins (16-33).

These considerations raised the following questions: Do human chondrosarcomas contain proteoglycan monomer and aggregate forms analogous to those found in normal cartilages? If so, do proteoglycan monomer and aggregate synthesized by human chondrosarcomas differ in structure from the proteoglycan species of normal human articular cartilage? Are these structural differences present in the glycosaminoglycan components of proteoglycans (chondroitin sulfate, keratan sulfate, and hyaluronic acid) or in the protein components of proteoglycans (proteoglycan monomer core-protein and link protein)? Are these structural differences in proteoglycans from chondrosarcomas related to the degree of malignancy of the tumors? To examine these questions, we isolated and characterized proteoglycans from human chondrosarcomas and from normal human articular cartilages.

METHODS

Proteoglycans were extracted from human chondrosarcomas by slow stirring for 48 h at 5°C in 5.5 M guanidinium hydrochloride and 0.15 M potassium acetate (pH 6.3) containing 5 mM concentrations of iodoacetate, benzamidine hydrochloride, and phenylmethanesulfonyl fluoride as protease inhibitors. Extracts were filtered, and the filtrates dialyzed at 5°C for 16 h against 20 vol. of 0.15 M potassium acetate (pH 6.3) containing 5 mM concentrations of the protease inhibitors. The extracts were then fractionated by equilibrium density gradient centrifugation under associative conditions (34) in 3.5 M cesium chloride, 0.15 M potassium acetate (pH 6.3). The initial density of the solution at 10°C was 1.46 to 1.47 g/ml. Density gradient centrifugations were carried out at 40 000 r.p.m. (128 000g_{av}) for 50 h at 10°C in 2.5 x 8 cm polyallomer tubes with high-force cap assemblies, in a Beckmann 50.2 fixed-angle rotor. The gradients were divided into six equal fractions. These were dialyzed against 0.15 M potassium acetate (pH 6.3) at 5°C, precipitated with three volumes of ethanol, washed with ethanol and ether, and dried in a vacuum. The fractions from the bottom to the top of this associative gradient are called A1 through A6. Matrix proteins and glycoproteins are of low buoyant density, and were separated into Fractions A6 and A5 at the top of the gradient. A mixture of proteoglycan monomer and proteoglycan aggregate was separated free of extraneous matrix proteins and glycoproteins into Fraction A1 at the bottom of this gradient.

Proteoglycan monomer (D1) was isolated from proteoglycan Fraction A1 by equilibrium-density gradient centrifugation under dissociative conditions. Fraction A1 was dissolved in 5.5 M guanidinium hydrochloride, 0.15 M potassium acetate (pH 6.3), containing 5 mM concentrations of the three protease inhibitors, and stirred for 16 h at 5°C to insure complete dissociation of the proteoglycan aggregate. Cesium chloride was added to give a final solution containing 4 M guanidinium hydrochloride, 3 M cesium chloride, and 2 mg of proteoglycan per ml, the density of which at 5°C was 1.51 g/ml. Density-gradient centrifugation was carried out at 40 000 r.p.m. for 60 h at 5°C. The gradient was cut into six equal fractions. The fractions from the bottom to the top of this dissociative gradient are called D1 through D6. Proteoglycan monomer was recovered, free of link protein and hyaluronic acid, in Fraction D1 at the bottom of this gradient. Fraction D1 was recycled through a second dissociative gradient. The proteoglycan monomer present in Fraction D1 was then chemically and physically characterized.

The analytical methods for uronic acid, galactosamine, glucosamine, galactose, glucose, sialic acid, protein, hydroxyproline, and amino acid content have previously been described (34). Sedimentation-velocity studies were carried out at 48 000 r.p.m. and 20°C in 0.15 M KCl, 0.01 M MES, pH 7.0 (ρ = 1.061 g/ml, η = 1.001), as previously described (34), or in 0.15 M NaCl, 0.01 M MES, pH 7.0 (ρ = 1.046 g/ml).

RESULTS AND DISCUSSION

To determine whether human chondrosarcomas contain proteoglycan monomer and aggregate species analogous to those found in normal cartilages, proteoglycans were extracted from human chondrosarcomas, and separated from extraneous matrix proteins and glycoproteins by equilibrium-density gradient centrifugation under associative conditions in 3.5 M cesium chloride. In the case of normal cartilages, this procedure separated a mixture of proteoglycan monomer and proteoglycan aggregate into Fraction A1 at the bottom of the associative gradient (ρ = 1.6 g/ml); extraneous matrix proteins and glycproteins were separated into the top one-third of the gradient (ρ < 1.45 g/ml). Aliquots of Fraction A1 were then fractionated by equilibrium-density gradient centrifugation under dissociative conditions in 4 M guanidinium hydrochloride - 3 M cesium chloride. In the case of normal cartilages, this procedure results in the isolation of proteoglycan monomer in pure form, free of hyaluronic acid and link proteins, in Fraction D1 at the bottom of the dissociative gradient (ρ = 1.6 g/ml). Link proteins were separated into Fractions D5 and D6 at the top of the gradient, and hyaluronic acid was separated into the middle of the gradient. Proteoglycan Fractions A1 and D1 were then chemically characterized and examined in sedimentation-velocity experiments (34).

Representative Schlieren patterns of proteoglycan Fractions A1 and D1 from a low-grade hyaline chondrosarcoma, and from a more highly malignant hyaline chondrosarcoma are shown in Fig. 1. Proteoglycan Fraction A1 from the human chondrosarcomas (Fig. 1B and 1D) showed proteoglycan aggregate and monomer species analogous to those from normal cartilages. Proteoglycan Fraction D1 from the human chondrosarcomas (Fig. 1A and 1C) showed a sharply unimodal proteoglycan monomer species analogous to that from normal cartilages. Fraction D6 from six human chondrosarcomas, and from one normal human articular cartilage, each showed on sodium dodecyl sulfate-polyacrylamide gel electrophoresis two link proteins having a mol. wt. of approximately 40 000 and 49 000 (Table I).

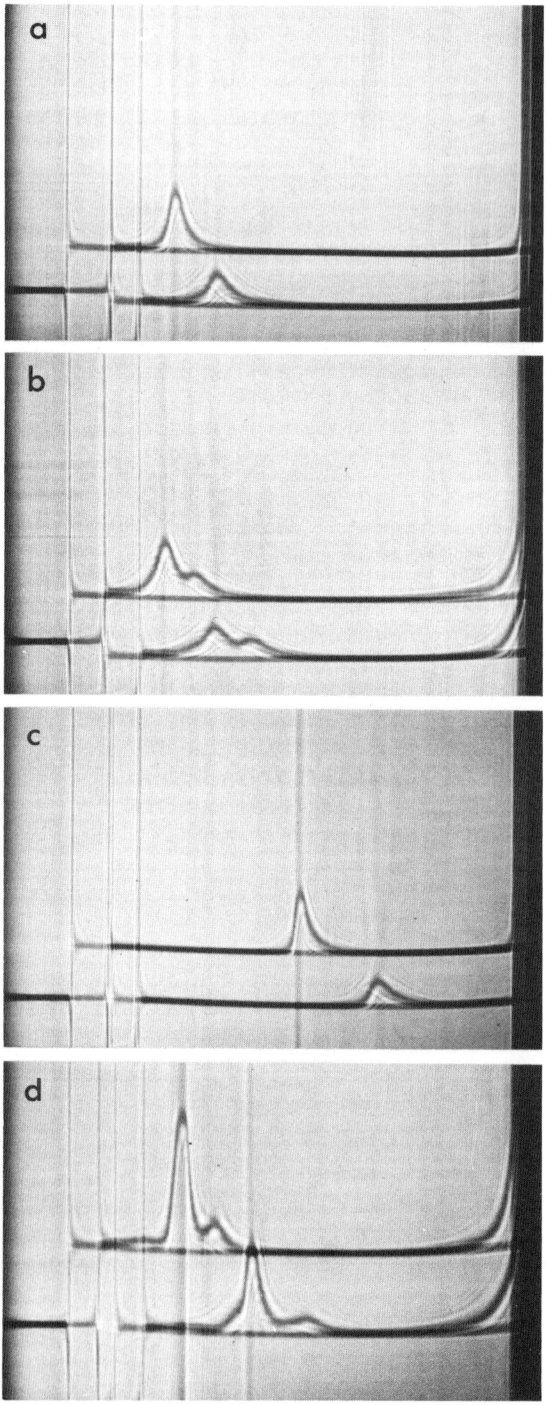

Fig. 1. Sedimentation velocity patterns of proteoglycan monomer in Fraction D1, and/or proteoglycan monomer and aggregate in Fraction A1 from two human chondrosarcomas. Runs were made at 48 000 r.p.m. and 20°C in 0.15 M NaCl, 0.01 M MES (pH 7.0). The direction of sedimentation is from left to right. (a) Proteoglycan monomer in Fraction D1 from a low-grade hyaline chondrosarcoma, Case 1, G.F.; 1.88 mg/ml (top) and 2.48 mg/ml (bottom) at 16 min. (b) Proteoglycan monomer and aggregate in Fractions A1 from Case 1, G.F.; 2.97 mg/ml (top) and 2.47 mg/ml (bottom) at 16 min. (c) Proteoglycan monomer in Fraction D1 from a more highly malignant hyaline chondrosarcoma, Case 6, F.B.; 2.04 mg/ml (top) and 1.55 mg/ml (bottom) at 44 min. (d) Proteoglycan monomer and aggregate in Fraction A1 from Case 6, F.B.; 3.06 mg/ml (top) and 2.54 mg/ml (bottom) at 24 min.

Table I. Sedimentation Coefficients of Proteoglycan Monomer and Aggregate Species and Molecular Weights of Link-Protein Species from Human Chondrosarcomas

Grade of malignancy	Case	Age	Sedimentation coefficient[a] Fraction D1 Monomer	Sedimentation coefficient[a] Fraction A1 Monomer	Sedimentation coefficient[a] Fraction A1 Aggregate	Mol. wt. ($\times 10^{-3}$) of link proteins	
Low (Hyaline)	1, G.F.	48	20.1	21.4	65	38.7	47.8
	2, E.E.	27	22.3				
	3, O.R.	19	19.4				
Intermediate or high (Hyaline)	4, P.D.	48	17.7	18.6	51	40.9	49.0
	5, J.E.	55	18.3	19.0	35	38.1	47.4
	6, F.B.	36	23.2	28.5	135		
	7, D.F.	47	20.8	29.7	118		
High (Myxoid)	8, M.K.	74	16.2	19.5	49.2	40.2	49.2
	10, J.T.	51	16.4				
	11, E.C.	44		17.7	39.2	41.2	49.5
Normal articular cartilage	E.C.	44				40.2	48.9
	L.L.	12	15.3	20.9	58		

[a] Values for $s^{o}_{20,\ \mathrm{solvent}}$ are in Svedbergs

The malignant chondrocytes of each human chondrosarcoma have synthesized each of the elaborate macromolecular species required for aggregate formation (proteoglycan monomer, hyaluronic acid, and link protein), and each species possesses the functional capacity essential for non-covalent association into proteoglycan aggregates.

Sedimentation coefficients of the proteoglycan monomer species in Fraction D1, and of the proteoglycan monomer and aggregate species present in Fraction A1 were determined, to establish whether there were differences in the size of these proteoglycan species related to the degree of malignancy of the chondrosarcomas. The concentration dependence of the sedimentation coefficients of the proteoglycan aggregate and monomer in Fraction A1, and of the proteoglycan monomer in Fraction D1 from Case 1, G.F., is shown in Fig. 2. The sedimentation coefficients of the proteoglycan species from ten human chondrosarcomas, and the mol. wts. of the link-protein species from six human chondrosarcomas are given in Table I. There is no apparent relationship between the size of a proteoglycan monomer or aggregate and the degree of malignancy of the chondrosarcoma from which it was isolated.

Chemical Composition of Proteoglycan Monomer from Human Chondrosarcomas. In a previous study, we examined the chemical composition of proteoglycan monomer from six human chondrosarcomas (34). Because proteoglycan monomers from only six tumors were available, proteoglycan monomers from both hyaline and myxoid chondrosarcomas were included in a single group. We have now isolated and characterized proteoglycan monomers from over a dozen human chondrosarcomas. Hyaline and myxoid chondrosarcomas are being considered as separate groups, and several important differences between the hyaline and myxoid chondrosarcomas have become apparent.

The chemical composition of proteoglycan monomer from seven hyaline chondrosarcomas is given in Table II. On the left of Table II is shown the chemical composition of proteoglycan monomer of articular cartilage from the youngest (age 12 years) of normal, human distal femur so far examined. Each of the proteoglycan monomers in Table II showed a sharply unimodal single component on analytical ultracentrifugation and contained little or no glycogen as indicated by glucose content. In proteoglycan monomer, glucuronic acid is found only in chondroitin sulfate, and glucosamine and sialic acid are found only in keratan sulfate; the chondroitin sulfate content of the monomer can be calculated from its uronic acid content, and the keratan sulfate content can be calculated from its glucosamine content.

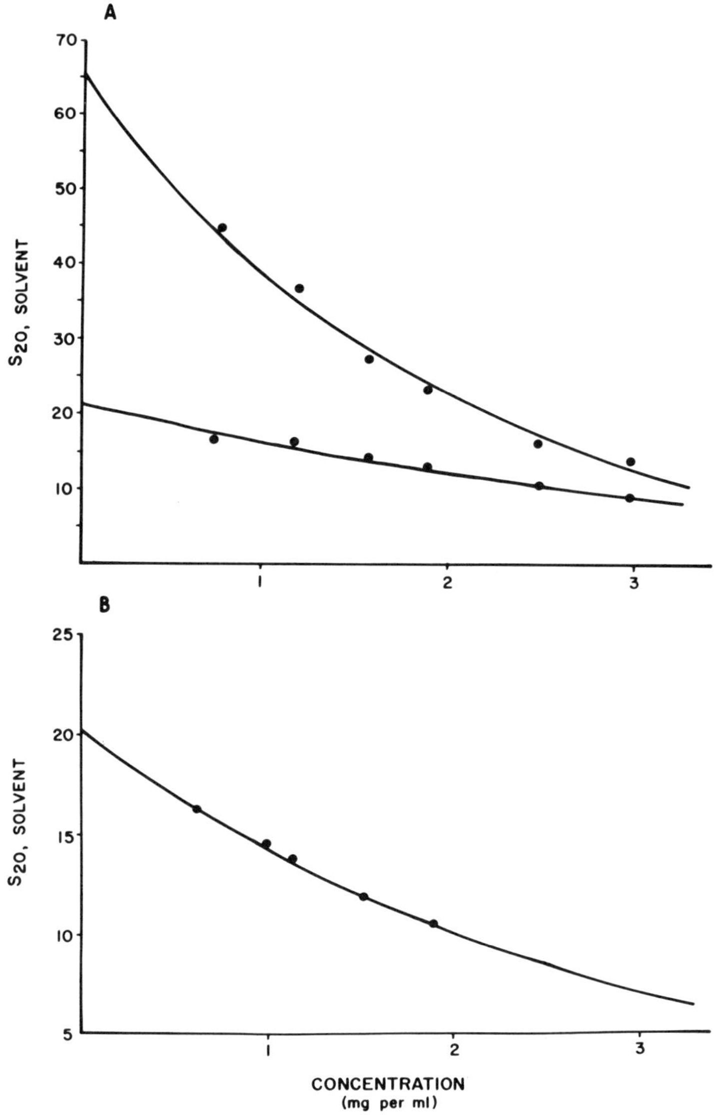

Fig. 2. Concentration dependence of sedimentation coefficients of: (a) proteoglycan aggregate and monomer in Fraction A1, and (b) proteoglycan monomer in Fraction D1 from human chondrosarcoma, Case 1, G.F.

Table II. *Chemical Composition and Physical Properties of Proteoglycan Monomer from Seven Human Hyaline Chondrosarcomas*[a]

Composition and properties	Normal	Grade of malignancy						
		Low grade			Intermediate to high grade			
		Case						
		1	2	3	4	5	6	7
	Patient							
	L.L.	G.F.	E.E.	O.R.	P.D.	J.E.	F.B.	D.F.
	Age							
	12	48	27	19	48	55	36	47
Uronic acid (%)	21.4	23.3	26.0	25.8	25.8	27.1	28.1	28.5
Galactosamine (%)	19.6	21.3	23.4	22.2	22.5	23.3	24.2	24.5
Glucosamine (%)	5.07	3.17	3.50	2.60	2.69	1.88	1.61	0.67
Galactose (%)		4.99	5.19	4.63	4.19	3.53	2.85	2.41
Sialic acid (%)	1.20	0.93	1.35	1.58	2.45	1.90	2.12	1.59
Protein (%)	8.3	6.2	7.0	7.5	7.8	7.8	6.0	6.1
Hexose (%)	9.4	11.2	10.4	8.7	7.2	6.6	10.5	7.5
Glucose (%)		2.54	1.36	0.40	0.10	0.35	2.20	1.21
Molar ratios								
GlcN/NeuAc	7.29	5.88	4.47	2.84	1.90	1.72	1.31	0.73
Gal/NeuAc		6.09	3.95	2.64	1.05	0.98	0.18	0.07
GalN/GlcN	3.9	6.7	6.7	8.5	8.4	12.4	15.1	36.4
s^o_{20} (Sved.)		20.1	22.3	19.4	17.7	18.3	23.2	20.8

[a] One previously reported case (ref. 34, Case D.P.) on chemotherapy with vincristine, methotrexate, and 5-(3,3-dimethyl-1-triazino)imidazole-4-carboxamide has been deleted. None of the cases included in this Table have received chemotherapy

The chemical compositions of the hyaline chondrosarcoma proteoglycan monomers differ from that of the young, normal human articular cartilage in the following ways: The chondroitin sulfate contents of the chondrosarcoma proteoglycan monomers are higher than that of the proteoglycan monomer from the young, normal cartilage, as indicated by their glucuronic acid or galactosamine contents. The keratan sulfate contents of the chondrosarcoma proteoglycan monomers are lower than that of the normal cartilage, as indicated by their glucosamine contents.

Of greater interest are the differences between the chemical compositions of the proteoglycan monomers from the individual tumors in relation to the differences in the degree of malignancy of these tumors. As indicated at the top of Table II, the chondrosarcomas have been classed as low-grade, or as intermediate to high-grade degrees of malignancy, based on clinical, radiographic, and histopathologic findings. As the degree of malignancy of the chondrosarcomas increases, there is a progressive increase in the chondroitin sulfate content of the proteoglycan monomers (as indicated by glucuronic acid or galactosamine content), and a parallel decrease in keratan content (as indicated by glucosamine content). The changes in glucosamine content are large, relative to the glucosamine content of a proteoglycan monomer from young, normal cartilage, and are particularly striking. The question was raised whether the decrease in keratan sulfate content with increasing degree of malignancy results from a decrease in the number of keratan sulfate chains per proteoglycan monomer, or from a decrease in the average chain length of keratan sulfate with increasing malignancy. To examine these possibilities, keratan sulfate chain-lengths were determined from glucosamine-to-sialic acid and from galactose-to-sialic acid molar ratios. Keratan sulfate chain-lengths were calculated from glucosamine-to-sialic acid molar ratios, assuming one sialic acid residue per chain, and using a mol. wt. for glucosamine of 179.2 and for sialic acid of 309.2. That is, the assumption was made that the short chain-length keratan sulfates of human chondrosarcomas contain a single sialylgalactosyl disaccharide linked to the *N*-acetylgalactosamine residue within the linkage region of keratan sulfate. As indicated in Table II, the average chain length of keratan sulfate from normal, young cartilage is 7.29 repeating units. The chain length of keratan sulfate in one proteoglycan monomer from a chondrosarcoma of low-grade malignancy (Case 1, G.F.) is 5.88 repeating units, and most nearly approaches that of the young, normal cartilage. Keratan sulfate chain-length in the low-grade chondrosarcomas then falls to *ca.* 2.5 repeating units (Case 3, O.R.). In the intermediate- to high-grade chondrosarcomas, keratan sulfate chain-length is in each case less than 2 repeating units and falls, in Case 7,

D.F., to less than 1 repeating unit. Keratan sulfate chain-lengths were also calculated from the galactose-to-sialic acid molar ratios by the following procedure: (a) The total number of mol of galactose per g of proteoglycan monomer was calculated from the percentage of galactose given in Table II; this value includes the galactose residues within the keratan sulfate repeating units, the galactose residues in the chondroitin sulfate linkage-regions, and the galactose residues linked to *N*-acetylgalactosamine residues in the keratan sulfate linkage regions. (b) The number of mol of galactose in the chondroitin sulfate linkage-regions was calculated, on the basis of potassium salt of chondroitin sulfate and assuming an average mol. wt. of 20 000 for the chain. (c) The number of mol of galactose in the keratan sulfate linkage-regions was calculated by assuming 1 mol of galactose of the keratan sulfate linkage-region per mol of sialic acid. (d) The number of mol of galactose within the keratan sulfate repeating-units was calculated from the values just obtained, giving the molar ratio of galactose to sialic acid. As indicated in Table II, the keratan sulfate chain-lengths calculated from the galactose-to-sialic acid molar ratios are similar to those calculated from the glucosamine-to-sialic acid molar ratios.

Some comment should be made regarding the alignment of the seven hyaline chondrosarcomas as to the degree of malignancy, in the order shown in Table II. The distinction between a low-grade chondrosarcoma, and a chondrosarcoma of intermediate- to high-grade malignancy is usually easy to make based on histologic and clinical findings. There is agreement based on independent evaluations by different bone pathologists that the tumors included in Table II are either low-grade chondrosarcomas, or intermediate- to high-grade chondrosarcomas. However, within the group of chondrosarcomas of intermediate- and high-grade malignancy (Cases 4 through 7), the alignment of the tumors in order of increasing degree of malignancy is subjective; differences of opinion exist as to whether the degree of malignancy of the four tumors increases in the order shown in Table II. In the case of the low-grade chondrosarcomas, each of the tumors showed similar histopathologic features (little cellularity; high ratio of matrix to cells; no pleomorphism; small, dark round nuclei; no nuclear detail or chromatin pattern discernible; few or no double nucleated cells; and no mitoses), and it is impossible to align the low-grade tumors in terms of degree of malignancy from histologic findings. The data presented in Table II suggest that changes in keratan sulfate chain-length may provide a more sensitive reflection and precise indicator of the degree of malignancy of human chondrosarcomas than histopathologic findings.

Chemical Composition of Proteoglycan Monomer from Fetal Cartilage. The observations just described suggest that we have identified a change in the chemical composition of the proteoglycan monomer from human chondrosarcomas that indicates the degree of malignancy of these tumors more precisely than is possible from histopathologic findings. The question now arises, what is the biologic significance of this apparent decrease in keratan sulfate content and chain length with increasing degree of malignancy in the human chondrosarcomas? Does this change in keratan sulfate with malignancy relate to any known change in cartilage composition or proteoglycan structure during normal development? Studies over the years (35,37) have shown that, in human and animal cartilages, keratan sulfate content is relatively low in early childhood, and increases during adolescence and with aging in adult life. Until recently, the changes in keratan sulfate content and chain length during fetal development have not been studied. Recently, however, De Luca *et al.* (38) have examined the chemical and physical changes in proteoglycans during development in chondrocytes, grown *in vitro,* of the chick limb bud, and have shown that keratan sulfate content and chain-length increase progressively from day 3 to day 8 during development in culture. In our own laboratory, we have recently compared the chemical composition of proteoglycan monomer (D1) from mature (1- to 2-year old) and fetal (age 220 days gestation) bovine articular cartilage. The results are shown in Table III.

The chemical composition of the proteoglycan monomer from adult bovine articular cartilage, and the keratan sulfate chain-length (6.8 repeating units) tentatively based on the GlcN/NeuAc molar ratio, is very close to that of the proteoglycan monomer from normal human articular cartilage, age 12 years (Table II, patient L.L.). In the near-term, fetal proteoglycan monomer, the chondroitin sulfate content is increased and the keratan sulfate content is greatly decreased. The keratan sulfate chain-length is *ca.* 2 repeating units at the end of bovine fetal development. The decrease in keratan sulfate content and chain-length in proteoglycan monomers from human chondrosarcomas appears analogous to the decrease in keratan sulfate content and chain-length seen in proteoglycan monomer from fetal cartilage.

Biochemical Basis for Histologic Characteristics of Myxoid Chondrosarcomas. Human chondrosarcomas are classified as hyaline, myxoid, clear cell, mesenchymal, and dedifferentiated chondrosarcomas. Hyaline and myxoid chondrosarcomas are commonly encountered forms of human chondrosarcomas. Clear cell (39), mesenchymal (40-42), and dedifferentiated (43,44) chondrosarcomas are rare. Hyaline and myxoid chondrosarcomas are distinguished from one another mainly by striking differences in their gross appearance; differences in their microscopic appearance have not been fully utilized. Grossly, hyaline chondrosarcomas consist of lobules of firm, rubbery, and yellowish-white tissue, enclosed in fibrous tissue. Myxoid chondrosarcomas consist of irregular lobules of much softer tissue, or gelatinous material, or cystic collections of liquid, surrounded by thin, tenuous fibrous pseudocapsules. One pathologist has stated, "...myxoid tissue is a soupy, glistening, liquid-like tissue which often oozes into the wound if the surgeon is not careful".

Table III. Chemical Composition and Physical Properties of Proteoglycan Monomer from Fetal and Adult Bovine Articular Cartilage

Composition and properties	*Adult*	*Fetal*
Uronic acid (%)	*21.0*	*28.1*
Galactosamine (%)	*18.5*	*24.2*
Glucosamine (%)	*5.7*	*1.8*
Galactose (%)	*8.5*	*3.7*
Sialic acid (%)	*1.44*	*1.46*
Protein (%)	*10.1*	*7.3*
Molar ratio		
GlcN/NeuAc	*6.8*	*2.1*
Gal/NeuAc	*7.9*	*1.7*
GalN/GlcN	*3.2*	*13.4*
Density (g/ml)	*1.62*	*1.61*
s^{o}_{20} *(Sved.)*	*17.6*	*22.5*

The distinction between hyaline and myxoid chondrosarcomas is of more than academic interest, since some observers believe that myxoid and hyaline chondrosarcomas differ in degree of malignancy. Our own experience indicates that myxoid chondrosarcomas are generally highly malignant tumors, which grow rapidly, are highly invasive, and metastasize early. Dahlin believes that myxoid chondrosarcomas are unusually aggressive (15). Schiller has suggested that a myxoid stroma represents a less differentiated area of cartilage and that myxoid tumors may tend to behave in a more malignant fashion. However, no systematic study has been carried out in which hyaline and myxoid chondrosarcomas have been separated into two distinct groups, and the degree of malignancy of the hyaline and myxoid tumors compared in terms of growth rate, invasiveness, incidence of metastasis, and survival rate. What are the reasons for this lag in development of knowledge? Such a study would require that the tumors be diagnosed as hyaline or myxoid from their gross and microscopic pathologic features. While the term myxoid is frequently used, precise detailed descriptions of the microscopic appearance and histologic characterization of myxoid chondrosarcomas have not yet been given in the literature (1-15; 39-45). It is as though the term myxoid, like a picture, is worth a thousand words, so that any definition or description of what actually is meant by myxoid would be superfluous. In fact, any meaningful description of the microscopic appearance of the myxoid chondrosarcomas must be based on some knowledge of the structural alterations in cells or matrix that lead to the peculiar histologic changes. In the absence of a biochemical and structural explanation for the peculiar histologic changes, pathologists have been justifiably reluctant to make detailed statements about the histopathology of myxoid chondrosarcoma. Until now, no information was available on the differences in the chemical compositions of intercellular matrix or the structures of proteoglycans and collagen in myxoid and hyaline chondrosarcomas.

The histologic features of hyaline chondrosarcoma (Fig. 3) resemble those of normal hyaline cartilages. The tissue consists of chondrocytes separated by broad areas of homogeneous matrix. The chondrocytes are contained within lacunae, formed by a dense network of collagen fibers. The intercellular matrix of the hyaline chondrosarcoma is tough and cohesive, and holds together on histologic processing (Fig. 3a,b,c,d). This is because the intercellular matrix of the hyaline chondrosarcoma contains adequate collagen (22 to 27%, compared with 60% in normal human articular cartilage). On light microscopy, the integrity of the matrix is preserved and the matrix appears homogeneous.

Myxoid chondrosarcomas are more highly cellular and show a higher ratio of cells to matrix. Lacunae are absent. The cells are distributed throughout a loosely textured, delicate, and tenuous intercellular matrix. During histologic processing, the intercellular matrix of the myxoid chondrosarcoma tends to shrink, coalesce, and tear. Clefts, fissures, and vacuoles are formed, of the kind shown in Fig. 3e,f,g, and h.

What is the biochemical basis for these peculiar histologic characteristics of the myxoid chondrosarcomas? They do not appear to be related to differences in the chemical composition of the proteoglycans, or in the proteoglycan content of the myxoid and hyaline chondrosarcomas. The proteoglycan contents of the two types are similar, based on analyses of whole dry tissue and on the results of histochemical staining. The chemical composition of the proteoglycan monomer from four highly malignant myxoid chondrosarcomas is given in Table IV.

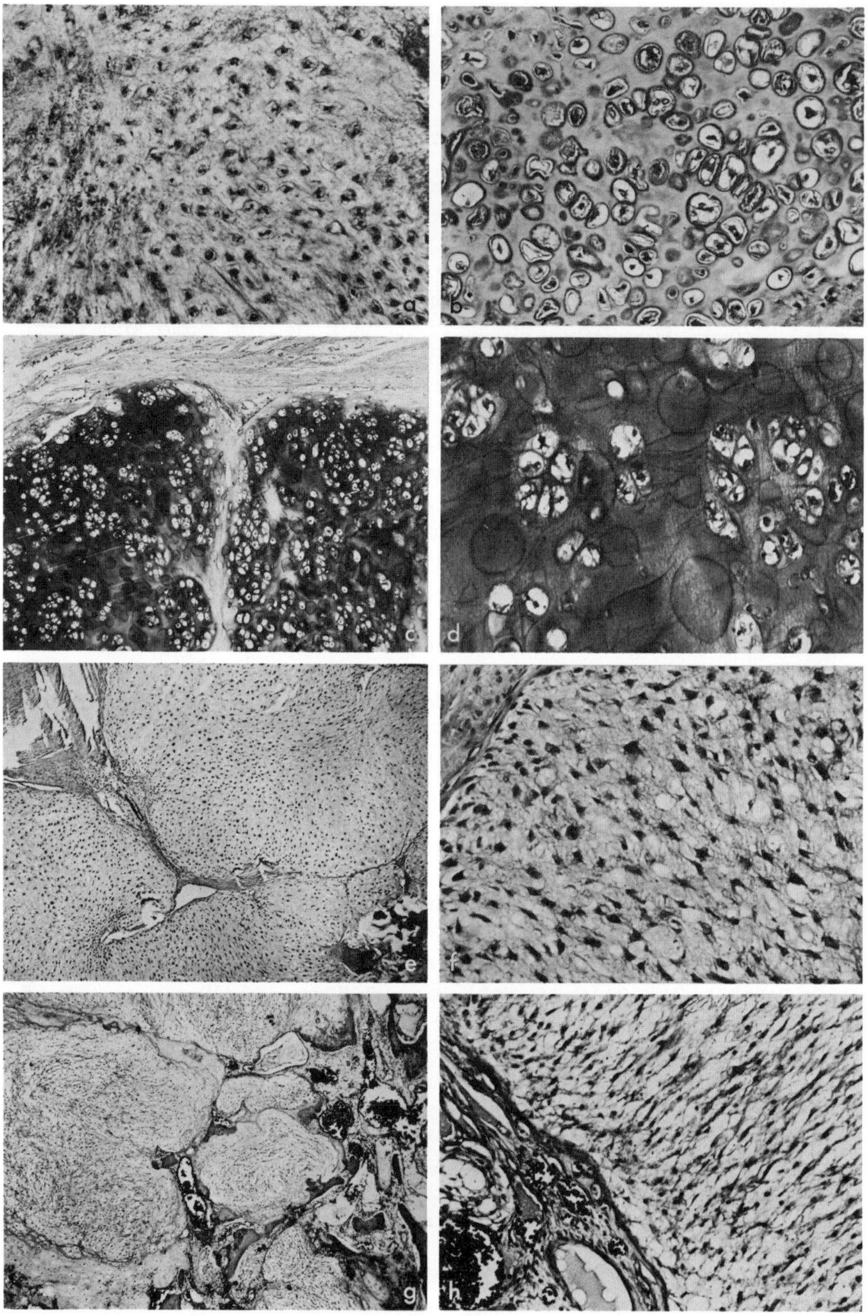

Fig. 3. Histologic features of hyaline and myxoid human chondrosarcomas. In the hyaline chondrosarcomas (a,b,c,d),

Each proteoglycan monomer shows an extremely low keratan sulfate content and chain-length. The D1 Fractions from myxoid chondrosarcomas also show a variable and sometimes high content of glucose. However, in Case 8, M.K., where little glucose is present, the chemical composition of the proteoglycan monomer is similar to those of the more highly malignant hyaline chondrosarcomas.

The absence of lacunae and delicate intercellular matrix of the myxoid chondrosarcomas are a direct result of the reduced collagen content of the myxoid chondrosarcomas. The collagen content of normal human articular cartilage (age 55 years) is 60%. The collagen content of the whole dry tissue in the hyaline chondrosarcomas ranges from 22 to 27%. In the myxoid chondrosarcoma from case 13, E.C., the collagen content of the whole dry tissue was only 6%. While both highly malignant hyaline chondrosarcomas and myxoid chondrosarcomas show greatly reduced keratan sulfate contents and chain-lengths, myxoid chondrosarcomas also show evidence of sharply reduced collagen

---

chondrocytes lie in lacunae formed by a dense network of collagen fibers. The chondrocytes and lacunae are separated by a homogeneous matrix. The intercellular matrix, because of its collagen content, is tough and cohesive, and its integrity is preseved after histologic processing. (a) Hematoxylin- and eosin-stained section (x 150) of the low-grade chondrosarcoma from Case 1, G.F. The chondrocytes contain small, round dark staining nuclei with no discernible chromatin pattern, and there is no pleomorphism. No binucleate cells are present. (b) Section (x 150) from a chondrosarcoma of intermediate-grade malignancy (Case 5, J.E.) showing increased cellularity, moderate pleomorphism, and some binucleate cells. (c and d) Sections (x 25 and x 150) from a large hyaline chondrosarcoma (Case 6, F.B.). (e and f) Sections (x 25 and x 150) from a large, highly invasive, and highly malignant myxoid chondrosarcoma (Case 9, C.D.). No lacunae have been formed. The intercellular matrix, which contains abundant proteoglycan but very little collagen, is delicate and fragile. During histologic processing, the intercellular matrix of the myxoid chondrosarcoma tears and coalesces, forming clefts of the kind seen in Section f. (g and h) Sections (x 25 and x 150) of another large, highly invasive, and highly malignant myxoid chondrosarcoma (Case 11, E.C.). The collagen content of the whole dry tissue was 6%, compared to 60% in normal human articular cartilage. No lacunae have been formed. During histologic processing, the non-cohesive intercellular matrix of the myxoid chondrosarcoma tears and coalesces, forming clefts and vacuoles of the kind seen in Section h.

synthesis. This, together with the clinical features of the myxoid chondrosarcomas, supports the concept that myxoid chondrosarcomas are less differentiated than hyaline chondrosarcomas.

Table IV. Chemical Composition and Physical Properties of Proteoglycan Monomer from Four Highly Malignant Myxoid Chondrosarcomas

Composition and properties	*Case*			
	8	*9*	*10*	*11*
	Age			
	74	*48*	*51*	*44*
Uronic acid (%)	*27.0*	*23.2*	*21.9*	*18.4*
Galactosamine (%)	*22.4*	*21.9*	*17.7*	*10.9*
Glucosamine (%)	*1.88*	*1.85*	*1.05*	*0.50*
Galactose (%)	*4.35*	*3.62*	*2.73*	*2.40*
Sialic acid (%)	*1.79*	*2.33*	*2.83*	*1.30*
Protein (%)	*9.2*	*7.7*	*11.2*	*5.2*
Hexose (%)	*7.2*	*19.1*	*36.4*	*62.3*
Glucose (%)	*0.27*	*5.64*	*17.3*	*43.4*
Molar ratios				
GlcN/NeuAc	*1.81*	*1.37*	*0.64*	*0.66*
Gal/NeuAc	*1.79*	*0.82*	*0.00*	*0.95*
GalN/GlcN	*11.9*	*11.8*	*16.9*	*21.8*
s^{o}_{20} *(Sved.)*	*16.2*		*16.4*	

The observations described in this report suggest that a biochemical examination of proteoglycans from human cartilage tumors may provide a useful method for assessing the degree of malignancy and differentiation of these tumors. Biochemical study of the chemical composition of the whole tissue from cartilage tumors, and of the alterations in structure of the macromolecular species comprising their intercellular matrix should also provide a rational basis for understanding and interpreting the dramatic variations in histologic appearance seen in human cartilage tumors. The results of such studies should lead to improvements in histopathologic classification and diagnosis, and in the assessment of the degree of malignancy of human cartilage tumors, which have value in management of patients with cartilage tumors.

ACKNOWLEDGMENT

This work was ssupported by Grants RR05499-14 and AM01431.

REFERENCES

1. Roberg, O. T., Jr., *Surg. Gynecol. Obstet. 61*, 68 (1935).
2. O'Neal, L. W., and Ackerman, L. V., *Cancer 5*, 551 (1952).
3. Lindbom, A., Soderberg, G., and Spjut, H. J., *Acta Radiol. 55*, 81 (1961).
4. Barnes, R., and Catto, M., *J. Bone Joint Surg. 48-B*, 729 (1966).
5. Dahlin, D. C., and Henderson, E. D., *J. Bone Joint Surg. 38-A*, 1025 (1956).
6. Henderson, E. D., and Dahlin, D. C., *J. Bone Joint Surg. 45-A*, 1450 (1963).
7. Gilmer, W. S., Jr., Kilgore, W., and Smith, H., *Clin. Ortop. 26*, 81 (1963).
8. Marcove, R. C., and Huvos, A. G., *Cancer 27*, 794 (1971).
9. Marcove, R. C., Mike, V., Hutter, R. V. P., Huvos, A. G., Shoji, H., Miller, T. R., and Kosloff, R., *J. Bone Joint Surg. 54-A*, 561 (1972).
10. Reiter, F. B., Ackerman, L. V., and Staple, T. W., *Radiology 105*, 525 (1972).
11. Evans, H. L., Ayala, A. G., and Romsdahl, M. M., *Cancer 40*, 818 (1977).
12. Erlandson, R. A., and Huvos, A. G., *Cancer 34*, 1642 (1974).
13. Jaffe, H. L., *in* "Tumors and Tumorous Conditions of the Bones and Joints", Lea and Febiger, Philadelphia, (1958).
14. Lichtenstein, L., "Bone Tumors", C. V. Mosby Company, St. Louis, (1972).
15. "Bones and Joints: Monograph in Pathology", (Ackerman, L. V., Spjut, H. J., and Abell, M. R., eds.), Williams and Wilkins, Baltimore, (1976).
16. Sajdera, S. W., and Hascall, V. C., *J. Biol. Chem. 244*, 77 (1969).
17. Hascall, V. C., and Sajdera, S. W., *J. Biol. Chem. 244*, 2384 (1969).
18. Gregory, J. D., *Biochem. J. 133*, 383 (1973).
19. Heinegård, D., *Biochim. Biophys. Acta 285*, 181 (1972).
20. Heinegård, D., *Biochim. Biophys. Acta 285*, 193 (1972).
21. Hascall, V. C., and Heinegård, D., *J. Biol. Chem. 249*, 4232 (1974).
22. Hascall, V. C., and Heingård, D., *J. Biol. Chem. 249*, 4242 (1974).
23. Heinegård, D., and Hascall, V. C., *J. Biol. Chem. 249*, 4250 (1974).
24. Hardingham, T. E., and Muir, H., *Biochim. Biophys. Acta 279*, 401 (1972).

25. Hardingham, T. E., and Muir, H., *Biochem. J. 135*, 905 (1973).
26. Hardingham, T. E., and Muir, H., *Biochem. Soc. Trans. 1*, 282 (1973).
27. Rosenberg, L., Schubert, M., and Sandson, J., *J. Biol. Chem. 242*, 4691 (1967).
28. Rosenberg, L., Pal, S., Beale, R., and Schubert, M., *J. Biol. Chem. 245*, 4112 (1970).
29. Rosenberg, L., Hellmann, W., and Kleinschmidt, A. K., *J. Biol. Chem. 245*, 4123 (1970).
30. Rosenberg, L., Pal, S., and Beale, R., *J. Biol. Chem. 248*, 3681 (1973).
31. Rosenberg, L., Hellman, W., and Kleinschmidt, A. K., *J. Biol. Chem. 250*, 1877 (1975).
32. Strider, W., Pal, S., and Rosenberg, L., *Biochim. Biophys. Acta 379*, 271 (1975).
33. Rosenberg, L., Wolfenstein-Todel, C., Margolis, R., Pal, S., and Strider, W., *J. Biol. Chem. 251*, 6439 (1976).
34. Pal, S., Strider, W., Margolis, R., Gallo, G., Lee-Huang, S., and Rosenberg, L. C., *J. Biol. Chem. 253*, 1279 (1978).
35. Rosenberg, L., Johson, B., and Schubert, M., *J. Clin. Invest. 44*, 1647 (1965).
36. Kaplan, D., and Meyer, K., *Nature 183*, 1267 (1959).
37. Inerot, S., Heinegård, D., Audell, L., and Olsson, S.-E., *Biochem. J. 169*, 143 (1978).
38. De Luca, S., Heinegård, D., Hascall, V. C., Kimura, J. H., and Caplan, A. I., *J. Biol. Chem. 252*, 6600 (1977).
39. Unni, K. K., Dahlin, D. C., Beabout, J. W., and Sim, F. H., *J. Bone Joint Surg. 58-A*, 676 (1976).
40. Lichtenstein, L., and Bernstein, D., *Cancer 12*, 1142 (1959).
41. Salvador, A. H., Beabout, J. W., and Dahlin, D. C., *Cancer 28*, 605 (1971).
42. Steiner, G. C., Mirra, J. M., and Bullough, P. G., *Cancer 32*, 926 (1973).
43. Dahlin, D. C., and Beabout, J. W., *Cancer 28*, 461 (1971).
44. Kahn, L. B., *Cancer 37*, 1365 (1976).
45. Fu, Y.-S. , and Kay, S., *Cancer 33*, 1531 (1974).

Structure–Function Relationships

Plants Respond Defensively to a Microbial Oligosaccharide which Possesses Pheromone-like Activity

Peter Albersheim and Barbara S. Valent

Plants are exposed to attack by an immense array of microorganisms, and yet plants are resistant to almost all of these potential pests. Many plant tissues have been observed to respond to an invasion by a pathogenic or nonpathogenic microorganism, whether a fungus, a bacterium or a virus, by accumulating phytoalexins, low-molecular-weight compounds which inhibit the growth of microorganisms. The production of phytoalexins appears to be a widespread mechanism by which plants attempt to defend themselves against pests (1-3). The molecules of microbial origin that trigger phytoalexin accumulation in plants have been called elicitors (4). Plants recognize and respond to elicitors as foreign molecules. It is highly improbable that plants have evolved separate recognition systems for every bacterial species and strain, and every fungal race, and every virus that plants are exposed to and respond to defensively. Thus, elicitors are likely to be molecules common to many microbes and, in fact, the one to be described in this paper is a structural polysaccharide of the mycelial walls of many fungi (5).

Most plants produce several structurally related phytoalexins (Scheme 1). The most studied phytoalexin of soybeans is glyceollin (6). Lyne *et al.* (7) have characterized two additional soybean phytoalexins which are structural isomers of glyceollin and which appear to have similar antibiotic characterisitics. The synthesis of glyceollin, a phenylpropanoid derivative, is probably initiated from phenylalanine *via* the reaction catalyzed by phenylalanine ammonia-lyase, but, as yet, no biosynthetic pathway for the production of a phytoalexin has been completely described.

ISBN 0-12-301301-1

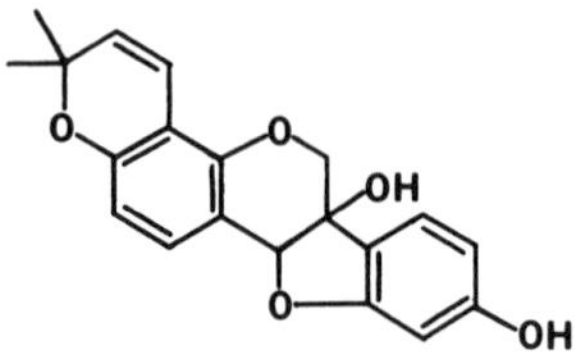

GLYCEOLLIN

A PTEROCARPAN FROM
SOYBEANS (*Glycine max*)

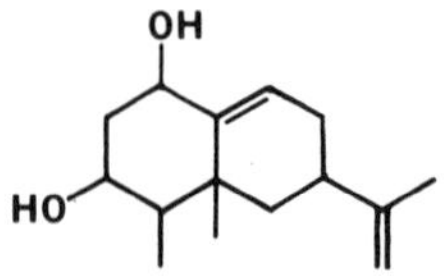

CAPSIDIOL

A SESQUITERPENE FROM
PEPPERS (*Capsicum frutescens*)

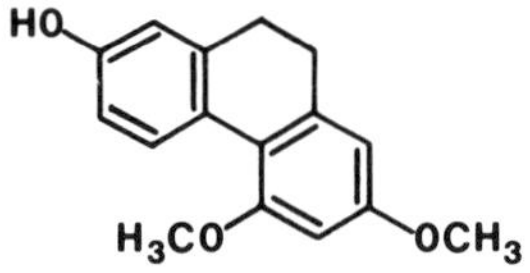

ORCHINOL

A PHENANTHRENE FROM
ORCHIDS (*Orchidaceae*)

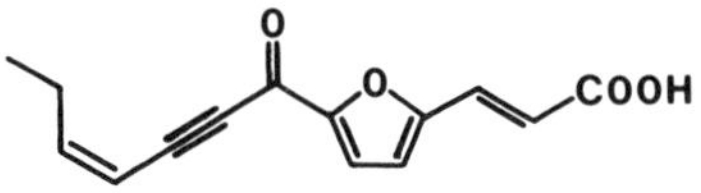

WYERONE ACID

A POLYACETYLENE FROM
BROAD BEANS (*Vicia faba*)

Scheme 1. Typical phytoalexins

The mechanism by which phytoalexins work is unknown. Glyceollin is a static agent rather than a toxic agent, a trait which seems to be common to many phytoalexins. Glyceollin inhibits the growth, *in vitro*, of the soybean pathogen, *Phytophthora megasperma* var. *sojae* (Pms), the causal agent of root and stem rot. In addition, S. Thomas, in our laboratory, has found that glyceollin will stop the growth of three Gram-negative bacteria, *Pseudomonas glycinea, Rhizobium trifolii*, and *Rhizobium japonicum*, of the Gram-positive bacterium, *Bacillus subtilis*, and of baker's yeast, *Saccharomyces cerevisiae*. Interestingly, it requires about 25 μg/ml of glyceollin to inhibit by 50% and 100 μg/ml to inhibit by 100% the growth of all of these different organisms. Thus, it appears that a plant's phytoalexins can potentially protect the plant from a broad spectrum of microorganisms.

Glyceollin is accumulated by soybean tissues in response to infection by Pms, the soybean pathogen. Glyceollin accumulates in soybean hypocotyls, within 9 hours of infection with Pms mycelia, to levels that are inhibitory to the growth of Pms *in vitro*. A component of Pms mycelial walls has been demonstrated to stimulate glyceollin accumulation at the same rate as live Pms mycelia (8). This observation and other data have convinced us that this mycelial wall component is responsible for triggering glyceollin accumulation during infection by the live fungus; and, therefore, we believe that the mycelial wall component is the natural elicitor of this system.

Three different soybean tissues respond to the Pms elicitor by accumulating glyceollin, and these have been used for biological assays of elicitor activity. An assay using 8-day-old cotyledons (seed leaves) was used for the purification of the elicitor since this was the least laborious assay developed (9). A second bioassay uses the hypocotyls (upper stems) of 5-day-old soybean seedlings (9), and a third assay uses suspension-cultured soybean cells (10). In all three assays, the production of glyceollin is proportional to the amount of elicitor applied. The time course of elicitor-stimulated glyceollin accumulation and the amount of elicitor required is very similar in all three soybean tissues.

Soybean tissues are sensitive to extremely small amounts of Pms elicitor. About 10^{-12} mol of elicitor applied to a single hypocotyl stimulates quantities of glyceollin sufficient to prevent the growth of Pms and other microorganisms *in vitro*. It is impressive, too, to observe the effects on the growing suspension-cultured soybean cells caused by the addition of submicromolar quantities of the polysaccharide elicitor. These cells respond to the small amount of elicitor even though the cells are growing in the presence of 50 mM sucrose. Within a few hours, the cells turn light brown. At the same time, the activity in the cells of at least one of the enzymes believed to be involved in the synthesis of glyceollin, phenylalanine ammonia-lyase, is greatly increased. The increase in activity of the phenylalanine ammonia-lyase precedes the accumulation of glyceollin both in the cells and in the culture medium. The growth of the suspension-cultured cells, as measured by fresh weight, stops upon addition of the elicitor (10). The cells also stop taking up ions from the media, which is another indication of the lack of growth of these cells (10).

THE CHEMICAL NATURE OF THE PMS ELICITOR

The component of the Pms mycelial walls that stimulates glyceollin accumulation by soybean tissues is a structural polysaccharide. The elicitor was first found in the fluid of old cultures of Pms, probably being released into the culture fluid by autolysis. It was later demonstrated that elicitor-active molecules with the same properties as the culture fluid elicitor could be isolated from the mycelial walls of Pms by a heat treatment similar to that used to solubilize the surface antigens from the cell walls of *S. cerevisiae* (8). The best method for obtaining large amounts of Pms elicitor is partial acid hydrolysis of the mycelial walls. The series of polysaccharides and oligosaccharides so obtained are extremely active as elicitors and possess characteristics identical to the culture fluid elicitor. All of the Pms-produced, elicitor-active molecules examined have been found to be α-D-glucans. Methylation analysis, periodate oxidation, and enzymic hydrolysis of the purified elicitor-active glucan have demonstrated that the glucan is largely a 3-linked polymer with glucosyl branches to C-6 of about one out of every three of the backbone glycosyl residues. Approximately 90% of the elicitor-active glucan is hydrolyzed by an *exo*-β-(1→3)-glucanase isolated from *Euglena gracilis* (11), indicating that the Pms mycelial wall glucan is a β-D-linked polymer. Optical rotation and n.m.r. studies have confirmed that the glucan is β-D-linked. This is not surprising as other *Phytophthora* cell walls have a quantitatively dominant structural component which is a (1→3)-β-D-linked glucan with some branches to C-6. Indeed, it appears that as much as 60% of the mycelial walls of Pms is composed of this polymer.

The Pms mycelial wall elicitor has been well characterized (12, and unpublished results of this laboratory). The portion of the elicitor that is released from the walls by aqueous extraction at 121°C is heterogeneous in size, with an average mol. wt. of *ca*. 100 000. The *E. gracilis* enzyme hydrolyzes D-glucans from the nonreducing end and is capable of hydrolyzing the glycosidic bond of 3-linked D-glucosyl residues that have other D-glucosyl residues attached to O-6. The product of the *exo*-glucanase-degraded, mycelial wall-released elicitor is still heterogeneous in size, but shows an average mol. wt. of approximately 10 000. This highly branched glucan fragment retains as much activity as the undegraded elicitor. The predominant glycosidic linkages remaining after extensive *exo*-glucanase treatment are 3-linked, 3,6-linked, and terminal glucosyl residues in a ratio of 1:1:1. Small amounts of 4-linked and 6-linked glucosyl residues are also present.

The evidence that demonstrated that the Pms elicitor is a polysaccharide included the findings that the elicitor is stable to autoclaving at 121°C for several hours, lacks affinity for both anion- and cation-exchange resins, is completely stable to treatment by Pronase, and is size-heterogeneous. Periodate treatment of the wall-released elicitor confirmed the polysaccharide nature of the active component and demonstrated the essential role of a branched oligosaccharide having terminal glycosyl residues. Exposing the elicitor to periodate eliminated almost all of the elicitor activity. On the other hand, if the periodate-degraded polymers were reduced with sodium borohydride and then were subjected to mild acid hydrolysis, a considerable portion of the elicitor activity was regained. Since the 3- and 3,6-linked glucosyl residues are resistant to periodate degradation, it seems likely that the periodate oxidation destroyed the elicitor activity by modifying the terminal glucosyl residues or the quantitatively minor, but periodate-susceptible, 4- or 6-linked glucosyl residues, or both. However, recovery of elicitor activity, by mild acid hydrolysis of the periodate-inactivated elicitor, points to periodate attack on the terminal glucosyl residues as the cause for periodate inactivation of the elicitor. The degradation of 4- and 6-linked glucosyl residues would lead to splitting of the glucan chain, whereas periodate destruction of terminal glucosyl residues followed by mild acid hydrolysis could lead to the exposure of new terminal glucosyl residues, which might provide the proper structure of an active elicitor.

The requirement for elicitor activity of a branched oligosaccharide is supported by our observation that (1→3)-linked β-D-glucans, which lack branches to C-6 or have only a single 6-branched glucosyl residue, such as laminarin, have little or no elicitor activity (less than 1/1000 of the Pms elicitor). Indeed, a series of commercially available polysaccharides, oligosaccharides, methyl glucosides, and simple sugars have been tested for elicitor activity, and, besides laminaran, the only commercially available product found with detectable elicitor activity was nigeran, a mycelial wall-component from the fungus, *Aspergillus niger* (10).

A major goal of our research is the determination of the detailed molecular structure of the active-site of the Pms elicitor. It is expected that this goal will be achieved by the isolation and structural characterization of the smallest possible elicitor-active oligosaccharide that can be derived from the glucan elicitor. A relatively small elicitor-active oligosaccharide has been produced by partial acid hydrolysis of Pms mycelial walls. The series of oligosaccharides obtained by this partial hydrolysis have been partially resolved by, first, low resolution (Fig. 1), and then by high-resolution (Fig. 2) Bio-Gel P-2 gel-permeation chromatography. It was found that oligosaccharides containing as few as 8 or 9 glucosyl residues still retain elicitor activity. Glucose is the only detected component of these oligosaccharides.

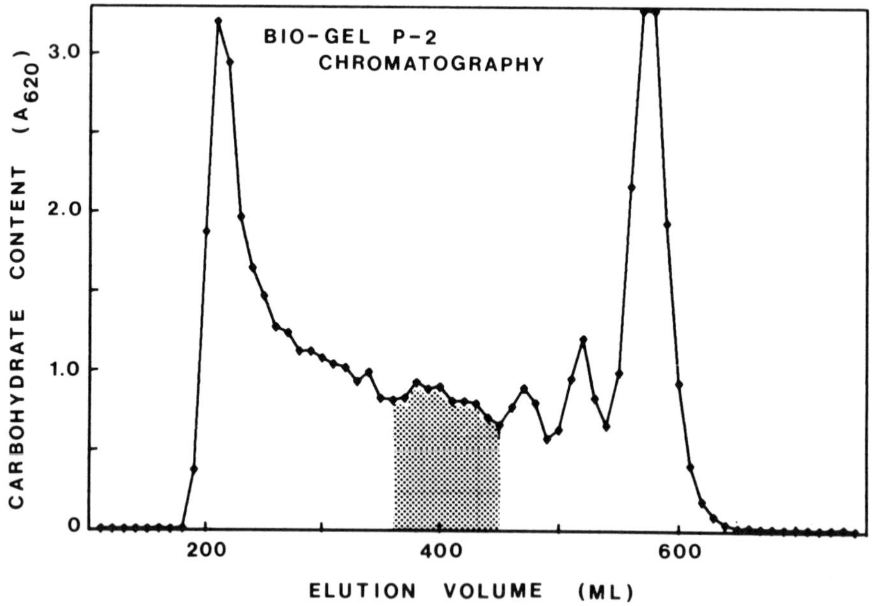

Fig. 1. Low-resolution gel-permeation chromatography of the crude elicitor obtained by partial acid hydrolysis of Phytophthora megasperma *var.* sojae *mycelial walls. The shaded fractions, which contain the smallest elicitor-active oligosaccharide, were combined and chromatographed on a high resolution Bio-Gel P-2 column (see Fig. 2).*

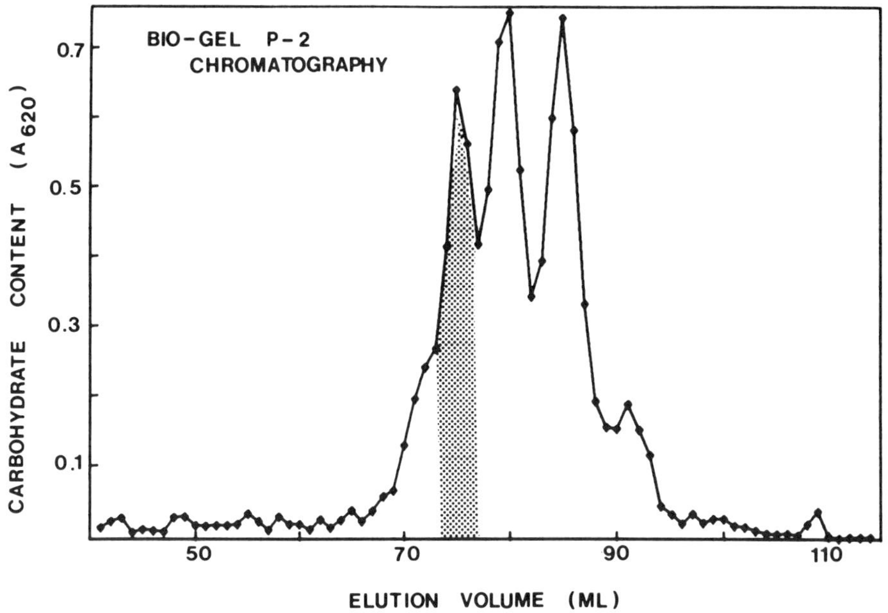

Fig. 2. High-resolution gel-permeation chromatography of the material from the shaded fractions in Fig. 1. The fractions shaded in this high-resolution chromatograph contain the smallest elicitor-active oligosaccharide. These fractions were subjected to high-pressure liquid chromatography (see Fig. 3).

The smallest elicitor-active, oligosaccharide-containing fraction from the high resolution P-2 column (Fig. 2) has been subdivided by high-pressure liquid chromatography (Fig. 3) into at least five oligosaccharide fractions. Two of the five oligosaccharide fractions obtained by high-pressure liquid chromatography can be eliminated by treatment of the mixture of oligosaccharides with the *E. gracilis exo*-glucanase. There appears to be little loss of elicitor activity after treatment with the *exo*-glucanase. Of the three oligosaccharide fractions remaining, two actively elicit soybean tissues to accumulate glyceollin. The purest elicitor-active oligosaccharide fraction obtained by high-pressure liquid chromatography still contains at least two distinct oligosaccharides.

Methylation analysis of this purest elicitor-active oligosaccharide indicates an approximate composition of three 6-linked glucosyl residues, two terminal glucosyl residues, and a single residue each of 3-, 4-, and 3,6-linked glucose. Indirect evidence suggests to us that the 4-linked glucosyl residues are part of an inactive oligosaccharide. Our best guess for the composition of the elicitor-active oligosaccharide is that it contains two residues each of 3-, 6-, and 3,6-linked glucose, and three residues of terminal glucose. Further purification techniques are required to obtain a pure active oligosaccharide.

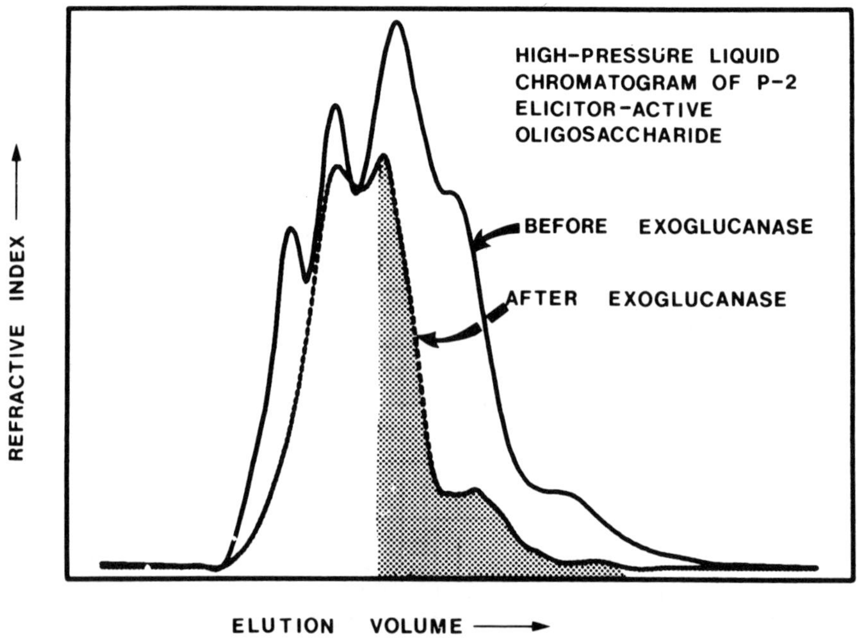

Fig. 3. High-pressure chromatography in acetonitrile-water (on the Water's Associates "carbohydrate column") of the high-resolution Bio-Gel P-2 fractions (see Fig. 2) containing the smallest elicitor-active oligosaccharide. These fractions were chromatographed before and after digestion with a Euglena gracilis *exo-(1→3)-β-D-glucanase. The elicitor-active oligosaccharides are not digested by the exoglucanase. The shaded fractions contain the smallest elicitor-active oligosaccharide.*

ELICITORS LACK SPECIES SPECIFICITY

Our experiments have shown that the elicitors of phytoalexin accumulation are not the specificity-determining factors in the Pms-soybean system. Three Pms races (races 1, 2, and 3) are distinguished by their differing abilities to infect various soybean cultivars. The elicitor obtained from each of the three Pms races may be purified in exactly the same manner, and all the discernible structural features of the elicitors from the three races are identical (12). The activities of the elicitors purified from the three Pms races were carefully examined by use of the three separate bioassays: the cotyledon assay (9), the hypocotyl assay (9), and the cell suspension-culture assay (10). All three assays gave the same results, that is, the activities of the elicitors from different Pms races are identical. These findings indicate that the three races of Pms are equally effective at stimulating phytoalexin accumulation in their host soybean tissues.

The results of another type of experiment support our conclusion that elicitors are not responsible for race-specific resistance in the Pms-soybean system. Soybean hypocotyls accumulate glyceollin when inoculated with living mycelia of Pms. The response, which is characteristic of natural infections with either an infective or a noninfective race of Pms, is retained with this inoculation technique. We have compared the relative effectiveness of live mycelia and purified elicitor in stimulating glyceollin accumulation. The result is the following: the onset and the rate of glyceollin accumulation in seedlings inoculated with infective mycelia was indistinguishable from the onset and rate of glyceollin accumulation in seedlings inoculated with either noninfective mycelia or purified elicitor. These results demonstrate that differences in rates of glyceollin accumulation in response to different races of Pms do not account for the resistance or susceptibility of various soybean cultivars to the Pms races.

The available evidence does indicate that elicitors have a role in resistance even though they are not determinants of race specificity. The elicitor isolated from Pms is capable of protecting soybean hypocoyls from infection by a normally infective race of Pms if the elicitor is applied to the hypocotyls 6 hours prior to inoculation with Pms. The elicitor cannot protect soybean tissue when applied simultaneously with an infective race of Pms.

PHYTOALEXINS ARE NOT CAPABLE, BY THEMSELVES, OF PROTECTING PLANTS FROM THEIR PATHOGENS

A microorganism that has evolved the ability to grow successfully on a plant and thus has become pathogenic to that plant must also have evolved a mechanism of avoiding the toxic effects of phytoalexins. There are several plausible mechanisms for such avoidance by successful pathogens. One such mechanism might be simply the ability of an infective strain of a pathogen to grow away from the areas in which the plant is accumulating toxic levels of phytoalexin. This possibility seems likely as an explanation for the avoidance of the effects of glyceollin in soybean plants by infective races of Pms.

There are other mechanisms by which a pathogen might prevent a plant from stopping the growth of the pathogen by accumulation of phytoalexins. For example, a pathogen might secrete a toxin that kills the plant cells in the region of the pathogen before those cells are capable of synthesizing the enzymes necessary for synthesis of phytoalexins. Still another mechanism by which a successful pathogen might prevent a plant from accumulating sufficient phytoalexins might be to repress synthesis of one or more enzymes involved in phytoalexin synthesis or else to inhibit the enzymes once they are synthesized. A known mechanism by which pathogens overcome phytoalexin inhibition is the enzymatic conversion of the phytoalexins to less toxic or unstable compounds (13-17). Another possibility might be the production by the pathogen of proteins or other molecules that specifically inhibit the enzymes of the host that solubilize elicitors from the mycelial walls of the pathogen. Evidence suggestive of this type of mechanism has been obtained (18). Finally, pathogens might produce a carbohydrate that effectively but innocuously competes with the elicitor for binding to the elicitor's receptor.

ELICITORS ARE WIDESPREAD IN NATURE

Soybean plants have evolved the ability to recognize and respond to the structural β-D-glucan of *Phytophthora* mycelial walls. Similar β-D-glucans are found in the walls of a wide range of fungi (5). One fungus containing such β-D-glucans is brewer's yeast, *S. cerevisiae*, a nonpathogen of plants. An elicitor has now been purified from a commercially available extract of brewer's yeast (Difco) (19).

The 80% ethanol-insoluble fraction of the yeast extract contains a very active elicitor of glyceollin accumulation in soybeans. Most of the polysaccharide in this 80% ethanol-insoluble fraction is mannan. However, yeast extract does contain small amounts of the β-D-glucan. The glucan could be almost completely separated from the mannan by binding the latter polysaccharide to an affinity column consisting of Concanavalin A covalently attached to Sepharose. The glucan could be separated from glycoproteins by binding the proteins to sulfopropyl-Sephadex. Both the purified mannan and purified glucan remained contaminated by small amounts (approx. 2%) of arabinogalactan. Ribose-containing polymers, which contaminate the 80% ethanol-insoluble fraction, were removed on a DEAE-cellulose column. The elicitor activity of the precipitate from the crude, 80% ethanol extract of yeast resides in the glucan component. The small amount of residual activity remaining in the mannan fraction can be attributed to the minor contamination of this fraction by glucan. The glucan is composed of the same glucosyl linkages found in the Pms elicitor. The same quantities of the yeast and Pms elicitor are required to stimulate glyceollin accumulation in soybeans.

Our laboratory has obtained other evidence that elicitors have the ability to stimulate phytoalexin accumulation in a wide variety of plants. For example, the Pms elicitor stimulates suspension-cultured cells of sycamore and parsley to produce large amounts of phenylalanine ammonia-lyase activity (10). In addition, we have obtained evidence that the Pms elicitor stimulates *Phaseolus vulgaris*, the true bean, and *Pisum sativum*, the pea, to accumulate their phytoalexins (K. Cline, unpublished results). We have already reported (20) that a wall glucan from *Colletotrichum lindemuthianum*, a pathogen of *P. vulgaris*, stimulates the *P. vulgaris* to produce its phytoalexin. And, finally, we have demonstrated that the Pms elicitor stimulates potato tubers to accumulate their phytoalexins (M. Wade, unpublished results).

In summary, elicitors appear to be general in nature, and diverse plants are able to respond to a single elicitor. Elicitors can be considered to be pheromones, for elicitors are synthesized by one organism and control the metabolism of another organism. Elicitors may provide a new way of protecting plants against their pests, since elicitors might be used to activate the plant's own defense mechanism and thereby eliminate some of the need for spraying agricultural crops with poisonous pesticides.

ACKNOWLEDGMENT

Supported by USDA # 616-15-73; Department of Energy EY-76-S-02-1426; and the Rockefeller Foundation RFGAAS 7510.

REFERENCES

1. Ingham, J. L., *Bot. Rev. 38*, 343 (1972).
2. Kuc, J., *Annu. Rev. Phytopathol. 10*, 207 (1972).
3. Deverall, B. J., *Proc. R. Soc. London, Sec. B. 181*, 233 (1972).
4. Keen, N. T., Partridge, J. E., and Zaki, A. I., *Phytopathology 62*, 768 (1972).
5. Bartnicki-Garcia, S., *J. Gen. Microbiol. 42*, 57 (1966).
6. Burden, R. S., and Bailey, J. A., *Phytochemistry 14*, 1389 (1975).
7. Lyne, R. L., Mulheirn, L. J., and Leworthy, D. P., *J. Chem. Soc. Chem. Commun.* 497 (1976).
8. Ayers, A. R., Ebel, J., Valent, B., and Albersheim, P., *Plant Physiol. 57*, 760 (1976).
9. Ayers, A. R., Ebel, J., Finelli, F., Berger, N., and Albersheim, P., *Plant Physiol. 57*, 751 (1976).
10. Ebel, J., Ayers, A. R., and Albersheim, P., *Plant Physiol. 57*, 775 (1976).
11. Barras, D. R., and Stone, B. A., *Biochim. Biophys. Acta 191*, 342 (1969).
12. Ayers, A. R., Valent, B., Ebel, J., and Albersheim, P. *Plant Physiol. 57*, 766 (1976).
13. Higgins, V. J., *Physiol. Plant Pathol. 6*, 5 (1975).
14. Higgins, V. J., Stoessl, A., and Heath, M. C., *Phytopathology 64*, 105 (1974).
15. Van Den Huevel, J., and VanEtten, H. D., *Physiol. Plant Pathol. 3*, 327 (1973).
16. Van Den Huevel, J., VanEtten, H. D., Coffen, D. L., and Williams, T. H., *Phytochemistry 13*, 1129 (1974).
17. Ford, J.E., McCance, D. J., and Drysdale, R. W., *Phytochemistry 16*, 545 (1977).
18. Albersheim, P. and Valent, B. S., *Plant Physiol. 53*, 684 (1974).
19. Hahn, M., and Albersheim, P., *Plant Physiol.*, in press.
20. Anderson-Prouty, A., and Albersheim, P., *Plant Physiol. 56*, 286 (1975).

Studies on Glycoconjugates by F. Egami and His Former Co-workers with Special Reference to Sulfated Carbohydrates and Glycosidases*

Fujio Egami and Tatsuya Yamagata

The contributions to glycoconjugates of one of the authors (F. Egami), consist of two parts: one part on carbohydrate sulfates and sulfatases, and the other on glycosidases.

CARBOHYDRATE SULFATES AND SULFATASES

Glucosulfatase and charoninsulfuric acid. The study of these subjects was initiated in 1932 when F. Egami began his first work as an undergraduate student in the laboratory of Professor T. Soda of the Department of Chemistry at the Faculty of Science of the University of Tokyo. The theme of the research was "glucosulfatase of a large marine snail, *Charonia lampas*." Now we should like to briefly summarize the history of the theme.

* *Dr. L. Rodén, Program Chairman of this symposium, suggested to Professor Egami that he give a special lecture on his work from a more personal point of view. Professor Egami was very sorry to be absent because of business reasons, but Dr. Rodén kindly offered Dr. Yamagata, one of Egami's former coworkers, the possibility of delivering the present report. Professor Egami and Dr. Yamagata were very honored to have this opportunity.*

ISBN 0-12-301301-1

T. Soda, before being appointed associate professor of biochemistry, had worked for two years (1921-1923) under the direction of Professor C. Neuberg at the Kaiser-Wihelm Institute für Biochemie in Berlin-Dahlem. The work on sugar sulfates was suggested by Neuberg; we do not know why he was interested in sugar sulfates at that time. The reason may be that, shortly before, the existence of polysaccharide sulfates, such as chondroitin sulfates and the fact that they were analogs of sugar phosphates* had been recognized. Soda synthesized in Neuberg's laboratory glucose 6-sulfate, analogous to glucose 6-phosphate (Robison ester), and demonstrated that in contrast to the latter compound, it could not be fermented by yeast systems (1).

Soda then returned to his mother country and continued the studies on sugar sulfates. In 1931 he found a new enzyme, glucosulfatase, in a land snail, and a year later after an extensive search for better sources for the enzyme, he found that the hepatopancreas of a large marine snail, *Charonia lampas*, was the best source. In 1932, Egami came to Soda's laboratory and began to study the enzymatic properties of "glucosulfatase." It was found to be a hydrolase giving equimolar amounts of glucose and sulfate (2).

"*Charonia lampas* must contain a carbohydrate sulfate as the natural substrate of the enzyme." Indeed, from the mucous gland (hypobranchial gland) of the snail, a highly viscous polysaccharide sulfate was isolated and named charonin sulfate (3). Not long before, E. Jorpes had demonstrated that an anticoagulant, heparin, was a polysaccharide sulfate, and it was suggested and found that charonin sulfate had anticoagulant activity (3).

After the war, Egami reinvestigated charonin sulfate (1955-1961). It consisted of low-sulfated and high-sulfated fractions. The former fractions contained both α- and β-D-glucan sulfates, and the latter mainly cellulose sulfates: the higher the sulfate content, the more prevailing the cellulose structure. The most highly sulfated fraction was

* *In this connection, Egami and his co-workers later synthesized nucleoside sulfates, such as adenosine 5'-sulfate in 1953 and 1955, and riboflavine 5'-sulfate in 1957.*

indeed a cellulose polysulfate in which all of the hydroxyl groups were sulfated (4,5). The mechanism of such a high degree of sulfation and the biochemical relations between low-sulfated glucans and high-sulfated celluloses remain to be elucidated. The physiological function of charonin sulfate has not been studied, but it should be pointed out that the highly viscous product can be found only in carnivorous gastropods, such as *Charonia*. Moreover, it was observed cinematographically that *Charonia* releases a cloudy material when attacking animals. Therefore, Egami suggested that the highly viscous material (charoninsulfuric acid) inhibits the movement of animals serving as food. The following degradation of charonin sulfate by the hepatopancreas extract of the animal was suggested (6,7):

Cellulose polysulfate (SO_4-rich) —*Cellulose polysulfatase* (splits mainly 2- and 3-sulfate)→ Cellulose sulfate (SO_4-poor)

—*Polysaccharidases*→ Glucose sulfates and glucose (mainly 6-sulfate)

—*Glucosulfatase*→ Glucose

Other works carried out before 1940. Three findings from this period should be mentioned: (a) Some sugar sulfates having a reducing group can be catalytically desulfated by hydrazine (8). The sulfate groups attached to primary hydroxyl groups, such as in glucose 6-sulfate, cannot be hydrolyzed by hydrazine. The method can be applied to the structural analysis of sugar sulfates. (b) Trypaflavine and related acridine bases form, with sulfate esters, salt-like complexes that precipitate. It is well-known that acidic polymers, such as nucleic acids, precipitate with acridine bases, but in the case of sulfate esters not only sulfated polysaccharides but also low-mol. wt. sugar sulfates and arylsulfates are precipitated with acridine base (9). This observation can be used for histochemical or chromatographic detection of sulfate esters. These findings, published in Japanese, were mostly confirmed after 1950 by Dodgson and his co-workers (10,11). (c) The precipitation reaction with trypaflavine was used for the purification and fractionation of chondroitin sulfates. By this procedure the existence of oversulfated chondroitin sulfate (S/N > 1) in shark cartilage was observed (12). This important finding was overlooked until it was reconfirmed by many specialists on mucopolysaccharides (13-15). S. Suzuki and T. Yamagata, Egami's former co-workers, have carried out extensive studies on chondroitin sulfates and related substances, and these results will be summarized in a later section.

Other studies on mucopolysaccharides and glycoproteins. As an extension of the studies on sugar sulfates, the discovery and characterization of horatin sulfate from the hepatopancreas of *C. lampas* (16,17), the enzymatic degradation of keratan sulfates by enzyme systems from the hepatopancreas of the organism (18), and the isolation of sulfated glycoprotein from bovine brain (19) should be noted. Horatin sulfate is a unique compound. It is highly sulfated (sulfate content 4-10% as S) and contains fucose, mannose, glucose, galactose, glucosamine, galactosamine, and sialic acid as sugar components; it may be regarded as a highly sulfated glycoprotein. The chemical nature of sulfated glycoprotein from bovine brain was not fully studied. A similar substance was found later in rat brain by Brunngraber *et al.* (20).

Studies on ascorbate 2-sulfate and reinvestigation of glycosulfatase. It is generally accepted that sulfate groups in sugar sulfates are transferred from PAPS by specific sulfotransferases. Egami suggested a long time ago the alternative possibility that in certain cases sulfatases might play the role of sulfotransferases (21,22). He has not excluded this possibility, although the probable participation of PAPS in the biosynthesis of charonin sulfate was demonstrated (23).

Discovery of ascorbate 2-sulfate by Mead and Finamore (1969) led Egami to two lines of study: its probable role as a sulfate donor and its enzymatic degradation. The sulfate group was found to be incorporated from ascorbate 2-sulfate into chondroitin sulfates by embryonic chick cartilage epiphyses. Based upon the difference in the time courses of sulfation and the slight difference in the chondroitin sulfates produced from inorganic sulfate and ascorbate 2-sulfate, we concluded that the sulfate group was incorporated from ascorbate 2-sulfate without the intermediate formation of inorganic sulfate. It was suggested that in certain cases, ascorbate 2-sulfate may play the role of sulfate donor in the synthesis of sulfated mucopolysaccharides, either directly or indirectly through PAPS. In the latter case, it is comparable to phosphagens in phosphate transfer. Further studies are required to elucidate the problem (24,25).

The discovery of ascorbate 2-sulfate led Egami to re-investigate the specificity of sulfatases of *C. lampas*. "Arylsulfatase" and "Glycosulfatase" of the organism were separated from each other and highly purified. The aryl-sulfatase hydrolyzed ascorbate 2-sulfate, cerebroside 3-sulfate, and sulfoglycerogalactoside besides arylsulfates, and the glycosulfatase hydrolyzed D-glucose 6-sulfate, D-glucose 3-sulfate, *N*-acetyl-D-glucosamine 6-sulfate, but not D-galactose sulfates, UDP-*N*-acetyl-D-galactosamine 4-sulfate, nor cerebroside 3-sulfate. These results show that the relationship between the established nomenclature of the two enzymes and their respective specificities is inconsistent. The purified "glycosulfatase" should rather be called "glucosulfatase" as Soda had originally named it in 1931.* The specificity of the arylsulfatase seems to be essentially similar to that of mammalian arylsulfatase A. We hope that these studies will be useful for the structural analysis of sulfated glycoproteins (26).

GLYCOSIDASES

When Egami studied *C. lampas* for the first time in 1932, he observed small fishes and star fishes in the digestive tracts of the organism and concluded that the digestive enzymes of this organism must be very strong. He made up his mind at that time to study these digestive enzymes in the future. Since then more than 30 years have passed, and in 1965 he began to investigate the glycosidases of *C. lampas*. Why did he begin these studies at that time?

In 1957 Egami had discovered RNase T_1, which opened the way to the elucidation of the nucleotide sequence of RNA. This remarkable result stimulated him to study, after RNases,

* *Crude extracts of the hepatopancreas of C. lampas contain various carbohydrate sulfatases besides glucosulfatase.*

glycosidases which would be certainly useful for the determination of the sequence of sugar residues in heteropolysaccharides and glycoconjugates. The importance of these compounds has been recognized especially in connection with the cell membrane structure and the mechanism of cell recognition. Thus, he remembered his observation of 33 years ago.

Egami and Muramatsu confirmed at first the release of mannose and *N*-acetylglucosamine from ovalbumin glycopeptide by the crude extract of the hepatopancreas of the organism (27). Thus, with the assurance that the glycosidases of the extract were active on natural glycoproteins, a systematic survey of glycosidases of *C. lampas* and of a similar marine gastropod, *Turbo cornutus*, was undertaken. These two organisms were selected because the first is carnivorous and the second herbivorous, and the authors hoped to obtain glycosidases that act on substrates of both animal and plant origins. Besides these enzymes a few fungal glycosidases (α-L-rhamnosidase and α-D-fucosidase) were studied (28,29).

Most of the studies on gastropod glycosidases were presented at the Third International Symposium on Glycoconjugates in Lille in 1973 (30), and will be summarized only briefly here. So far, α-D- and β-D-mannosidase (31), β-D-xylosidase (32), α-L-fucosidase (33,34), *N*-acetyl-α-D-glucosaminidase, *N*-acetyl-α-D-galactosaminidase, *N*-acetyl-β-D-glucosaminidase (*N*-acetyl-β-D-galactosaminidase) (35), and sialidase (unpublished) were purified enough to be used for structural analysis, and a few examples may be cited. The enzymes were used for elimination of the protein part from chondroitin sulfate proteoglycan (36), structural analysis of the carbohydrate moiety of thyroglobulin (37), and in collaboration with other laboratories, structure determination of the carbohydrate moiety of stem bromelain (38,39), study of the receptor for Japanese encephalitis virus (40), and determination of the anomeric configuration of the mannose residues in *Salmonella* O-antigens (41). These findings show that the glycosidases are very useful for the structural analysis of glycoconjugates.

Egami has not continued his studies on glycosidases, but his former co-workers, I. Yamashina, T. Muramatsu, and N. Takahashi have extended these studies, including endoglycosidases and enzymes splitting the bonds between carbohydrate and protein in glycoproteins.

ENZYMES RELATED TO GLYCOCONJUGATES STUDIED BY EGAMI'S FORMER CO-WORKERS (T. MURAMATSU, I. YAMASHINA, AND N. TAKAHASHI).

The first reliable report on the discovery of endoglycosidase was made by Muramatsu in 1971. While incubating IgG glycopeptides with a glycosidase fraction obtained from *Diplococcus pneumoniae*, he noticed the appearance of two oligosaccharides, in addition to galactose and mannose. One oligosaccharide consisted of mannose residues with one *N*-acetylglucosamine residue at the reducing end, and the other of fucose, *N*-acetylglucosamine, and amino acid residues. The enzyme was found to split the linkage between vicinal *N*-acetylglucosamine residues and was thus an endoglycosidase (42).

The following year, Tarentino and associates found an endoglycosidase in the extract of *Streptomyces griseus* (1972), and Huang and Aminoff found an endoglycosidase activity in the culture medium of *Clostridium perfringens* (1972). Muramatsu and his collaborators purified these endoglycosidases and studied extensively their specificities, using a number of glycopeptides as substrates. A series of glycopeptides was prepared by splitting sugar residues one by one from such glycopeptides, as those obtained from ovalbumin or IgG, with exoglycosidases. The endoglycosidase of *D. pneumoniae* was found to be an endo-*N*-acetyl-β-D-glucosaminidase having a strict substrate specificity. It needs the presence of an α-D-mannosyl residue linked to the D-mannose residue next to an *N*-acetylglucosamine residue that is the substrate for the enzyme. If the α-D-mannosyl residue is substituted with another sugar, such as D-mannose in the case of the glycopeptide of ovalbumin, the glycopeptide will not undergo enzyme degradation. This enzyme was named endo-*N*-acetyl-β-D-glucosaminidase D (43).

During the course of study on the specificity of endo-*N*-acetyl-β-D-glucosaminidase, the entire structure of IgG glycopeptide from bovine serum was determined (44). The enzyme obtained from *Streptomyces*, that was named endo-*N*-acetyl-β-D-glucosaminidase H, was found to act on the glycopeptides of the Man-GlcNAc type, but not on complex glycopeptides: the presence of either one or two mannosyl residues attached to the second mannosyl residue was found to be necessary (45). Thus, these two endo-*N*-acetyl-β-D-glucosaminidases, with complementary specificities, could be used for the study of asparagine-linked carbohydrate chains, such as the cell-surface glycopeptides from growing and nongrowing human diploid cells. Large oligomannosyl cores were found to be preponderant in the glycopeptides from growing cells, whereas the amount of a small oligomannosyl core, consisting of 3 mannosyl residues, increased in the glycopeptides from non-growing cells (46). The introduction of endo-*N*-acetyl-β-D-glucosaminidases is surely a breakthrough in the field of glycopeptide research, and they will be widely used to prepare oligosaccharides from glycopeptides.

Another enzyme, amidase, which splits the aspartylglycosylamine bond, was isolated by Yamashina *et al*. (47).* Very recently, Takahashi, a former co-worker of Egami, demonstrated for the first time a new type of amidase that cleaves the aspartylglycosylamine linkage in glycopeptides having both an oligosaccharide and an oligopeptide moiety (48). She obtained this amidase from the extract of almond emulsin and showed that a glycopeptide obtained from stem bromelain was cleaved by the amidase to give an oligopeptide and an oligosaccharide:

Asn-Asn(oligosaccharide)-Glu-Ser-Ser →
Asn-*Asp*-Glu-Ser-Ser + oligosaccharide.

This enzyme is expected to be a valuable tool for structural studies of glycoproteins.

CHONDROITIN SULFATES AND RELATED SUBSTANCES STUDIED BY EGAMI'S FORMER CO-WORKERS (S. SUZUKI AND T. YAMAGATA)

As the last chapter in our presentation, we will describe the studies on chondroitin sulfates and related materials that have been carried out mostly by Suzuki, Yamagata, and their collaborators. At first our attention was focused mainly on chondroitin sulfate-degrading enzymes.

Chondroitin sulfate-degrading enzymes. Enzymes were purified from two sources, namely *P. vulgaris* and *Flavobacterium heparinum*, as their occurrence had been reported in these two bacteria. We found that the chondroitin sulfate-degrading

* *Yamashina and his co-workers have isolated several useful exoglycosidases (49-51).*

enzymes from *P. vulgaris* attack any of the chondroitin sulfates, including dermatan sulfates, and produces unsaturated disaccharides corresponding to the repeating unit of the mother structures. Since it degraded chondroitin 4- and 6-sulfate and dermatan sulfate, it was called CHase ABC. The chondroitin sulfate-degrading enzyme from *F. heparinum* was called CHase AC, as it degraded chondroitin 4- and 6-sulfate but not dermatan sulfate (52). Furthermore, we found and purified from *P. vulgaris* two chondrosulfatases, namely chondroitin 4- and 6-sulfatase, which act on unsaturated disaccharide 4-sulfate and 6-sulfate, respectively. Neither chondrosulfatase degrades intact chondroitin sulfate molecules. These enzymes made possible (a) to determine quantitatively, by their specificities, the amounts of chondroitin sulfate isomers in any source (53); (b) to identify chondroitin sulfates in any source, especially living tissues that have been labeled radioisotopically and could never have been shown to contain chondroitin sulfates without chondroitinases; and (c) after paper chromatographic separation of the enzymatic digests, to establish a number of new repeating units of chondroitin sulfates as well as of dermatan sulfates. Furthermore, the specificities of the chondroitinases showed the occurrence of a copolymer of chondroitin sulfate and dermatan sulfate.

Studies on fine structures revealed by chondroitinases. A dermatan sulfate preparation from pig skin gave an oversulfated, unsaturated disaccharide, ΔGlcUA(-S)-GalNAc(-4S) (ΔDi-diS$_B$), in addition to ΔGlcUA-GalNAc(-4S) (ΔDi-4S), when digested with CHase ABC but not with CHase AC, suggesting the oversulfation at iduronic acid residues (54).

Chondroitin sulfate of shark cartilage was found long ago by Egami to be oversulfated. Digestion with CHase ABC established that the oversulfation was caused by sulfation of part of the repeating units of chondroitin 6-sulfate (54). A chondroitin sulfate preparation from endocranial cartilage of squid gave rise to a new repeating unit that consists of β-D-glucosyluronic acid-*N*-acetylgalactosamine 4,6-disulfate (54). Biosynthetic studies revealed that squid cartilage has a unique sulfotransferase that catalyzes the introduction of sulfate from PAPS at C-6 of an *N*-acetylgalactosamine residue already bearing a sulfate group at C-4. Though various mono-, di-, and poly-saccharides were available as sulfate acceptors, the presence of a sulfate ester at C-4 of their *N*-acetylgalactosamine moiety was prerequisite for the acceptor activity (55). The same repeating unit, β-D-glucosyluronic acid-*N*-acetylgalactosamine 4,6-disulfate, was later found in bovine brain (1970), human aorta (1969), and other sources (1969).

A mucopolysaccharide preparation from hagfish notochord was found to contain an L-idosyluronic acid residue. When it was digested with CHase ABC, ΔDi-diS_E was detected along with ΔGlcUA-GalNAc(-4S) as the enzymatic products. The dermatan sulfate of hagfish notochord was not degraded by the action of CHase AC. We concluded that the main repeating unit of the hagfish notochord dermatan sulfate was composed of L-idosyluronic acid-*N*-acetylgalactosamine 4,6-disulfate residues (56). This newly found repeating unit was not restricted to hagfish, but it was also detected in the sea urchin embryo (57). This was the first report of the occurrence of a dermatan sulfate in an invertebrate, and it was shown to be a part of a unique proteoglycan, the side-chains of which consist of dermatan sulfates and chondroitinase-resistant mucopolysaccharides. The linkage between side-chains and the core protein appeared not to be an O-glycosyl bond (58).

During the course of our studies on the composition of chondroitin sulfates of various materials, we observed the appearance of oligosaccharides when a chondroitin sulfate fraction from meniscus was degraded with CHase AC. The oligosaccharides disappeared when the original sample was digested with CHase ABC. It was established that these oligosaccharides were derived from the iduronic acid-containing portions of dermatan sulfate-chondroitin sulfate copolymer (59). The occurrence of dermatan sulfate-chondroitin sulfate copolymer was also observed by Fransson.

Studies on proteoglycans. During the 1960's it was established that the mucopolysaccharide chains are, except for hyaluronic acid, side-chains of proteoglycan molecules. After the breakthrough of Sajdera and Hascall on the method of isolation of proteoglycan molecules (1969), our interest shifted from mucopolysaccharides to proteoglycans.

The proteoglycans of epiphyseal cartilage of a 13-day old chick embryo were separated into two fractions by sucrose density-gradient centrifugation. The proteoglycan which sedimented faster was called proteochondroitin sulfate H, and the lighter one was named proteochondroitin sulfate L. Neither of these was an intermediate of the other, and neither was shown to be a degradation product of the other.

Kitamura and Yamagata reported last year that undifferentiated mesenchymal cells synthesized another unique type of proteochondroitin sulfate, type M (60). Type M is distinctly different from type H and type L proteochondroitin sulfates in three ways: first, its position in a sucrose density gradient is between types H and L; second, the amino acid compositions differ, and third, the lengths of the bound mucopolysaccharide chains differ.

At present, we do not know the function, in development, of each proteochondroitin sulfate, especially that of type M. The proteochondroitin sulfate synthesized by cells shifts from type M to type H during chondrogenesis. We also noticed that the synthesis by chondrocytes of type L was quickly stopped when the cells were cultured in suspension in a medium containing F12X-10% FCS (61). However, cartilage-specific type H continued to be synthesized under the same conditions.

By adding chick embryo extract to the medium in which chondrocytes were cultured in suspension, the type of proteochondroitin sulfates synthesized by the cells could be switched from H to M. The effect of embryo extract was reversible; the ability to synthesize type H was restored when the embryo extract was removed from the medium (62).

Thus, we know the occurrence of three proteochondroitin sulfates during the differentiation of chick embryos. The analytical data of each could allow an analysis of the control mechanism of the differentiation of chondrocytes. However, the difficulties of these problems are still formidable. The further the research on proteoglycans advances, the more are we confronted by unsolved problems: (a) we do not know how the protein cores of types H, L, and M proteochondroitin sulfates differ; (b) we do not know whether one given type of proteochondroitin sulfate consists of an identical core protein or not; and (c) we do not know the function, localization, biosynthesis, and catabolism of these proteochondroitin sulfates. Further extensive work is required to elucidate these problems.

CONCLUSION

Following the tradition of Neuberg and Soda, Egami extended largely the domains of glycoconjugate research, and his co-workers have realized discontinuous but often rapid progress in related domains as well. It may be recognized that his group has contributed remarkably to the progress of glycoconjugate research, especially in sulfated carbohydrates and enzymes related to glycoconjugates.

REFERENCES

1. Soda, T., *Biochem. Z. 135*, 46 (1923).
2. Soda, T., and Egami, F., *Bull. Chem. Soc. Jpn. 8*, 148 (1933).
3. Soda, T., and Egami, F., *Bull. Chem. Soc. Jpn. 13*, 652 (1938).
4. Egami, F., Asahi, T., Takahashi, N., Suzuki, S., Shikata, S., and Nishizawa, K., *Bull. Chem. Soc. Jpn. 28*, 685 (1955).
5. Nakanishi, K., Takahashi, N., and Egami, F., *Bull. Chem. Soc. Jpn. 29*, 434 (1956).
6. Takahashi, N., and Egami, F., *Biochim. Biophys. Acta 38*, 375 (1960).
7. Takahashi, N., and Egami, F., *Biochem. J. 80*, 384 (1961).
8. Egami, F., *Nippon Kagaky Kaishi 59*, 1034 (1938); *60*, 849 (1939); *61*, 377 (1940); *61*, 592 (1940); *62*, 274 (1941); *63*, 763 (1942).
9. Egami, F., *Nippon Kagaki Kaishi 60*, 853 (1939); *61*, 855 (1940); *63*, 763 (1942).
10. Dodgson, K. S., and Spencer, B., *Biochem. J. 57*, 310 (1954).
11. Dodgson, K. S., Rose, F. A., and Spencer, B., *Biochem. J. 60*, 346 (1955).
12. Soda, T., Egami, F., and Horigome, T., *Nippon Kagaku Kaishi 61*, 43 (1940).
13. Seno, N., and Meyer, K., *Biochim. Biophys. Acta 78*, 258 (1953).
14. Mathews, M. B., *Nature (London) 181*, 421 (1958).
15. Furuhashi, T., *J. Biochem. (Tokyo) 50*, 546 (1961).
16. Inoue, S., and Egami, F., *J. Biochem (Tokyo) 54*, 557 (1963).
17. Inoue, S., *Biochim. Biophys. Acta 101*, 16 (1965).
18. Nishida-Fukuda, M., and Egami, T., *Biochem. J. 119*, 39 (1970).
19. Arima, T., Muramatsu, T., Saigo, K., and Egami, F., *Jpn. J. Exp. Med. 39*, 301 (1969).
20. Brunngraber, E. G., Hof, H., Susz, J., Brown, B. D., Aro, A, and Chang, I., *Biochim. Biophys. Acta 304*, 781 (1973).

21. Suzuki, S., Takahashi, N., and Egami, F., *Biochim. Biophys. Acta 24,* 444 (1957).
22. Suzuki, S., Takahashi, N., and Egami, F., *J. Biochem. (Tokyo) 46,* 1 (1959).
23. Yoshida, H., and Egami, F., *J. Biochem. (Tokyo) 57,* 215 (1965).
24. Hatanaka, H., Yamagata, T., and Egami, F., *Proc. Jpn. Acad. 50,* 747 (1974).
25. Hatanaka, H., and Egami, F., *J. Biochem. (Tokyo) 80,* 1215 (1976).
26. Hatanaka, H., Ogawa, Y., and Egami, F., *Biochem. J. 159,* 445 (1976).
27. Muramatsu, T., and Egami, F., *Jpn. J. Exp. Med. 35,* 171 (1965).
28. Kurosawa, Y., Ikeda, K., and Egami, F., *J. Biochem. (Tokyo) 73,* 31 (1973).
29. Iwashita, S., and Egami, F., *J. Biochem. (Tokyo) 73,* 1217 (1973).
30. Egami, F., *Colloq. Int. C.N.R.S. 221,* 289 (1973).
31. Muramatsu, T., and Egami, F., *J. Biochem. (Tokyo) 62,* 700 (1967).
32. Fukuda, M., Muramatsu, T., and Egami, F., *J. Biochem. (Tokyo) 65,* 191 (1969).
33. Iijima, Y., Muramatsu, T., and Egami, F., *Arch. Biochem. Biophys. 145,* 50 (1971).
34. Iijima, Y., and Egami, F., *J. Biochem. (Tokyo) 70,* 75 (1971).
35. Muramatsu, T., *J. Biochem. (Tokyo) 64,* 521 (1968).
36. Fukuda, M., and Egami, F., *J. Biochem. (Tokyo) 66,* 157 (1969).
37. Fukuda, M., and Egami, F., *Biochem. J. 123,* 415 (1971).
38. Fukuda, M., Muramatsu, T., Egami, F., Takahashi, N., and Yasuda, Y., *Biochim. Biophys. Acta 159,* 215 (1968).
39. Yasuda, Y., Takahashi, N., and Murachi, T., *Biochemistry 9,* 25 (1970).
40. Yasui, K., Nazima, T., Homma, K., and Ueda, S., *Acta Virol. 13,* 158 (1968).
41. Fukuda, K., Egami, F., Hammerling, G., Lüderitz, O., Bagdian, B., and Staub, A. M., *Eur. J. Biochem. 20,* 438 (1971).
42. Muramatsu, T., *J. Biol. Chem. 246,* 5535 (1971).
43. Koide, N., and Muramatsu, T., *J. Biol. Chem. 249,* 4897 (1974).
44. Tai, T., Ito, S., Yamashita, K., Muramatsu, T., Iwashita, S., and Kobata, A., *Biochem. Biophys. Res. Commun. 65,* 968 (1975).
45. Arakawa, M., Muramatsu, T., *J. Biochem. (Tokyo) 76,* 307 (1974).

46. Muramatsu, T., Koide, N., Ceccarini, C., and Atkinson, P. H., *J. Biol. Chem. 251,* 4673 (1976).
47. Yamashina, I., in "Glycoproteins" (Gottschalk, A. ed.) Ed. 2, p. 1187, Elsevier, Amsterdam, (1972).
48. Takahashi, N., *Biochem. Biophys. Res. Commun. 76,* 1194 (1977).
49. Makino, M., Kojima, T., Ohgushi, T., and Yamashina, I., *J. Biochem. (Tokyo) 63,* 186 (1968).
50. Sugahara, K., Okumura, T., and Yamashina, I., *Biochim. Biophys. Acta 268,* 488 (1972).
51. Okumura, T., and Yamashina, I., *J. Biochem. (Tokyo) 68,* 561 (1970).
52. Yamagata, T., Saito, H., Habuchi, O., and Suzuki, S., *J. Biol. Chem. 243,* 1523 (1968).
53. Saito, H., Yamagata, T., and Suzuki, S., *J. Biol. Chem. 243,* 1536 (1968).
54. Suzuki, S., Saito, H., Yamagata, T., Anno, K., Seno, N., Kawai, Y., and Furuhashi, T., *J. Biol. Chem. 243,* 1543 (1968).
55. Habuchi, O., Yamagata, T., and Suzuki, S., *J. Biol. Chem. 246,* 7357 (1971).
56. Anno, K., Seno, N., Mathews, M. B., Yamagata, T., and Suzuki, S., *Biochim. Biophys. Acta 237,* 173 (1971).
57. Yamagata, T., and Okazaki, K., *Biochim. Biophys. Acta 372,* 469 (1974).
58. Oguri, K., and Yamagata, T., *Abstr. Int. Cong. I.S.D.B. 8,* 110 (1977).
59. Habuchi, H., Yamagata, T., Iwata, H., and Suzuki, S., *J. Biol. Chem. 248,* 6019 (1973).
60. Kitamura, K., and Yamagata, T., *FEBS Lett. 71,* 337 (1976).
61. Yasumoto, S., Yamagata, T., Kondo, S., Oguri, K., Yamazaki, K., and Kato, Y., *Tissue Culture 2,* 189 (1976).
62. Yamagata, T., Yasumoto, S., Hayasaka, M., and Oguri, K., *Abstr. Int. Cong. I.S.D.B. 8,* 52 (1977).

GLYCOCONJUGATE RESEARCH / Volume I
Proceedings of the Fourth International Symposium on Glycoconjugates

The Lipid-Linked Oligosaccharide and Its Role in Glycoprotein Synthesis

Phillips W. Robbins, S. J. Turco, S. Catherine Hubbard, Dyann Wirth, and Theresa Liu

Oligosaccharide chains bound to protein *via* asparagine linkage frequently have a common inner core structure (1). The inner core consists of three D-mannose residues linked to *N,N'*-diacetylchitobiose. Structures vary from this point and may be classified into two types: The "complex" type contains sialic acid, galactose, *N*-acetylglucosamine, and fucose, while the "high mannose" type contains only additional mannose residues. There is considerable evidence that the terminal sugar residues of the complex structure are added to a growing oligosaccharide chain by the direct transfer from the appropriate nucleotide sugars (2). In contrast, the *N*-acetylglucosamine and mannose residues of either type are believed to be preassembled in an activated state bound to the isoprenoid lipid dolichol pyrophosphate.

The involvement of lipid-linked oligosaccharides in the glycoprotein of asparaginyl oligosaccharides of membrane and secreted glycoproteins has been under intensive investigation during the last six years. Lipid-linked oligosaccharides containing 2 *N*-acetylglucosamine and up to five mannose residues have been shown to serve as intermediates for transfer in several microsomal preparations (for review, see ref. 3). However, recent studies have shown that the major lipid-linked oligosaccharide synthesized in tissue slices (4) and in other *in vitro* systems (5) is larger, containing 8-11 mannose, 2 *N*-acetylglucosamine, and 1-3 glucose residues. The role of the D-glucose-containing oligosaccharide lipid was not established.

ISBN 0-12-301301-1

In this paper, we report two types of experiments designed to investigate the role of the glucose-containing, lipid-linked oligosaccharide in the glycosylation of proteins. First, we describe the *in vitro* synthesis in Nil 8 microsomes of similar glucose-containing oligosaccharide-lipid and the kinetics of its transfer to protein acceptor in an *in vitro* reconstitution system. Secondly, we describe the *in vivo* synthesis of a lipid-linked oligosaccharide similar to the *in vitro* product and examine the processing of the oligosaccharide after it is transferred to protein. These results also suggest that processing of the glucose and probably some mannose residues occurs after transfer to protein, since glucose is not commonly found as a constituent of asparaginyl-linked oligosaccharides.

ENZYMATIC SYNTHESIS OF THE LIPID OLIGOSACCHARIDE

When microsomes of Nil-8 hamster cells were incubated with GDP-[^{14}C]mannose or UDP-[^{3}H]glucose or both, radioactivity was incorporated into the lipid-linked oligosaccharide which can be isolated by appropriate extraction procedures and chromatography on DEAE-cellulose. Mild acid hydrolysis converted this material to water-soluble oligosaccharides, which have been fractionated by chromatography on Bio-Gel P-6. A typical set of profiles for singly- and doubly-labeled materials prepared from samples incubated for 10 min is shown in Fig. 1. The striking result is the alteration in the mannose-oligosaccharide profile produced by UDP-glucose addition. Although the total yield of [^{14}C]mannose-oligosaccharide material is the same in A and B, approximately half of the mannose label appeared in a larger oligosaccharide when 0.2 μM UDP-glucose was added to the incubation mixture. Radioactive glucose from UDP-glucose was incorporated into this same larger oligosaccharide. Some ^{3}H was incorporated into the larger oligosaccharide area in the absence of GDP-mannose, but this incorporation was stimulated 2- to 3-fold by GDP-mannose addition. Addition of higher levels of UDP-glucose lead to almost exclusive synthesis of the larger oligosaccharide.

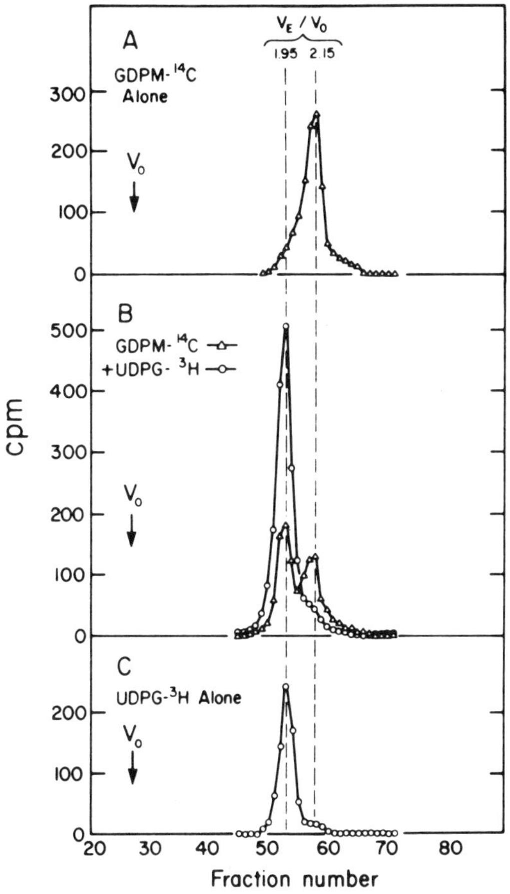

Fig. 1. Bio-Gel P-6 chromatograms of oligosaccharides released from the chloroform-methanol-water (1:1:0.3) fraction by mild acid hydrolysis. Acid hydrolysis and Bio-Gel P-6 chromatography were carried out as described in Robbins et al. *(12). GDPM-^{14}C, GDP-[^{14}C]mannose; UDPG-^{3}H, UDP-[^{3}H]glucose.*

PREPARATION OF THE TWO OLIGOSACCHARIDE-LIPIDS

Attempts to separate the glucose-containing oligosaccharide lipid from the glucose-free oligosaccharide-lipid by DEAE cellulose, silicic acid chromatography, Sephadex LH-20 gel filtration, and thin-layer chromatography in several solvent systems were unsuccessful. It was possible, however, to obtain each oligosaccharide-lipid in radiolabeled pure form by adjusting the UDP-glucose concentration in the preparative incubation mixture. Thus in the presence of 20 μM UDP-*N*-acetylglucosamine and 2 μM GDP-[^{14}C]-mannose and in the absence of UDP-glucose, only the glucose-free oligosaccharide-lipid was labeled with ^{14}C. Conversely, in the presence of 1 mM UDP-glucose, 85% of the ^{14}C radioactivity resided with oligosaccharide-lipid I. Furthermore, when UDP-[^{3}H]glucose was substituted for unlabeled UDP-glucose, all of the tritium radioactivity resided with the larger oligosaccharide-lipid.

TRANSFER OF OLIGOSACCHARIDE TO PROTEIN

In order to evaluate the relative rates of transfer of the glucose-containing and glucose-free oligosaccharide-lipid, equimolar amounts (9.7 nmol) of each [^{14}C]mannose-labeled substrate were added to separate microsomal preparations, as shown in Fig. 2. The glucose-containing oligosaccharide was transferred to endogenous acceptors at an initial rate 9 times greater than the glucose-free material. Also, up to 41% of the glucose-containing oligosaccharide was incorporated, as compared to 5% of the glucose-free oligosaccharide. Since in this experiment the glucose-containing oligosaccharide was labeled with [^{14}C]mannose, we were unable to determine whether the oligosaccharide was transferred *en bloc* or whether the glucose residues were cleaved during transfer. However, as shown in Fig. 3, addition of [^{3}H]glucose-oligosaccharide-lipid I and [^{14}C]mannose-oligosaccharide-lipid II to the same microsomal preparation again resulted in a greater initial rate (at least 4-fold) and in a greater extent of incorporation (4-fold) of the glucose-containing oligosaccharide.

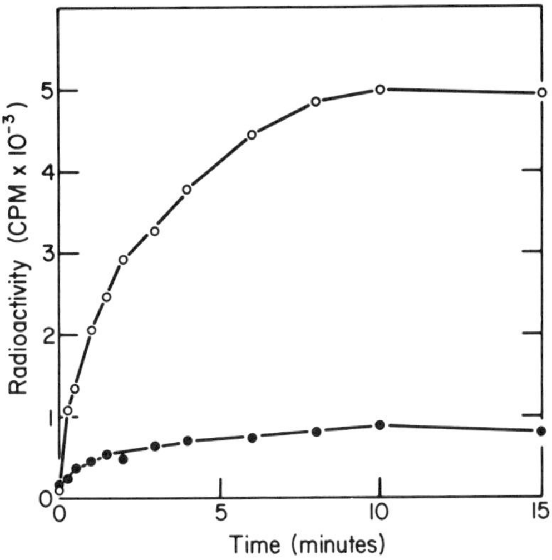

Fig. 2. Kinetics of incorporation of [^{14}C]mannose-labeled oligosaccharides by the endogenous acceptors of NIL cell microsomes. [^{14}C]mannose-oligosaccharide-lipid I (220 000 cpm) and [^{14}C]mannose-oligosaccharide-lipid II (350 000 cpm) were dried in separate tubes under a stream of nitrogen. To each tube was added 0.4 ml of 64 mM $MnCl_2$ and 0.4 ml Tris buffer saline (TBS). Following sonication, 0.4 ml of 3% sodium deoxycholate was added. Each reaction was initiated by the addition of 2.0 ml of microsomal protein (84 mg), and stirred at 25°C. At each time point, a 160 µl aliquot was removed from each tube, and the reaction was terminated by addition of 2.0 ml of chloroform-methanol (3:2) and 200 µl of water. The residue fraction was extracted as described. [^{14}C]Mannose-labeled oligosaccharide-lipid I -O-; [^{14}C]mannose-labeled oligosaccharide-lipid II -●-.

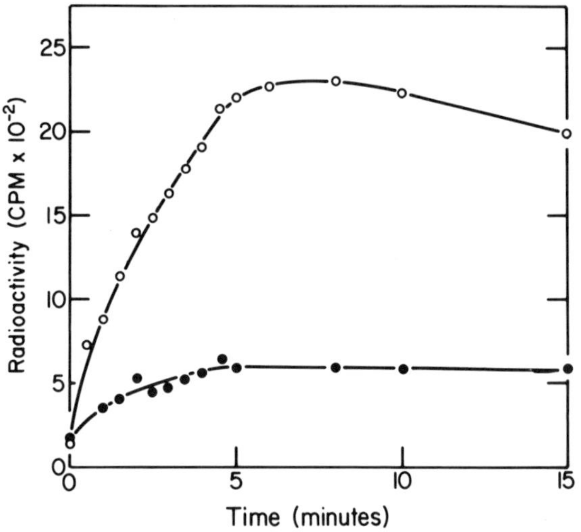

Fig. 3. Kinetics of incorporation of [^{3}H]glucose- and [^{14}C]mannose-labeled oligosaccharides by the endogenous acceptors of NIL cell microsomes. [^{3}H]Glucose-oligosaccharide lipid I (200 000 cpm) and [^{14}C]mannose-oligosaccharide-lipid II (200 000 cpm) were combined and dried under a stream of nitrogen. $MnCl_2$ (0.28 ml of a 64 mM solution) and 0.28 ml of TBS, pH 7.4 were added, and the mixture was sonicated. Following the addition of 0.28 ml of 3% sodium deoxycholate, the reaction was initiated by the addition of 1.4 ml of microsomal protein (42 mg) and was stirred at 28°C. At the indicated times, 160 μl aliquots were removed and the reaction was terminated by the addition of 2 ml of chloroform-methanol (3:2) and 200 μl of water. The residue fraction was extracted as described. [^{3}H]Glucose-labeled oligosaccharide-lipid I -O-. [^{14}C]Mannose-labeled oligosaccharide-lipid II -●-.

CHARACTERIZATION OF PRODUCT OF TRANSFER

The endogenous radiolabeled acceptor(s) were characterized as protein by the following methods: The products of transfer with either [^{3}H]glucose or [^{14}C]mannose-oligosaccharide-lipid I were 99% insoluble in trichloroacetic acid and greater than 95% sensitive to exhaustive Pronase digestion. In contrast, 92% of the product from [^{14}C]mannose-oligosaccharide-lipid was insoluble in trichloroacetic acid and only 61% sensitive to Pronase digestion. Gel filtration profiles on Bio-Gel P-6 (Fig. 4) of the products of Pronase digestion are very similar to corresponding profiles obtained in experiments with nucleotide-sugars (not shown).

COMPARISON OF LIPID-LINKED OLIGOSACCHARIDE *in vivo* AND *in vitro* BY A VARIETY OF CULTURED CELLS

We were interested in comparing the *in vitro* synthesized glucose-containing oligosaccharide with the *in vivo* synthesized oligosaccharides of Nil 8 cells and of other cultured cells in order to determine whether there exists one common oligosaccharide-lipid or a heterogeneous population of lipid-linked oligosaccharides. Gel-filtration profiles of *in vivo* and *in vitro* oligosaccharides isolated from the lipid carrier are shown in Fig. 5. The *in vivo* labeled oligosaccharides were obtained from the lipid-linked oligosaccharide synthesized by incubating whole cells with radioactive mannose. In all cell lines tested, a major oligosaccharide was obtained which emerged from the column at the same place as the glucose-containing oligosaccharide synthesized *in vitro* by Nil 8 microsomes (Fig. 5, left panels). The results of α-mannosidase digestion of the major species revealed further structural similarities. In all instances, approximately half of the total radioactive mannose residues were removed by the action of the enzyme.

The distribution of lipid-linked oligosaccharides obtained by incubating microsomal preparation from a variety of cells with 2 μM GDP-[^{14}C]mannose and 0.2 μM UDP-[^{3}H]-glucose are shown in Fig. 5, right panels. As originally observed with Nil 8 microsomes, the largest oligosaccharide was labeled with both [^{3}H]glucose and [^{14}C]mannose while the smaller oligosaccharide was labeled with [^{14}C]mannose only. The proportion of the two oligosaccharides differed, but this could be explained by differences in the concentration of the immediate endogenous lipid acceptors of nucleotide sugars or from different optimal conditions required by the particular cells for labeling.

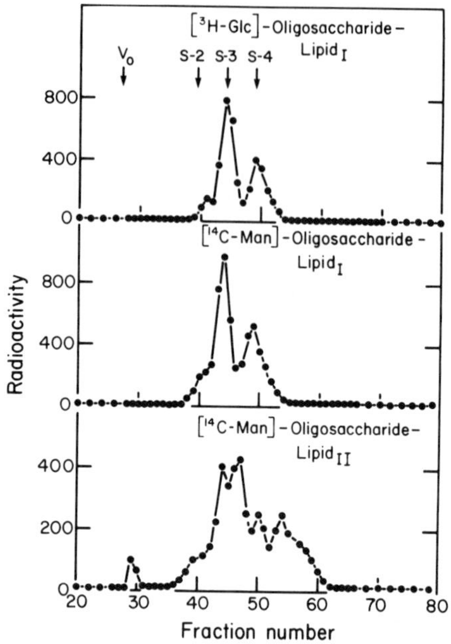

Fig. 4. Bio-Gel P-6 chromatogram of glycopeptides derived by Pronase digestion of labeled microsomal proteins. The non-extractable residue fractions prepared from 10-min incubations with oligosaccharide-lipids were treated with Pronase as follows: The residue fractions were suspended in 500 µl of 0.1 M Tris HCl, pH 8.0, 10 mM $CaCl_2$. A 1% solution of pre-digested Pronase in the same medium (Calbiochem) was added in 200 µl of aliquots at the following times: 0, 22, 48, and 72 h). The digestion was carried out at 50°C, and the preparations were protected from bacterial contamination by periodic addition of a drop of toluene. After 92 h the samples were concentrated to 400 µl under a stream of nitrogen, and applied directly to a Bio-Gel P-6 column. The vertical arrows indicate the positions of the exclusion volumes and Sindbis virus glycopeptides S-2, S-3, and S-4. The carbohydrate portions of S-4 and S-3 have molecular weights of 1550 and 1800, respectively. S-2 and S-1 are similar to S-3, except that they contain, respectively, 1 and 2 residues of sialic acid.

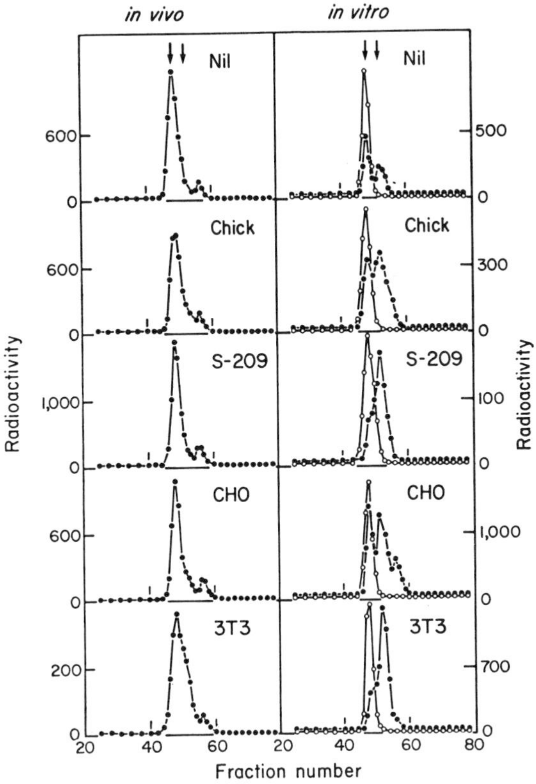

Fig. 5. Comparison of lipid-linked oligosaccharide derived from a number of sources. "Left Panels", [^{3}H]mannose-oligosaccharides synthesized in vivo *by the various cell lines were applied to a column of Bio-Gel P-6 (200--400 mesh, 1 x 115 cm) equilibrated in 0.1 M Tris HCl, pH 8.0, 0.02% sodium azide; 1-ml fractions were collected. "Right Panels", [^{14}C]mannose- and [^{3}H]glucose-labeled oligosaccharides synthesized by microsomal preparations from various sources were applied to the same column as described above. [^{3}H]Glucose-labeled oligosaccharide, -O-; [^{14}C]mannose-labeled oligosaccharides -●-. The arrows designate the positions of the two [^{14}C]mannose oligosaccharide synthesized* in vitro *by NIL 8 microsomes. These two oligosaccharides served as markers. The cell lines are: Nil 8 (hamster embryo), chick embryo, S-209 (human), BALB 3T3 (mouse embryo), and CHO (Chinese hamster ovary).*

Similar to observations made with *in vivo* labeled oligosaccharides, α-mannosidase digestion of the glucose-containing oligosaccharide removed approximately half of the [^{14}C]mannose, resulting in the formation of a [^{3}H]glucose- and [^{14}C]mannose-labeled, α-mannosidase-resistant component. These preliminary results indicate that the major lipid-linked oligosaccharides may have a similar or identical structure in a number of species.

EXPERIMENTS WITH SINDBIS AND VSV INFECTED CELLS

Both Sindbis and VSV are lipid-enveloped RNA viruses containing glycoprotein molecules in their membranes. They serve as appropriate model systems for the study of membrane-glycoprotein biosynthesis, since normal host-cell protein as well as RNA synthesis are inhibited during viral infection. Neither virus has enough coding capacity for the enzymes involved in oligosaccharide biosynthesis. Hence, cellular machinery must be employed. Sindbis virus has two envelope glycoproteins, E1 and E2, each containing one complex and one high-mannose oligosaccharide (6, 7). The proposed structures of these oligosaccharides are shown in Fig. 6. The mature VSV envelope protein (G) carries two complex glycopeptides similar in structure but probably not identical to the complex Sindbis glycopeptide shown in Fig. 6 (8).

A useful tool for the analysis of intracellular forms of these proteins and their oligosaccharide side-chains is the enzyme endo-β-*N*-acetylglucosaminidase H (endo H) described by Maley and coworkers (9, 10). As illustrated in Fig. 6, the enzyme cleaves between the *N*-acetylglucosamine residues in large high-mannose, asparagine-oligosaccharides. Complex chains are resistant and small oligosaccharides with three mannose residues are much less sensitive (10). Since the lipid-linked oligosaccharide and high-mannose glycopeptides were excellent substrates for the enzyme, as described below, we have used endo H to examine the carbohydrate chains attached to pulse-labeled normal and virus-infected cell proteins.

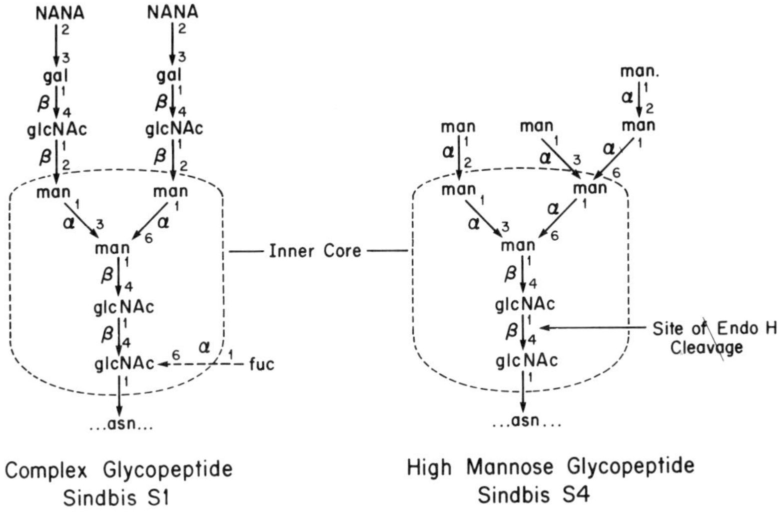

Fig. 6. Structures of Sindbis Virus Glycopeptides S1 and S4. The structures are those suggested by Burke (6). They are typical members of the "complex" and "high-mannose" classes of oligosaccharides. As pointed out by Burke and Keegstra (7), "S4" is heterogeneous and may contain components with fewer or more mannose residues than the structure shown here. This heterogeneity can be seen clearly in Fig. 8. The structure with seven mannose residues shown here is considered by Burke to be a major component of the S4 mixture. Two other complex glycopeptides of Sindbis virus are S2 and S3, which differ from S1 only in that they contain 1 and 0 residue of sialic acid (NANA) respectively.

The VSV G protein glycopeptide has a complex type side-chain and is resistant to the action of endo H (11). However, when chick cells infected with VSV were labeled for 10 min or less with [^{3}H]mannose, treatment with endo H released all the protein-bound label as free oligosaccharides. Gel filtration of this material (Fig. 7) revealed that, in cells labeled for 2.5 min, the oligosaccharide released from the glycoprotein fraction was very similar in size to the oligosaccharide from the lipid-linked intermediate. However, the endo H-released glycoprotein-oligosaccharide from cells labeled for 5 min or 10 min was significantly smaller.

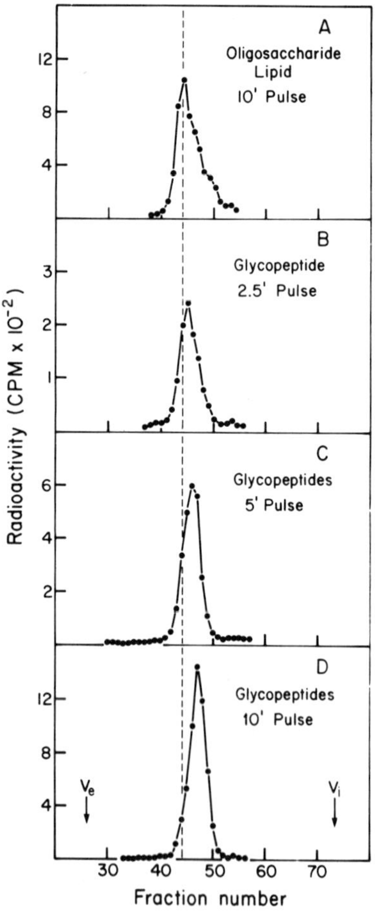

Fig. 7. Gel-filtration profiles of endo H-treated glycopeptides and lipid-linked oligosaccharides from VSV-infected

The mature Sindbis viral proteins each have one complex and one high-mannose side-chain. If mature virions are treated with endo H, about 60% of the mannose label is released; however, if newly synthesized proteins are examined, all of the mannose label is susceptible to cleavage by endo H. [^{3}H]-Mannose pulse-labeling of Sindbis-infected cells gave similar results (Fig. 8).

---

cells pulse-labeled with [^{3}H]mannose. Chick cells (1.5 x 10^6 60 mm culture plate) were labeled briefly with [2-^{3}H]mannose 5.5–6.5 h after infection with VSV (50 PFU/cell). One μg/ml actinomycin D was added 1 h after infection, and the labeling medium was Minimal Essential Medium without glucose (MEM-glucose) containing 5 mM sodium pyruvate and 800 μCi/ml (5 min sample) or 100 μCi/ml (10 min sample) [2-^{3}H]mannose (2 Ci/nmol, Amersham). After each labeling period, the cells were washed and scraped off the plates into ice-cold phosphate-buffered saline (PBS), and centrifuged at 0°C. The pellets were extracted twice with chloroform-methanol (2:1 v/v) and washed extensively with water. The lipid-linked oligosaccharide was extracted with chloroform-methanol-water (10:10:3, v/v); this material was subjected to mild acid hydrolysis to release free oligosaccharide. The chloroform-methanol-water-insoluble material was treated with Pronase to release glycopeptides. The glycopeptides and the oligosaccharide from the lipid-linked material were treated with endo H and mixed with a marker (endo H-treated, GDP-[^{14}C]mannose labeled, and lipid-linked oligosaccharide prepared in vitro *from Nil-8 microsomes) before application to a column of Bio-Gel P-4 (Bio-Rad, 200--400 mesh, 1 cm x 115 cm, 1-ml fractions, eluting buffer 0.1 M Tris, pH 8.0, containing 0.5 mM NaN_3). On this column, endo H-resistant material was eluted before fraction 43; the elution position of the marker is indicated by the dotted line. (A) Endo H-treated lipid-linked oligosaccharide from cells labeled for 10 min (4% sample). (B) Endo H-treated glycopeptides from cells labeled for 2.5 min (50% of sample). (C) Endo H-treated glycopeptides from cells labeled for 5 min (25% of sample). (D) Endo H-treated glycopeptides from cells labeled for 10 min (4% of sample).*

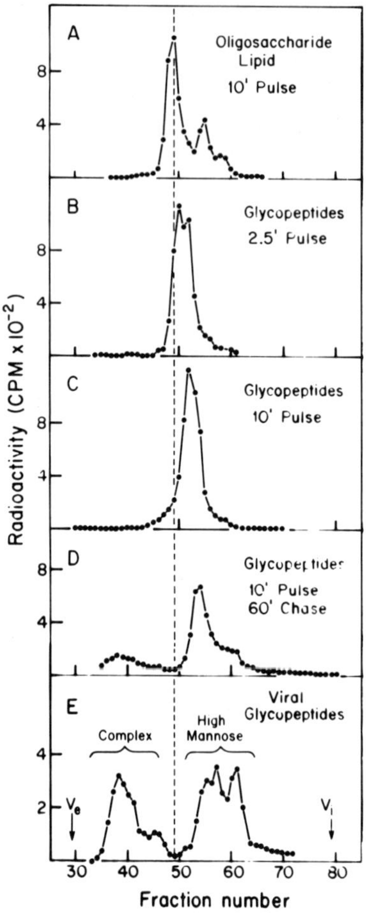

Fig. 8. Gel-filtration profiles of endo H-treated glycopeptides and lipid-linked oligosaccharides from Sindbis-virus-infected cells, pulse-labeled with [^{3}H]mannose. Chick cells were labeled briefly with [2-^{3}H]mannose 4.5-6.5 h after infected with Sindbis virus, and oligosaccharide from the lipid-linked oligosaccharide and glycopeptides from the protein fraction were prepared. Details of the procedure were as described in the legend to Fig. 7, except that after the labeling period, some cultures were washed and chased in Dulbecco's Modified Eagel's medium (DME), supplemented with 5% calf serum and 1 mM mannose, and containing 10/ml actinomycin D. The oligosaccharides and glycopeptides obtained, as well as glycopeptides from mature, [2-^{3}H]mannose-labeled Sindbis virions, were treated with endo H, mixed with a marker, and applied to

Again, glycopeptides from cells labeled for 2.5 min, 5 min, or 10 min were completely susceptible to endo H. Examination of the endo H-released oligosaccharides by gel filtration showed that they were indistinguishable from the corresponding material in the VSV experiment. Again, the protein-bound oligosaccharide from cells labeled for 2.5 min had approximately the same size as the lipid-linked oligosaccharide, while that from cells labeled for 5 min or 10 min was significantly smaller. All these released oligosaccharides were sensitive to α-mannosidase, which converted them to smaller oligosaccharides and free mannose (not shown). In cells labeled with [^{3}H]mannose for 10 min and then chased for various amounts of time, endo H- and α-mannosidase-resistant, protein--bound oligosaccharides first appeared following a 10 min chase (not shown), and the proportion of resistant material increased thereafter. After a 60 min chase, the remaining labeled, endo H-sensitive, protein-bound oligosaccharide was somewhat smaller than that from cells labeled for 10 min, corresponding approximately in size to the material released by endo H from [^{3}H]mannose-labeled, mature Sindbis virus glycoproteins.

The oligosaccharide initially attached to the protein seems to have about the same size as the lipid-linked oligosaccharide (containing 8-12 mannose residues), while the final complex oligosaccharides in mature VSV or Sindbis contain only three mannose residues. This processing must result in the removal of mannose and presumably glucose residues form the protein-attached oligosaccharide. This should be evident from quantitative pulse-chase studies. We have pulse-labeled infected cells simultaneously with [^{35}S]methionine and [^{3}H]-mannose. Following the pulse and varying periods of chase, the cells were lysed and aliquots were subjected to endo H treatment, followed by trichloroacetic acid precipitation. In each

---

a column of Bio-Gel P-4. The marker and gel filtration procedure are described in the legend to Fig. 7. On this column, oligosaccharides and glycopeptides not treated with or resistant to endo H were eluted before fraction 47; the elution position of the marker is indicated by the dotted line. (A) Endo H-treated lipid-linked oligosaccharide from cells labeled for 10 min (2% of sample). (B) Endo H-treated glycopeptides from cells labeled for 2.5 min (50% of sample). (C) Endo H-treated glycopeptides from cells labeled for 10 min (2% of sample). (D) Endo H-treated glycopeptides from cells labeled for 10 min and chased for 60 min (2% of sample). (E) Endo H-treated glycopeptides from mature Sindbis virions labeled with [2-^{3}H]mannose.

case, mannose recovery in treated and untreated samples was normalized to recovery of methionine. The methionine level remained constant after the beginning of the chase period, and the recovery of [^{35}S]methionine varied by less than 20%. Fig. 9 shows the total incorporation and endo H sensitivity of such samples from Sindbis-infected cells. Mannose incorporation into protein continued for 15-30 min following the beginning of the chase period. This is presumably the result of high levels of label in precursor pools. There is high degree of endo H sensitivity at early time points. The total level of acid-precipitated mannose decreased from 30 to 180 min, and the level of endo H resistance increased during that same period. The final, endo H-resistant mannose accounts for 40% of the total acid-precipitated label. This is similar to the value for mature virions. In the case of VSV similar results were obtained. The endo H-resistant material accounted for approximately 70% of the total mannose label present after 180 min of chase.

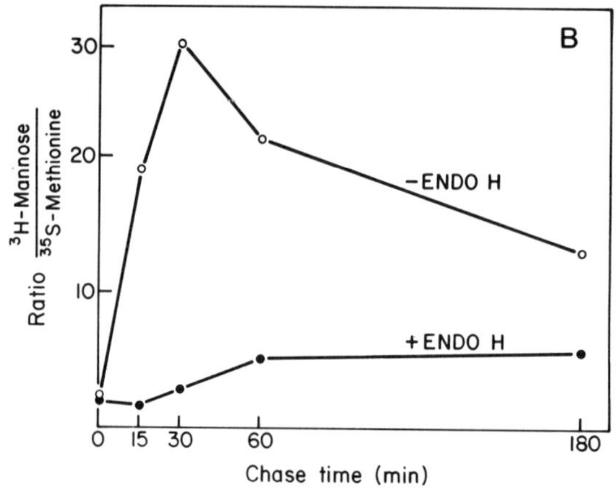

Fig. 9. Quantitative determination of endo H-sensitive and -resistant compounds containing [^{3}H]mannose, after a pulse label and chase. Chicken embryo cells infected with Sindbis virus were pulse-labeled with [^{3}H]mannose and [^{35}S]methionine. Aliquots of dodecyl sodium sulfate-lysed cells were treated with endo H. Parallel aliquots were incubated in the same buffer, but without the enzyme. At the end of the incubation, the samples were treated with 0.5 M NaOH and subsequently with 4 vol. of 10% trichloroacetic acid. Under these conditions, [^{35}S]methyl RNA is hydrolyzed as is the dolichol pyrophosphate oligosaccharide (not shown). The trichloroacetic-insoluble radioactivity was determined.

DISCUSSION

We have shown that microsomal preparations of Nil 8 cells, as well as other cultured cells, catalyze the synthesis in the presence of GDP-mannose and UDP-glucose of two major lipid-linked oligosaccharides. The smaller oligosaccharide is composed of 2 *N*-acetylglucosamine and 8-12 mannose residues, whereas the larger oligosaccharide contains, in addition, 1-3 glucose residues. In preliminary structural studies, the latter oligosaccharide is shown to have features similar to the *in vivo* labeled lipid-linked oligosaccharide with regard to elution profiles on Bio-Gel P-6 and susceptiblility to the action of α-mannosidase. Furthermore, these properties are similar to those reported for the major lipid-linked oligosaccharide obtained from tissue slices (4) and from microsomes from calf pancreas (5).

The importance of the presence of glucose residues on lipid-linked mannosyl-oligosaccharides was shown in comparative studies on the kinetics of transfer to protein. With substrates of comparable specific activities, the glucose-containing oligosaccharide was transferred from the lipid carrier to endogenous protein acceptors at a greater initial rate and to a greater extent than the glucose-free oligosaccharide. We have not conclusively demonstrated whether the two oligosaccharide-lipids compete for the same protein acceptors, although possible competition for transfer was suggested when the two oligosaccharide-lipids were incubated together (Fig. 3).

The observations that the major lipid-linked oligosaccharides from different sources have similar features and that glucose-containing oligosaccharide-lipid is a preferential substrate suggested the possiblity that asparaginyl-linked oligosaccharides originate from the transfer of a common lipid-linked intermediate. Subseqently, the protein-bound oligosaccharide undergoes processing. Further support for this proposal was obtained from the results of pulse-chase experiments with viral-infected cells. In these studies, it was observed that during pulse periods nearly all of the oligosaccharide bound to protein was susceptible to endo H. Initially, the released material is nearly identical in size to the product of endo H treatment of the lipid-linked oligosaccharide present in these cells. During chase periods, the oligosaccharide portion of the protein became increasingly less sensitive to the enzyme and the material released during this period was smaller in size than the oligosaccharide from the lipid.

ACKNOWLEGMENTS

S. Turco was a fellow of the Cystic Fibrosis Foundation and is now a fellow of the Leukemia Society of America, Inc. This investigation was supported by Grants CA14142 and CA14051 awarded by the National Cancer Institute.

REFERENCES

1. Kornfeld, R. and Kornfeld, S., *Annu. Rev. Biochem. 45*, 217 (1976).
2. Schacter, H., and Roden, L., *in* "Metabolic Conjugation and Metabolic Hydrolysis" (W. H. Fishman, ed.), vol. 3, p. 1, Academic Press, New York, 1973.
3. Waechter, C. J., and Lennarz, W. J., *Annu. Rev. Biochem. 45*, 95 (1976).
4. Spiro, M. J., Spiro, R. G., and Bhoyroo, V., *J. Biol. Chem. 251*, 6420 (1976).
5. Herscovics, A., Bugge, B., and Jeanloz, R. W., *J. Biol. Chem. 252*, 2271 (1977).
6. Burke, D., Ph.D. Thesis SUNY (Stony Brook), 1976.
7. Burke, D. J., and Keegstra, K., *J. Virol. 20*, 676 (1976).
8. Etchison, J. R., Robertson, J. S., and Summers, D. F., *Virology 78*, 375 (1977).
9. Tarentino, A. L., and Maley, F., *J. Biol. Chem. 249*, 811 (1974).
10. Tarentino, A. L., Trimble, R. B., and Maley, F., *Methods Enzymol. 50*, 574 (1978).
11. Reading, C. L., Penhoet, E. E., and Ballou, C. W., *J. Supramolec. Struct., Supplement 1*, p. 7 (1977).
12. Robbins, P. W., Krag, S. S., and Liu, T., *J. Biol. Chem. 252*, 1780 (1977).

Possible Functions of Lectins in Microorganisms, Plants, and Animals

Nathan Sharon

It has been known for many years that cell-agglutinating proteins are widely distributed in Nature. Such proteins, named lectins, (for recent review, see refs. 1 and 2) are further characterized by their ability to specifically bind carbohydrates, from simple monosaccharides to complex oligosaccharides, and may thus be classified together with other sugar-binding proteins (Table I).

Table I. Carbohydrate-binding Proteins

Cell-agglutinating	*Non-agglutinating*
Lectins	*Enzymes*
Antibodies	*Hormones*
	Toxins
	Interferon

ISBN 0-12-301301-1

Lectins are similar to antibodies not only in their ability to agglutinate cells, but also in other biological properties (Table II). There are, however, striking differences between these two classes of protein. Most importantly, antibodies are the product of the immune system of animals, and are formed in a response to an antigenic stimulus, whereas lectins are formed without such a stimulus, since many of them are found in organisms, such as plants, which do not posses the capacity of immunologic response. Antibodies and lectins differ in several other respects. Thus, although each antibody is specific for the antigen that elicited its formation, the range of specificity of antibodies is very wide, encompassing a broad spectrum of organic structure; the specificity of lectins, on the other hand, is restricted to carbohydrates. Also, antibodies are structurally similar; lectins are diverse, their only common feature being that they are all proteins.

Table II. Biologic effects of lectins on cells[a]

1.	*Agglutination of erythrocytes and other types of cell*
2.	*Mitogenic stimulation of lymphocytes*
3.	*Inhibition of phagacytosis*
4.	*Inhibition of vacuole formation in macrophages*
5.	*Immunosuppressive effects*
6.	*Toxicity*
7.	*Inhibition of growth of tumor cells*
8.	*Inhibition of migration of tumor cells*
9.	*Inhibition of fertilization of ovum by sperm*
10.	*Induction of insulin release from pancreatic islets*
11.	*Insulin-like effects on fat cells*
12.	*Induction of platelet-release reaction*
13.	*Induction of histamine release from basophils and mast cells*
14.	*Inhibition of fungal growth*

[a] *From Lis and Sharon (2)*

In this article, I briefly review some properties of lectins and summarize the main current ideas about their possible roles in Nature as determinants of intercellular recognition between bacteria on the one hand, and plants and animal tissues on the other, and as protective agents of plants against phytopathogenic fungi. Some of the practical implications of these ideas, in particular in medicine, will be pointed out.

SUGAR SPECIFICITY AND BIOLOGICAL ACTIVITIES

Many lectins combine preferentially with a single sugar structure, for example D-galactopyranose or L-fucopyranose. For some lectins, the specificity is broader and includes several closely related sugars, *e.g.*, D-mannopyranose, D-glucopyranose, and D-arabinopyranose; other lectins interact only with complex carbohydrate structures, such as those that occur in glycoproteins, glycolipids, or on cell surfaces. The sugars with which lectins combine best are those that are typical constituents of glycoproteins or glycolipids. Perhaps this is a reflection of the way in which lectins are detected (namely, by hemagglutination), as a result of which lectins specific for sugars other than those present on animal cell surfaces might be overlooked.

A wide range of specificities and biological activities has been observed in the interaction of lectins with cells (Table II). Lectins show selectively in their agglutination of erythrocytes of different animal species, and with human erythrocytes some of them are even blood-type specific. This ability to distinguish between different human blood groups provided the basis for the name "lectins", coined by Boyd and Sharpleigh in 1954, which was derived from the Latin "*legere*" (to select or choose). Both species and class (T or B) specificity have also been demonstrated in the interaction of lectins with lymphocytes; moreover, certain lectins may distinguish between lymphocyte subpopulations from the same animal or organ (3,4). Another intriguing property of lectins is their ability to agglutinate malignantly transformed cells much better than normal cells (reviewed in 5-7).

The ready availability of lectins, their ease of preparation in purified form, the fact that they are amenable to chemical manipulation, and that many of them are inhibited by simple sugars, has made them a most attractive tool in biological research.

DISTRIBUTION IN NATURE

The early studies seemed to indicate that lectins are confined to plants and they were, therefore, known as phytohemagglutinins or phytoagglutinins. However, during the last decade many lectins have been isolated from diverse groups of organisms including bacteria, slime molds, snails, crabs, and fishes. The isolation of the first mammalian lectin, the hepatic-binding protein from rabbit liver (8,9), and the demonstration (10) that extracts of a variety of mammalian tissues contain a lectin specific for β-D-galactosides, have served to emphasize the wide distribution of these macromolecules. Several other mammalian lectins have now been purified, from calf heart and lung (11) and from chick embryo muscle (12,13), all specific for β-D-galactosides. Recently, an *N*-acetyl-D-glucosamine-specific lectin has been isolated from chicken liver (14), and there is evidence for the presence in bovine liver of a lectin with similar specificity (15).

Lectins are not confined to a single site in the organisms in which they occur. Thus, the rabbit liver lectin has been recovered in good yield not only from the liver plasma membrane, but also from membranes of the Golgi complex, the microsomes, and the lysosomes (16).

In plants, where they may contribute as much as 3-4% of the dry matter of the seeds (*e.g.*, of the jack bean), lectins have now been shown to occur in the cell walls and in a variety of membrane fractions of different tissues. They are present in mitochondrial membranes of castor bean (17-19) and in mung bean cell-walls (20,21). Preliminary experiments carried out in our laboratory have demonstrated that soybean agglutinin and peanut agglutinin are found in plant tissue other than seeds, being present in roots, shoots, and leaves where they are membrane components (22). High levels of these lectins are maintained throughout development from the seedlings to the mature plant (measured at 2, 5, and 7 weeks). The highest amounts of the lectins were recovered by treatment of the isolated membranes with detergents, indicating that these proteins are integral membrane components.

In another study of the levels of soybean agglutinin in different tissues of soybean plants, the bulk (>90%) of the lectin was extracted by buffer solutions from the cotyledons, but it was also detected in extracts of the embryo axis and the seed coat (23). Soybean agglutinin was present in all tissues of the young seedlings but its level decreased as the plants matured and was not detected in plants older than 2 to 3 weeks. However, as in these experiments detergents were not used for extraction of the tissues, the lectin present in cellular and intracellular membranes might have been missed.

PROPERTIES OF PURIFIED LECTINS

To date, over 50 lectins have been obtained in purified form. Several examples are listed in Table III, where their properties are also summarized. Three of these, soybean agglutinin, peanut agglutinin, and wheat germ agglutinin have been under intensive investigation in our laboratory (for review, see ref. 2 and references therein). As can be seen from this Table, lectins vary considerably in composition, sugar content, mol. wt., subunit structure, number of carbohydrate binding sites per molecule, and metal requirement. Many lectins contain covalently bound sugar and are therefore glycoproteins. However, concanavalin A, wheat germ agglutinin, and peanut agglutinin, which are amoung the best characterized proteins of this class, are devoid of sugar. Soybean agglutinin, the most thoroughly investigated glycoprotein lectin, contains 6% of sugar comprised of D-mannose and *N*-acetyl-D-glucosamine, and consists of 4 subunits (mol. wt. 30 000), each of which carries one carbohydrate chain, $Man_9(GlcNAc)_2$, linked to the protein *via* an *N*-acetyl-D-glucosaminyl-asparagine linkage (24-26). We have now elucidated the structure of the carbohydrate side-chain of soybean agglutinin and have demonstrated for the first time the presence, in a plant glycoprotein, of the branched core α-Man-(1→3)-[α-Man-(1→6)]-β-Man-(1→4)-β-GlcNAc-(1→4)-β-GlcNAc, previously found in many animal glycoproteins as well as in those from fungi and yeasts (H. Lis and N. Sharon, in this volume; 27).

In spite of their apparent structural diversity (28), lectins from closely related sources do exhibit considerable similarities in amino acid composition and molecular properties. Extensive homologies exist between the first 25 residues of the amino terminal sequence of the β chains of the lentil and pea lectins, of soybean and peanut agglutinins, and the R and L subunits of the phytohemmagglutinin (PHA) from red kidney bean (all from leguminous plants), ranging from near identity in the case of the β chains of the lentil and pea lectins, to 24% of identity between soybean agglutinin and L-PHA (assuming two deletions in the latter) (Table IV) (29). In the case of peanut and soybean agglutinins, 11 out of the first 25 amino acids are identical, and 8 additional ones (Asn-7, Ala-17, Ile-18, Asn-19, Phe-20, Gln-21, Val-24, and Thr-25) in peanut agglutinin could have arisen each by a single nucleotide substitution. Lectins derived from plants of families other than the legumes will most likely prove to possess different primary structures, as can be predicted, for example, from the very unusual amino acid composition of wheat germ agglutinin (30) and the potato lectin (31).

Table III. Properties of Some Purified Lectins[a]

Source: Phylogenetic name	Source: Common name	Abbreviated name of lectin
PLANTS		
Legumes		
Arachis hypogaea	*peanut*	*PNA*
Bandeiraea simplicifolia		
Canavalia ensiformis	*jack bean*	*Con A*
Dolichos biflorus	*horse gram*	
Glycine max	*soybean*	*SBA*
Lens culinaris	*lentil*	
Lotus tetragonolobus	*winged pea*	
Phaseolus limensis	*lima bean*	
Phaseolus vulgaris	*red kidney bean*	*PHA*
Other		
Agaricus campester	*meadow mushroom*	
Ricinus communis	*castor bean*	*RCA*
Solanum tuberosum	*potato*	
Triticum vulgaris	*wheat*	*WGA*
BACTERIA		
Pseudomonas aeruginosa		
LOW ANIMALS		
Anguilla anguilla	*eel*	
Electrophorus electricus	*electric eel*	
Helix pomatia	*garden snail*	
Limulus polyphemus	*horseshoe crab*	

[a] *For references, see (2)*

			Specificity	
No. of sugar-binding sites	*Mol. wt.*	*Sugar*	*Human-blood type*	*Mitogenic activity*
2	110 000	GalNAc		+
4	114 000	Gal	B	
4	102 000	Man, Glc		+
4	110 000	GalNAc	A	
2	120 000	Gal		+
	ca. 60 000	Man, Glc		+
	ca. 120 000	L-Fuc	O	
4	230 000	GalNAc	A	+
	120 000	GalNAc		+
	64 000	?		+
2	120 000	Gal		
2	120 000	$(GlcNAc)_2$		
4	36 000	$(GlcNAc)_2$		
	65 000	Gal		
4	130 000	L-Fuc	O	
	33 000	Gal		
6	79 000	GalNAc	A	
	335 000	NANA		+

Table IV. Amino Terminal Sequences of the β Chains from Lentil and Pea Lectins, the Agglutinins of Soybean and Peanut, and of the R and L Subunits of PHA[a]

	1									10	
Lentil β	Thr	Glu	Thr	Thr	Ser	Phe	Ser	Ile	Thr	Lys	Phe
Pea β					Leu						
Soybean	Ala			Val				Trp	Asn		
Peanut	Ala			Val			Asn	Phe	Asn	Ser	
R-PHA	Ala	Ser	Glu					Phe	Glu	Arg	
L-PHA	Ser	Asn	Asp	Ile	Tyr		Asn	Phe	Glu	Arg	

[a] *Homologies are indicated by a solid line. Deletions [ ]*

								20					25
Ser	Pro	Asp	Gln	Gln	Asn	Leu	Ile	Phe	Gln	Gly	Asp	Gly	Tyr
											Asn		
Val		Lys	Glu	Pro	Asp	Met		Leu	Glu			Ala	Ile
	Glu	Gly	Asn	Pro	Ala	Ile	Asn					Val	Thr
Asn	Glu	Thr	[		]			Leu		Arg		Ala	Ser
Asn	Glu	Thr	[		]			Leu		Arg		Ala	Ser

were introduced to maximize homology (30)

The extensive homologies between different lectins obtained from a single plant family, the legumes, strongly suggest a common genetic origin for these proteins (29). It would appear, therefore, that lectins can be grouped in families, which have conserved their primary structure even through their carbohydrate specificity and some of their biological properties may be different. Moreover, the homologies suggest a common ancestry for the genes coding for these lectins, and indicate that these proteins may have an important physiological role in plants. The latter probably holds true for lectins found in other organisms, from bacteria to animals.

DETERMINANTS OF RECOGNITION

In the past there have been many speculations on the role that lectins may have in Nature (32,33). Some recent proposals were discussed in a Symposium on "Cell Wall Biochemistry Related to Specificity in Host-Plant Pathogen Interactions" held at the University of Tromsö, Norway, in August 1976 (34).

According to an early suggestion, lectins in plants may be involved in sugar transport and storage. It was, however, shown that the most potent inhibitors of the lectin present in the sieve tube sap of *Robinia pseudoacacia* were *N*-acetyl-D-galactosamine and glycosides containing D-galactose, even though galactosides are not transported in the sieve tubes of this plant, leading to the conclusion that most of the lectin in the plant is not directly involved in sugar transport mechanisms (35,36).

In view of the mitogenic properties of lectins, the possibility has been raised that their function is to control cell division and germination in plants. Indeed, it was reported that Con A and PHA, both of which stimulate lymphocytes to grow and divide, also stimulate pollen germination *in vitro*, by reducing the lag-period before the emergence of the pollen tube (37). However, when the effect of soybean agglutinin and PHA on the growth of soybean root and tobacco pith segments *in vitro* was examined, no increase or induction of cell division was observed (38). It was suggested (38) that the slight and transient increase in mitotic acitivities, previously reported (39,40) with PHA, might have been due to the presence of contaminants (possibly plant growth substances) in the impure lectin preparations used.

Currently, there are increasing indications that lectins function in recognition phenomena of microorganisms, plants, and animals, both intercellular and intracellular. They may be responsible for a variety of intercellular interactions, from the adhesion of bacteria to animal cells, to the attachment of sperm to egg.

Bowles and Kauss (15,36) have suggested that the action of exogenous lectins on animal cells *in vitro* may mimic that of carbohydrate-binding proteins normally present at the surface of adjacent cells. According to Ashwell and Morell (41,42), carbohydrate-mediated cellular and intracellular recognition phenomena may be regarded as a direct manifestation of native endogenous lectins. Just as enzymes represent a class of ubiquitous proteins that recognize specific substrates in the course of catalyzing specific reactions, or as antibodies act in neutralizing foreign agents, so lectins may be perceived as a unique category of proteins with the capacity to recognize subtle differences in cell-surface carbohydrate sequences as a means of influencing and regulating a variety of normal physiological functions.

Since recognition also implies distinction between self and nonself and between friend and enemy, these ideas are in line with earlier suggestions that lectins play an important role in host-parasite relationships, both in animals and in plants, and may serve as part of the defense mechanisms of plants against pathogenic microorganisms, whether fungi or bacteria. Recognition by lectins may also be the basis of another important biological process, the association between legumes and their symbiotic nitrogen-fixing bacteria.

BINDING OF NITROGEN-FIXING BACTERIA TO LEGUMES

The association between legumes and nitrogen-fixing bacteria, such as rhizobia, is highly specific. Even though many other microorganisms live in the soil around the plant roots, each species of legume is usually infected by only one rhizobial species and, conversely, most rhizobia infect only one kind of legume. For example, rhizobia that infect and nodulate soybeans cannot nodulate garden beans or white clover, and *vice versa*. The basis for this specificity may reside in the capacity of the bacterial cell to be recognized and bound by some component, most likely a lectin (or lectins), found in the roots of the plants (summarized in ref. 44). Much of the evidence implicating lectins in the establishment of symbiotic nitrogen-fixing relationships is based on the finding that a lectin from a particular legume binds only to the corresponding rhizobial species and not to bacteria that infect other legumes (44,45). Thus, it was reported that fluorescein-labeled soybean agglutinin binds to 22 out of 25 strains of *Rhizobium japonicum*, which infect soybeans, but does not bind to any of the 23 strains from 5 species of rhizobia that infect only other legumes (45). Similar results were obtained in another study on the binding of soybean agglutinin to different strains of *Rhizobium*, although it was observed that the lectin-binding properties of the lectin-positive strains changed substantially with the age of the culture (46,47). Whenever tested, binding of the lectin to the bacteria was reversible and competitively inhibited by monosaccharides for which soybean agglutinin is specific (*N*-acetyl-D-galactosamine and D-galactose) showing that binding is not due to nonspecific adsorption of the lectin to the bacteria. Also, the lectin has been detected in soybean root extracts (22,45), and it is released by soybean seeds during water uptake (48).

In a somewhat different approach to this problem, lectins from the seeds of 4 legumes (soybean, pea, red kidney bean, and jack bean) and lipopolysaccharides from the 4 corresponding symbiotic rhizobial species were isolated (49). In all cases, the bacterial lipopolysaccharide bound only to the lectin from the legume with which the bacterium forms a symbiotic relationship, although the proportion of lipopolysaccharide bound was low, between 5-35%.

Other researchers, however, found no correlation between the binding of soybean agglutinin to rhizobia isolates and their ability to nodulate soybeans (50,51). Also, no evidence was found for the preferential binding of Con A, the lectin of jack bean seeds, to *Rhizobium* strains capable of nodulating jack bean (52).

Lectins may function in a different way in plant-rhizobia interactions, as suggested on the basis of a study of the symbiotic associations between *Rhizobium trifolii* and white clover (53,54). It was shown that the clover roots and *R. trifolii* possess a common antigen, which is present on the cell surfaces of infective strains of *R. trifolii*, but is absent, inaccessible, or present in reduced quantities on noninfective strains. Furthermore, evidence was obtained for the presence in clover seed extracts of a lectin capable of binding to the isolated capsular polysaccharide antigen and to infective, but not the noninfective, strains of rhizobia. It was proposed that the clover lectin forms a bridge between common antigenic structures for the preferential adsorption of infective strains of *R. trifolii* to the root surface of the host (Fig. 1).

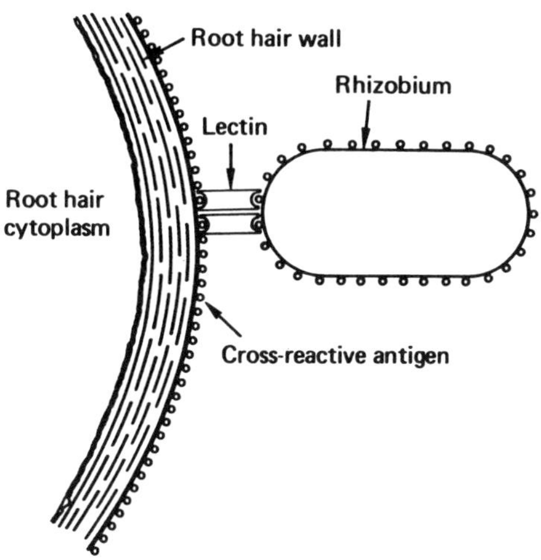

Fig. 1. Diagram of the proposed cross-binding of the cross-reactive antigens of Rhizobium trifolii *and clover root hairs with a clover lectin (redrawn from ref. 53).*

PROTECTION AGAINST PLANT PHYTOPATHOGENS

Interactions of lectins with fungal hyphae were first demonstrated by our group at the Weizmann Institute (55), where it was found that wheat germ agglutinin, a lectin specific for chitin oligosaccharides, binds to hyphal tips and hyphal septa of *Trichoderma viride* (Fig. 2) and that this binding is specifically inhibited by tri-*N*-acetylchitotriose, $(GlcNAc)_3$.

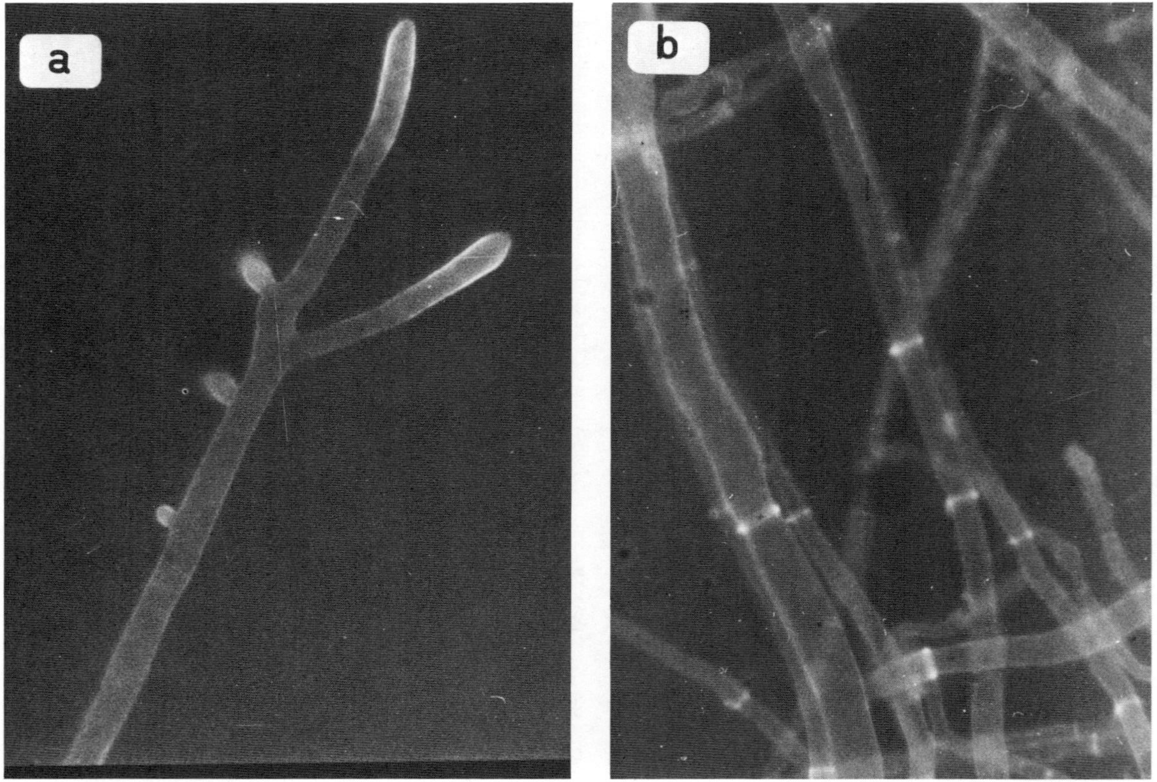

Fig. 2. Binding of fluorescein isothiocyanate-labeled wheat germ agglutinin to the tips (a) and septa (b) of Trichoderma viride *hyphae (from ref. 55).*

We have also demonstrated that binding of the lectin inhibited hyphal growth and spore germination of this chitin-containing fungus (Fig. 3). Based on these findings, we have suggested that wheat germ agglutinin may function in protecting wheat seedlings against chitin-containing phytopathogens during seed imbibition, germination, and early seedling growth. It was further postulated that lectins with sugar specificities different from those of wheat germ agglutinin act similarly, as natural inhibitors of the growth of fungi, the surfaces of which are covered by polysaccharides that react with these lectins.

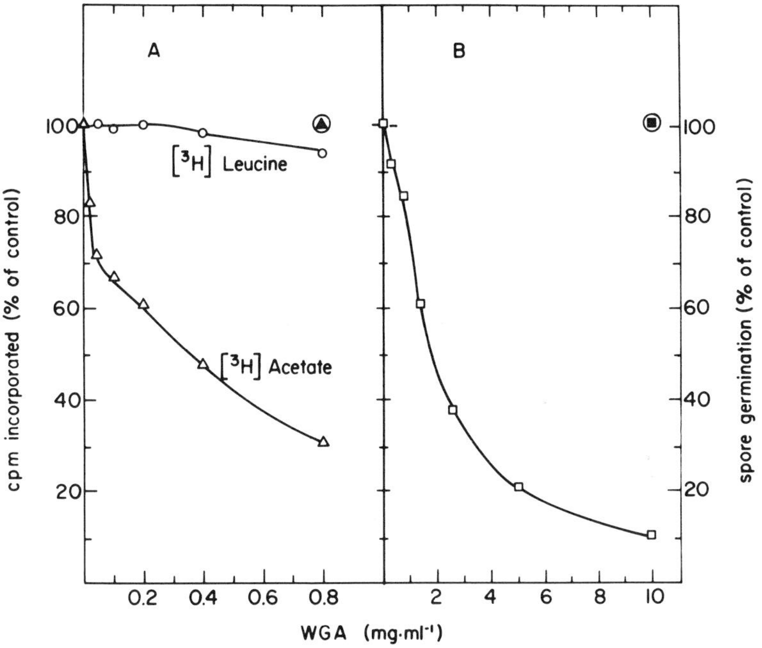

Fig. 3. (A) Effect of wheat germ agglutinin on the incorporation of sodium acetate (△) and L-leucine (o) into young hyphae of T. viride*; ▲, experiments done in the presence of 2 mM (GlcNAc)$_3$. (B) Inhibition by wheat germ agglutinin of* T. viride *spore germination in the absence (□) and presence (■) of 2 mM (GlcNAc)$_3$ (from ref. 55).*

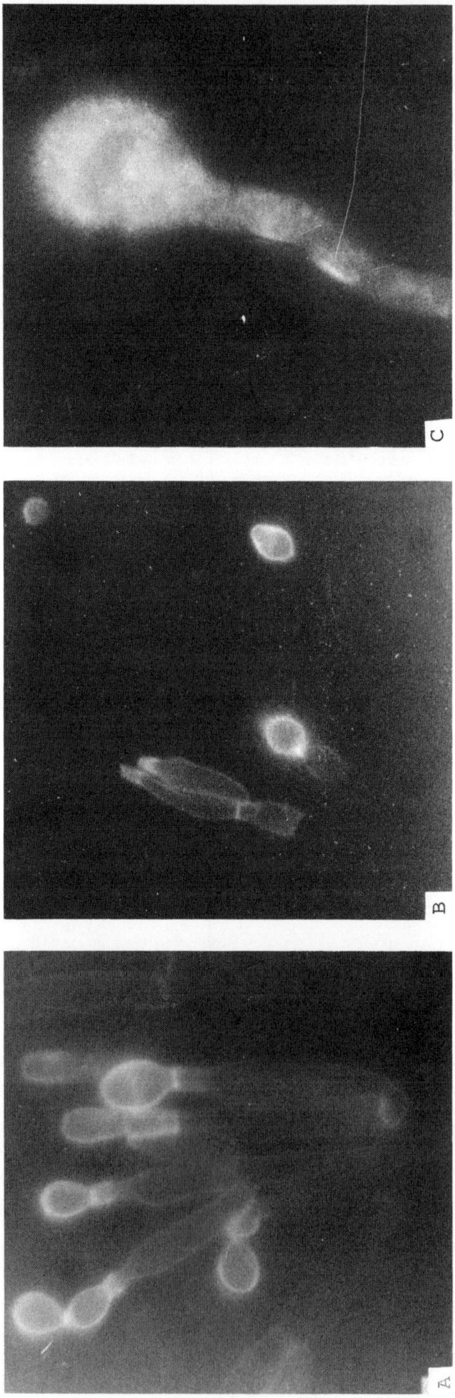
C
B
A

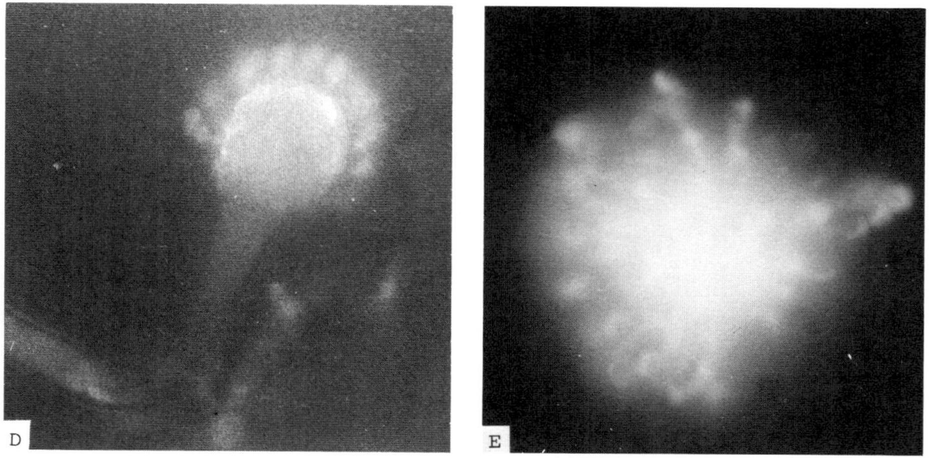

Fig 4. Binding of fluorescein isothiocyanate-labeled lectins to fungi: Wheat germ agglutinin to sterigmata and young spores of P. italicum (x1913) (A), and to sterigmata and young spores of P. expansum (x1913) (B). Peanut agglutinin to conidiophores and vesicle of A. flavus (x 600) (C), to a conidiophore and a head of A. flavus with a single series of sterigmata (x 600) (D), and to a head of A. ochraceus (upper view) (x1262) (E).

To assess the extent to which binding of lectins to fungi and their growth-inhibiting effects are indeed general phenomena, and to obtain additional information on the possiblity of applying lectins to probe fungal surfaces, we have recently investigated the interaction of several lectins with a large number of fungi belonging to different taxonomic groups (56). The lectins used in this study were soybean agglutinin, peanut agglutinin, and wheat germ agglutinin.

As a first step, the ability of the various fungi to bind the fluorescein derivatives of the lectins was examined (Fig. 4). Of the fungi that exhibited binding, representative species of *Penicillium* (*P. italicum*) and *Aspergillus* (*A. niger* and *A. ochraceus*) were selected for detailed examination of the effect of lectins on their growth and uptake of various nutrients.

Wheat germ agglutinin, peanut agglutinin, and soybean agglutinin inhibited the incorporation of sodium [^{3}H]acetate, *N*-acetyl-D-[1-^{3}H]glucosamine, and D-[1-^{14}C]galactose into young hyphae of *A. ochraceus* (Fig. 5). Incorporation was not affected when the lectins were preincubated with their specific inhibitors. With *A. niger* and *Stemphylium botryosum*, chitin fungi that bind soybean agglutinin and peanut agglutinin poorly, or not at all, only wheat germ agglutinin caused a marked inhibition of acetate incorporation.

The inhibition of incorporation of various precursors into the fungal hyphae by lectins of different specificities seems to be the result of coating of the hyphal surfaces, by binding of available surface receptors to the corresponding lectins.

Growth inhibition by the various lectins was indicated also by their effect on fungal spore-germination (Fig. 6). All lectins tested caused marked inhibition at 5 mg lectin/ml or more, which is a much higher concentration than that needed for the inhibition of precursor incorporation (Fig. 5). Inhibition of spore germination occurs probably at a very early stage of the germination process, after the spores swell and before initiation of detectable germ tubes. This inhibition was mainly expressed by the prolongation of the latent period which precedes germination (Fig. 7). However, once initiated, the rate of germination appeared to be normal and after 18 h of incubation the percent of lectin-treated spores that germinated was almost the same as that of the untreated ones.

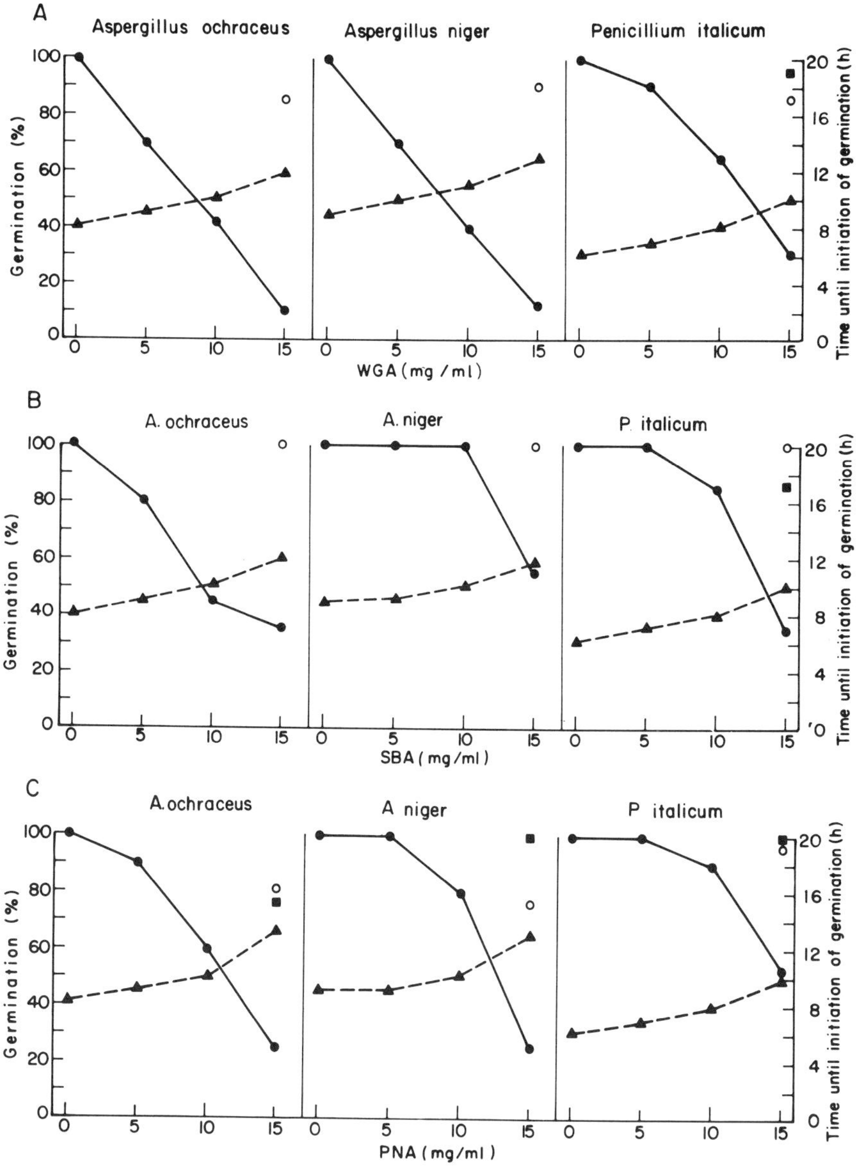

Fig. 5. Inhibition by lectins of precursor incorporation into young fungal hyphae. Precursors: (A) [^{3}H]acetate; (B) N-acetyl-D-[^{3}H]glucosamine; and (C) D-[^{14}C]galactose. Lectins: Δ, wheat germ agglutinin (WGA); □, soybean agglutinin (SBA); ○, peanut agglutinin (PNA). Percentage of incorporation obtained when lectins preincubated with specific inhibitors: ●, PNA + 0.2 M D-galactose; ▲, WGA + 2 mM (GlcNAc)$_3$; ■, SBA + 0.2 M GalNAc (from ref. 56).

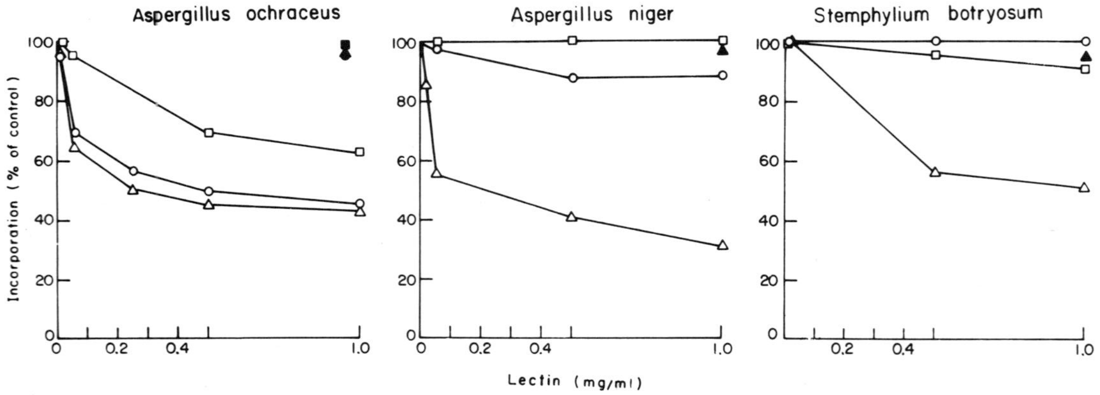

Fig. 6. Effect of lectins on spore germination in potato dextrose broth at 23°C. (A) WGA; (B) SBA; (C) PNA. Percentage of germination after 14 h, ● ●; after 14 h in the presence of specific sugar inhibitor, ; after 18 h, o. Specific inhibitors for lectins: 2 mM $(GlcNAc)_3$ for WGA; 0.2 M GalNAc for SBA; 0.2 M D-galactose for PNA. Time of initiation of germination, ▲------▲ (from ref. 56).

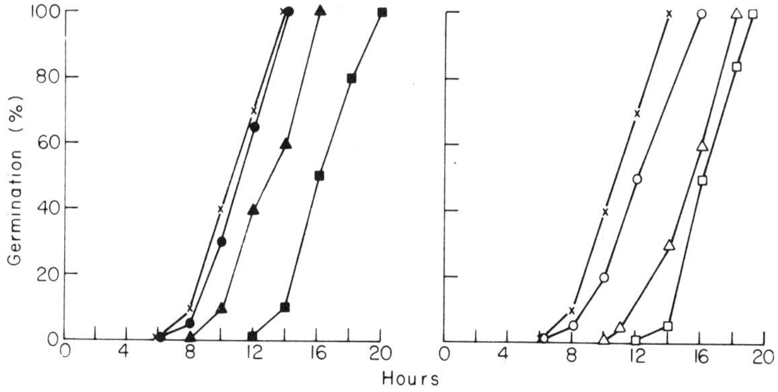

Fig. 7. Latent period of spore germination of A. ochraceus *in the presence of peanut agglutinin (●—● 5 mg/ml; ▲—▲ 10 mg/ml; ■—■ 15 mg/ml) or wheat germ agglutinin (o—o 5 mg/ml; Δ—Δ 10 mg/ml; □—□ 15 mg/ml), compared with untreated spores (x—x) (from ref. 56).*

These results clearly indicate that lectins may interfere with fungal growth. This lends support to our suggestion that lectins in plants are part of their protection system, helping them to combat attack by fungal pathogens. Further studies on the distribution of lectins in plants, in relation to plant resistance to fungal infection, and on the effect of lectins on different fungi are required in order to validate this hypothesis.

There have also been suggestions that lectins are involved in the defense of plants against invasion by pathogenic bacteria (57). When tobacco leaves are inoculated with avirulent strains of *Pseudomonas solanacearum*, a pathogen of tobacco and potato plants, the bacteria rapidly attach to the plant cell-walls and are then enveloped by the walls. Virulent bacteria, however, remain free in the spaces between the cells, where they multiply and spread. It has been therefore postulated that avirulent strains may be attaching to lectins in the plant. Accordingly, 55 virulent and 34 avirulent strains of *P. solanacearum* from different geographic regions and representing all major races and biotypes were tested for their interaction with the potato lectin (specific for *N*-acetyl-D-glucosamine oligomers). All avirulent isolates bound the lectin and agglutinated strongly, whereas virulent isolates did not. Binding of the lectin to avirulent cells was hapten-specific and could be inhibited by chitin oligomers.

Failure of the virulent bacteria to bind the lectin was correlated with the presence on their surface of an extracellular polysaccharide not formed by avirulent cells. Indeed, when most of the extracellular polysaccharide was removed from virulent cells by washing, the cells were agglutinated strongly by the lectin.

Studies with the red kidney bean have shown that the saprophytic bacterium *Pseudomonas putida* is immobilized and encapsulated in the intercellular spaces of the leaves of the plant, whereas the phytopathogenic bacteria *Pseudomonas phaseolicola* and *Pseudomonas tomato* did not adhere to the plant cell-walls, nor were they encapsulated (58). It was concluded that plant lectins may be involved in the immobilization and encapsulation, since only *P. putida* cells were agglutinated by PHA, the lectin from the red kidney bean.

Thus, the attachment of avirulent cells to host cell-walls in plants presumably involves recognition between plant lectins and carbohydrates on the bacterial surface. The attachment leads to envelopment of the avirulent bacteria by the host mesophyll cell-walls, thereby preventing infection. Virulent bacterial cells cannot be recognized and bound by the host lectin; they remain free in the intercellular spaces and multiply rapidly subsequent to invasion, becoming systemic within the host.

Because slight changes in chemical composition or accessibility of the bacterial polysaccharide and plant lectin may affect binding to each other, the potential for variability in the system is very high. Thus, mutations that affect any of these constituents could alter their specificity, and this may be expressed by the ability or inability of the bacteria to multiply in the plant tissue (57).

BACTERIAL ADHERENCE TO ANIMAL CELLS

Adherence or attachment to host tissues may play a crucial role not only in bacterial infection of plants but also in that of mammals. Recent studies in our laboratory and elsewhere strongly suggest that bacterial lectins are involved in what is perhaps the most important property of virulence possessed by microorganisms pathogenic to animals, namely their ability to adhere to and colonize the epithelial and endothelial cells of the host. The initial attachment of bacteria to the nasal or intestinal mucosa, for example, prevents them from being washed away by secretions bathing these surfaces. Paradoxically, however, it would appear that such lectins may also be responsible for the attachment of bacterial cells to macrophages, as a prelude to the ingestion of the bacteria and their elimination from the body.

We have investigated the adherence of several strains of *Escherichia coli* to human buccal epithelial cells (59). Such attachment can be readily demonstrated *in vitro*. Only cells collected from cultures 26-72-h old adhered well to the epithelial cells; on the average, 50-70 bacterial cells per epithelial cell could be seen under the microscope in Gentian Violet-stained preparations (Fig. 8).

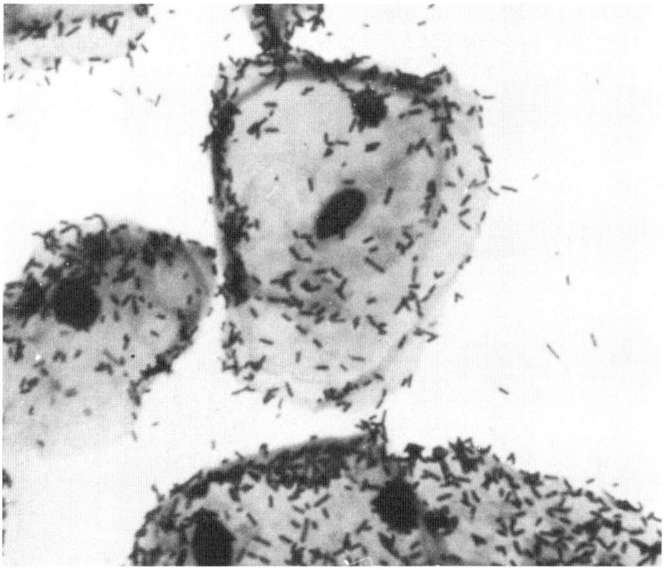

Fig. 8. Attachment of Escherichia coli *to human buccal epithelial cells. Preparation stained with Gentian Violet. Micrograph courtesy of I. Ofek and E. H. Beachey.*

Ten different sugars, as well as one polysaccharide (yeast mannan), were tested for their ability to inhibit the attachment; of the compounds tested, only D-mannose, methyl α-D-mannopyranoside, and yeast mannan were effective. At concentrations of 5 and 25 mg/ml, methyl α-D-mannopyranoside was 75% and 95% inhibitory, respectively, while D-mannose was somewhat less inhibitory (Fig. 9). The inhibition by methyl α-D-mannopyranoside of the adherence was dose related and was linear in the range of 2-6 mg of sugar per ml. Addition of D-mannose or methyl α-D-mannopyranoside (25 mg per ml) to epithelial cells, to which *E. coli* was preattached, caused displacement of the organisms from the epithelial cells within 2-5 min. Similarly, mannan (2 mg per ml) removed bacteria attached to the epithelial cells, whereas D-glucose, glycogen, or D-galactose failed to displace the bacteria. On remixing the washed cells, epithelial attachment of the bacteria was restored, indicating that the inhibition of adherence by the sugar is reversible and probably does not involve any irreversible alterations of either type of cell.

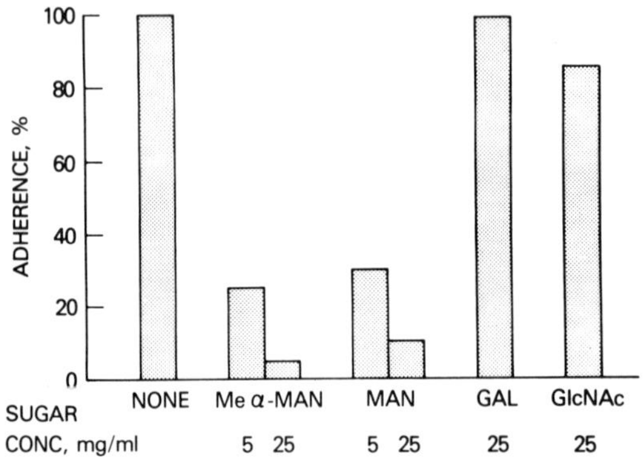

Fig. 9. Inhibition of adherence of Escherichia coli *to human oral epithelial cells by sugars (data from ref. 59).*

Additional experiments have shown that epithelial cells treated with sodium metaperiodate were no longer able to bind *E. coli*. Con A, which binds D-mannose and D-glucose residues, inhibited the attachment of the bacteria to the epithelial cell surfaces; the other lectins tested (wheat germ agglutinin specific for *N*-acetyl-D-glucosamine and peanut agglutinin specific for D-galactose) were not inhibitory. Both soluble yeast mannan and intact yeast cells (*Saccharomyces cerevisiae*), which are known to contain mannans on their cell surface, agglutinated *E. coli* cells. The agglutination was completely blocked by D-mannose or by methyl α-D-mannopyranoside at 25 mg per ml, but not by any other sugar tested at the same concentration.

We have now extended these findings, and have shown that the attachment of *E. coli* and *Salmonella typhi* to mouse peritoneal macrophages is specifically inhibited by D-mannose, methyl α-D-mannopyranoside, and yeast mannan (Table V) (60). D-Mannose and its derivatives also inhibited the attachment of *E. coli* to human polymorphonuclear leucocytes. The inhibition by D-mannose or methyl α-D-mannopyranoside of the attachment was linear in the range 0.1-2.5 mg of sugar per ml. The same pattern of binding and specific inhibition was obtained both at 37°C and 4°C, indicating that the sugars inhibited the attachment phase of phagocytosis, since the engulfment phase does not take place at the lower temperature. Mannan inhibited phagocytosis when preincubated with *E. coli*, but not when preincubated with leucocytes. In contrast to the non-opsonized bacteria, binding of opsonized bacteria to leucocytes was not inhibited by any of the compounds tested. The fact that opsonized bacteria bind to macrophages shows that, under the conditions used, sugars do not alter the normal biological behavior of phagocytes, and that recognition between both cells mediated by specific anti-*E. coli* antibodies and serum factors does not involve interaction between lectins and cell-surface sugars.

Table V. Inhibition of attachment of Escherichia coli *and* Salmonella typhi *to human and mouse phagocytes by carbohydrates*[a]

Bacterial strain	Inhibitor used[b]	Conc. in reaction mixture (mg/ml)	Attachment[c] to mouse peritoneal macrophages	Attachment[c] to human morpho-nuclear leucocytes
E. coli	D-Mannose	2.0	22	d
		20.0	10	27
	Methyl α-D-mannopyranoside	2.0	14	d
		20.0	6	29
	Mannan	2.0	12	d
		20.0	8	27
	Methyl α-D-glucopyranoside	2.0	74	d
		20.0		93
	L-Fucose	20.0	98	93
	Mannan	e	5	d
	Mannan	f	88	d
S. typhi	Methyl α-D-mannopyranoside	20.0	4	d
	D-Mannose	20.0	6	d
	L-Fucose	20.0	93	d

[a] *Data from ref. 60*

[b] *The percentage attachment in the presence of other sugars (20 mg/ml) was in all cases >90%. The sugars tested were: D-xylose, D-arabinose, D-galactose,* N-*acetyl-D-glucosamine, and* N-*acetyl-D-galactosamine*

[c] *Values were calculated as percentage attachment in the presence of carbohydrates compared with control (without inhibitor), which was taken as 100% attachment*

[d] *Not determined*

[e] *Pretreated bacteria*

[f] *Pretreated macrophage*

Our results suggest that the binding of *E. coli* to epithelial cells and macrophages is mediated by a mannose-specific, lectin-like substance present on the surface of *E. coli*, which binds to mannose-like receptor sites on the mammalian cells. Preliminary evidence for the presence in *E. coli* of a mannose-specific lectin has now been obtained by us. Extracts of *E. coli* K_{12} agglutinated yeast cells, and the agglutination was inhibited specifically by D-mannose or methyl α-D-mannopyranoside. Moreover, the extracts inhibited attachment of *E. coli* to epithelial cells, and the inhibitory activity was abolished by adsorption with yeast cells; also the agglutinating activity of the extract was abolished by heating or by incubation with trypsin. Further work on the purification and characterization of the *E. coli* lectin is in progress.

Production of mannose-specific lectins by members of Gram-negative organisms, including strains of *E. coli*, has been described (for example, see ref. 61). In most of these studies, the target host-cell was the erythrocyte. Duguid and Gillies (62) reported some years ago that D-mannose inhibited binding of *E. coli* to intestinal epithelial cells, but this was not investigated in any detail and was overlooked in subsequent studies (63, 64).

The findings just described show that lectins of certain bacteria may play a dual role in host-parasite relationship: the lectins enable the organisms to establish colonization on mucosal surfaces, by adhering to mannose residues on epithelial cells, as well as to recognize and attach to such residues on the surface of phagocytes, as a prelude to their ingestion and digestion by the latter cells. Increasing evidence has now become available for the generality of this concept, which may hold true both *in vivo* and *in vitro*.

Thus, it has been reported that adhesion of *Vibrio cholerae* to mucosal surfaces (*e.g.*, brush borders) and agglutination of human group O erythrocytes by these bacteria were specifically (although incompletely) inhibited by L-fucose and various glycosides of L-fucose, and to a lesser extent by D-mannose (65). *V. cholerae* adhered specifically to L-fucose covalently linked to agarose beads. It has been proposed that an L-fucose-specific lectin, to which the name adhesin was given, may be responsible, at least in part, for the initial attachment of *V. cholera* to mucosal surfaces, and may therefore be an important determinant of bacterial infectivity.

In vivo studies in intact animals carried out in our laboratory and elsewhere suggest that binding to cell-surface sugars may be a prerequisite for the colonization by *E. coli* of the urinary tract as well as of the intestinal tract.

Under suitable conditions, methyl α-D-mannopyranoside (but not methyl α-D-glucopyranoside) is effective in preventing experimental urinary tract infections by *E. coli* in mice (66). Thaler, Hirschberger, and Mirelman (67,68) have found that D-mannose and its derivatives inhibit specifically the adherence of *E. coli* to intestinal mucosa, not only in mucosal-cell suspensions, but also in closed loops of the intestine of young rabbits and rats. Moreover, addition of D-mannose 30 min after the bacteria had been added also inhibited attachment to the intestine. They have suggested that binding of the bacteria (*via* the bacterial lectin) to specific cell-surface sugars may play a role in rapid intestinal colonization after birth, and in persistences of infantile diarrhea. D-Mannose and its derivatives may thus be potentially useful agents in the treatment of such conditions.

These studies provide a novel approach to the prevention of bacterial colonization: rather than killing the pathogen with antibiotics and other drugs, or mobilizing the immune defenses of the body with vaccines, it may be possible to interrupt the infectious process at the crucial stage, when the pathogen attempts to adhere to its target cell.

ROLE IN SEXUAL MATING AND IN DIFFERENTIATION AND DEVELOPMENT

Additional proposals for the roles of lectins in Nature have recently been made, all of which support the hypothesis that their main function is that of recognition.

There are indications that lectins or lectin-like materials are involved in the initial attachment of opposite mating types in microorganisms, such as yeast (69) and chlamydomonas (70), of stigma to pollen in plants (71), and of sea urchin eggs to sperm (72,73).

Several lectins appear to be responsible for intercellular species-specific adhesion in the organisms from which they are isolated. For example, the cell aggregation factor from certain sponges is specifically inhibited by D-glucuronic acid (74,75). In the cellular slime-molds, *Dictyostelium discoideum* and *Polysphondylium pallidum*, lectins appear as the organisms differentiate from a vegetative to a cohesive state (76,77; reviewed in ref. 78). The carbohydrate specificities of the lectins from the two species are different, and it is therefore possible that the slime-mold lectins are involved in species-specific recognition.

The finding of a β-D-galactoside-specific lectin in the electric eel and the demonstration that lectins of similar specificity occur in relatively high concentrations in tissue cultures of nerve and muscle cells provided grounds for the hypothesis that such lectins may mediate the fusion of chick myoblasts. This hypothesis seems, however, to be incompatible with the finding that these lectins do not require Ca^{++} for activity, whereas myoblast fusion is a strictly Ca^{++}-dependent phenomenon (12).

CONCLUDING REMARKS

In this article, I have discussed possible roles of lectins in Nature, with special emphasis on work from our laboratory on the inhibition of fungal growth by lectins and on bacterial adherence to D-mannose residues on the surface of epithelial cells and macrophages. The survey is by no means exhaustive. For example, there have been suggestions that lectins protect plants from insect predators (79), serve as structural elements in membranes (sort of "glue" holding together glycoprotein constituents of membranes) (33,80), or mediate cell extension during growth in plants (20,21). These and other proposals have not been mentioned, mainly because the evidence available is very meager.

The elucidation of the function of naturally occurring compounds is often a difficult task, especially when the problem cannot be tackled with the aid of genetic mutants. There are many examples of well known and extensively characterized natural products whose role is still a mystery. Lectins are no exception; in spite of the marked increase of our knowledge on their occurrence and their molecular and biological properties, there is as yet no firm evidence for any of the roles postulated for them.

Studies of lectins are becoming all the more important, not only because of the great usefulness of these proteins in biological research and in the clinic, but because of their possible involvemnt in host-parasite relationships both in animals and in plants, and in symbiotic associations between leguminous plants and bacteria. It is not difficult to envisage that such research will lead to important practical applications in medicine and agriculture.

ACKNOWLEDGMENT

This article was written during tenure of a John E. Fogarty International Center Scholarship-in-Residence at the National Institutes of Health, from August 1, 1977 to March 31, 1978.

REFERENCES

1. Sharon, N., *Sci. Am. 236*, (6) 108 (1977).
2. Lis, H., and Sharon, N., *in* "The Antigens" (M. Sela, ed.), vol. 4, p. 429, Academic Press, New York, (1977).
3. Reisner, Y., Linker-Israeli, M., and Sharon, N., *Cell. Immunol. 25*, 129 (1976).
4. Reisner, Y., Ravid, A., and Sharon, N., *Biochem. Biophys. Res. Commun. 72*, 1585 (1976).
5. Lis, H., and Sharon, N., *Annu. Rev. Biochem. 42*, 541 (1973).
6. Rapin, A. M. C., and Burger, M. M., *Adv. Cancer Res. 20*, 1 (1974).
7. Nicolson, G. L., *Int. Rev. Cytol. 39*, 89 (1974).
8. Stockert, R. S., Morell, A. G., and Scheinberg, H. I., *Science 186*, 365 (1974).
9. Hudgin, R. L., Pricer, W. E., Jr., Ashwell, G., Stockert, R. S., and Morell, A. G., *J. Biol. Chem. 249*, 5536 (1974).
10. Teichberg, V., Silman, I., Beitsch, D. D., and Resheff, G., *Proc. Natl. Acad. Sci. U.S.A. 72*, 1383 (1975).
11. de Waard, A., Hickman, S., and Kornfeld, S., *J. Biol. Chem. 251*, 7581 (1976).
12. Den, H., and Malinzak, D. A., *J. Biol. Chem. 252*, 5444 (1977).
13. Novak, T. P., Kobiler, D., Roel, L. E., and Barondes, S. H., *J. Biol. Chem. 252*, 6026 (1977).
14. Kawasaki, T., and Ashwell, G., *J. Biol. Chem. 252*, 6536 (1977).
15. Bowles, D. J., and Kauss, H., *FEBS Lett. 66*, 16 (1976).
16. Pricer, W. E., and Ashwell, G., *J. Biol. Chem. 251*, 7539 (1976).
17. Bowles, D. J., and Kauss, H., *Plant Sci. Lett. 4*, 411 (1975).
18. Bowles, D. J., and Kauss, H., *Biochim. Biophys. Acta 443*, 360 (1976).

19. Bowles, D. J., Schnarrenberger, C., and Kauss, H., *Biochem. J. 160*, 375 (1976).
20. Kauss, H., and Glaser, C., *FEBS Lett. 45*, 304 (1974).
21. Kauss, H., and Bowles, D. J., *Planta (Berlin) 130*, 169 (1976).
22. Bowles, D. J., and Sharon, N., unpublished results.
23. Pueppke, S. G., Bauer, W. D., Keegotra, K., and Ferguson, A. L., *Plant Physiol. 61*, 779 (1978).
24. Lis, H., Sharon, N., and Katchalski, E., *J. Biol. Chem. 241*, 684 (1966).
25. Lis, H., Sharon, N., and Katchalski, E., *Biochim. Biophys. Acta 192*, 364 (1969).
26. Lotan, R., Siegelman, H. W., Lis, H., and Sharon, N., *J. Biol. Chem. 249*, 1219 (1974).
27. Lis, H., and Sharon, N., *J. Biol. Chem. 253*, 3468 (1978).
28. Sharon, N., Lis, H., and Lotan, R., *Colloq. Int. C.N.R.S. 221*, 693 (1974).
29. Foriers, A., Wuilmart, C., Sharon, N., and Strosberg, A. D., *Biochem. Biophys. Res. Commun. 75*, 980 (1977).
30. Allen, A. K., Neuberger, A., and Sharon, N., *Biochem. J. 131*, 155 (1973).
31. Allen, A. K., and Neuberger, A., *Biochem. J. 135*, 307 (1973).
32. Boyd, W. C., *Vox Sang. 8*, 1 (1963).
33. Sharon, N., and Lis, H., *Science 177*, 949 (1972).
34. "Cell Wall Biochemistry Related to Specificity in Host-Plant Pathogen Interactions" (B. Solheim and J. Raa, eds.), p. 487, Universitetsforlaget, Oslo, (1977).
35. Kauss, H, and Ziegler, H., *Planta (Berlin) 121*, 197 (1974).
36. Kauss, H., in ref. 34, p. 347.
37. Southworth, D., *Nature (London) 258*, 600 (1975).
38. Vasil, I. K., and Hubbel, D. H., in ref. 34, p. 361.
39. Nagl, W., *Planta (Berlin) 106*, 269 (1972).
40. Nagl, W., *Exp. Cell. Res. 74*, 599 (1972).
41. Ashwell, G., and Morell, A. G., *Trends Biochem. Sci. (TIBS) 2*, 76 (1977).
42. Ashwell, G., *Trends Biochem. Sci. (TIBS) 2*, N186 (1977).
43. Marx, J. L., *Science 196*, 1429 and 1478 (1977).
44. Hamblin, J., and Kent, S. P., *Nature New Biol. 245*, 28 (1973).
45. Bohlool, B. B., and Schmidt, E. L., *Science 185*, 269 (1974).

46. Bauer, W. D., Bhuvaneswari, T. V., and Pueppke, S. G., in ref. 34, p. 377.
47. Bhuvaneswari, T. V., Pueppke, S. G., and Bauer, W. D., *Plant Physiol. 60*, 486 (1977).
48. Fountain, D. W., Foard, D. E., Repolgle, W. D., and Yang, W. K., *Science 197*, 1185 (1977).
49. Wolpert, J. S., and Albersheim, P., *Biochem. Biophys. Res. Commun. 70*, 729 (1976).
50. Brethauer, T. S., and Paxton, J. D., in ref. 34, p. 381.
51. Chen, A. T., and Philips, D. A., *Physiol. Plant. 38*, 83 (1976).
52. Dazzo, F. B., and Hubbel, D. H., *Plant Soil. 43*, 713 (1975).
53. Dazzo, F. B., and Hubbel, D. H., *Appl. Microbiol. 30*, 1017 (1975).
54. Dazzo, F. B., and Brill, W. J., *Appl. Environmental Microbiol. 33*, 132 (1977).
55. Mirelman, D., Galun, E., Sharon, N., and Lotan, R., *Nature (London) 256*, 414 (1975).
56. Barkai-Golan, R., Mirelman, D., and Sharon, N., *Arch. Microbiol. 116*, 119 (1978).
57. Sequiera, L., and Graham, T. L., *Physiol. Plant Pathol. 11*, 43 (1977).
58. Sing, V. O., Schroth, M. N., *Science 197*, 759 (1977).
59. Ofek, I., Mirelman, D., and Sharon, N., *Nature (London) 269*, 623 (1977); *ibid.* "News and Views", p. 584.
60. Bar-Shavit, Z., Ofek, I., Goldman, R., Mirelman, D., and Sharon, N., *Biochem. Biophys. Res. Commun. 78*, 455 (1977).
61. Old, D. C., *J. Gen. Microbiol. 71*, 149 (1972).
62. Duguid, J. P., and Gillies, R. R., *J. Path. Bact. 74*, 397 (1957).
63. Gibbons, R. J., and van Houte, J., *Annu. Rev. Microbiol. 29*, 19 (1975).
64. Gibbons, R. J., *Microbiology 1977*, 395 (1977).
65. Jones, G. W., and Freter, R., *Inf. Immun. 14*, 240 (1976).
66. Aronson, M., Medalia, O., Mirelman, D., Sharon, N., and Ofek, I., manuscript in preparation.
67. Thaler, M. M., Hirschberger, M., and Mirelman, P., *Clin. Res. 25*, 469A (1977).
68. Hirschberger, M., Mirelman, D., and Thaler, M. M., *Gastroenterology 72*, 1069 (1977).
69. Yen, P. H., and Ballou, C. E., *Biochemistry 13*, 2428 (1974).

70. Wiese, L., and Wiese, W., *Develop. Biol. 43*, 264 (1975).
71. Knox, R. B., Clarke, A., Harrison, S., Smith, P., and Marchalonis, J. J., *Proc. Natl. Acad. Sci. U.S.A. 73*, 2788 (1976).
72. Schmell, E., Earles, B. J., Breaux, C., and Lennarz, W. J., *J. Cell Biol. 72*, 35 (1977).
73. Vacquier, V. D., and Moy, G. W., *Proc. Natl. Acad. Sci. U.S.A. 74*, 2456 (1977).
74. Turner, R. S., and Burger, M., *Nature (London) 244*, 509 (1973).
75. Muller, W. E. G., Muller, I., Zahn, R. K., and Kurelec, B., *J. Cell Sci. 21*, 227 (1976).
76. Rosen, S. D., Kafka, J. A., Simpson, D. L., and Barondes, S. H., *Proc. Natl. Acad. Sci. U.S.A. 70*, 2554 (1973).
77. Frazier, W. A., Rosen, S. D., Reitherman, R. W., and Barondes, S. H., *J. Biol. Chem. 250*, 7714 (1975).
78. Barondes, S. H., and Rosen, S. D., *in* "Neuronal Recognition" (S. H. Barondes, ed.), p. 331, Plenum Press, New York, (1976).
79. Janzen, D. H., Juster, H. B., and Liener, I. B., *Science 192*, 795 (1976).
80. Bowles, D. J., and Hanke, D. E., *FEBS Lett. 82*, 34 (1977).

Influence of Bovine Tendon Glycoprotein on Collagen Fibril Formation

John C. Anderson, Rhona I. Labedz, and Michael A. Kewley

It is possible that the variation in morphology of collagen fibres of the same collagen type in different tissues may be controlled by macromolecules of the extracellular matrix, such as proteoglycans and glycoproteins. Proteoglycans have been shown to have the ability to influence the formation of collagen fibrils *in vitro* (1).

A glycoprotein fraction of approximate mol. wt. 60 000 was isolated from bovine flexor tendon as described by Anderson (2). Alternatively, it was prepared by digestion of saline-extracted tendon with protease-free bacterial collagenase (3). The glycoprotein was obtained from the digest supernatant by affinity chromatography on Concanavalin A-Sepharose followed by chromatography on Sephadex G-200 (2).

The interaction between bovine tendon glycoprotein and calf skin acid-soluble collagen (1 mg/ml) was investigated by following fibril formation at physiological ionic strength, pH, and temperature (4). Typical results are shown in Fig. 1; the presence of glycoprotein (80 and 160 μg/mg of collagen) retards fibril formation and decreases the final absorbance of the collagen gel.

The influence of increasing concentrations of glycoprotein on the specific absorbance ΔA_{sp} and the specific retardation R_{sp} is shown in Figs. 2 and 3. Both $-\Delta A_{sp}$ and R_{sp} increase with increasing glycoprotein-to-collagen ratio, and the relationship between $-\Delta A_{sp}$ and glycoprotein-to-collagen ratio appears to be sigmoidal. Both curves would presumably level off at higher concentrations of glycoprotein, and indicate that there is only a weak interaction between glycoprotein and collagen. Such weak binding is confirmed by the observation that glycoprotein is extracted from tendon by physiological saline solution.

ISBN 0-12-301301-1

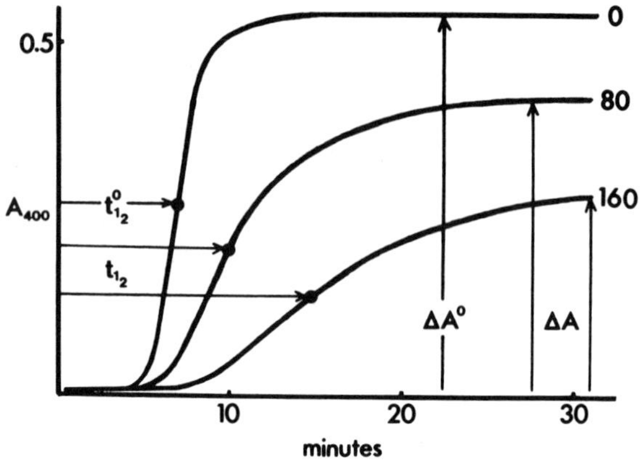

Fig. 1. Interaction between bovine tendon glycoprotein and calf skin acid-soluble collagen.

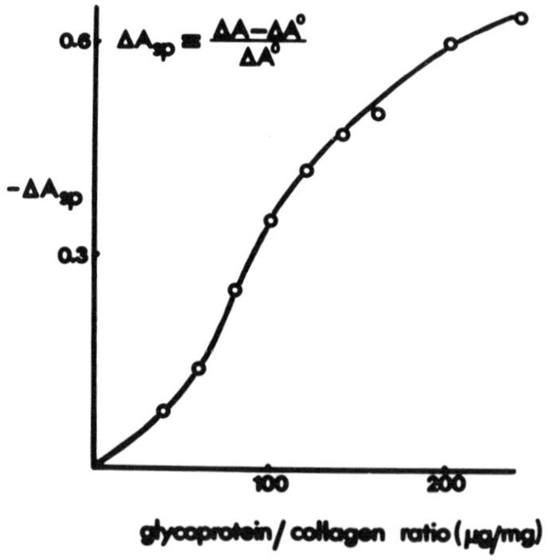

Fig. 2. Influence of glycoprotein concentration on specific absorbance.

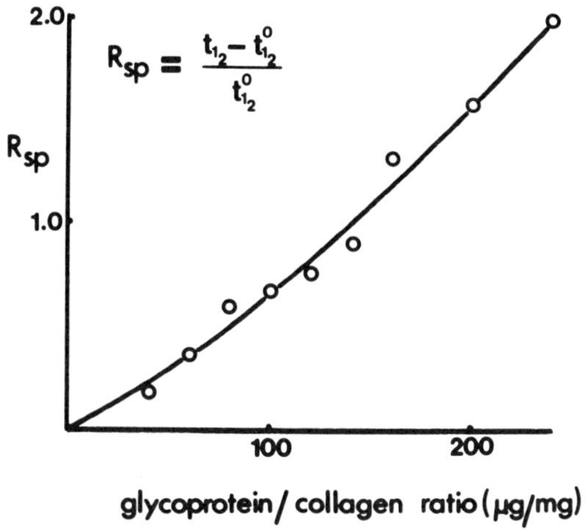

Fig. 3. Influence of glycoprotein concentration on specific retardation.

When glycoprotein (100 μg/ml of collagen) was added 2 min after the start of the incubation, it was still able to exert its effect fully (Fig. 4). Even addition at 4.3 min still produced full retardation, although the decrease in absorbance was less. It is thus evident that glycoprotein influences the later stages of fibril formation.

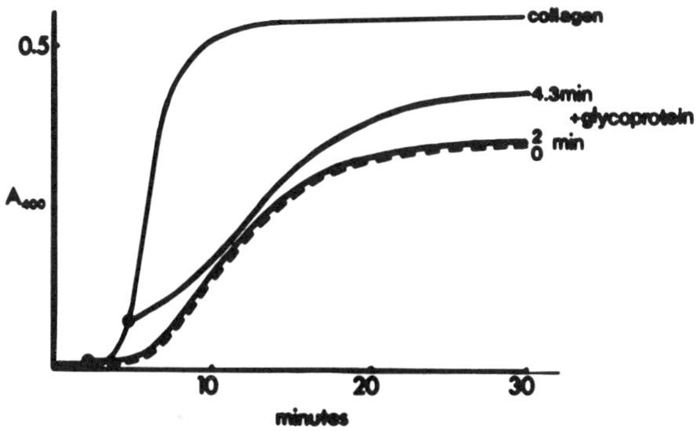

Fig. 4. Effect of lapse of time on fibril formation.

Electron micrographs (x 55 000) of fibrils formed in the absence and presence of glycoprotein are shown in Figs. 5 and 6, respectively. The average width of fibrils formed in the absence of glycoprotein was 33.5 ± 5.5 nm, while in the presence of glycoprotein fibrils had an average width of 26.6 ± 3.3 nm. Two different statistical treatments showed that the fibrils formed in the presence of glycoprotein were significantly thinner ($p<0.001$) than those in control collagen gels. This difference probably accounts for the decreased absorbance of collagen gels containing glycoprotein.

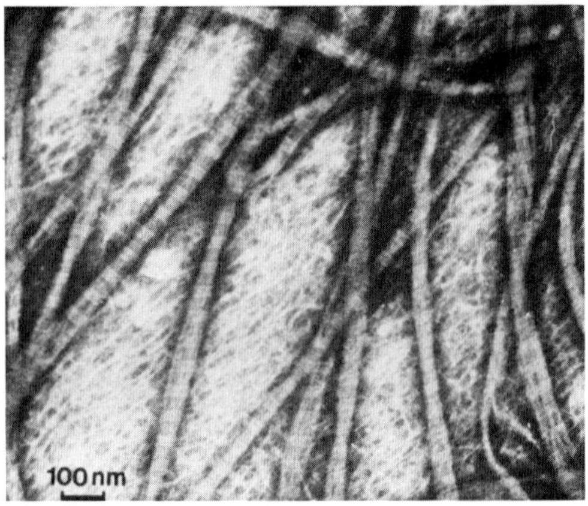

Fig. 5. E.m. of fibrils in the absence of glycoprotein.

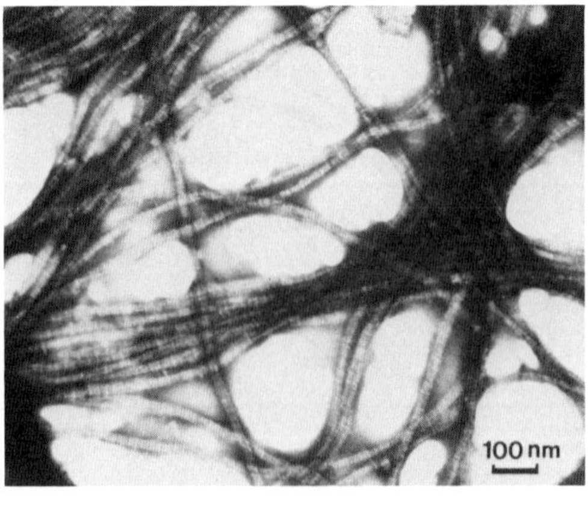

Fig. 6. E.m. of fibrils in the presence of glycoprotein.

We cannot say whether our observations have any relevance *in vivo* until we have investigated further tissues and we know the concentration of glycoprotein in each tissue and how its concentration varies during development of the tissue.

REFERENCES

1. Lowther, D. A., and Natarjan, M., *Biochem. J. 127*, 607 (1972).
2. Anderson, J. C., *Biochim. Biophys. Acta 379*, 444 (1975).
3. Lee-Own, V., and Anderson, J. C., *Prep. Biochem. 5*, 229 (1975).
4. Anderson, J. C., Labedz, R. I., and Kewley, M. A., *Biochem. J. 168*, 345 (1977).
5. Oegema, T. R., Laidlaw, J., Hascall, V. C., and Dziewiatkowski, D., *Arch. Biochem. Biophys. 170*, 698 (1975).

GLYCOCONJUGATE RESEARCH / Volume I
Proceedings of the Fourth International Symposium on Glycoconjugates

Glycosaminoglycans and Sialoglycopeptides Associated with Mammalian Cell Nuclei

Veerasingham P. Bhavanandan

The characterization of the glycosaminoglycans associated with the nuclei of B16 mouse melanoma cells has been described previously (1). The major nuclear glycosaminoglycan component is a chondroitin 4-sulfate, which was further characterized by use of the milligram quantities isolated from spent media. The o.r.d. curve and i.r. spectrum of the melanoma chondroitin sulfate resemble those of cartilage chondroitin 4-sulfate. However, the mol. wt. of the melanoma chondroitin sulfate was estimated to be in the range of 90 000-120 000 by sedimentation equilibrium analysis, whereas the corresponding values for rib cartilage chondroitin sulfate is 12 000-15 000 (2). The minor nuclear glycosaminoglycans were characterized as heparan sulfate and chondroitin sulfate of lower mol. wt.

The sialoglycopeptides associated with the mouse melanoma nuclei were fractionated into two classes (I and II) by use of cetylpyridinium chloride (Scheme 1). The class I glycopeptides differed from those of class II, which are not precipitated with cetylpyridinium chloride, in greater molecular size and charge (sialic acid content) and affinity for wheat germ agglutinin. The distribution of the radioactivity in these glycopeptides was as follows: sialic acid (50%), galactosamine (43%), and glucosamine (7%) for class I; and sialic acid (36%), galactosamine (23%), and glucosamine

ISBN 0-12-301301-1

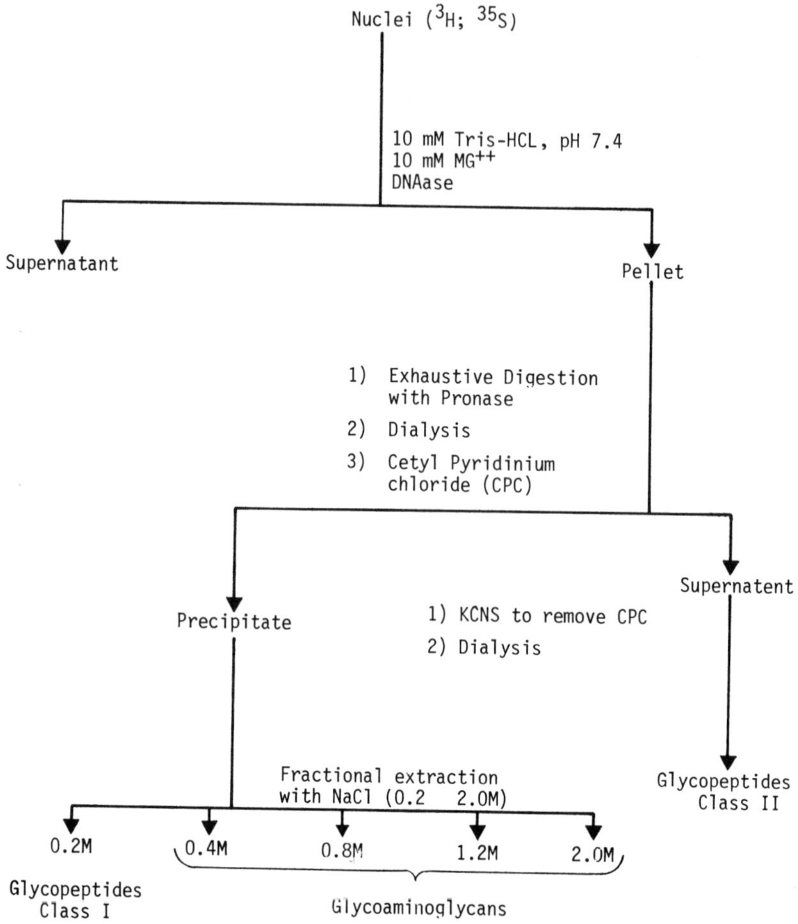
Nuclei (3H; 35S)
10 mM Tris-HCL, pH 7.4
10 mM MG++
DNAase
Supernatant
Pellet
1) Exhaustive Digestion with Pronase
2) Dialysis
3) Cetyl Pyridinium chloride (CPC)
Precipitate
Supernatent
1) KCNS to remove CPC
2) Dialysis
Fractional extraction with NaCl (0.2 2.0M)
Glycopeptides Class II
0.2M
0.4M
0.8M
1.2M
2.0M
Glycopeptides Class I
Glycoaminoglycans

Scheme 1.

(41%) for class II. About 85% of the radioactivity was in alkali-labile oligosaccharides in class I compared to 43% in class II. The sialoglycopeptides isolated from nuclei had a higher percentage of label in sialic acid than the sialoglycopeptides isolated from spent media. These preliminary data indicate that the nuclear sialoglycoproteins are richer in sialic acid and galactosamine as compared to the sialoglycoproteins present in the medium.

The nuclei isolated in our studies were free of obvious contamination by other organelles, as judged by phase-contrast and electron microscopy and by assays for marker enzymes (5'-nucleotidase, acid phosphatase, and succinic dehydrogenase). The nuclei, which were freed of the outer membrane, and therefore also of nonnuclear membrane contaminants, by treatment with 2.0% Triton X-100 (3), still retained between 70 to 80% of the initial radioactivity. Further, chromatin isolated from ^{3}H- and ^{35}S-labeled mouse melanoma nuclei yielded, on digestion with Pronase, the same complex saccharides as those obtained from the intact nuclei. These results indicate that glycosaminoglycans and sialoglycoproteins are integral components of the nuclei. In other control experiments, soluble glycoconjugates and membrane fractions prepared from cells labeled with [^{3}H]glucosamine and $Na_2{}^{35}SO_4$ were added to unlabeled cells prior to homogenization and isolation of nuclei. The results show that there was negligible adsorption of soluble glycoconjugates on the purified nuclei and that the contribution from nonnuclear membrane contaminants could not account for more than 10% of the radioactivity associated with the nuclei isolated by our procedures.

In preliminary studies, evidence for the presence of glycosaminoglycans associated with the nuclei of rat liver, Morris hepatoma 7777, human breast cells, and human breast cancer cells has been obtained. The presence of glycosaminoglycans associated with the nuclei of HeLaS_3 cells (4), rat brain (5), and human skin fibroblasts (6) has been reported. Although the significance of these findings remains to be determined, the present evidence strongly suggests that the association of glycosaminoglycans with the cell nuclei may be a general phenomenon.

ACKNOWLEDGMENTS

I am grateful to Mr. Jeff Kemper for excellent technical assistance. This work is supported in part by USPHS Grant CA 17686.

REFERENCES

1. Bhavanandan, V. P., and Davidson, E. A., *Proc. Natl. Acad. Sci. U.S.A. 72,* 2032 (1975).
2. Bhavanandan, V. P., and Davidson, E. A., *Carbohydr. Res. 57,* 173 (1977).
3. Blobel, A., and Potter, V. R., *Science 154,* 1662 (1966).
4. Stein, G. S., Roberts, R. M., Davis, J. L., Head, W. J., Stein, J. L., Thrall, C. L., van Veen, J., and Welch, D. W., *Nature (London) 258,* 639 (1975).
5. Margolis, R. K., Crockett, C. P., Kiang, W.-L., and Margolis, R. U., *Biochim. Biophys. Acta 451,* 465 (1976).
6. Fromme, H. G., Buddecke, E., v. Figura, K., and Kresse, H., *Exp. Cell Res. 102,* 445 (1976).

Effect of Blood Group Determinants on Binding of Human Salivary Mucous Glycoproteins to Influenza Virus

Thomas F. Boat, James Davis, Robert C. Stern, and PiWan Cheng

Sialoglycoproteins inhibit the agglutination of red blood cells by myxoviruses. The inhibitory activity is determined by the *N*-acetylneuraminic acid content and molecular size of the glycoprotein, and by the extent to which glycoprotein and viral surfaces are "complementary" (1-3). We have demonstrated that the influenza B virus hemagglutination inhibition (VHI) activity in human saliva is destroyed by neuraminidase and is associated with the mucous glycoprotein (void-volume fraction) after Bio-Gel A-5m fractionation of this secretion. This fraction contains more than 80% of the total nondialyzable sialic acid in saliva. The amount of sialoglycoprotein required to inhibit virus hemagglutination is greater for saliva from secretors of blood group A and B substances than for saliva from secretors of H and Le^a substances (Fig. 1). Because the mean sialic acid content of saliva samples from secretors of the 4 blood-group substances does not vary significantly (22-40 μg/ml), the observed differences in VHI activity cannot be attributed to differences of sialic acid content. Treatment of 4 saliva samples containing blood-group B substance with α-galactosidase from *Ruminococcus AB* (4) decreases blood-group B titers 2^6-2^7 fold, increases blood group H titers 2^2-2^3 fold, and increases VHI activity 2^2-2^3 fold. Further evidence for a reciprocal relationship between blood group titers and VHI activity is found by study of the saliva of individuals with cystic fibrosis (CF). Fig. 2 demonstrates that whereas more saliva from CF than from non-CF blood-group A secretors is required to inhibit agglutination of type A red blood cells with anti-A serum, CF saliva is a 20-fold more potent inhibitor of influenza virus hemagglutination.

ISBN 0-12-301301-1

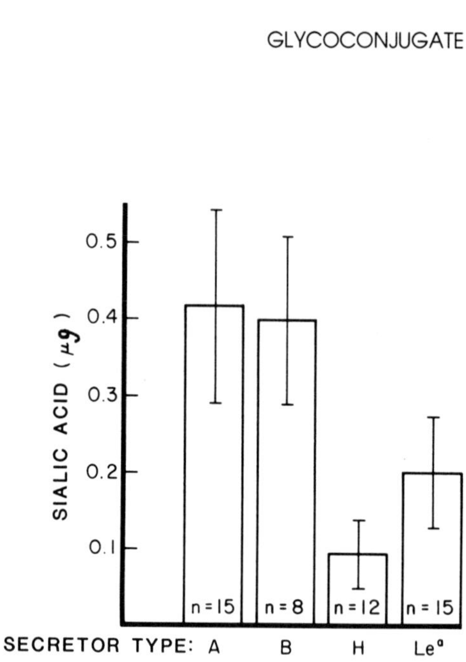

Fig. 1. Amount of glycoprotein sialic acid in 50 µl of saliva required to inhibit 2 hemagglutinating units of influenza B virus ($p<0.01$ for H vs. *A or B, and* Le^a vs. *A or B blood group substances). Sialic acid was determined by the method of Warren (6).*

The mean concentration of nondialyzable sialic acid is only 2-fold higher in CF saliva samples (72.8 *vs.* 35.0 µg/ml), indicating that increased sialic acid content alone cannot account for differences in VHI activities. Because the secretory rate of salivary proteins and glycoproteins is elevated in CF (5), the low blood-group titers and high VHI activities of CF salivary glycoprotein may be the consequence of incomplete oligosaccharide-chain synthesis.

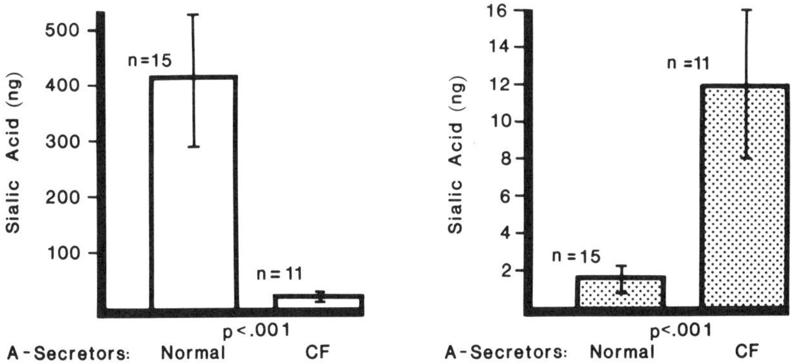

Fig. 2. (Left) Amount of glycoprotein sialic acid in 50 μl of saliva from normal and CF secretors of blood-group A substance required to inhibit 2 hemagglutinating units of influenza B virus. Donors ranged in age from 16 to 42 years; similar numbers of males and females were represented in both groups. (Right) Amount of glycoprotein sialic acid in 50 μl of saliva from normal and CF donors required to inhibit 2 hemagglutinating units of anti-A serum.

We conclude that terminal α-linked D-galactose, and probably α-linked *N*-acetyl-D-galactosamine residues, may interfere with the access of the sialic acid binding-sites of influenza virus to the sialic acid-containing oligosaccharide chains of salivary mucins. These studies provide evidence for a mechanism whereby "complementariness" of mucous glycoproteins and influenza virus surfaces may vary.

REFERENCES

1. Springer, G.F., Schwick, H.G., and Fletcher, M.A., *Proc. Natl. Acad. Sci. U.S.A. 64,* 634 (1969).
2. Morawiecki, A., and Lisowska, E., *Biochim. Biophys. Res. Commun. 18,* 606 (1965).
3. Fazekas de St. Groth, S., and Gottschalk, A., *Biochim. Biophys. Acta. 78,* 248 (1963).
4. Hoskins, L.C., and Boulding, E.T., *J. Clin. Invest. 57,* 63 (1976).
5. Chernick, W.S., Eichel, H.J., and Barbero, G.J., *J. Pediat. 65,* 694 (1964).
6. Warren, L., *J. Biol. Chem. 234,* 1971 (1959).

Fibronectin in Basement Membranes and Acidic Structural Glycoproteins of Lung and Placenta

Bonnie Anderson Bray

Fibronectin is a new name for a fibroblast surface antigen also known as LETS protein. Cold-insoluble globulin (CIG) of plasma (1) has been shown to be antigenically identical to fibronectin (2), and is called plasma fibronectin. The interaction of fibronectin with collagen (3) and fibrin (2,4), and the observation that plasma transglutaminase can cross-link it into high mol. wt. multimers (4,5) suggest that fibronectin may influence the organization of connective tissues and basement membranes. By immunofluorescence, it has been detected in the connective tissue and basement membranes of the developing chick embryo (6). In this study, fibronectin has been detected in isolated basement membranes and acidic structural glycoproteins (ASGS) from human placenta and lung by use of an antiserum to CIG.

MATERIALS AND METHODS

Human basement membranes from terminal villi of placenta, lung parenchyma, and renal cortex were prepared by sieving the blood-free tissues and sonicating the sieved portion (7,8). The membranes were digested with elastase or highly purified collagenase to provide soluble fragments for immunodiffusion tests (8). ASGS were extracted from isolated trophoblast basement membrane (TBM), as well as from terminal villi of placenta (7) and from lung parenchyma (9), with 0.3 M acetic acid at 4°C, and were recovered by isoelectric precipitation at pH 4.7 and dissolved in 0.2 M glycine, pH 8.0. ASGS from placental villi were chromatographed on DEAE-cellulose (7), and the material eluted with 0.1 M acetic acid was used here as ASG-DEAE (Fig. 1). Specific rabbit antiserum to human CIG (1) was a gift from Dr. Mosesson. Reference CIG was prepared from Cohn Fraction I by clotting out the fibrinogen with thrombin.

ISBN 0-12-301301-1

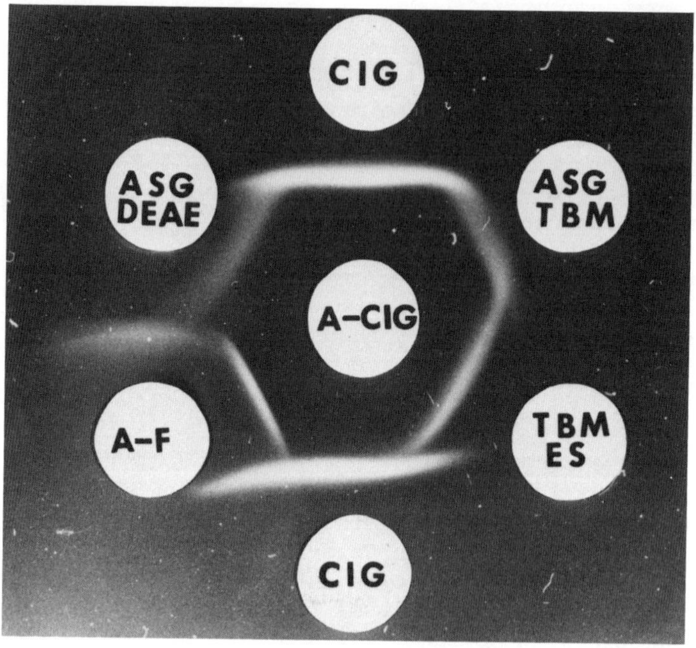

Fig. 1. Immunodiffusion reactions in agarose between anti-CIG antiserum (center well) and placental fractions. Clockwise from the top, the outer wells contained: Reference CIG; ASG from isolated TBM; elastase-solubilized (ES) fragments from TBM; reference CIG; antiserum to human fibrinogen (A-F); and ASG from terminal villi partially purified on DEAE-cellulose.

RESULTS AND DISCUSSION

Fibronectin was detected in isolated TBM and alveolar basement membrane (ABM) but not in glomerular basement membrane (GBM). The intact antigen could be extracted from TBM with 0.3 M acetic acid (ASG-TBM; Fig. 1), but was degraded when elastase was used to solubilize TBM (TBM-ES). Collagenase solubilized the intact antigen from ABM (not shown), but neither collagenase nor elastase produced reactive fragments from GBM. The failure to detect fibronectin in isolated GBM is one of several lines of evidence suggesting that fibronectin is part of the structure of microfibrils, since these are more prominent in TBM (7,10) and ABM (11) than in GBM.

ASG fractions from both placenta and lung contained fibronectin along with antigenic determinants of fibrinogen. The placental ASG fraction (ASG-DEAE; Fig. 1) gave precipitin lines with both anti-CIG (center well) and antifibrinogen (A-F). The lines fused, suggesting that both types of determinants (CIG as well as fibrinogen) were present in the same molecular species. [The anti-CIG contained fibrinogen as a result of absorption (1)]. An association of fibronectin with collagen in both placenta and lung was suggested by the observation that collagenase solubilized the antigen from both tissues.

Fibronectin is concentrated in amniotic fluid relative to maternal and fetal plasma (12), and is probably synthesized by the fetal trophoblast. The hypothesis is advanced that fibronectin may contribute to the formation of the immunological barrier (13) between mother and fetus.

ACKNOWLEDGMENT

Supported by USPHS Grant 15832 and aided by a grant from the New York Lung Association.

REFERENCES

1. Mosesson, M.W., and Umfleet, R.A., *J. Biol. Chem. 245,* 5728 (1970).
2. Ruoslahti, E., and Vaheri, A., *J. Exp. Med. 141,* 497 (1975).
3. Engvall, E., and Ruoslahti, E., *Int. J. Cancer 20,* 1 (1977).
4. Mosher, D.F., *J. Biol. Chem. 250,* 6614 (1975).
5. Keski-Oja, J., Mosher, D.F., and Vaheri, A., *Cell 9,* 29 (1976).
6. Linder, E., Vaheri, A., Ruoslahti, E., and Wartiovaara, J., *J. Exp. Med. 142,* 41 (1975).
7. Bray, B.A., Hsu, K.C., Wigger, H.J., and LeRoy, E.C., *Connect. Tissue Res. 3,* 55 (1975).
8. Bray, B.A., and LeRoy, E.C., *Microvasc. Res. 12,* 77 (1976).
9. Bray, B.A., and Turino, G.M., *Fed. Proc. 35,* 1741 (1976).
10. Verbeek, J.H., Robertson, E.M., and Haust, M.D., *Am. J. Obst. Gynecol. 99,* 1136 (1967).
11. Low, F.N., *Anat. Rec. 142,* 131 (1962).
12. Chen, A.B., Mosesson, M.W., and Solish, G.I., *Am. J. Obstet Obstet. Gynecol. 125,* 958 (1976).
13. Medawar, P.B., *Symp. Soc. Exp. Biol. 7,* 320 (1953).

Keratan Sulfate-like Substance as a Function of Age in the Brain and Eye

Moira Breen, Lidia B. Vitello, Hyman G. Weinstein, and Paul A. Knepper

Our aim in this study was to characterize and measure age-related changes in the glycosaminoglycan (GAG) substituents of rat cerebral cortex. Changes in the concentration and composition of the GAG in brain during development have been reported for the rat (1,2), and the human (3). At birth or shortly after, the concentration of the GAG increased to a peak and then declined. Two main GAG constituents were found in the brain: hyaluronic acid and chondroitin 4-sulfate (1-7).

EXPERIMENTAL

The experimental animal in this study was the Fischer-344 rat from the Charles River Breeding Laboratories, Wilmington, MA. These rats are a Caesarian-originated breed stock, specifically bred for senescent studies in a microbially controlled environment. Five different age groups were used: 1 (weanling), 3 (young adult), 12 (mature), 18, and 25 (senescent) months old.

The rats were decapitated and their brains removed, and the cerebral cortices were dissected, frozen immediately in liquid nitrogen, and stored at -85°C. After delipidation and proteolysis, the GAG were precipitated with alcohol containing 5% of potassium acetate (8,9). This glycosaminoglycan fraction was analyzed for uronic acid (10), glucosamine, and galactosamine (11,12). The data are shown in Table I.

ISBN 0-12-301301-1

Table I. Glycosaminoglycans in Rat Cerebral Cortex[a]

Content	Age (month)				
	1	3	12	18	25
Hexosamine (HexN)	9.2	14.7	12.3	9.8	6.2
Uronic acid (UA)	6.6	5.2	6.5	5.8	5.5
HexN in excess of UA (μmol)	2.6	9.5	5.8	4.0	0.7
UA:HexN mole ratio	0.72	0.35	0.53	0.59	0.89

[a] *Glycosaminoglycan substituents in the GAG fraction (μmol/g) of dry, delipidated weight*

At all ages studied, the hexosamine was in excess of the uronic acid content, indicating the presence of a keratan sulfate-like component in the GAG fraction of the cerebral cortex. This substance increased from weanling (1 month of age) to young adult (3 months of age), and then decreased steadily to a negligible quantity in the senescent rat (see Fig. 1).

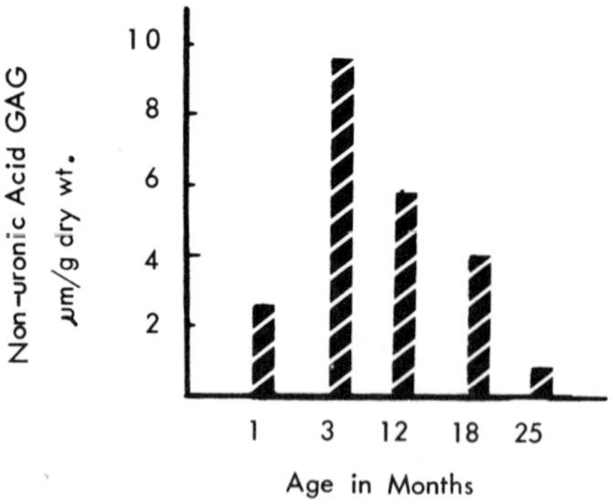

Fig. 1 The variation of the keratan sulfate-like component (non-uronic acid GAG) with age in the cerebral cortex.

The presence of keratan sulfate-like GAG in brain tissue was inferred by Van Hoof and Hers (13) who used an isolation procedure similar to ours. Most other investigators (1-3, 5-7) used cetylpyridinium chloride (CPC) to precipitate the GAG. Since keratan sulfate is soluble in the presence of CPC, it would not be detected under these conditions, if present.

Keratan sulfate is known to be a component of cartilage proteoglycan. Little is known about the role of proteoglycans in tissues. Earlier studies by us have shown that keratan sulfate appears in the late stages of human fetal development, associated with an increase in corneal thickness and accounting for 33% of the increase observed in GAG concentration (14,15).

We have also found that the ratio of keratan sulfate to the total GAG concentration in the anterior segment of the senescent rabbit (aged 3 years) decreased when compared with the young rabbit (aged 8 weeks).

In human tracheobronchial cartilage (16), keratan sulfate increased from trace amounts at birth to a plateau value in the fifth decade. Choi and Meyer (17) found that the concentration of keratan sulfate doubled in costal cartilage of aged humans (60-90 years), as compared with children (4-9 years), although the total amount of GAG decreased slightly.

The evidence indicates that keratan sulfate or a keratan sulfate-like component is present in a number of tissues and that its concentration varies with development, maturation, and senescence. A possible explanation for these apparently diverse observations may be found in the suggestion of Mathews and Cifonelli (18), who concluded that keratan sulfates from cornea and cartilage are "fragments of two different classes of macromolecules in vertebrate connective tissues".

ACKNOWLEDGMENT

This work was supported by the U.S. Veterans Administration and the Helfaer Foundation, Milwaukee, WI.

REFERENCES

1. Singh, M., and Bachawat, B.K., *J. Neurochem. 12,* 519 (1965).
2. Margolis, R.U., Margolis, R.K., Chang, L.B., and Preti, C., *Biochemistry 14,* 85 (1975).
3. Singh, M., and Bachawat, B.K., *J. Neurochem. 15,* 249 (1968).
4. Meyer, K., Hoffman, P., Linker, A., Grumbach, M.M., and Sampson, P., *Proc. Soc. Exp. Biol. Med. 102,* 587 (1959).
5. Szabo, M.M., and Roboz-Einstein, E., *Arch. Biochem. Biophys. 98,* 406 (1962).
6. Margolis, R.U., *Biochim. Biophys. Acta 141,* 91 (1967).
7. Margolis, R.U., and Margolis, R.K., *Biochemistry 13,* 2849 (1974).

8. Breen, M., Weinstein, H.G., Anderson, M., and Veis, A., *Anal. Biochem. 35,* 146 (1970).
9. Brunngraber, E.G., Brown, B.D., and Hof, H., *Clin. Chim. Acta 32,* 159 (1971).
10. Blumenkrantz, N., and Asboe-Hansen, G., *Anal. Biochem. 54,* 484 (1973).
11. Kremen, D.M., and Vaughn, J.G., *Fractions No. 2,* 8 (1972).
12. Lee, Y.C., Scocca, J., and Muir, L., *Anal. Biochem. 27,* 559 (1969).
13. Van Hoff, F., and Hers, H.G., *Eur. J. Biochem. 7,* 34 (1968).
14. Breen, M., Johnson, R.L., Sittig, R.A., Weinstein, H.G., Veis, A., and Marshall, R.T., *Connect. Tissue Res. 1,* 291 (1972).
15. Borcherding, M.S., Blacik, L.J., Sittig, R.A., Bizzell, J.W., Breen, M., and Weinstein, H.G., *Exp. Eye Res. 21,* 59 (1975).
16. Mason, R.M., and Wusteman, F.S., *Biochem. J. 120,* 777 (1970).
17. Choi, H., and Meyer, K., *Biochem. J. 151,* 541 (1975).
18. Mathews, M.B., and Cifonelli, J.A., *J. Biol. Chem. 240,* 4140 (1965).

GLYCOCONJUGATE RESEARCH / Volume I
Proceedings of the Fourth International Symposium on Glycoconjugates

Levels of Sialic Acid and L-Fucose in Human Cervical Mucus Glycoprotein during the Normal Menstrual Cycle

Eric Chantler and Eric Debruyne

The rheological properties of human cervical mucus show a distinct, hormonally dependent variation during the menstrual cycle. At the time of ovulation, a copious secretion of low-viscosity mucus occurs whilst during the rest of the cycle only scant secretion of a viscous mucus is observed (1). Previous studies of the carbohydrate content of human cervical mucus have been mainly based on pooled mucus (2,3), where it has been reported that the variation in rheology coincides with a change in the ratio of the terminal carbohydrate residues of the mucus glycoprotein. In the ovulatory phase, L-fucose predominates over sialic acid, whereas in the luteal phase, the level of sialic acid increases.

We have studied mucus samples, partially purified before sialic acid and L-fucose measurement, and taken at two-day intervals throughout the menstrual cycle. The highest level of sialic acid in the glycoprotein is seen as a peak at the time of ovulation (124 nmol/mg of dry mucus) with only a slight rise occurring during the luteal phase. In contrast, the level of L-fucose is relatively constant throughout the cycle (mean 110 nmol/mg of dry mucus) with a slight decrease in the luteal phase. The ratio of the two sugar residues shifts to favour sialic acid during the luteal phase but this is mainly caused by a drop in the level of L-fucose. The most significant change in the composition of the terminal

ISBN 0-12-301301-1

sugar residue is the increase in the sialic acid content at mid-cycle. The sialic acid to L-fucose ratio varies about one in the proliferative phase but rises to 2 in the luteal phase. The increase in the proportion of sialic acid coincides with an increase in the total sialyltransferase activity of the cervical epithelium at mid-cycle (measured by use of asialo human α_1-acid glycoprotein as an acceptor).

In contrast, no such mid-cycle change was seen in preliminary determinations of the total fucosyltransferase activity of the epithelium (by use of lactose as an acceptor). A factor, which may be significant in the decreased levels of L-fucose in the luteal-phase mucus glycoprotein, is the presence of L-fucosidase activity in the cervical mucus (with 4-methylumbelliferyl L-fucoside as a substrate). This activity increases throughout the luteal phase to reach a maximum of 30 nmol of substrate hydrolyzed/min/g of wet-weight mucus, however we have not yet measured the activity of this enzyme against the native mucus glycoprotein.

In conclusion, it can be seen that the explanation of the changes in the rheology of normal human mucus in terms of the ratio of the two terminal sugars of the glycoprotein probably is an oversimplification. The underlying mechanism for this change is probably related to a more complex alteration in the hydrated glycoprotein (4), which may be related to the change in the carbohydrate composition.

ACKNOWLEDGMENT

This investigation was supported by the World Health Organization Project no. 75049 and 75178.

REFERENCES

1. Lamar, J. K., Shettles, L. B., and Delfs, E., *Am. J. Physiol. 129,* 234 (1940).
2. Iacobelli, S., Garcea, N., and Angeloni, C., *Fertil. Steril. 22,* 727 (1971).
3. Gelle, P., Crepin, G., Roussel, P., Degand, P., and Harvez, R., *Gyn. Obstet. 68,* 279 (1969).
4. Wolf, D. P., Blasco, L., Khan, M. A., and Litt, M., *Fertil. Steril. 28,* 41 (1977).

GLYCOCONJUGATE RESEARCH/Volume I
Proceedings of the Fourth International Symposium on Glycoconjugates

Relationship between Allotransplantability and Cell-Surface Glycoproteins in TA3 Ascites Mammary Carcinoma Cells

John F. Codington, George Klein, Amiel G. Cooper, Nora Lee, Michel C. Brown, and Roger W. Jeanloz

It has been previously suggested that allotransplantability in the TA3-Ha mammary carcinoma ascites tumor of the strain A mouse was due to masking of cell surface histocompatibility antigens by large endogenous glycoprotein (epiglycanin) molecules (1,2). Evidence for masking was based upon the contrasting physical (3,4), chemical (5,6), and biological (2) properties of the nonstrain-specific TA3-Ha cell and the strain-specific ascites subline, TA3-St, that had been derived from the same tumor. In order to test this hypothesis further, a series of six hybrid ascites cell lines (TA3-Ha/A.CA), resulting from fusion *in vitro* of TA3-Ha cells with normal A.CA embryonic fibroblasts (7,8), were investigated.

Each of the hybrid lines exhibited non-strain specificity, and the capacities of the cells to grow in foreign mouse strains were intermediate between those of the TA3-Ha and the TA3-St lines, as shown in Table I. The capacity of each hybrid cell-line to absorb antibody (9) to histocompatibility ($H\text{-}2^a$) antigen (absorptive capacity) was significantly less than that of the strain-specific TA3-St cell (Table I). These results are consistent with a masking mechanism, which would explain the allotransplantability of the hybrid cells. One hybrid line, TA3-Ha/A.CA/7, absorbed even less antibody than the TA3-Ha cell.

ISBN 0-12-301301-1

Table I. Allotransplantabilities, Absorptive Capacities for Antibody to H-2[a] Antigen, and Amounts of Epiglycanin-like Material (mg/10^9 Cells) in Six TA3-Ha/A.CA Hybrid and TA3-Ha and TA3-St Cell Lines

Cell line	Allotransplantability[a]	Absorptive capacity[b] (log 1/G)	Epiglycanin (mg/10^9 cells)[c]
TA3-Ha/A.CA/3B	93	-7.31	5.6
4	85	-7.19	2.4
6	71	-7.00	1.4
7	83	-7.94	7.4
10	57	-7.32	5.9
11	15	-6.92	3.5
TA3-Ha	95	-7.65	4.2
TA3-St	5	-6.66	0.

[a] *Percent of mice killed within 30 days after intraperitoneal inocula of 10^6 cells/mouse into seven foreign mouse strains*
[b] *See ref. 9*
[c] *Values represent relative amounts, since results obtained with* Vicia graminea *lectin [criteria (a) and (b)] were obtained after neuraminidase treatment, and the exact purity of isolated epiglycanin [criteria (c)] was not determined*

On the basis of four criteria, epiglycanin-like material was present at the cell surface of each hybrid line: (*a*) Viable cells absorbed the *Vicia graminea* lectin at approximately the same level as the TA3-Ha cell (10); (*b*) inhibitory activity to hemagglutination by the same lectin was released from viable cells by proteolysis (10); (*c*) fractionation on gel filtration columns of the macromolecules released by proteolysis produced a peak at the same effluent volume as material (epiglycanin fragments) released by the same procedure from TA3-Ha cells (11); (*d*) material having a mol. wt. (*ca*. 500 000), carbohydrate and amino acid composition, and inhibitory activity to hemagglutination by *Vicia graminea* lectin (12), similar to those of epiglycanin, was shed *in vivo* from ascites cells into the ascites fluid. Average values of epiglycanin-like material, determined by criteria (*a*), (*b*), and (*c*) are listed in Table I. For (*a*) and (*b*), values determined after treatment with neuraminidase were used.

It is concluded that allotransplantability and the low level of antibody absorption in the TA3-Ha/A.CA hybrid cells may be due to masking of histocompatibility antigens by large epiglycanin-like molecules at the cell surface. Values presented in Table I suggest a possible correlation between the amount of epiglycanin-like material present and both the absorptive capacity (in the reverse order of magnitude) and the degree of allotransplantability of the hybrid cells.

ACKNOWLEDGMENTS

This work was supported by U.S. Public Health Service, National Institues of Health Grants CA 18600 (to J.F.C.); CA 08418 (to R.W.J); 5 R01 CA 14054-02 (to G.K.); CA 19987 (to A.G.C.); and a grant from the American Cancer Society BC-201 (to A.G.C.). This is publication 758 of the Robert W. Lovett Memorial Group for the Study of Diseases Causing Deformaties, Harvard Medical School and Massachusetts General Hospital. We wish to thank Ms. Marianne Jahnke and Mrs. Cyla Silber for technical assistance.

REFERENCES

1. Codington, J. F., Sanford, B. H., and Jeanloz, R. W., *J. Natl. Cancer Inst. 51,* 585 (1973).
2. Sanford, B. H., Codington, J. F., Jeanloz, R. W. and Palmer, P. D., *J. Immunol. 110,* 1233 (1973).
3. Slayter, H. S., and Codington, J. F., *J. Biol. Chem. 248,* 3405 (1973).
4. Miller, S. C., Hay, E. D., and Codington, J. F., *J. Cell Biol. 72,* 511 (1977).
5. Codington, J. F., Linsley, K. B., Jeanloz, R. W. Imimura, T. and Osawa, T., *Carbohyd. Res. 40,* 171 (1975).
6. Codington, J. F., Van Den Eijnden, D. H., and Jeanloz, R. W., *in* "Cell Surface Carbohydrate Chemistry" (R. E. Harman, ed.), p. 49, Academic Press, New York (1977).
7. Klein, G., Friberg, S., Jr., Wiener, F., and Harris, H., *J. Natl. Cancer Inst. 50,* 1259 (1973).
8. Friberg, S., Jr., Klein, G., Wiener, F., and Harris, H., *J. Natl. Cancer Inst. 50,* 1269 (1973).
9. Reif, A. E., *Immunochemistry 3,* 267 (1966).
10. Codington, J. F., Cooper, A. G., Brown, M. C., and Jeanloz, R. W., *Biochemistry 14,* 855 (1975).
11. Codington, J. F., Sanford, B. H., and Jeanloz, R. W., *Biochemistry 11,* 2559 (1972).
12. Cooper, A. G., Codington, J. F., and Brown, M. C., *Proc. Natl. Acad. Sci. U.S.A. 63,* 1418 (1974).

Immunochemical Studies on the Pr_{1-3} and MN Antigens

Werner Ebert, Hans Peter Geisen, Franz Nader, Dieter Roelcke, and Helmut Weicker

The antigens MN (1) and Pr_{1-3} (2) are known to be located on sialoglycoproteins of the human red cell membrane. The MN blood group antigens correspond to alloantibodies, although human anti-M and anti-N occasionally are found as auto-antibodies. Pr_{1-3} antigens react with autoantibodies which are cold agglutinins. The amino sugar *N*-acetylneuraminic acid (NeuAc) is the immunodominant component of the MN (1) and Pr_{1-3} antigens (3).

Oligosaccharides were split off the peptide backbone by alkaline borohydride treatment of the sialoglycoprotein according to the conditions described by Iyer and Carlson (4). In the hemagglutination inhibition assay, the isolated oligosaccharides clearly inhibited all the anti-Pr sera, whereas the inhibition of anti-M or -N turned out to be rather weak when compared to the original substance. Thus, we conclude that the alkali-labile oligosaccharides represent the Pr_{1-3} determinants. In t.l.c. they revealed a substantial heterogeneity (Fig. 1). The significance of this observation is supported by the results of Hun and Springer (5) on protease-labile glycopeptides of red cells. Their alkaline borohydride degradation also led to a mixture of reduced oligosaccharides. Multiple use of preparative t.l.c. allowed the isolation of one pure oligosaccharide, which had distinct Pr_2 activity. In collaboration with the group of Dr. Lüderitz (Freiburg, Germany) the structure of this compound was elucidated by g.l.c.-m.s. as the tetrasaccharide NeuAc-(2→3)-β-D-Gal-(1→3)-[NeuAc-(2→6)]-β-D-GalNAc. This compound had been previously described by Thomas and Winzler (6) and Adamany and Kathan (7).

ISBN 0-12-301301-1

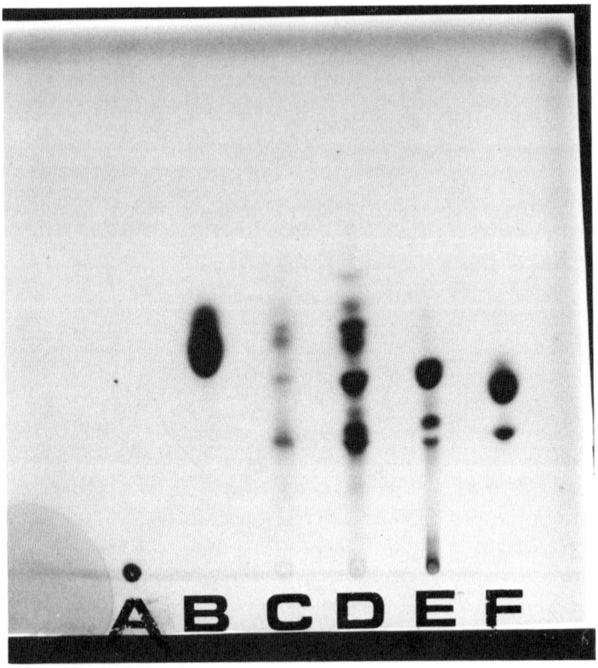

Fig. 1. T.l.c. on Silica gel 60 plates (Merck). Solvent: 1-propanol-25% NH_4OH-water (6:1:2, v/v). The plate was stained with Bial reagent. (A) Untreated sialoglycoprotein; (B) NeuAc; (C) and (D) alkali-labile, dialyzable oligosaccharides; (E) nondialyzable fraction; (F) N-*acetylneuraminosyl-D-lactose.*

Since NeuAc obviously plays a dominant role for the various antigenic activities, structural features of this molecule were systematically altered by diverse chemical reactions: (a) In the presence of carbodiimide, the carboxyl group of the NeuAc residues was amidated with the following amino components: glycinamide, and the methyl esters of glycine, 3-aminopropionic acid, 4-aminobutyric acid, and 5-aminovaleric acid. Subsequent alkaline hydrolysis of the ester functions yielded free carboxyl groups of the amino acid components. (b) By treatment of the sialoglycoprotein with carbodiimide in the absence of additional nucleophile groups, we have attempted to achieve inter- or intra-molecular coupling of NeuAc carboxyl groups with internal nucleophilic centers. (c) Periodate oxidation according to the method of Suttajit and Winzler (8) shortened the polyhydroxy side-chain of NeuAc resulting in a C-7-NeuAc derivative.

The influence of these structural modifications on the antigenic activities was tested by hemagglutination inhibition assays (Table I). The aminolysis of NeuAc carboxyl groups with glycinamide and the series of homologous amino acid methyl esters did not affect the Pr_{1-3} specificities, whereas the MN receptor sites recognized by rabbit antisera were destroyed. The subsequent hydrolysis of the ester group failed to restore the MN activity. Thus, an unmodified NeuAc carboxyl group seems to be essential for the MN determinants. In the absence of external nucleophile groups the carbodiimide-activated carboxyl groups reacted with nucleophilic centers of the sialoglycoprotein molecule, since after this treatment some NeuAc groups were no longer susceptible to neuraminidase. The dominant feature of this reaction is expressed by a cross-linking of sialoglycoprotein molecules; intramolecular linkages and the formation of stable isourea or imide derivatives or both may have occurred. The coupling reaction inactivated both MN and Pr_2 specificities. While Pr_1 was only slightly affected, an increased Pr_3 activity was observed. The results are similar to the degradation by periodate oxidation of the polyhydroxy side-chain of NeuAc which yielded the C-7 analogue. Here, a loss of MN, Pr_1, and Pr_3 antigenicities but a strong increase of the Pr_2 activity was observed (9).

Table I. Immunological Data of MN and Pr_{1-3} Antigens

		Sialoglycoprotein human red blood cell			
Antigen	*Antibody Ig-Class L-chain type*	*Untreated*	*Amidated*	*Carbo-diimide-treated*	*Periodate oxidized*
M	*Rabbit Ig*	+	-	-	-
N	*Rabbit Ig*	+	-	-	-
Pr_1	*IgA (κ)*	+	+	+	-
Pr_2	*IgM (κ)*	+	+	-	+++
Pr_3	*IgM (κ)*	+	+	+++	-

All chemical modifications applied here destroyed both the M and the N specificities indicating that the structural difference between M and N antigens resides neither in the carboxyl group nor in the polyhydroxy side-chain. To the contrary, the antigens of the Pr system showed an individual behavior.

ACKNOWLEDGMENTS

We are grateful to Mrs. J. Fey and Miss Ch. Gärtner for their skilled technical assistance. We also would like to thank Drs. O. Lüderitz, H. Mayer, and F. T. Rietschel, Freiburg, for their help with mass spectrometric analyses. The work was supported by the Deutsche Forschungsgemeinschaft, Bonn-Bad Godesberg.

REFERENCES

1. Springer, G. F., and Ansell, N. J., *Proc. Natl. Acad. Sci. U.S.A. 44*, 182 (1958).
2. Ebert, W., Metz, J., Weicker, H., and Roelcke, D., *Hoppe-Seyler's Z. Physiol. Chem. 352,* 1309 (1971).
3. Roelcke, D., *Vox Sang. 16,* 76 (1969).
4. Iyer, R. N., and Carlson, D. M., *Arch. Biochem. Biophys. 142,* 101 (1971).
5. Hun, J. Y., and Springer, G. F., *Fed. Proc. 35,* 1444 (1976).

Interaction of Small Solutes with a Hyaluronate Matrix that Facilitates their Movement

Nortin M. Hadler and Mary A. Napier

Our previous observations (1,2) suggest that hyaluronate (HA) domains in synovial fluids interact with some small solutes to facilitate their movement. The current corroborative studies were undertaken utilizing isolated human umbilical cord HA. Bulk diffusion coefficients were determined as previously described (2) utilizing methods initially developed by Redwood *et al.* (3).

The following conclusions can be drawn from Table I: (a) When compared with agarose, a 2.5% matrix of HA facilitates the movement of lysine and glucose but not glutamic acid or sucrose; (b) this facilitated movement is critically dependent on Ca^{2+}. It is abrogated at supraphysiologic concentrations; and (c) the effect of Ca^{2+} cannot be explained by ionic strength or pH.

The following conclusions can be drawn from Table II: (a) Facilitated movement of lysine and glucose is dependent on matrix concentration in buffered HA; (b) in the presence of serum proteins, the 0.5 and 1% HA matrices support enhanced diffusivity as effectively as the 2.5% matrix; and (c) the addition of serum proteins to agarose tends to impede solute movement in contrast to the phenomenon documented in HA.

We have previously suggested several hypotheses to explain such aberrant or facilitated movement within an HA matrix (4). One hypothesis borrows heavily from Metzner's considerations of diffusive movement in structured media (5). It is postulated that the molecular "packing" of the solvent adjacent to the polymer is different from that in the bulk solvent phase. This could provide a region with low resistance to mass transport analogous to the mechanism of conduction of electricity in disperse systems or diffusion in crystalline solids. It is the concept of adsorption of the solute to a surface with high intrinsic conductance.

ISBN 0-12-301301-1

With a ^{13}C-n.m.r. analysis (Napier and Hadler, in this volume) we have demonstrated the presence of interacting segments of adjacent chains in regions of the HA matrix at physiologic concentrations of calcium. At higher calcium concentrations, such regions become disordered and the ability of the matrix to enhance the translational diffusivity of glucose and lysine is lost. It is possible that these regions of interacting stiff segments retain the hexad motif documented in hydrated HA putty studied by x-ray diffraction (6). In that case, these regions could have the configuration of corrugated sheets with highly structured solvation shells and might well provide a high conductance surface.

Table I. The Effect of Calcium, pH, and Ionic Strength on Diffusion Coefficients in Matrices[a]

Matrix Solvent	Solute			
	Lysine	*Sucrose*	*Glucose*	*Glutamate*
2.5% HA				
H_2O	2.03	0.44	1.91	0.69
0.01 M Ca^{2+}	0.53	0.44	0.48	0.56
0.10 M Ca^{2+}	0.57	0.46	0.54	0.55
H_2O	2.19	0.50	2.16	0.69
0.15 M NaCl[b]	2.11	0.52	1.99	0.69
2.5% Agarose				
H_2O	0.85	0.68	0.91	0.96
0.01 M Ca^{2+}	0.94	0.75	0.92	0.98
0.10 M Ca^{2+}	0.96	0.78	0.96	1.00
0.15 M NaCl[b]	0.94	0.78	0.94	0.97

[a] *All data were derived from co-diffusion experiments measuring the ratios of [^{14}C]lysine to [^{3}H]sucrose and [^{14}C]glucose to [^{3}H]glutamic acid together. The solute is applied in a concentration less than 0.5 mM and the experiment prepared at 27°C. The mean of a minimum of three experiments is tabulated, D being expressed in 10^{-5} $cm^2 sec^{-1}$. The standard errors averaged 8% of the respective means*

[b] *These samples were buffered at pH 7.0 with 0.01 M sodium phosphate*

Table II. Diffusion Coefficients[a]

Matrix	Solvent[b]	Solute			
		Lysine	Sucrose	Glucose	Glutamate
HA					
0.5%	D-PBS	0.76	0.63	0.80	0.74
	2.5% BSA			1.48	1.13
	50% Serum	1.08	0.85	1.41	0.94
1.0%	D-PBS	0.61	0.48	0.64	0.66
	2.5% BSA	2.15	0.59	1.11	0.75
	50% Serum	1.80	0.95	1.71	1.30
2.5%	D-PHS	2.11	0.52	1.99	0.69
	2.5% BSA	2.24	1.02	2.23	0.89
Agarose					
0.5%	D-PHS	0.88	0.61	0.90	0.94
	2.5% BSA	0.70	0.58	0.96	9.69
	50% Serum	0.82	0.59	0.85	0.71
1.0%	D-PBS	0.87	0.56	0.73	0.75
	2.5%	0.79	0.61	0.70	0.71
	50% Serum	0.65	0.49	0.67	0.68
2.5%	D-PHS	0.94	0.78	0.94	0.97
	2.5% BSA	0.78	0.57	0.83	0.81

[a] *See Table I*

[b] *All solvents were buffered with Dulbecco's phosphate buffered saline (D-PHS). Where indicated, bovine serum albumin (BSA) or pooled normal human serum was added*

ACKNOWLEDGMENT

N. M. H. is a recipient of the Established Investigatorship from the American Heart Association.

REFERENCES

1. Hadler, N. M., *Fed. Proc. 36*, 1069 (1977).
2. Napier, M. A., and Hadler, N. M., *Proc. Natl. Acad. Sci. U.S.A. 75*, 2261 (1978).
3. Redwood, W. R., Rall, E., and Perl, W., *J. Gen. Physiol. 64*, 706 (1974).
4. Hadler, N. M., and Napier, M. A., *Semin. Arthritis Rheum. 7*, 141 (1977).
5. Metzner, A. B., *Nature (London) 224*, 240 (1965).
6. Sheehan, J. K., Atkins, E. D. T., and Nieduszynski, I. A., *J. Mol. Biol. 91*, 153 (1975).

GLYCOCONJUGATE RESEARCH/Volume I
Proceedings of the Fourth International Symposium on Glycoconjugates

Studies on a Human Salivary Glycoprotein with Specific Bacterial Adhesive Properties

Kathleen M. Guilmette and Shelby Kashket

Work in our laboratory has been devoted to a study of the mechanisms by which specific salivary macromolecules bring about the aggregation of certain oral microorganisms (1). The phenomenon of bacterial aggregation is considered to be of importance in regulating the nature of the microbial population in the oral environment (2). *In vivo,* aggregation can lead to the clearance of microorganisms from the oral cavity, although there is evidence that the same salivary factors become adsorbed to hydroxyapatite of enamel and act as substrates for the attachment of specific microorganisms to the tooth surface (3). The purpose of the present study was to isolate the specific salivary factor for *Streptococcus sanguis,* strain H7PR3, so that its role in the oral environment, as well as the mechanisms of its interaction with the bacterial cell surface, can be investigated in detail.

Human submandibular-submaxillary saliva was collected with an appropriate collecting device, with sour lemon drop stimulation. The saliva was dialyzed and mixed with cetyltrimethylammonium bromide to 1% final concentration. The resulting precipitate was collected, resuspended in 50% $CaCl_2$, and dialyzed overnight. Sufficient 95% ethanol was added to the nondialyzable solution to give a final concentration of 80%, and this precipitate was collected, dissolved

ISBN 0-12-301301-1

in water, and dialyzed overnight against water at 4°C. The ethanol-precipitable material was layered onto an 8–15% sucrose density gradient and centrifuged for 3 h at 144 000*g*. One-ml fractions were collected, dialyzed against water, and assayed for aggregating activity by the electronic particle-counting procedure of Kashket and Donaldson (1). The results are shown in Fig. 1. Although it was found that the ethanol-insoluble material exhibited high aggregating activity, it can be seen that none of the density gradient fractions exhibited activity. However, if a trace of unfractionated submandibular saliva was added to each assay tube, three regions of aggregating activity became apparent (Peaks I,

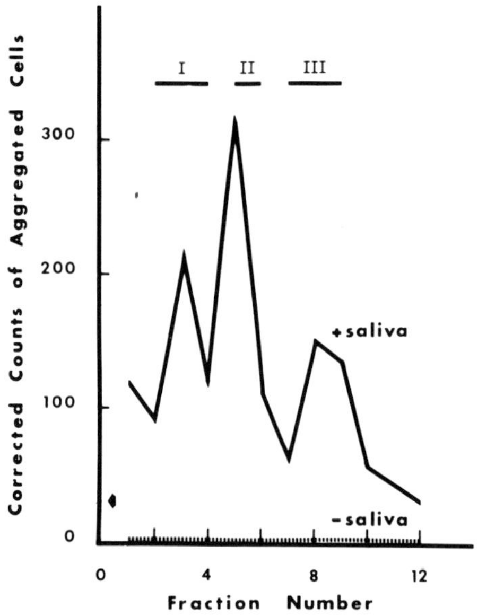

Fig. 1. Sucrose density gradient separation of ethanol-precipitable, aggregating factor. Fractions were assayed with or without the addition of trace amounts of unfractionated saliva. The arrow indicates aggregating activity of the latter.

II, and III). These findings are consistent with our earlier observations on the multimeric nature of the aggregating factor (4,5). The fractions within each peak region were pooled and concentrated, and when these were combined, they showed aggregating activity in the absence of unfractionated saliva. It can be seen from Table I that Peak I, when combined with Peak III, gave good activity, Peaks I plus II were moderately active, while Peaks II plus III were inactive. The peak constituents did not enter into 5-7% polyacrylamide gels during electrophoresis, which is consistent with their supposed high mol. wt. (6). Preliminary results with isoelectric focusing at pH 3.5-6.0 showed Peak I to contain at least 5 u.v.-absorbing bands between pH 5.0 and 6.0. The region that exhibited aggregating activity (in the presence of trace amounts of unfractionated saliva) was found to contain 66% of reducing sugars, and the amino acid analysis showed a high content of serine, glutamic acid, glycine, and alanine, some hexosamine, and low levels of proline and basic amino acids. These studies are being continued.

Table I. Reactivation of Aggregating Factor Activity of Isolated Glycoprotein Fractions

Sucrose density gradient peaks	*Aggregating activity (Corrected counts of aggregated cells)*
I	*14*
II	*16*
III	*15*
I + II	*70*
I + III	*147*
II + III	*17*

In conclusion, the results show that the specific salivary factor, which is able to bring about the aggregation of a strain of the oral microorganism *S. sanguis*, is glycoprotein in nature. The high mol. wt. factor appears to be made up of at least two subunits, and it has been shown that, while each of the subunits is biologically inactive, they can reassemble to give active aggregating factor.

ACKNOWLEDGMENT

This research was supported in whole by the National Institute of Dental Research, N.I.H., Research Grant No. R01 DE-03430 and RR-05483.

REFERENCES

1. Kashket, S., and Donaldson, C. G., *J. Bact. 112,* 1127 (1972).
2. Gibbons, R. J., and van Houte, J., *Ann. Rev. Microbiol. 29,* 19 (1975).
3. Hillman, J. D., van Houte, J., and Gibbons, R. J., *Arch. Oral. Biol. 15,* 899 (1970).
4. Kashket, S., and Hankin, S. R., *Arch. Oral Biol. 22,* 49 (1977).
5. Kashket, S., Skobe, Z., and Garant, P. R., *Arch. Oral Biol. 23,* 125 (1978).
6. Hay, D. I., Gibbons, R. J., and Spinell, D. M., *Caries Res. 5,* 111 (1971).

GLYCOCONJUGATE RESEARCH/Volume I
Proceedings of the Fourth International Symposium on Glycoconjugates

Proteoglycan Structure and Ca Release by Enzymatic Proteolysis

Nobuhiko Katsura, Hiroko Takita, Noriyuki Kasai, Masaki Shiono, and Ken-ichi Notani

CsCl equilibrium density-gradient centrifugation, which is widely used for the preparation and purification of proteoglycan (PG), does not give a preparation of PG that gives a single spot on two dimentional electrophoresis (2E) (1). Although the density-gradient fraction designated "AIDI" (2) prepared from bovine nasal cartilage was further purified by repeated (4X) CsCl centrifugation, no improvement was evident on 2E. In addition, this fraction showed decreasing solubility and a steadily increasing uronic acid to protein ratio. PG or free chondroitin sulfate (CS) does not give a buoyant band in the CsCl system, but is distributed in the density range 1.5-1.9 g/ml. In contrast, a sharp buoyant band of PG at 1.2-1.3 g/ml was obtained when the centrifugation was carried out in Metrizamide. This PG band had a higher uronic acid to protein ratio but still did not exhibit a single spot on 2E. The efficacy of the CsCl method insofar as purification of PG is concerned remains questionable.

An improved procedure for the preparation of PG subunit was developed as follows: 3 M $MgCl_2$ extraction of sperm whale intermaxillar cartilage slices, CTAB purification (P-PG), complete removal of hyaluronic acid by streptomyces hyaluronidase digestion in the presence of benzamidinium chloride, ultracentrifugation, dithiothreitol reduction in 4 M guanidinium chloride, modification of sulfhydryl residues with DTNB, concentration and dialysis with Diaflo XM 300, ultracentrifugation in 0.98 M acetic-0.15 M Na (pH 4), gel chromatography with AcA 22, and lyophilization. The preparation thus obtained was designated PGUS. PGUS showed a single spot on 2E just behind that of standard CS (Fig. 1); it contains about 6% of amino acids, 30% of galactosamine, and 5% of glucosamine.

ISBN 0-12-301301-1

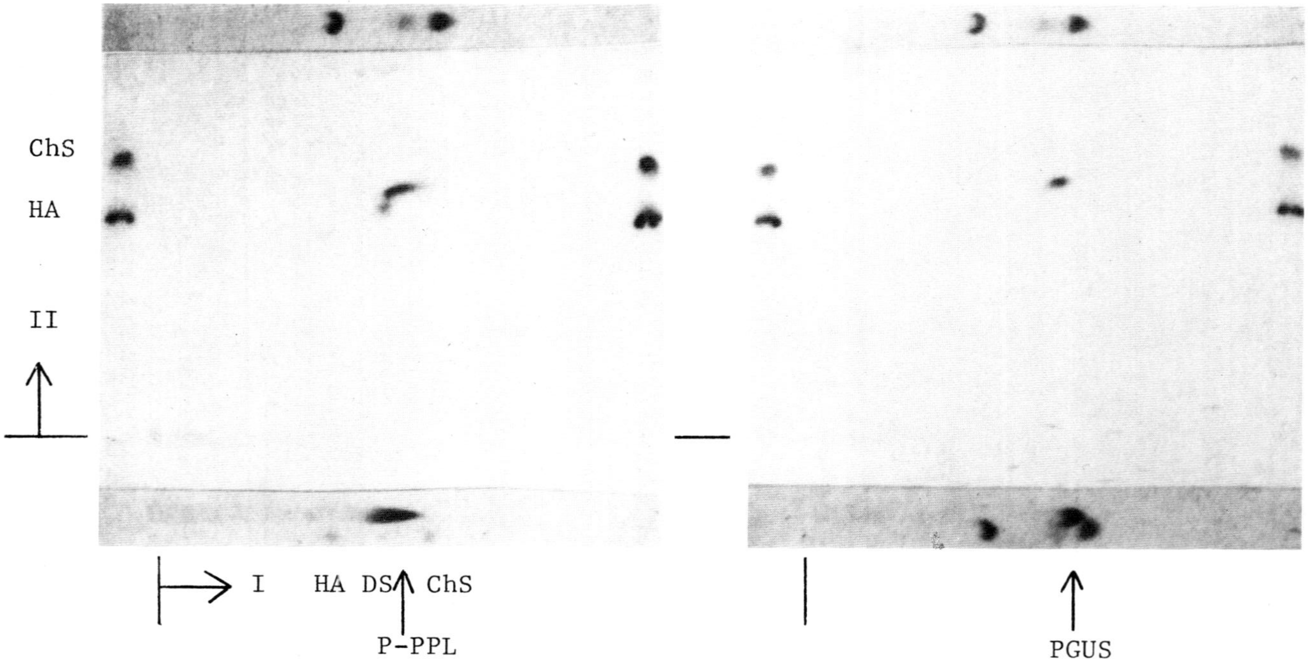

Fig. 1. Two-dimensional electrophoresis on Sepraphore III of P-PPL and PGUS.

The sedimentation coefficient is approximately 7.0 and the mol. wt. 1.0-1.25 x 10^5. At least one cysteine residue is present, but attempts to detect a N-terminal amino acid were not successful. The PGUI fraction that was eluted at the void volume of the exclusion column had a higher cysteine content, as much as 30 per 1000 residues, and lower solubility. A speculative structural model (Fig. 2) of PG is shown as a stiff rod, about 300 nm in length.

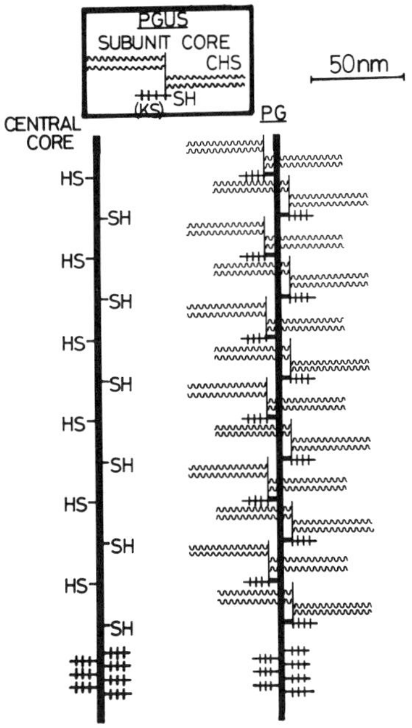

Fig. 2. A speculative structural model of PGUS and PG: CS, chondroitin sulfate; KS, keratan sulfate; and SH, sulfhydryl group.

The central core-protein (PGUI) carries 10-16 PGUS subunits through disulfide, ionic, and hydrophobic bonds. The PGUS unit is about 24 nm in length with the subunit core-protein carrying one keratan sulfate and two doublets of chondroitin sulfate chains. This model is in excellent agreement with both the analytical data and electron microscopic images.

A hollow fiber dialyzer (Minibeaker b/HFD 1/20, Dow Chemical Co.) filled with Ca salt solution was washed through hollow fibers with a constant flow of physiological saline, and the diffused Ca concentrations were determined. Diffusible Ca decreased exponentially in the same manner as a radioactivity decay curve, $C=C_oe^{-\lambda t}$. The half lives of various Ca salt solutions were determined (see Table I). They indicate that the protein cores of PG or PGUS are essential to immobilization of Ca ions, and that PG plays some important roles in the process of calcification.

Table I. Half-lives of Various Calcium Salts

		Half-life (min)	
Ca salt	*Treatment*	*Before treatment*	*After treatment*
Chloride		*33*	
1% Proteoglycan	*Trypsin*	*90*	*52.5*
	Test. hyal.[a]		*54*
1% Chondroitin 4-sulfate	*Test. hyal.*	*69*	*51*
1% Chondroitin 6-sulfate		*64.5*	
1% Hyaluronic acid	*Test. hyal.*	*79.5*	*70.5*
1% PGUS	*Trypsin*	*78*	*57*

[a] *Abbreviation: Test. hyal., testicular hyaluronidase*

REFERENCES

1. Hata, R., and Nagai, Y., *Anal. Biochem.* 45, 462 (1972).
2. Hascall, V. C., and Sajdera, S. W., *J. Biol. Chem.* 245, 4920 (1970).

GLYCOCONJUGATE RESEARCH/Volume I
Proceedings of the Fourth International Symposium on Glycoconjugates

Binding Studies on the Liver Receptor for Asialoglycoproteins

Leif Jansson and Nils E. Nordén

In 1966, the Ashwell group discovered that several glycoproteins were rapidly removed from the blood circulation after desialylation (1). This removal is effected by a liver membrane receptor, which was isolated from rabbit and found to be a glycoprotein. The affinity between the receptor and various glycoproteins *in vitro* varied in parallel to the rate of elimination from the blood *in vivo*. We have now isolated the receptor from rabbit and pig liver, using the same method as previously described (2), except that in the affinity chromatography asialoorosomucoid was replaced by asialofetuin.

Antibodies were raised against the porcine receptor, and the rabbit and pig receptor preparations were studied by immunodiffusion with the antiserum in the central well, and the pig and rabbit receptors alternating in the outer wells. Complete fusion without spurs was observed, indicating immunochemical identity between the two receptors.

In order to study the homogeneity of the preparations, they were first treated with a buffer containing SDS (dodecyl sodium sulfate), and then run on polyacrylamide gel electrophoresis. Two main components, as well as three minor components, were eluted separately from the gel. They all contained material reacting with the antiserum, indicating that the minor components are aggregates of the main components. When the receptor preparations were studied by cross immunoelectrophoresis in the presence of the non-ionic detergent Triton X-100, most of the material remained on the starting line in the first dimension. As we thought that this result was due to aggregation, we used a mild treatment with SDS before cross immunoelectrophoresis in the next experiment. This resulted in a higher mobility in the first dimension. Only a single line of precipitation was seen in both experiments, which means that the preparation can be

ISBN 0-12-301301-1

considered immunochemically homogeneous.

The binding of different ligands to the receptor was studied by the use of radiochemical binding and inhibition assays described previously (2). It was possible to inhibit the binding with the antiserum, thus indicating the specificity of the antiserum towards the receptor. The concentration (w/v) needed for 50% of inhibition in the inhibition assay was used as a measure of the affinity of different ligands to the receptor protein. Thus, it was confirmed that asialofetuin has an affinity three times lower than that of the asialo-orosomucoid used as a standard in all experiments (3). The bacterial polysaccharide S14 (4) had an affinity that was 9 times lower than that of asialo-orosomucoid. The polysaccharide has a large number of nonreducing galactose terminal residues. On the other hand, asialotransferrin and an oligosaccharide isolated from a patient with GM_1-gangliosidosis (5) did not show any measurable affinity to the receptor.

Since some of the experiments described have indicated that the receptor protein tends to aggregate, even in the presence of detergents, the binding of a constant amount of asialo-orosomucoid was studied as a function of the receptor concentration. Thus, it was shown that the binding first increases and reaches a maximum, and then decreases with increasing receptor concentration. It is possible to explain these results by the aggregation of the receptor molecules, if one assumes that the binding of the ligand is sterically hindered in the aggregates. Preliminary thermochemical results (obtained in collaboration with Dr. J. Suurkuusk, Thermochemistry Laboratory, Chemical Center, University of Lund, Sweden) show that the rate of heat production changes when the receptor solution (without ligand) is diluted, indicating that there is aggregation of the receptor molecules and also that the degree of aggregation changes when the solution is diluted. It is desirable to find conditions in which the aggregation is eliminated, before determination of the binding constants is attempted.

As a conclusion, the binding to the liver receptor was found to be not as specific as previously thought, since it can bind not only certain asialoglycoproteins and carbohydrate fragments from asialoglycoproteins, but also a polysaccharide with a carbohydrate structure that is completely different from the structures seen in asialoglycoproteins, except that it contains nonreducing galactose terminal residues. On the other hand, an oligosaccharide having two nonreducing galactose terminal residues and a structure closely related to the carbohydrate structures of several asialoglycoproteins had no detectable binding, indicating a very low affinity. Thus, the number and possibly the sterical arrangement of the terminal galactose residues, rather than the structure of the interior of the carbohydrate chain of the ligand, seem to be of primary importance for the degree of binding. Therefore, the ability of the receptor to recognize carbohydrate structures is a quantitative rather than an all or none phenomenon.

REFERENCES

1. Ashwell, G., and Morell, A.G., *Trends Biochem. Sci. 2,* 76 (1977).
2. Hudgin, R.L., Pricer Jr., W.E., and Ashwell, G., *J. Biol. Chem. 17,* 5536 (1974).
3. Ashwell, G., and Morell, A.G., *Adv. Enzymol. 41,* 99 (1974).
4. Lindberg, B., Lönngren, J., and Powell, D.A., *Carbohydr. Res. 58,* 177 (1977).
5. Wolfe, L.S., Senior, R.G., and Ng Ying Kin, N.M.K., *J. Biol. Chem. 244,* 1828 (1973).

GLYCOCONJUGATE RESEARCH / Volume I
Proceedings of the Fourth International Symposium on Glycoconjugates

Affinity of Lectins for Human Bronchial Mucosa and Secretions

Michel Lhermitte, Annie-Claude Roche, Philippe Roussel, and Marc Mazzuca

The specific affinity of a battery of lectins for different carbohydrate structures might allow a comparative study of the glycoproteins from a mucosal secretion and the cells involved in their synthesis. It is well known that, by use of the periodic acid-Schiff reagent (PAS), different cells involved in the tracheobronchial secretion are stained: goblet cells and mucous cells of the submucosal glands are strongly stained, whereas serous cells stain more weakly (1). Bronchial secretion contains different glycoproteins; some of them have the mucin-type with GalNAc-Ser(Thr) linkages and no mannose, whereas others have the serum-type and contain mannose. The aim of this study was to compare the affinity of four lectins [*Ricinus communis* lectins (RCA I and RCA II), concanavalin A (Con A), and limulin] for the secreted bronchial glycoproteins, and for the glandular cells of the human bronchial epithelium and of the submucosal glands.

AFFINITY OF CONCANAVALIN A, R.C. LECTINS, AND LIMULIN FOR HUMAN BRONCHIAL MUCOSA

A biopsy of human bronchial mucosa, obtained in a macroscopically healthy area from a subject belonging to blood group A, was fixed, dehydrated, embedded in paraffin, and cut (2). *Ricinus communis* lectins (RCA I and RCA II) and limulin were labeled with peroxidase (2,3). Bronchial mucosa sections were stained either with each of the peroxidase-labeled lectins (2,3) or with Con A and subsequently with peroxidase (4). After the washing, the peroxidase was localized with 3,3'-diaminobenzidine and fresh hydrogen peroxide. The sections were washed, dehydrated, and mounted with Euckitt.

ISBN 0-12-301301-1

Con A did not bind to goblet cells or mucous cells and reacted strongly with serous cells. The peroxidase-labeled RCA I or RCA II reacted strongly with the goblet cells and the mucous cells, whereas the cytoplasm of the serous acini was also stained with less intensity (3). With the peroxidase-labeled limulin, there was a diffuse staining of the apical part of goblet cells and, in the serous cells, the labeling appeared as small granules localized either at the apical pole or in the whole cytoplasm (2).

AFFINITY OF BRONCHIAL GLYCOPROTEINS FOR CON A, R.C. AGGLUTININS, AND LIMULIN

Sepharose chromatography of solubilized fibrillar mucus. Bronchial fibrillar mucus from a patient with blood group A who suffered from chronic bronchitis was solubilized and subjected to chromatography on Sepharose 4 B (3). Three fractions were obtained. Fraction 1, corresponding to high-molecular-weight constituents, was excluded from the column. Fraction 2 was a population of mucins and reacted with R.C. lectins and limulin but not with Con A. Fraction 3 corresponded to a mixture of proteins and glycoproteins and reacted with Con A, R.C. lectins, and weakly with limulin.

Affinity chromatography of Fraction 2 on R.C. lectins-Sephadex G-25. Affinity chromatography of Fraction 2 resulted in the separation of two mucin fractions, a minor fraction having no affinity for R.C. lectins, and a major fraction reacting with R.C. lectins and limulin. The minor fraction recovered an affinity for R.C. lectins after delipidation (3).

Fractionation, by affinity chromatography, of the proteins and glycoproteins contained in Fraction 3. Fraction 3 was separated by affinity chromatography on Con A-Sepharose into two fractions devoid of affinity for Con A, and two fractions having an affinity for this lectin. The latter two fractions contained glycoproteins, such as IgA, α_1-antitrypsin, α_1-antichymotrypsin, and bronchotransferrin. The fractions having no affinity for Con A were separated, by affinity chromatography on R.C. lectins-Sephadex G-25, into two groups of components. The first group having no affinity for R.C. lectins comprised albumin and antigenic determinants related to α_1-antitrypsin and α_1-antichymotrypsin. The second group corresponded mainly to a small amount of mucins.

DISCUSSION

Most of the bronchial mucins were separated from other proteins or glycoproteins by chromatography on Sepharose 4 B. These molecules, contained in Fraction 2, did not react with Con A, but reacted with limulin and R.C. lectins before or after prior delipidation. Mucous cells and goblet cells were strongly stained by peroxidase-labeled R.C. lectins but, under our conditions, have no visible affinity for Con A. Their affinity for limulin was different: mucous cells had no affinity for this lectin, whereas goblet cells had a strong affinity. The glycoproteins of the solubilized fibrillar mucus that have a small molecular weight correspond to mucins in minor amount that have an affinity for R.C. lectins, and no affinity for Con A, and to a group of glycoproteins that have an affinity for Con A and related either to plasma proteins or to molecules already described in the serous cells, such as broncho-transferrin (5). Serous cells, which were weakly stained by PAS and by peroxidase-labeled R.C. lectins, had a strong affinity for Con A and limulin.

These results demonstrate that, by use of four lectins, the various cell types involved in the biosynthesis of bronchial secretion may be characterized and that affinity chromatography of solubilized bronchial mucus is useful for the separation of the mucin-type and of the serum-type glycoproteins. They suggest that serum-type glycoproteins are related to serous cells and that various mucin-type glycoproteins are related to goblet cells and mucous cells.

ACKNOWLEDGMENTS

This work was supported by Contrat CECA 1977 No. 7246-22-3-00-3 and Contrat de Recherches UER III des Sciences Médicales de Lille.

REFERENCES

1. Spicer, S., Chakrin, L. W., and Wardell, J., *in* "Sputum" (M. Dulfano, ed.), vol. 1, p. 22, Thomas, Illinois (1973).
2. Mazzuca, M., Roche, A. C., Lhermitte, M., and Roussel, P., *J. Histochem. Cytochem. 25,* 470 (1977).
3. Lhermitte, M., Lamblin, G., Degand, P., Roussel, P., and Mazzuca, M., *Biochimie 59,* 611 (1977).
4. Bernhard, W., and Avrameas, S., *Exp. Cell. Res. 64,* 232 (1971).
5. Masson, P. L., Heremans, J. F., Prignot, J. J., and Wauters, G., *Thorax 21,* 538 (1966).

GLYCOCONJUGATE RESEARCH/Volume I
Proceedings of the Fourth International Symposium on Glycoconjugates

Calcium Ion Binding to Glycosaminoglycans and Corneal Proteoglycans

Michael A. Loewenstein and Frederick A. Bettelheim

Data on calcium ion binding to glycosaminoglycans (GAGs) and proteoglycans (PGs), which are reported in the literature (1-3), suggest either a non-specific type of electrostatic interaction or a chelation-type mechanism for binding. To further establish one type or the other, Ca^{+2} binding data were obtained for two sulfated GAGs of different charge densities, as well as the PGs that contain these polysaccharides in various proportions. The (bovine) cornea provides a convenient source for such PGs whose two main GAGs are chondroitin 4-sulfate (C-4-S) and keratan sulfate (KS).

MATERIALS

Samples of C-4-S and C-6-S were obtained commercially, and KS was previously isolated from bovine corneas by Plessy and Bettelheim (4). The PGs (three fractions plus a fourth, new one) were isolated from bovine corneas according to the procedure of Bettelheim and Plessy (5).

PROCEDURE FOR CA^{+2} BINDING DETERMINATION

Binding data were obtained by molecular filtration (the apparatus was purchased from Millipore) at constant volume. A membrane filter that retained polymers of nominal mol. wt. of 1000 was used. The filtrate was collected in 0.7-ml increments with a fraction collector, and each fraction was analyzed for Ca^{+2} by atomic absorption. Filtration proceeded until a steady-state concentration in Ca^{+2} was achieved. This value was taken to be the free Ca^{+2} concentration in the binding cell. After filtration, a sample of the cell solution was analyzed for the total Ca^{+2} concentration. The difference between the total and

ISBN 0-12-301301-1

the free Ca^{+2} is the concentration of bound Ca^{+2}.

RESULTS

The composition of PG fractions IA, IB, and II, as reported in ref. 5, is presented in Table I along with an analysis of the new fraction I'.

Table I. Composition of Isolated Corneal Proteoglycans

	Fractions			
Compounds (%)	*I'*	*Ia*	*Ib*	*II*
Protein	*60*	*41*	*48*	*41*
Galactose	*9.2*	*3.1*	*16.3*	*13.1*
Uronic Acid	*6.6*	*16.3*	*3.3*	*5.2*

Binding data are expressed as the binding constant $K = \gamma/\alpha\beta$, defined for the reaction: Ca^{+2}+ Binding $site^{-2} \rightleftarrows$Complex, and as a binding capacity, *B*, defined by $B = f/w$ where $f = \gamma/\alpha+\gamma$. Here, α and γ represent the molarities of the free and bound Ca^{+2} respectively, *w* is the weight concentration (g/liter) of either GAG or PG and is fixed by solution preparation, and β represents the molarity of free sites and is calculated as the difference between the total concentration of sites and the concentration of bound Ca^{+2}. A site is assumed to be a saccharide unit that presents two negative charges. For the GAGs, this corresponds to binding site mol wts. of 503 g/mol for CS and 934 g/mol for KS. The total number of binding sites on the PGs is calculated from the CS and KS concentrations present as side chains.

Binding capacities and constants for the various GAGs and PGs are listed in Table II, along with the concentration ratio Ca^{+2} (mol/liter)/*w*(g/liter) at which they were measured. There is an uncertainty of approximately ± 30 μM in the Ca^{+2} concentration as determined by atomic absorption.

DISCUSSION

From Table II, *B* and *K* for each GAG and PG are seen to be independent of the conditions (Ca^{+2}/w) at which the determinations were made. The *B* values for the PGs and GAGs in general are very similar. Since there are no correlations between *B* and the different charge densities of the various polyelectrolytes, a non-specific type of electrostatic interaction is not indicated by the data.

Table II. Ca^{+2} *Binding*[a]

Compounds (sodium salt)	Ca^{+2}/w	B $(g/liter)^{-1}$	$K(M^{-1})$
Chondroitin 4-sulfate	*1.70*	*0.015*	*8.2*
	0.78	*0.016*	*9.0*
Chondroitin 6-sulfate	*1.60*	*0.017*	*9.6*
	0.86	*0.015*	*8.4*
Keratan sulfate	*1.66*	*0.020*	*22*
	0.64	*0.020*	*20*
I'	*0.38*	*0.022*	*42*
IA	*1.22*	*0.017*	*22*
	0.35	*0.022*	*29*
IB	*0.72*	*0.043*	*122*
II	*0.64*	*0.021*	*48*
	1.65	*0.019*	*52*

[a] *All measurements were carried out at* ca. *22°C and on solutions in the buffer 0.05* M *Tris-HCl, 0.25* M *NaCl, pH 7.1. The nominal macromolecular concentration was 2.5 g/liter for all substances except I' and IA, where it was 4.0 g/liter*

On the other hand, *K* values show that the PGs bind calcium more strongly than the GAGs. The difference may be attributed in part to the neglect of possible binding-sites on the protein core of the PG, which constitutes 40-60% of its dry weight. The two parameters *B* and *K* for the PG are based upon models of extreme assumptions. For *B*, it was assumed that the protein core and the GAG side-chains bind CA^{+2} to the same extent and, therefore, the interaction is on a per unit-mass basis. In calculating *K*, the assumption was that binding sites were provided only by the GAG side-chains. Clearly, the actual binding must be between these two models. In any case, the PG binds the cation more strongly than the GAG.

This difference in binding strength together with a view of the bottlebrush structure (6) of PG suggests a chelation-type mechanism for Ca^{+2} binding to the PG. The protein core aligns the GAG side-chains so that interchain calcium bridges form. Such a representation would be consistent with the findings of Winter and Arnott (7), that cations exert strong influences on both conformation and packing of GAG and PG.

ACKNOWLEDGMENT

This research was supported in part by a grant EY00501-09 of the National Eye Institute (National Institutes of Health, USPHS).

REFERENCES

1. Woodward, C, and Davidson, E.A., *Proc. Natl. Acad. Sci. U.S.A. 60,* 201 (1968).
2. MacGregor, E.A., and Bowness, J.M., *Can. J. Biochem. 49,* 417 (1971).
3. Dunstone, J.R., *Biochem. J. 85,* 336 (1962).
4. Plessy, B., and Bettelheim, F.A., *Mol. Cell. Biochem. 6,* 85 (1975).
5. Bettelheim, F.A., and Plessy, B., *Biochem. Biophys. Acta 381,* 203 (1975).
6. Mathews, M.B., and Lozaityte, I., *Arch. Biochem. Biophys. 74,* 158 (1958).
7. Winter, W.T., and Arnott, S., *J. Mol. Biol. 117,* 761 (1977).

GLYCOCONJUGATE RESEARCH / Volume I
Proceedings of the Fourth International Symposium on Glycoconjugates

Changes in Gastric Mucosal Blood Group ABH and I Activities in Association with Cancer

Jean Picard and Ten Feizi

It is well known that blood group A, B, and H antigens are deficient at the site of many human adenocarcinomas (reviewed in ref. 1). The aim of the present studies has been to investigate whether the diminution of these antigens is associated with an increase in incomplete or precursor-like substances which are known to express the Ii antigens (2-4).

Although the precise sugar sequences recognized by the majority of anti-I and -i antibodies are not yet known, one anti-I antibody (from patient Ma) has been shown to recognise the type 2 chains of the precursors of secreted blood group substances (3,4). This antibody has proved valuable for the detection of incomplete A, B, and H antigens by quantitative precipitation assays, and we have used it as a reagent in the studies described below.

Assays of the ABH and I (Ma type) activities of gastric mucins and mucosal extracts of apparently healthy persons and of those with benign gastric ulcers and chronic gastritis have shown two main patterns of antigen expression, which correlate with secretor status. In secretors the ABH antigens are usually well expressed, and there is little precipitating I activity of the type recognized by anti-I Ma. In non-secretors on the other hand, there is usually little or no detectable ABH activity, but there is substantial I activity

ISBN 0-12-301301-1

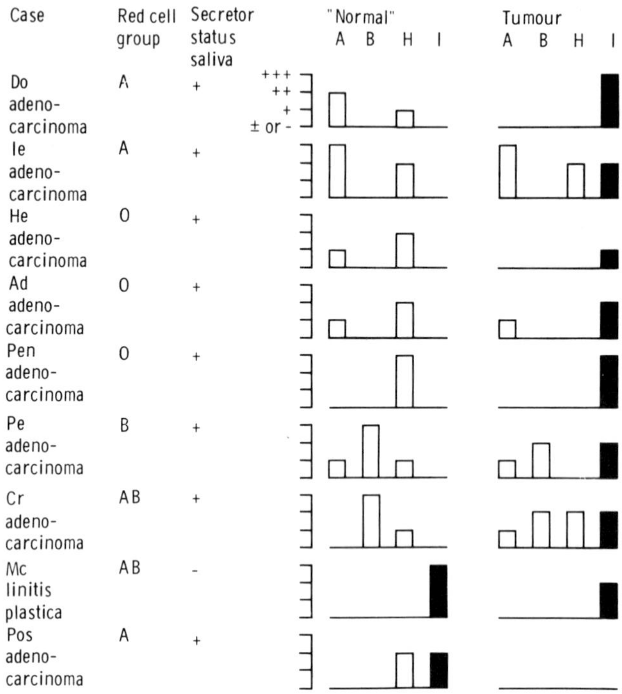

Fig. 1. Blood-group ABH and I (Ma type) activities of glycoprotein-rich extracts (prepared by pepsin digestion and ethanol precipitation) of gastric carcinomas and apparently uninvolved tissues. The A, B, and H activities were measured by haemagglutination inhibition of human anti-A and anti-B sera, and Ulex europaeus lectin, respectively, and the I activities by quantitative precipitation assays with anti-I Ma. The results were expressed as the minimum concentration (μg/ml) required to inhibit 8 haemagglutinating units of anti A, B, or H, and the amount (μg) required to precipitate 2 μg of antibody nitrogen: +++, <3-10 μg; ++, >10-50 μg; +, >50-100 μg; and ±, >100-500 μg.

of Ma type. Only exceptionally are substantial ABH and I activities simultaneously present.

Fig. 1 shows the antigen patterns observed in the tumours and the adjacent apparently normal mucosae of 9 patients with gastric carcinoma. In the uninvolved mucosae, with one exception, the secretor/non-secretor pattern is seen, *i.e.*, those whose extracts are rich in ABH are lacking I, and the case without ABH possesses I. Only one case had substantial amounts of H together with I. In those whose adjacent tissue extracts were of secretor type, I antigen regularly appeared in the tumours. This was usually associated with a loss or diminution of ABH. In the two cases that had I activity in their uninvolved mucosae this activity disappeared or diminished in the tumour extract. These findings and our previous observations on the presence of increased Ii activities in certain metastatic colon tumours (5) are compatible with the accumulation of glycoproteins with shortened oligosaccharide chains in the tumours.

A finding of uncertain significance was the detection of weak A-like activity in approximately one third of the mucosal extracts from apparently healthy or cancer-bearing persons of blood groups O and B (examples are seen in Fig. 1). Such anomalous expression of A-like activity has previously been described in gastric cancer patients and occasionally in patients with peptic ulcers (6). Preliminary investigations suggest that this is not due to the presence of Forssman antigen.

REFERENCES

1. Hakomori, S., and Kobata, A., in "The Antigens," (Sela, M., ed.) Vol. 2, p. 79, Academic Press, New York (1974).
2. Feizi, T., Kabat, E. A., Vicari, G., Anderson, B., and Marsh, W. L., *J. Exp. Med. 133,* 39 (1971).
3. Feizi, T., Kabat, E. A., Vicari, G., Anderson, B., and Marsh, W. L., *J. Immunol. 106,* 1578 (1971).
4. Feizi, T., and Kabat, E. A., *J. Exp. Med. 135,* 1247 (1972).
5. Feizi, T., Turberville, C., and Westwood, J. H., *Lancet ii,* 391 (1975).
6. Hakkinen, I. P. T., Raunio, V., Virtanen, S., and Kohonen, J., *Clin. Exp. Immunol. 4,* 149 (1969).

Protein–Sugar Interactions: Gangliosides and Limulin (*Limulus polyphemus* Agglutinin)

Annie-Claude Roche, Régine Maget-Dana, Angèle Obrenovitch, and Michel Monsigny

Limulin (*Limulus polyphemus* lectin, NeuAc → GalNAc > NeuAc, NeuGl) was purified from the hemolymph of horseshoe crab. Limulin, as many other lectins, agglutinates red blood cells (1), lymphocytes, and transformed cells. Its activity is inhibited by some glycoproteins, such as bovine submaxillary mucin (BSM), by horse erythrocyte gangliosides, and by some low mol. wt. carbohydrates containing sialic acid.

ISOLATION

Soluble proteins from the hemolymph were passed through Sephadex G-50 and DEAE-Sephadex columns (2). The agglutinating fractions, free of hemocyanin, were applied onto a column of Sepharose 4B substituted by BSM. A nonagglutinating material was eluted with a high ionic-strength buffer containing Ca^{2+} ions, and the agglutinin, limulin, was eluted with the same buffer but free of Ca^{2+} ions (3). Limulin requires calcium for its activity. The purity of the lectin was controlled by cross immunoelectrophoresis using rabbit antisera against active fractions eluted from DEAE-Sephadex, or against the pure limulin.

SPECIFICITY

Horse and cat erythrocytes were agglutinated by a very low concentration of limulin (10 ng/ml); rabbit, human, and mouse red blood cells required higher concentrations. The difference could be related to the high content of sialoglycolipids in horse and cat erythrocytes, as gangliosides may participate in the agglutination process (incorporated into vesicles, they inhibit the effect of limulin).

ISBN 0-12-301301-1

By use of oligosaccharides or glycoproteins as inhibitors of agglutination, it was found that: (a) *N*-Acetylneuraminic acid and *N*-glycolylneuraminic acid are equally slightly inhibitors; (b) α-*N*-acetylneuraminyl-(2→3 or 6)-*N*-acetyl-galactosamine, its derivatives, and glycoproteins containing these oligosaccharides are strong inhibitors; (c) α-*N*-acetylneuraminyl-(2→3 or 6)-galactose and glycoconjugates containing this disaccharide are not inhibitors; and (d) all the desialylated glycoconjugates have no inhibitory power.

MITOGENIC ACTIVITY

Limulin was found to induce the transformation of 50% of human peripheral lymphocytes (4), mouse T lymphocytes (peripheral, node, and thymus) but not mouse B lymphocytes (spleen of nude mice). The dose response curve showed a maximal activity for a concentration of about 20 µg/ml. The mitogenic activity, measured by [^{3}H]thymidine incorporation, was inhibited by BSM, but not by desialylated BSM. The mitogenic activity of PHA was not modified by BSM. Lymphocytes incubated with *Vibrio cholerae* neuraminidase (1.5 u/µl) were not transformed.

GANGLIOSIDE-LIMULIN INTERACTIONS

Gangliosides obtained from horse erythrocytes membrane were embedded in vesicles of hen egg lecithin with cholesterol (4:1) by sonication. The small vesicles were isolated by centrifugation (140 000*g*, 180 min). Upon adding limulin, the absorbance (500 nm) of the suspension increased and reached an equilibrium after about 30-60 min (5). The higher the ganglioside to lecithin ratio was, the higher was the turbidity increase. The enhancement of absorbance was also proportional to the lectin concentration, and was reversed after addition of inhibitors (BSM or EDTA), showing that the absorbance change was due to an aggregation and not to a fusion process. Purified ganglioside vesicles were not aggregated by wheat germ agglutinin (WGA). Brain gangliosides did not bind limulin.

The binding of limulin to the embedded gangliosides change the apparent fluorescence polarization of diphenyl-1,3,5-hexatriene (DPH) incorporated into the vesicles. However, the observed decrease of *p* was shown to be related to a light-scattering depolarization, which is parallel to the absorbance increase. After the appropriate corrections, it was found that the fluorescence polarization of DPH did not change upon limulin addition.

Limulin binds specifically glycoproteins containing the disaccharide NeuAc→GalNAc and certain gangliosides, but does not bind glycoconjugates containing NeuAc→Gal. Furthermore, the *N*-acetyl group of sialic acid may be replaced by a *N*-glycolyl group without change of binding. Because of its restricted specificity, this "sialic acid-binding lectin" may be used to visualize glycoconjugates containing NeuAc→GalNAc (6) and to identify the glycoconjugates involved in biological process, such as the transformation of lymphocytes.

REFERENCES

1. Cohen, E., Rose, A.W., and Wissler, F.C., *Life Sci. 4,* 2009 (1965).
2. Roche, A.C., and Monsigny, M., *Biochim. Biophys. Acta. 371,* 242 (1974).
3. Roche, A.C., Schauer, R., and Monsigny, M., *FEBS Lett. 57,* 245 (1975).
4. Roche, A.C., Perrodon, Y., Halpern, B., and Monsigny, M., *Eur. J. Immunol. 7,* 263 (1977).
5. Maget-Dana, R., Roche, A.C., and Monsigny, M., *FEBS Lett. 79,* 305 (1977).
6. Mazzuca, M., Roche, A.C., Lhermitte, M., and Roussel, P., *J. Histochem. Cytochem. 25,* 470 (1977).

GLYCOCONJUGATE RESEARCH/Volume I
Proceedings of the Fourth International Symposium on Glycoconjugates

Demonstration of *O*-Acetyl Groups in Ganglioside-Bound Sialic Acids and Their Effect on the Action of Bacterial and Mammalian Neuraminidases

Rüdiger W. Veh, Michael Sander, Johan Haverkamp, and Roland Schauer

The occurrence of a variety of *O*-acyl-substituted *N*-acylneuraminic acids in Nature is well documented (1,2), while information concerning the functional significance of these groups is still scant. One of the established facts, however, is the resistance of glycoconjugate-bound 4-*O*-acetyl-*N*-acetylneuraminic acid toward the action of bacterial neuraminidases (3). The highest activity of mammalian neuraminidases is found in brain tissue, and this enzyme shows some specificity for ganglioside substrates (4). The purpose of this study was to find gangliosides containing 4-*O*-acetyl-*N*-acylneuraminic acid and to obtain information about the effect of this 4-*O*-acetyl group on the release of acylneuraminic acid by mammalian neuraminidases.

As a first screening for the occurrence of 4-*O*-acetyl-*N*-acylneuraminic acids, ganglioside fractions were prepared from the brains of codfish, chicken, rabbit, rat, horse, pig, sheep, cow, cat, and man by conventional methods. T.l.c. of the ganglioside fractions gave no indications for the presence of *O*-acyl groups. However, after mild acid or enzymic hydrolysis, t.l.c. of the liberated and purified acylneuraminic acids demonstrated the presence of 10-20% of mono-*O*-acetyl-*N*-acetylneuraminic acid in all species investigated. G.l.c. revealed a corresponding peak with the R_{NeuNAc} value of 9-*O*-acetyl-*N*-acetylneuraminic acid. The position of the acetyl group at O-9 could be proved by g.l.c.-m.s. The presence of 9-*O*-acetyl-*N*-acetylneuraminic acid was unexpected, especially in the gangliosides from horse brain, because in the other tissues of this animal 4-*O*-acetyl-*N*-acetylneuraminic acid is the most abundant *O*-acylated neuraminic acid. Thus, no brain ganglioside containing 4-*O*-acetyl-*N*-acetylneuraminic acid has been found.

ISBN 0-12-301301-1

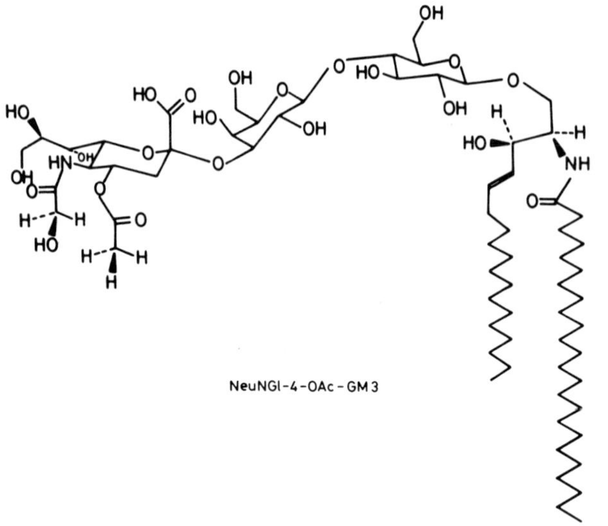

Scheme 1.

Table I. Effect of 4-O-Acetyl Group on the Activity of Neuraminidase[a]

Source of neuraminidase	*NeuNG1-GM3*	*NeuNG1-4-OAc-GM3*
Clostridium perfringens	5.5	0.0
Human brain (particulate)	10.6	0.8
Human brain (solubilised)	10.6	1.4
Human heart (particulate)	2.3	0.0
Horse liver (particulate)	2.1	0.0

[a] *The results are expressed as nmol of acylneuraminic acids released per 2 h at a substrate concentration of* $10^{-3}M$

As another possible source of 4-*O*-acetyl-*N*-acetylneuraminic acid, the sphingolipid fraction of horse erythrocyte membranes was investigated by use of a new t.l.c. procedure (two-dimensional chromatography on silica gel HPTLC-plates with an intermediate 12-h treatment in ammonia vapor), the occurrence of an *O*-acetylated hematoside in these membranes (5) could be confirmed. By comparing two-dimensional chromatograms of this hematoside before and after mild acid hydrolysis, it was found that removal of *O*-acetyl groups by the ammonia treatment altered only the $R_F$value of the hematoside and not the $R_F$value of the lactosylceramide produced by the acid treatment. Therefore, the *O*-acetyl group is bound to the neuraminic acid part of the hematoside molecule. After liberation and purification of the acylneuraminic acid, an $R_F$value corresponding to that of a mono-*O*-acetyl-*N*-glycolylneuraminic acid was obtained on t.l.c. G.l.c. showed a major peak with an R_{NeuNAc}value corresponding to that of 4-*O*-acetyl-*N*-glycolylneuraminic acid and a small peak corresponding to *N*-glycolylneuraminic acid. The structure of this *O*-acetylated glycolylneuraminic acid was established by g.l.c.-m.s. (6). Thus, the hematoside isolated by Hakomori and Saito (5) was shown to be a 4-*O*-acetyl-*N*-glycolylneuraminic acid-containing hematoside (NeuNGl-4-OAc-GM3), and therefore is a possible substrate for mammalian ganglioside-specific neuraminidases. Consequently, NeuNGl-GM3 as well as NeuNGl-4-Oac-GM3 (see Scheme 1) were labeled specifically in the acylneuraminic acid side-chain (7) and used as substrates for radioactive neuraminidase assays (8). The effect of the 4-*O*-acetyl group on the activity of bacterial and different mammalian neuraminidases is shown in Table I. The data indicate clearly that the activity of all neuraminidases investigated is almost or completely abolished by the presence of a 4-*O*-acetyl group on the substrate. Only in the case of human brain neuraminidase, a small amount of neuraminic acid was released, which may be due to de-esterification, during incubation, of some sialic acid residues before hydrolysis of the glycosidic bond.

REFERENCES

1. Buscher, H.-P., Casals-Stenzel, J., and Schauer, R., *Eur. J. Biochem. 50,* 71 (1974).
2. Schauer, R., *Methods Enzymol. 50,* 64 (1978).
3. Schauer, R., and Faillard, H., *Hoppe-Seyler's Z. Physiol. Chem. 349,* 961 (1968).
4. Leibovitz, Z., and Gatt, S., *Biochim. Biophys. Acta 152,* 136 (1968).
5. Hakomori, S., and Saito, T., *Biochemistry 8,* 5082 (1969).
6. Kamerling, J.P., Vliegenthart, J.F.G., Versluis, C., and Schauer, R., *Carbohydr. Res. 41,* 7 (1975).
7. Veh, R.W., Corfield, A.P., Sander, M., and Schauer, R., *Biochim. Biophys. Acta 486,* 145 (1977).
8. Veh, R.W., and Schauer, R., *Adv. Exp. Med. Biol. 101,* 497 (1978).

GLYCOCONJUGATE RESEARCH / Volume I
Proceedings of the Fourth International Symposium on Glycoconjugates

Isolation of the *Amphicarpaea bracteata* Lectin Using Epoxy-Activated Sepharose 6B

Hyman G. Weinstein, Lawrence J. Blacik, and Moira Breen

An anti-A_1 lectin in the seeds of *Amphicarpaea bracteata* has been isolated from the extract of the seeds by affinity chromatography on a column of Epoxy-activated Sepharose 6B coupled to *N*-acetyl-D-galactosamine (1). The use of Epoxy-activated Sepharose 6B in the isolation of lectins was first reported by Vretblad (2). In this paper, we give the details of the coupling procedure.

The reagents used in the procedure are described below: Tris-HCl buffer, pH 8.0, was prepared by titrating 0.05 M Tris base in 0.5 M NaCl with 1.0 M HCl. Sodium formate buffer, pH 4.0, was made by titrating 0.05 M formic acid in 0.5 M NaCl with 3 M NaOH. Isotonic phosphate-buffered saline, pH 7.4, I 0.15, was prepared by titration of a solution of 0.05 M KH_2PO_4 and 0.10 M NaCl with a 0.05 M Na_2HPO_4 solution. The final concentrations were: 0.010 M KH_2PO_4, 0.040 M Na_2HPO_4, and 0.020 M NaCl.

Epoxy-activated Sepharose 6B (Pharmacia Fine Chemicals, Piscataway, NJ) (7.14 g) was washed on a medium porosity sintered-glass filter with 1200 ml of twice-distilled H_2O during a 2-h period, then washed on the filter with 20 ml of 0.1 M NaOH, and washed from the filter into a 150 ml Corex centrifuge tube with 20-30 ml of the same solution. The tube was centrifuged at 13 200*g* for 10 min. The supernatant solution was drawn off and discarded, and a solution of 875 mg of *N*-acetyl-D-galactosamine (Pfanstiehl Laboratories, Waukegan, IL) in 20 ml of 0.1 M NaOH was added to the resin. After swirling of the content, the tube was capped and heated for 16 h at 45°C in a shaking water-bath. The resin was collected on a medium porosity sintered-glass filter and washed with 3 liters of twice-distilled H_2O, washed 5 times with 20 ml of Tris-HCl buffer, pH 8.0, and 5 times with 20 ml of sodium formate buffer, pH 4.0.

ISBN 0-12-301301-1

The coupled resin was washed twice with 30 ml of isotonic phosphate-buffered saline, and then washed with the same buffer into a suction flask so that the slurry could be de-aerated before being poured onto a Pharmacia K-15 column. The coupled resin was light yellow. The column bed volume was 3 ml per g of dry Epoxy-activated Sepharose 6B.

The *A. bracteata* lectin was isolated from the extract in one passage through the resin. Most of the agglutination activity appeared in the 0.1 M *N*-acetyl-D-galactosamine eluate. At higher concentrations of eluent (0.2-0.4 M), additional material absorbing at 280 nm was eluted, but these fractions had little agglutination activity. By the method just described, the agglutination activity was increased 146-fold compared to the crude extract. Dodecyl sodium sulfate-polyacrylamide gel electrophoresis of the lectin fraction showed essentially one band.

Our initial attempts to isolate the *A. bracteata* lectin by means of polyleucine-hog A+H substance, as reported by Etzler and Kabat (3) for the isolation of *Dolichos biflorus* lectin were unsuccessful. The hog A+H blood-group substance obtained from hog gastric mucosa did indeed inhibit the agglutination of A_1 red cells by the extract of *A. bracteata* seeds. However, the *A. bracteata* lectin was bound so strongly to the polyleucine-hog A+H medium that no material having agglutination activity (following dialysis) was eluted from the column, even at an *N*-acetyl-D-galactosamine concentration as high as 0.1 M. In contrast, the use of Epoxy-activated Sepharose 6B, as described above, proved to be a simple and effective means of isolating the *A. bracteata* lectin by one passage through the affinity column.

ACKNOWLEDGMENT

This work was supported by the U.S. Veterans Administration and the Helfaer Foundation, Milwaukee, Wisconsin.

REFERENCES

1. Blacik, L.J., Breen, M., Weinstein, H.G., Sittig, R.A., and Cole, M., *Biochim. Biophys. Acta 538*, 225 (1978).
2. Vretblad, P., *Biochim. Biophys. Acta 434*, 169 (1976).
3. Etzler, M.E., and Kabat, E.A., *Biochemistry 9*, 869 (1970).

Small Glycopeptides and Oligosaccharides with Human Blood Group M- and N-Specificities

Hung-Ju Yang and Georg F. Springer

We obtained low mol. wt. blood group M- and N-specific glycopeptides (GP) from the NH_2-terminal region of α-1 (1,2, *cf.* 3). α-1 Glycopeptides were homogeneous by PAGE and thin-layer electrophoresis (t.l.e.) at various pHs (1). GP IV fractions obtained by **Pronase** treatment of glycopeptides (1) were re-chromatographed on Sephadex G-50 followed by preparative t.l.e. M-derived GP IV (GP IV-M), the smallest, highly active GP fraction had up to 50% of the specific activity of intact glycoprotein with 27/27 human and animal sera. N-derived GP IV had *ca.* 20% of the activity of the intact antigen; it was active with 9/17 human and animal sera. Mol. wt. of GP IVs was between 3500-4500 (2).

Highly purified GP IV-M had 2 Ser, 2 Thr, 1 Gly, and no Leu, **Glx**, Lys or Met (also determined by Dr. A. Bezkorovainy). Carbohydrate measured as previously described (1), as well as immunochemistry and stepwise degradation of desialized α-1 glycopeptides-M (with Dr. B. Friedenson), yielded results compatible with the tentative structure illustrated in Fig. 1. After desialization, t.l.c. showed oligosaccharides (OS) moving like tri- and disaccharides. GP IV-M appears to contain one more large OS then GP IV-N. Partially purified GP IV-N had Leu, Glu, Ser, Thr, and no Gly, Lys or Met.

Determination of NH_2-terminal on intact, highly purified, electrodialyzed M and N antigens was unsuccessful. PTH revealed no NH_2-termini (see also 2,4), and fluorodinitrobenzene and dansyl assays gave nonstoichiometric quantities of Ser (2).

ISBN 0-12-301301-1

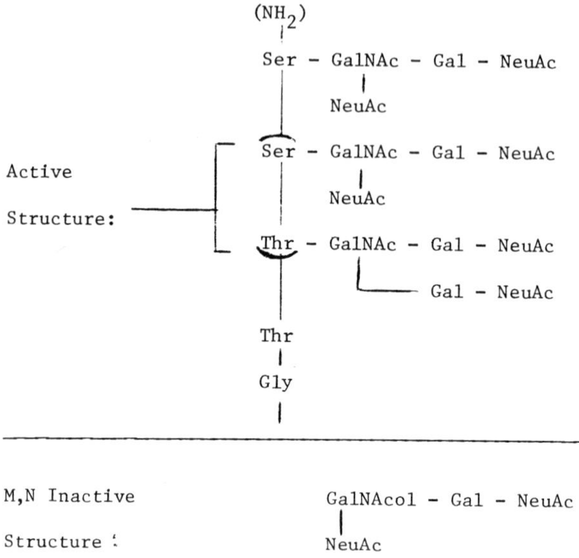

Fig. 1. Possible chemical structure of highly M-active glycopeptide (GP-IV-M). It has not yet been determined if the proposed pentasaccharide is attached to amino acid 2 or 4 rather than 3.

Desialized α-1 glycopeptides had only Ser as NH_2-terminus for α-1 M and only Leu for α-1 N (see also 2,5). The α-1 M is richer in all carbohydrates listed in Fig. 1, which indicates larger and/or more carbohydrate chains than α-1 N, and contains considerably more Ser and Gly, and less Leu and Glu.

OS obtained by β-elimination of α-1 glycopeptides (1) showed 4 major fractions after Sephadex G-50 elution. Repeated t.l.c. of fraction B (1) revealed a series of resorcinol-positive bands of higher mol. wt. than NeuAc-(2→3)-β-Gal-(1→3)-[NeuAc-(2→X)]-GalNActitol; 3 more bands were obtained from α-1 M than from α-1 N. OS fractions were up to 10% as active as glycoproteins with the majority of antisera tested (17/27 anti-M and 12/18 anti-N), with specificities corresponding to their origin.

The effect of chemical modification on the serologic activity of M- and N-active preparations varied greatly, depending only on the antiserum used. The reactivity with anti-M sera is given as example:

(a) *Human sera:* All 7 antisera, *well* inhibited by 1-5 mg/ml of OS obtained by β-elimination of C and D fractions (1), were also well inhibited by carbodiimide-treated (CIIT) and carbodiimide-catalyzed amidated (Am) glycoproteins, and partially inhibited by mildly IO_4-oxidized ($M-IO_4^-$) glycoprotein, and by α-1, GP II, and GP IV. They were not inhibited by carbamylated (CM) and strongly IO_4^--oxidized ($S-IO_4^-$) glycoproteins.

Three out of 15 human antisera *poorly* inhibited by OS (10-20 mg/ml) and 5/15 uninhibited at 20 mg/ml were rarely inhibited by CIIT- and Am-glycoproteins. They were inhibited well by $M-IO_4^-$- and by $S-IO_4^-$-glycoproteins. Carbamylation led to inactivation towards 6 of these antisera. GP II usually did not inhibit at 5 mg/ml or less, but GP IV inhibited well.

(b) *Animal sera:* 12/12 antisera were well inhibited (0.5-1 mg/ml) by α-1, GP II, and GP IV. Of these, 3 were well inhibited by OS (0.3-3 mg/ml), CIIT-, $M-IO_4^-$-, and CM-, but not by $S-IO_4^-$-glycoprotein. Four of the animal antisera were inhibited by OS at 5-15 mg/ml and were also inhibited by CIIT-glycoprotein, but not by $M-IO_4^-$-, $S-IO_4^-$-, and CM-glycoproteins. The 5 antisera uninhibited by OS (> 20 mg/ml) were generally not inhibited by CIIT-, $S-IO_4^-$-, and CM-glycoproteins, but remained active towards IO_4^--glycoprotein as measured with 3 of the antisera.

ACKNOWLEDGMENTS

Aided by NIH Grant CA 22540. We thank Drs. J. Finne and A. Gauhe for NeuAc-containing oligosaccharides and Drs. W. Ebert and D. Roelcke for chemically modified glycoproteins.

REFERENCES

1. Springer, G. F., Yang, H. J., *Immunochemistry 14,* 497 (1977).
2. Springer, G. F., Yang, H. J., and Huang, I. Y., *Naturwissenschaften 64,* 393 (1977).
3. Tomita, M., and Marchesi, V. T., *Proc. Natl. Acad. Sci. U.S.A. 72,* 2964 (1975).
4. Winzler, R. J., *in* "Blood and Tissue Antigens" (D. Aminoff, ed.), p. 124, Academic Press, New York (1970).
5. Wasniowska, K., Drzeniek, Z., and Lisowska, E., *Biochem. Biophys. Res. Commun. 76,* 385 (1977).

GLYCOCONJUGATE RESEARCH/Volume I
Proceedings of the Fourth International Symposium on Glycoconjugates

Interaction of Saccharides with Ricin: Microcalorimetric Study

Christian Zentz, Jean-Pierre Frénoy, and Roland Bourrillon

Ricin (RCA II, RCA_{60}), one of the two lectins of *Ricinus communis* binds specifically galactose and carbohydrates containing a terminal nonreducing galactose residue (1). In the present work, the interaction of ricin with its specific ligands, galactose, lactose (4-*O*-β-D-galactopyranosyl-D-glycopyranose), and lactulose (4-*O*-β-D-galactopyranosyl-D-fructofuranose) has been studied by microcalorimetry, equilibrium dialysis, and analytical ultracentrifugation, and the behaviour of the interaction is discussed.

A normal calorimetric titration curve, for an exothermic reaction, is observed for the ricin-galactose interaction (Fig. 1). The calorimetric curves of the binding of ricin to disaccharides present two parts: a first, ascending part as typical curves, from which the thermodynamic quantities of the interaction can be calculated, and a second, descending part, which does not correspond to a plateau as expected when the saturation of protein by ligand is reached.

Thermodynamic quantities (ΔG, ΔH, ΔS) of the interaction, calculated from the "double reciprocal" plots of the curves of Fig. 1 (Table I) indicate that galactosides contribute by the same large exothermic enthalpy value (ΔH from -10 to -14 Kcal/mol) to the Gibbs free energy term which, from a thermodynamic point of view, is responsible for specificity. Moreover, considering the negative entropy change, hydrophobic interactions probably do not play a dominant role in the complex formation (2).

ISBN 0-12-301301-1

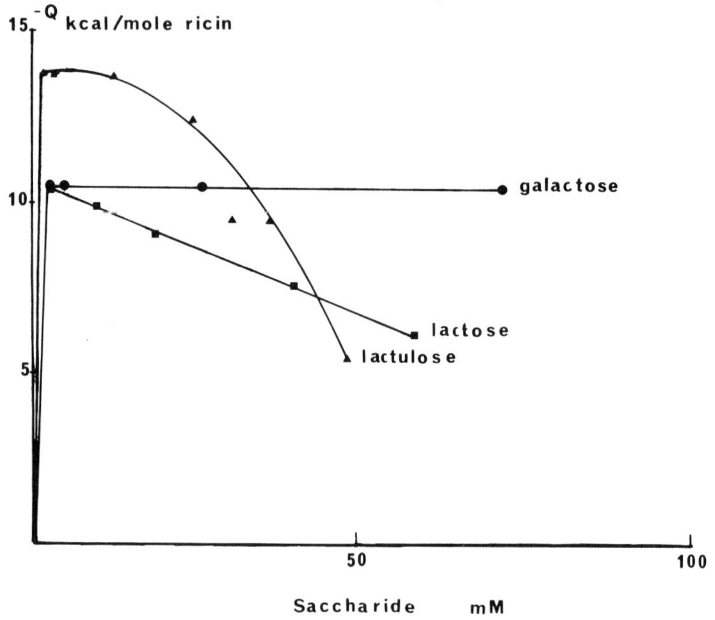

Fig. 1. Experimental heats of binding of galactose, lactose, and lactulose to ricin as a function of final saccharide concentration. Experiments were performed at 25°C, pH 6.9, and at a final protein concentration of 1.96 mM: (●) galactose, (■) lactose, and (▲) lactulose.

Table I. Thermodynamic Values for the Binding of Specific Oligosaccharides to Ricin

Sugar	K_A (M^{-1})	ΔG ($Kcal.mol^{-1}$)	ΔH ($Kcal.mol^{-1}$)	ΔS ($cal.mol^{-1}.deg^{-1}$)	Experimental method
Galactose	3 900	-4.9	-10.5 $\pm$0.7	-18.6	Calorimetry, pH 6.9, 25°C.
	8 000	-4.9			Equilibrium dialysis, pH 6.9, 4°C.
	4 600	-4.6			Equilibrium dialysis, pH 6.9, 20°C.
Lactose	6 500	-5.2	-10.4 $\pm$0.8	-17.3	Calorimetry pH 6.9, 25°C.
	35 000	-5.7			Equilibrium dialysis,
	2 800	-4.4			pH 6.9, 4°C.
Lactulose	7 900	-5.3	-14 $\pm$1	-29.1	Calorimetry, pH 6.9, 25°C.

The heat of ricin-glucose interaction is very low (≤ 1 Kcal/mol) and difficult to measure under our experimental conditions, but is not negligible, suggesting a low specificity of binding of glucose on ricin molecule. Moreover, when a mixture of 2 mM lactose (corresponding to the end point of the ricin-lactose titration curve) and 0.18 mM glucose was allowed to react with ricin in the calorimeter, an exothermic reaction with low ΔQ was observed (one third of ricin-lactose ΔQ), indicating that the excess of glucose interfered with ricin-lactose binding. In order to study the binding of saccharides to ricin, equilibrium dialysis with radioactivity-labeled galactose and lactose were carried out. From the data in Fig. 2 and Table I, it is deduced that the lectin can bind two lactose molecules per protein monomer of 60 000 daltons. The affinity of the second sugar molecule appears to be lower than that of the first molecule.

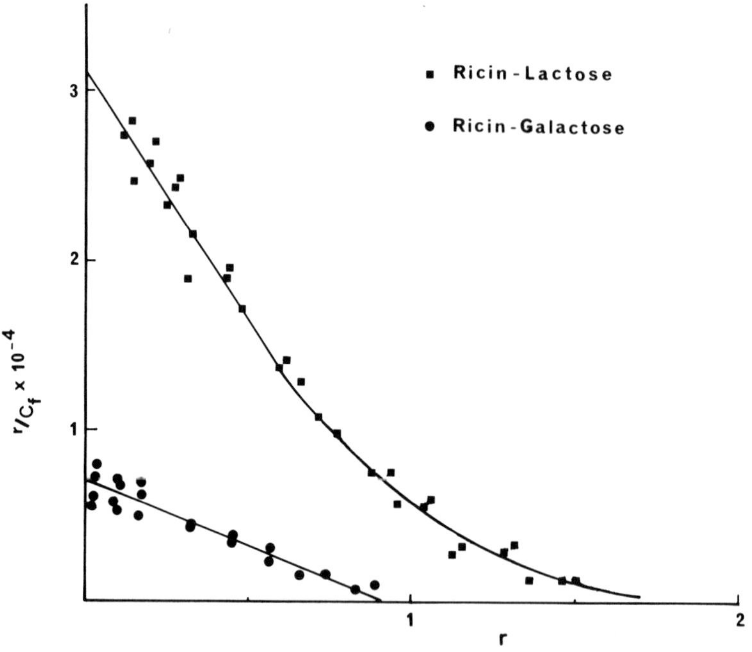

Fig. 2. Scatchard plots of the binding of [^{3}H]galactose and [^{14}C]lactose to ricin. Experiments were performed at 4°C, pH 6.9, and at a final protein concentration of 17.2 mM: (O) ricin-galactose, (■) ricin-lactose.

Of considerable interest is the second type of interaction between ricin and disaccharides, as revealed by microcalorimetric and equilibrium dialysis methods. A possible explanation was the superimposition of a second chemical reaction to the ricin-sugar binding itself. But the hypothesis of a modification, in the polymerisation state, of the ricin molecules in the presence of various lactose concentrations can be ruled out since, in the ultracentrifuge, no change of the mol. wt. in the presence of sugar was observed.

At present, two hypotheses can be drawn to explain the shape of the ricin-lactose titration curve and the influence of the excess of glucose on ricin-lactose binding: (a) The ricin molecule contains a unique sugar binding-site. This site has to be an extended site, larger than the size of a monosaccharide, as already suggested by Nicolson *et al.* from results of inhibition studies (3,4). In the presence of high lactose concentrations (>2 mM), a second type of interaction, similar to inhibition by excess substrate, would appear between ricin and lactose through its glucose moiety, altering the original interaction and giving a less exothermic total energetic evaluation. (b) As revealed by equilibrium dialysis experiments, the ricin molecule contains two sugar binding-sites. Analysis of the data cannot permit to distinquish between: (i) the existence of two independent sites and (ii) two identical sites with negative interaction, since the two models are formally equivalent (5).

ACKNOWLEDGMENTS

This work was supported by grants from the C.N.R.S. (ERA 321) and the I.N.S.E.R.M. (contrat libre 77.1.102.3).

REFERENCES

1. Olsnes, S., Saltvedt, E., and Pihl, A., *J. Biol. Chem. 249,* 803 (1974).
2. Zentz, C., Frénoy, J.P., and Bourrillon, R., *FEBS Lett. 81,* 23 (1977).
3. Nicolson, G.L., and Blaustein, J., *Biochim. Biophys. Acta 266,* 543 (1972).
4. Nicolson, G.L., Blaustein, J., and Etzler, M.E., *Biochemistry 13,* 196 (1974).
5. Ferguson, R.N., Edelhoch, H., Saroff, H.A., and Robbins, J., *Biochemistry 14,* 282 (1975).